Blockchain Technology

"In simple terms, Chandramouli and the team have explained complex technology with very relevant examples and illustrations. The need for more developers will continue to increase as Blockchain Technology becomes mainstream, and this book will address a critical need in that supply chain by educating anyone wanting to become a Blockchain developer—a must-read for all Blockchain enthusiasts."

Rajesh Balaji
Senior Vice President
Cognizant Technology Solutions

"Amidst the pandemonium created by the current information overload on Blockchain, this book is an excellent treatise on the subject that covers concepts to case studies and takes its readers and the learners gradually to a point where they are at ease with the principles and in readiness to deploy Blockchain in real-world use cases across industry verticals: banking, insurance, retail, manufacturing, and logistics. Incorporating the AICTE guidelines makes this a much-needed curriculum aid for university students."

A N Rao
Founder, Learning Sherpa
Former SVP and Global Head of Learning, Cognizant Technology Solutions

"One of the most disruptive technologies of the 21st century, the Blockchain, has been lucidly presented by the authors who have interspersed their discussion with lots of examples and case-studies from different industry domains for the benefit of both novices and experts in equal measure. Notably, the suggestions given on the adoption of Blockchain in life sciences and healthcare industry are pragmatic and futuristic."

V Samuel Gnana Prakash
Professor and Head
Centre for Marine Science and Technology
Manonmaniam Sundaranar University
Tamil Nadu, India

Blockchain Technology

S Chandramouli

Associate Director
Cognizant Technology Solutions

Asha A George

PPM and Strategy Consultant
Verbat Technologies

Abhilash K A

Solution Architect
6d Technologies

Meena Karthikeyan

Enterprise Solutions Consultant
Former Vice President, Cognizant Technology Solutions

Universities Press

BLOCKCHAIN TECHNOLOGY

UNIVERSITIES PRESS (INDIA) PRIVATE LIMITED

Registered Office
3-6-747/1/A & 3-6-754/1, Himayatnagar, Hyderabad 500 029, Telangana, India
info@universitiespress.com; www.universitiespress.com

Distributed by
Orient Blackswan Private Limited

Registered Office
3-6-752 Himayatnagar, Hyderabad 500 029 Telangana, India

Other Offices
Bengaluru, Chennai, Guwahati, Hyderabad, Kolkata,
Mumbai, New Delhi, Noida, Patna

First published 2021
Reprinted 2022, 2023, 2024

ISBN 978-93-89211-63-4

Cover and book design:

502156

Typeset in Bembo MT Pro 12 points *by*
Cameo Corporate Services Limited, Coimbatore

Printed in India by
B.B. Press, Tronica City, Ghaziabad 201 102

Published by
Universities Press (India) Private Limited
3-6-747/1/A & 3-6-754/1, Himayatnagar, Hyderabad 500 029, Telangana, India

My sincerest thanks and appreciation go to several people...
My parents Subramanian and Lalitha
My wife Ramya
My son Shri Krishna
My daughter Shri Siva Ranjani
And my colleagues and friends

S Chandramouli

This book is dedicated to…
My professional guides Ramanujam Ramaswami, Anura Boteju, Craig Ryall
My favourite people Nargisse Skalante, Lakshmi Mittra, Abenla Mar
My forever support system, my loving husband George and my son Steven

Asha A George

My sincerest thanks and appreciation go to several people...
My parents
My wife
My daughters Kanvika and Kanshika
And my colleagues and friends

Abhilash K A

My sincerest thanks and appreciation go to several people...
My parents M Subba Lakshmi and Late S Krishnan
My husband M Karthikeyan
My daughters Kanchana and Lavanya
My son-in-law Bharath Murali
And my colleagues and friends

Meena Karthikeyan

Contents

Preface

Today we are at the cusp of the Fifth Industrial Revolution, with over 50% of the world connected via the internet and technology progressing at its most sophisticated phase with Artificial Intelligence, Big Data, Virtual Reality, and one of the most disruptive technologies—the Blockchain, forming the bedrock of such development. This exciting new technology is the proverbial 'elephant in the room' that is now redefining how we store, update, and move data.

Blockchain Technology is based on the latest guidelines to present the subject concepts in an easily understandable way. Due care has been taken to present the concepts in a simple, comprehensive manner. The pictorial representations for explaining the hard concepts as well as the multiple examples that facilitate easy assimilation of each topic have been created to ensure that no aspects of the subject are left unexplained. This book is primarily designed for use as a course textbook on blockchain technologies. However, it can serve as a good point of reference for software practitioners as well. We have provided a blend of practical as well as theoretical concepts of Blockchain technology in this book so that the reader can easily relate the concept to the emerging real-world applications.

Why this Book

There are many aspects of Blockchain technology that require critical reading and in-depth understanding. We endeavour to cover as much area as possible and present it in a concise and clear language. All the complex topics are decomposed into subtopics so that the reader can grasp the concepts quickly.

Through a study of this book, the reader will first gain fundamental knowledge of how Blockchain works and be led, in stages, into the more technical aspects before finally learning about the vast scope of this nascent revolutionary technology for in-depth practical applications.

Who is this Book for?

Readers of this book will gain a thorough understanding of the principles of Blockchain technology, both theoretical and practical. Students, developers, technical managers, and anyone with a basic background in computer science or basic technology awareness will find the material in this book easy to read, study, and apply. Technical managers and practitioners will get an insight into how blockchain technology can be woven into the overall organization value management.

S Chandramouli
Asha A George
Abhilash K A
Meena Karthikeyan

Acknowledgements

The journey through traditional Project Management, Agile Project Management, Program and Portfolio Management, facilitated by the use of artificial intelligence, machine learning and blockchain technologies has been very rewarding, as it has given me the opportunity to work for some of the best organizations in the world and learn from some of the best minds. Along the way, several individuals have been instrumental in providing me with the guidance, opportunity and insights needed to excel in project and portfolio management. I wish to personally thank my colleagues at Cognizant Technology Solutions: Mr Rajesh Balaji Ramachandran, Senior Vice President; Mr Pradeep Shilige, Executive Vice President; Mr Alexis Samuel, Senior Vice President; Mr Hariharan Mathrubutham, Vice President; Mr Krishna Prasad Yerramilli, Vice President; and Mr Balasubramanian Narayanan, Senior Director for their inspiration and help in creating this book. They have immensely contributed to improve my skills.

This book is the result of my association with many highly reputed professionals in the industry. I was fortunate to work with them and, in the process, acquire knowledge that helped me in molding my professional career. I am obliged to Mr Chandrasekar, Ex CIO, Standard Chartered Bank, for demonstrating how the lives, thoughts and feelings of others in professional life are to be valued. He is a wonderful and cheerful man who inspired me and gave me a lot of encouragement when he launched my first book, *Virtual Project Management Office*.

My parents (Mr Subramanian and Ms Lalitha) have always been enthusiastic about this project. Their unconditional love and affection provided the much-needed moral support. My son Shri Krishna and daughter Shri Siva Ranjani constantly added impetus to my motivation to work hard. This book would not have been possible without the constant inspiration and support of my wife, Ramya. She was unfailingly supportive and encouraging during the long months that I had spent glued to my laptop while writing this book. Last but not least, I beg forgiveness of all those who have been with me over the course of the years and whose names I have failed to mention here.

S Chandramouli

I am obliged to my co-authors for their deliberation on the subject and sharing of their knowledge with great enthusiasm. We had multiple discussions on various topics related to blockchain development before we finally brought it out in the present book form with an aim to shorten the learning curve for anyone interested in this subject, whether it is a college student or an IT professional. I am grateful to the members of my family for their cooperation during the period of writing this book.

Asha George

I wish to thank Dr S Chandramouli for taking this initiative and inviting me to work on this project and for his unfailing support and guidance, throughout the project.

I am grateful to my parents (Late Mr KK Aravindakshan and Mrs Kanchana TG), who worked hard for us.

Last but not least, I am obliged to my wife Ms Rasmi, who has been supportive of my actions.

Abhilash K A

Writing a book is harder than I thought and is more rewarding than I could have ever imagined. Therefore, I would like to thank Almighty God for giving me the patience and perseverance to complete this book with conviction.

My father Late S Krishnan, who was an integral part of my life, empowered me in so many different ways, giving me the freedom and courage to express my opinion and assert myself in whatever I do. This book is a result of that freedom, which has allowed me to express my technical expertise and research work without any inhibitions. I am thankful to my family – my loving mother M Subbalakshmi, my beloved husband M Karthikeyan who has been a constant support and guide to me, my doting daughters and pillars of my life – Kanchana and Lavanya, who encouraged me to pursue my dreams and my son-in-law Bharath Murali with whom I could have long hours of discussion on any technical topic of interest.

I would like to thank my team of co-authors, Mouli, Asha and Abhilash who have been an absolute joy to work with. Lastly, I would like to thank all the mentors and friends in my professional life of 35 years.

Meena Karthikeyan

We are grateful to Universities Press, who came forward to publish this book. Messrs Thomas Mathew Rajesh, Kallol Das and Ramesh of Universities Press were always kind and understanding. Mr Ramesh reviewed this book with abundant patience and helped us say what we had wanted to, improvising each and every page of this book with care.

S Chandramouli
Asha A George
Abhilash K A
Meena Karthikeyan

About the Authors

Dr S Chandramouli, PhD, PMP, PMI-ACP, is an alumnus of Indian Institute of Management Kozhikode (IIMK) and Certified Global Business Leader from Harvard Business School. He is a prolific writer of business management articles dealing with technology management, delivery management, competitiveness, IT, organizational culture and leadership. He is a certified 'Green Belt' in six-sigma methodology and a certified master practitioner in Neuro Linguistic Programming (NLP).

Chandramouli has a good record of delivering large-scale, mission-critical projects including those that use blockchain technology, AI and machine learning, on time and within budget to the customer's satisfaction. A formerly active member of PMI's Organization Project Management Maturity Model (OPM3) and Project Management Competency Development Framework (PMCDF) assignments, he has been an invited speaker at various technology and management conferences. In addition, he has also addressed more than 10,000 software professionals worldwide on a variety of themes associated with Blockchain, AI, machine learning, delivery management, competitiveness and leadership.

Asha A George, CGEIT, PfMP, PMP, is a Certified Blockchain Expert and Certified Cryptocurrency Expert from Blockchain Council. She is a consultant with over 20 years of project program portfolio management experience, including 10 years as a portfolio manager running the Program Management Office (PMO). A blockchain enthusiast, learner, educator and mentor, she is passionate about developing the next generation of talents in areas of innovative technology, P3M, leadership and communication. She received her B.Tech. degree from Mar Athanasius College of Engineering, Kothamangalam, and holds recognized certifications in Six-sigma, COBIT, ITIL, Scrum and other project management and governance fields.

Asha is a specialist in enterprise program project management office and project management methodologies. She has established PMO in major organizations, formalizing and aligning the methodologies, frameworks, matrices and KPIs with organizational maturity to ensure measurable business results and continuous improvement. Throughout her professional career, she has successfully led complex cross-country transformation, systems integration, software development and process improvement projects.

Abhilash K A operates from Bengaluru, as a solution architect.

His expertise includes working with enterprise systems and business transformation. Abhilash has delivered complex projects, using Service Oriented Architecture (SOA) and microservices. He has proven expertise in business process modelling and telecom frameworks. He uses open source technologies and open API for product development.

His interests include cryptography and distributed computing.

Abhilash earned his degree in engineering from University of Madras (VMKV Engineering College). He learnt his management lessons from IIMK and computing lessons from C-DAC.

Meena Karthikeyan has over 35 years of global business experience in Enterprise Solutions Consulting in the areas of CRM, Supply Chain and ERP, and Process Automation and Manufacturing. She is a known global leader, visionary and strategist, being an intrapreneur and entrepreneur in the world of digital business.

An alumnus of Government College of Engineering, Thiruvananthapuram, Meena started her professional life as an electronics and communication engineer and worked with organizations like Bharath Electronics Limited, Tata Consultancy Services and Cognizant Technology Solutions during her journey to evolve as a digital enterprise solutions expert. Her areas of special interest include enterprise application space and emerging digital technologies, enhancing customer experience, analytical reporting, enterprise process management and enterprise integration.

Meena has been instrumental in helping global clients in their digital transformation of customer-critical business units and defining their overall digital roadmap. In IT consulting, she has expertise in areas of strategic process and technology consulting, global program/project delivery management, creating centres of excellence and setting up offshore delivery centres for global customers. She has helped numerous customers in building intelligent automation of their business processes using the niche IoT, AI/ML and Cloud Computing technologies. An avid reader and analyst of technical topics, she has reviewed several publications and published papers on thought leadership. She is an active NASSCOM member and spearheads innovation within the enterprise.

Meena is committed to community services and is a strong supporter of diversity at the workplace in terms of language, gender and physically challenged workforce.

CHAPTER 1

Fundamentals of Blockchain

LEARNING OBJECTIVES

Readers generally find it difficult to understand the philosophy of blockchain when they are new to the topic. Also, they do not know where to start learning about the concept or the types of problems which could be solved using blockchain. This chapter intends to give the readers a starting point to understand the Blockchain and the cutting-edge technology behind its functions. It commences with a historical look at the timeline in this field and proceeds to give a brief background into the different aspects of blockchain.

1.1 Introduction

Welcome to the world of Information Technology! You are about to explore a vibrant subject in the arena of innovative technology. Computer and information technology are a vital part of any industry. The need for humankind to perform a process with perfection and accuracy commenced with the origin of computers. The digitalized computing machines, initialized in the 1940s, merely carried out arithmetic operations marking the first step towards the innovative computers of today. Computers are part of our day-to-day life now. Computer networks connect not only computers around the globe but also human beings across the world.

Today we are at the Fifth Industrial Revolution with over 50% of the world connected via the internet and technology progressing at its most sophisticated phase with Artificial Intelligence, Big Data, virtual reality, and one of the most disruptive technologies—the Blockchain, forming the bedrock of such development.

Blockchain is fast becoming one of the most sought-after technologies of today. This exciting new technology is redefining how we store, update, and move data. In this chapter, we will see what Blockchain is and how the technology works, including the concepts of cryptography, mining, and its essential components.

1.2 Origin of Blockchain

1.2.1 What Is Blockchain

Technically, Blockchain is defined as a distributed, replicated peer-to-peer network of databases that allows multiple non-trusting parties to transact without a trusted intermediary and maintains an ever-growing, append-only, tamper-resistant list of time-sequenced records.

In short, Blockchain is a type of distributed ledger that sits on the internet for recording transactions and maintaining a permanent and verifiable record-set of information.

This innovative technology was first published in 2008 by Satoshi Nakamoto (pseudonym of a person/persons as-of-yet unknown) in a white paper titled "A Peer-to-Peer Electronic Cash System." The thought behind the design was to create a decentralized digital currency that is free from government regulation whereby two people can confidently trade directly with one another without the need for mediators or intermediaries.

In 2009, the concept became a reality when Satoshi Nakamoto implemented the first application of the Blockchain we all know as Bitcoin.

Though Bitcoin and Blockchain are often referred to interchangeably, they are not the same. Blockchain is the underpinning technology that the Bitcoin was built on. Now aside from Bitcoin blockchain, we have the Ethereum blockchain and other private blockchains that are adapted to cater to various industry applications such as supply chain, cross-border payments, and many others.

1.2.2 The Bitcoin and the Blockchain

To better understand the difference between Bitcoin and Blockchain, let us delve a little bit into what they are and what they are not.

1) Meaning

Bitcoin is a cryptocurrency, while Blockchain is a ledger or database of information. Bitcoin was created to reduce the government's control over cross-border transactions and to speed up the transaction process by removing the need for third-party intermediaries. Blockchain, on the other hand, provides a secure environment that Bitcoin needs for peer-to-peer transactions. In other words, blockchain acts as bitcoin's ledger and maintains all the transactions of bitcoin.

2) Scope of Usage

Bitcoin is limited to currency transactions, while the blockchain has numerous applications. It can not only trade currencies, property rights of stock but also be used for identity management, records management, research management, as well as many other areas that cover all aspects of the business.

3) Transparency

Bitcoin has a high degree of anonymity. Though the transactions are visible, it is close to impossible to identify the user. Until today, the identity of Satoshi Nakamoto is a mystery even though Satoshi has roughly one million bitcoins (BTC) worth around 8 billion dollars as of 2019.

On the other hand, Blockchain is quite transparent as it is expected to work across multiple industry applications. However, with Smart contracts and various consensus mechanisms, Blockchain can assure compliance to KYC and other industry standards.

Though Bitcoin will continue to hold the coveted position of the world's first decentralized cryptocurrency, its operational significance is considerably reducing with the blockchain technology taking the limelight into various industry sectors like Healthcare, Travel, Education, Government, and others.

1.2.3 The Evolution of Blockchain

The first Bitcoin was mined in 2009. Table 1.1 outlines a brief history on the journey of blockchain from the Genesis Block to the cusp of the Blockchain fourth generation.

Table 1.1 History of blockchain

Pre-blockchain – The Early Years	The 1950s	– First computers developed and adopted
	The 1960s	– 1969: **Arpanet**, the early Internet on the peer-to-peer network
	The 1970s	– 1973: **Public-key cryptography** implemented by Clifford Cocks – 1977: RSA, the public-key cryptosystem that is widely used for secure data transmissions, is released. – 1979: Ralph Merkle patents the concept of hash trees now called **Merkle tree**
	The 1980s	– 1982: IBM Personal Computer launched with DOS operating system
	The 1990s	– 1991: Stuart Haber and W Scott Stornetta work on a cryptographically secure chain – 1997: **Proof-of-work** with Hashcash – 1996: Nick Szabo introduced bit gold as a mechanism for a decentralized digital currency and smart contracts – 2000: Stefan Konst introduced a general cryptographic theory of secured chains

(*Continued*)

Table 1.1 (*Continued*)

Blockchain 1.0: Origin of Bitcoin	2008	– Oct 31: **Satoshi Nakamoto** releases Bitcoin white paper – a concept on the peer-to-peer payment system – Bitcoin.org registered in August
	2009	– Jan 03: **Bitcoin Genesis block** mined – Jan 12: Hal Finney receives first Bitcoin transaction, thus launching the first application of a public blockchain – Oct 12: Bitcoin registered open source code – Oct 31: Bitcoin Market – Bitcoin recognized as a digital currency
	2010	– May 22: First Bitcoin purchase - 10,000 BTC for a $25 pizza – Nov 06: Bitcoin marketplace surpasses $1 million
	2011	– **Namecoin**, the first Bitcoin fork – **Litecoin** released as an alternative to Bitcoin with different mining algorithm and faster transaction speed – Bitcoin reaches parity with the US dollar (1BTC=1USD)
	2012	– Diaspora, the first decentralized social network – **Ripple**, a permissioned blockchain, is launched. A payment protocol focusing on integration with banking systems – The Bitcoin Foundation launched in September
Blockchain 2.0: Ethereum and Smart Contracts	2013	– Mar 28: Bitcoin marketplace surpasses $1 billion – May 02: First Bitcoin ATM unveiled – The University of Nicosia in Cyprus accepts Bitcoin – **Mastercoin** (the first Altcoin) is one of the earliest – Vitalik Buterin releases **Ethereum** white paper
	2014	– Establishment of **R3**: a consortium of over 40 financial institutions committed to implementing Blockchain technology – Ethereum Blockchain is funded by crowdsale – PayPal announces Bitcoin integration – Microsoft accepts Bitcoin
	2015	– **Genesis block** in Ethereum created – Linux Foundation unveils **Hyperledger** to boost blockchain development. – Visa, Citi, Nasdaq, Capital One and Fiserv invest $30M in Blockchain startup Chain.com

(*Continued*)

Table 1.1 (*Continued*)

Blockchain 3.0: Distributed Applications	2016	– Bug in Ethereum DAO code exploited, causing theft of $50 M in ether
	2017	– **EOS** unveiled by Block.one as a new Blockchain protocol for industry-scale decentralized applications.
Blockchain 4.0: The Future	2018 - future	– Current Bitcoin marketplace between $10-$20 billion – TRON, a blockchain platform for the entertainment industry – Business-oriented hybrid blockchain projects – Integration with IoT, AI and Big Data

The concept of digital computing and cryptography started to appear in the 1900s and early 2000. It is these concepts and implementation that inspired the Blockchain 1.0 – the Bitcoin.

a) First Generation Blockchain (2008-2013): The Origin of Bitcoin

In 2009, Bitcoin became the first application of Blockchain technology. Satoshi Nakamoto formed the Genesis block. On 12 Jan 2009, the first successful bitcoin transaction on the blockchain takes place between Nakamoto and Hal Finney.

The first generation of blockchain promised transparency, immutability, accountability, and security in transactions. However, the protocols used (Proof-of-work consensus, explained later in this chapter) necessitated the use of heavy mining hardware and significant resources leading to problems in scalability, interoperability, and speed. To date, Bitcoin is one of the slowest cryptocurrency, taking about 10 minutes to confirm a transaction.

b) Second Generation Blockchain (2013-2015): Transactions with Smart Contracts

The second-generation sought to overcome the limitations of the Bitcoin. Thus emerged Ethereum in 2013, when it was realized that the underlying technology of Bitcoin could be used for all kinds of B2C and other general applications. It still used the Proof-of-work algorithm and had less-than-optimal speed. Though Ethereum required lesser energy to maintain, there were still concerns about future scalability. However, with the hosting of Smart Contracts that made available new functionalities, Ethereum was the go-to blockchain for Enterprise use.

c) Third Generation Blockchain (2015-2018): Distributed Applications

This generation saw the arrival of Hyperledger from the Linux Foundation and Decentralized Applications (DApps) of Ethereum. Though the Hyperledger is a platform that can plug in any consensus mechanism, Smart Contracts of Ethereum opened up the possibility of a Proof-of-Stake consensus mechanism. The focus of the generation was consensus mechanisms that can bring greater interoperability and boost network speeds.

Promoting cross-chain transactions, using sharding (a type of database partitioning that separates vast databases into smaller, faster, and more easily managed parts) and establishment of parallel chains are only some of the approaches being taken by the third generation blockchain solutions. However, EOS and TRON have taken over the DApps market due to the scaling issues still inherent in the Ethereum platform.

d) Fourth Generation Blockchain (2018–Future)

Blockchain technology's future appears optimistic as many governments and organizations are investing heavily in innovations and applications.

A significant innovation on the horizon is called blockchain scaling. A scaled blockchain is expected to accelerate the processing speeds, without sacrificing security significantly. This is done by assessing the computing power required to validate each transaction and then dividing the work efficiently. Progress in this front is yet to be seen, but the outlook has been positive and more is expected on this front.

1.3 Blockchain Solution

> "*A purely peer-to-peer version of electronic cash would allow online payments to be sent directly from one party to another without going through a financial institution. Digital signatures provide part of the solution, but the main benefits are lost if a trusted third party is still required to prevent double spending. We propose a solution to the double-spending problem using a peer-to-peer network.*"
>
> – Part of the abstract from the paper "A Peer-to-Peer Electronic Cash System."

It is evident from the paper that Satoshi Nakamoto was proposing a digital currency trading or payment solution that addressed the double-spending problem that is unique to digital currency transactions. But what is double-spending, and how is a blockchain transaction different from a traditional transaction?

1.3.1 Traditional vs. Blockchain Transactions

Business transactions are recorded in ledgers since time immemorial. History credits the Mesopotamians (now Iraq) as the first record keepers with evidence dating back to about 7000 years. Clay tablets were used to record the expenditures of goods received and traded. Paper-based ledger entries or bookkeeping can be traced back to the 13th century. It wasn't till the mid to late 20th century, until the widespread adoption of computers, that the digital ledger replaced the physical ledger. The digital ledger is a digital file or files or database that can be manipulated only using a computer program as it does not have a physical form. Although the medium of entry has changed, the fundamental principles of the recording of sales and purchases have remained the same throughout the centuries. The general ledger records the assets, liabilities, income, expenses, and capital of the company or business. While digital transactions resolved the disadvantages of manual paper-based transaction entry in terms of time-consuming recording and maintenance, human error, lack of security and limited copies that could not cater to large organizations, it, however, brought in its own set of complexities related to data inconsistencies, potential frauds, and other technical issues with respect to outages and virus attacks.

Let us understand transactions with the following basic example. Joe purchases a painting from Ann for $50.

Scenario 1: The simple transaction would be where Joe hands a physical $50 note to Ann, who will then hand over the painting to Joe (refer Fig. 1.1). In short, Joe has received goods worth 50 dollars.

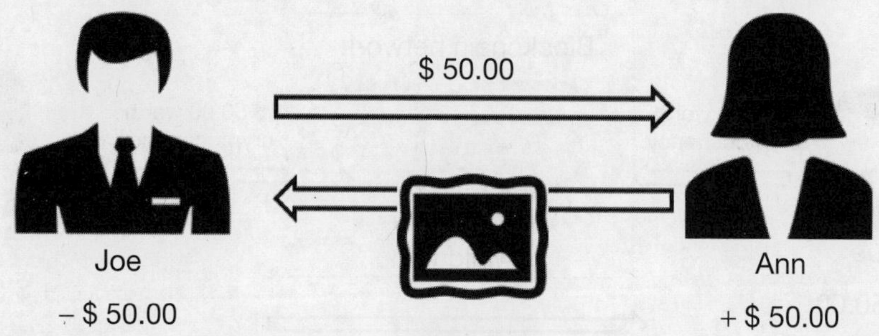

Figure 1.1: Physical transaction

Scenario 2: Assume a modern-day scenario where Joe and Ann are unable to meet. They would like to do the transaction online (refer Fig. 1.2). In such a case, the bank acts as the central authority or intermediary. In the online transaction scenario, the bank debits Joe's account by 50 dollars and adds 50 dollars to Ann's account. Once Ann verifies that her account is credited, she sends Joe the painting. Though no real money exchanges hands, a credit and debit entry is passed in the bank's centralized database or ledger. Only the bank has access to this ledger. It acts as a trusted third party for which the bank may collect a transaction fee. In our example, the bank has taken a further 1 dollar from Joe's account as a transaction fee.

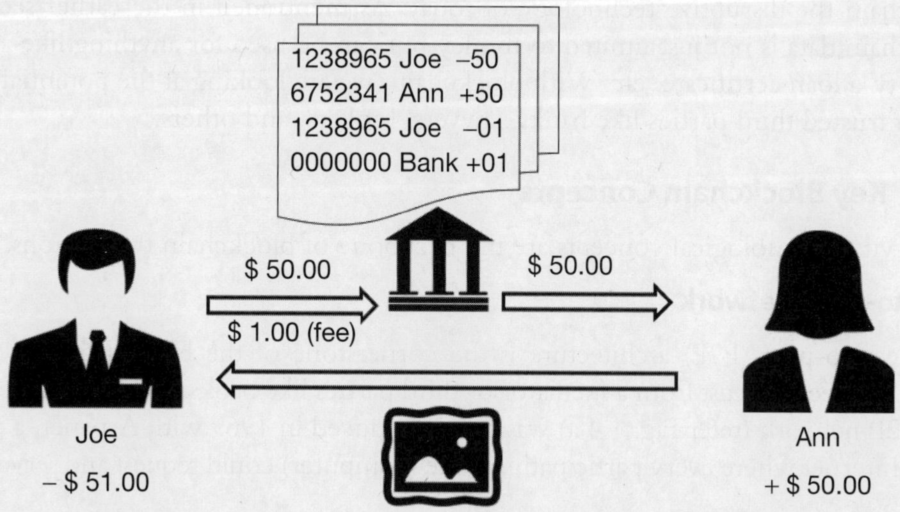

Figure 1.2: Traditional online transaction

However, this is a straightforward example. In a real-life scenario, we are talking of multiple banks or exchanges for cross-border transactions involving a large amount of money. With the advent of commerce on the internet, financial institutions have become indispensable to verify ownership, transaction maintenance, and any dispute mediation. For the services, the financial institutions need to charge fees, which increases the overall cost to the consumer.

Scenario 3: the Blockchain transaction

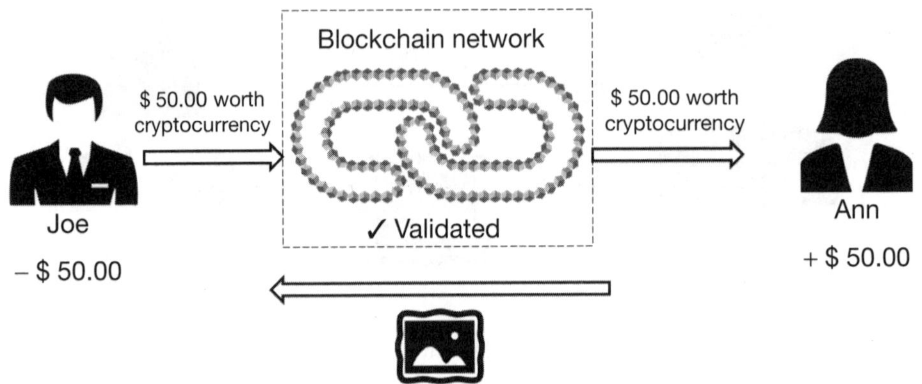

Figure 1.3: Blockchain transaction

In the blockchain scenario, Joe and Ann are both members of a blockchain network (refer Fig. 1.3). In the blockchain, transactions are done using cryptocurrency like Bitcoin. Joe initiates the transaction with 50-dollar worth of bitcoin or related cryptocurrency say. Once it is validated within the blockchain network, Ann receives 50 dollars equivalent in crypto. The whole transaction is done without any intermediary or fees. There is, of course, the reward fees that go to the miner, but that is insignificant compared to the fees charged by banks. This inherent trust in the peer-to-peer network is what makes blockchain the disruptive technology of today. As mentioned in the earlier section, the blockchain data is not just limited to money but can be used for anything like a proof of property, a loan certificate, etc. With blockchain, we are looking at the potential elimination of trusted third parties like banks, lawyers, brokers, and others.

1.3.2 Key Blockchain Concepts

Three vital technological concepts are the backbone of blockchain transactions.

Peer-to-peer Network

The peer-to-peer (P2P) architecture is the cornerstone of the blockchain network that makes the deed of trust from a mediator or third parties like banks, lawyers, etc. redundant. The P2P network (refer Fig. 1.4 a) was first introduced in 1969 with Arpanet, a precursor to the Internet where every participating node (computer) could request and serve content.

The standard network model traditionally used is the client–server model (refer Fig. 1.4 b), where communication is typically to and from a central server. For example, in File Transfer Protocol (FTP) services, the client requests the transfer, and the server executes it.

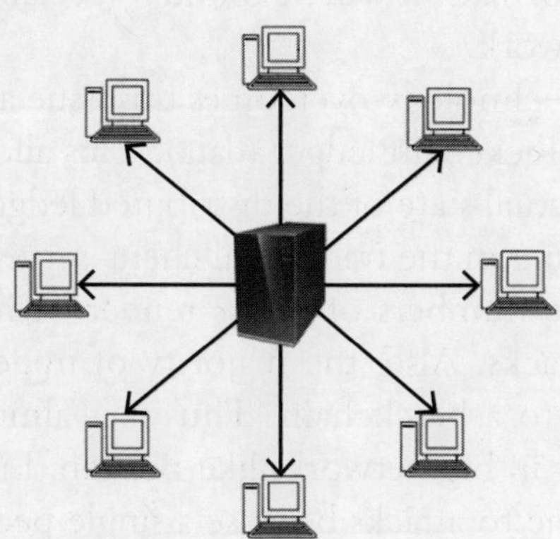

(a) A **client-server network** where individual clients request services from centralized servers

(b) A **peer-to-peer (P2P) network** where interconnected nodes or peers share resources amongst each other without any centralized administrative system

Figure 1.4

The P2P network, however, works on the principle of equal peer nodes acting as both clients and servers to the other nodes within the network. The P2P architecture was popularized in 1999 by the file-sharing system Napster designed for sharing digital music files in mp3 format within its music-sharing application network.

While in the traditional network, the information is stored in one central location, in the P2P network, multiple copies of the same data are stored in different locations/computer devices on the network. This prevents a single point of failure. Even if one node or server is inactive/lost/destroyed, multiple copies remain safe and secure elsewhere. Besides, if one piece of information is changed or is inaccurate, there are other exact copies with peers (majority) in existence, making the false record obsolete. Gnutella, BitTorrent, and IPFS (InterPlanetary File System) network and protocols use the P2P architecture.

Did you know?

Arpanet (Advanced Research Projects Agency Network) of the United States Department of Defense was the first packet-switching network using the TCP/IP protocol suite, which forms the technical foundation of the Internet. The Arpanet program aimed to work on time-sharing, to enable research institutions to use the CPU (Central Processing Unit) power of other institutions for doing research.

While P2P architecture is inherently distributed, in the real-world scenario it is not necessarily fully decentralized. There is a central authority (system of servers and administrators) to guide the network activity as a single data query could flood a network to reach as many peers as possible. This results in high CPU and memory usage. Even then, there is no guarantee that the required piece of information will be found, especially if the information sits with only a few nodes in the network.

The peer-to-peer architecture of blockchain technology overcomes this issue as every full node maintains a complete updated copy of blockchain ledger (data). This allows the nodes to collectively participate in verifying the actual state of the distributed ledger, thus assuring decentralization and security that is wanting in the traditional client-server models. The distribution of the blockchain over large numbers of nodes renders it resistant to cyber-attacks like Denial-of-Service (DoS) attacks. Also, the majority of nodes must establish consensus for a data block to be added to a blockchain. Thus it is almost impossible for an attacker to alter the data, especially in big networks like Bitcoin, Litecoin, Ethereum, etc. Smaller blockchains are more prone to attacks because a single peer node or group of peers with common interests could eventually gain control over a majority of nodes. This is referred to as a 51 percent attack.

Thus an extensive distributed peer-to-peer network, paired with a majority consensus requirement, gives blockchains a relatively high degree of resistance to malicious activity.

Public Key Cryptography

Also called asymmetric cryptography, it was discovered in 1976 by two Stanford mathematicians, Whitfield Diffie and Martin Hellman. Used in PKI (Public Key Infrastructure), two keys, called the private key and the public key, are generated, which can be used to encrypt/decrypt a message. They need to be used in combination. The public key cannot decrypt the message encrypted by the public key, nor can the private key decrypt message encrypted by the private key. This enables two essential cryptographic capabilities, i.e., confidentiality and integrity.

More details on cryptography and its utilization in Blockchain are detailed in Chapter 09: Security in Blockchain.

Did you know?

Though Diffie and Hellman discovered the concept of asymmetric cryptography first, it was the RSA algorithm published in 1978 by mathematicians Ron Rivest, Adi Shamir, and Leonard Adleman that monetized the concept more effectively. This ultimately resulted in RSA Security LLC, now part of the Dell Technologies family of brands.

Distributed Consensus

The distributed consensus protocols are fundamental in avoiding double-spending and other internet attacks. In 1992, researchers Cynthia Dwork and Moni Naor presented an approach to combat junk emails through a protocol known as the pricing function. Users would be able to assess their email service once they can compute a function or puzzle by engaging their processing power. In 1997, cryptographer Adam Back proposed a similar function called Hashcash. Hashcash utilized the cryptographic hash function SHA-1 (Secure Hash Algorithm 1) that would help email recipients to identify spam. Hashcash served as the inspiration behind the PoW consensus mechanism used within the Bitcoin distributed ledger. Once a transaction is verified, it is broadcasted over the network for a unified agreement or consensus.

More on Consensus Protocols are detailed in Chapter 2.

None of the above three technologies are new. But bring all the three together, and we have the birth of a cutting-edge technology called Blockchain.

1.3.3 How Blockchain Technology Works

One of the features in the blockchain is digital trust. In the digital world, trust relates to two questions: "Are you whom you say you are?" and "Do you have the right to do what you want to do?" This boils down to proving identity (authentication) and proving permissions (authorization).

Authentication is addressed with cryptography. Blockchain utilizes public-key or asymmetric cryptography that involves a public key and a private key (refer Fig. 1.5). When a private key is created, its public key pair is also created using a complex algorithmic process. Both the public key and private keys are generated and stored in a place referred to in the Blockchain world as "wallet." The Wallet is a software program that not only stores your private and public keys but also facilitates the sending and receiving of cryptocurrency through the blockchain, while also monitoring your balance. More on the wallet is explained in Chapter 3: Cryptocurrency.

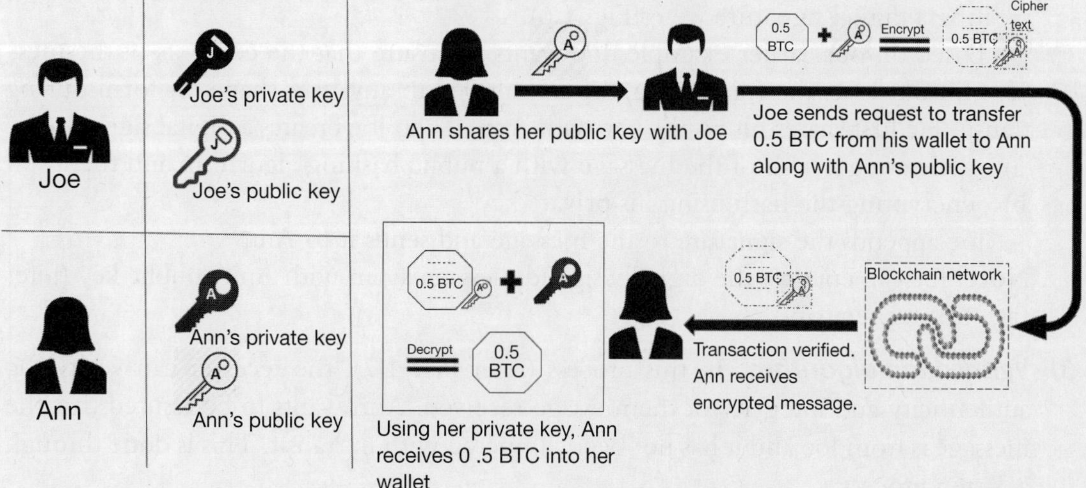

Figure 1.5: Public-key cryptography

While the private key must be maintained confidential, the public key may be distributed to anyone. You can relate the public key to your bank account number that you would need to share to receive money while your private key is similar to your bank account password or PIN that only you can access. So if anyone sends you a message with your public key, only you can open it with your private key provided no one else has access to your private key. Thus it goes without saying that the private key should not be shared with anyone.

The public key is shared across. Therefore, how can we guarantee the authenticity of the data and ensure that it will not be tampered with in transit? The answer lies in the digital signature. A digital signature provides validation and authentication in the same way signatures do but in digital form.

The digital signature algorithm generally comprises of three parts:

1) ***Generation of the private and public key:*** The keys are used to encrypt and decrypt the message in PKI.

   Let's say Joe wants to send some bitcoins, of say 0.5 BTC, to Ann (refer Fig. 1.5). Ann shares her public key with Joe. Joe takes 0.5 BTC from his wallet and, along with Ann's public key (encryption), sends the request. The transaction is routed through the blockchain network, where it is verified and sent to Ann. Ann can decrypt using her private key and receive 0.5 BTC into her wallet.

   Thus with public-key cryptography, blockchain can ensure the **security** of the transactions. Also, the transactions are time-stamped, thus making each transaction **unique**.

2) ***Signing algorithm:*** A digital signature produced through public-key cryptography safeguards the integrity of the data that is sent. This is done by combining the private key with the data that they wish to certify, through a mathematical algorithm, thus creating a digital signature (refer Fig. 1.6).

   Based on our earlier example, Joe wants to ensure that the contents of the message (in our case, the transaction) are not altered in any way, shape, or form during transit. He first shares his public key with Ann. Then Joe creates a digital signature by
   a) generating a hash of the message with a public hashing algorithm and then
   b) encrypting the hash using his private key.

   Joe appends the signature to the message and sends it to Ann.

   Note: Joe can encrypt the digitally signed message again with Ann's public key (refer Fig. 1.6) for security.

3) ***Verification algorithm:*** In this process (refer Fig. 1.7), the receiver can verify the authenticity and integrity of the message received. Ann wants to be assured that the message is from Joe and it has not been tampered with in transit. This is done through a 3-step process:

   Step 1: Ann decrypts the digital signature using Joe's public key to get the hash.

   Note: Only Joe's public key can decrypt the hashed message that was encrypted with Joes' private key.

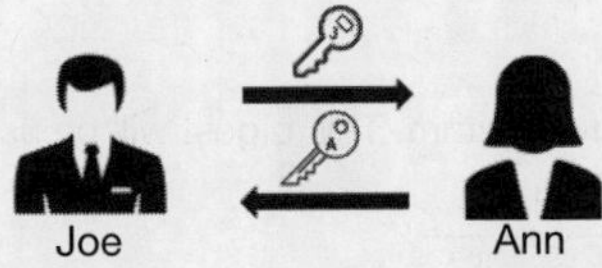

1. Joe and Ann share their public key with each other.

2. Joe generates a hash of the transaction and using his private key encrypts the hash thus creating his digital signature.

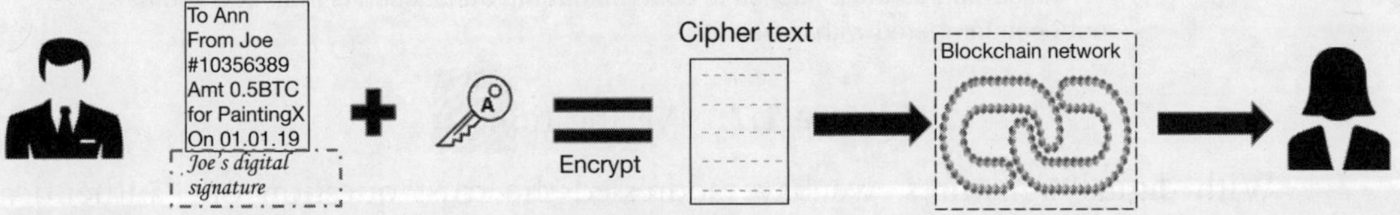

3. Joe encrypts the digitally signed transaction file with Ann's public key and sends it to Ann via the blockchain network.

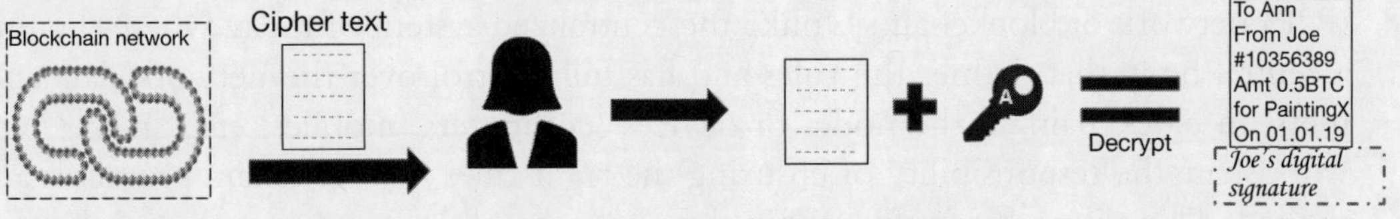

4. Ann receives the encrypted transaction file. She decrypts the file using her private key to access Joe's digitally signed document

Figure 1.6: Digital signature

Step 2: Ann runs the hash function on the message she received to generate the hash
Step 3: Ann compares the two hashes. If both are the same, Ann is assured that the message is from Joe (as Joe's private key decrypted the message).

Although the verification process seems like much work, it is generally done by a software application, and the process is typically invisible to the end-user.

Digital signatures are what gives the data recorded on a blockchain its **immutability**. Thus encryption and digital signatures are the de facto ways blockchain technology guarantees the security, authentication, integrity, and non-repudiation requirements of the data or transaction.

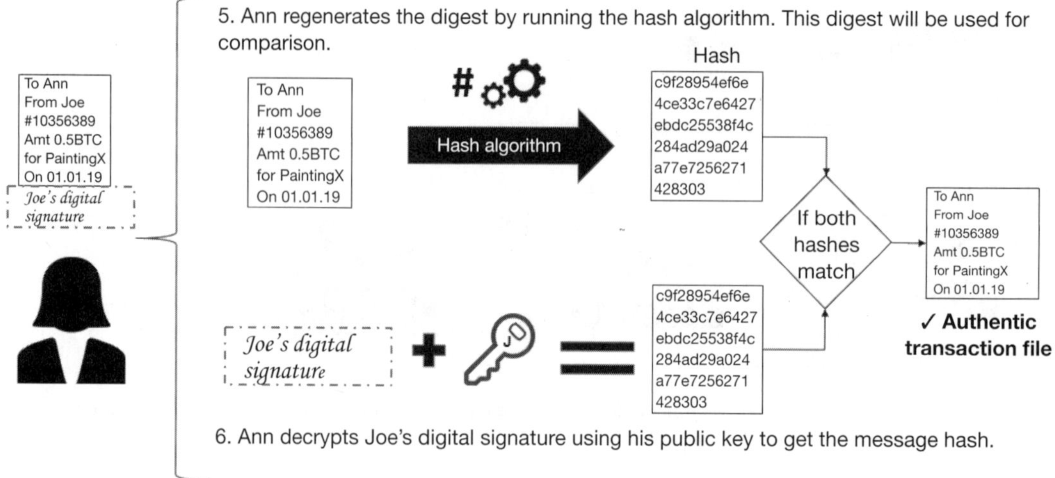

Figure 1.7: Verification

With digital signatures, we have addressed the trust question on "authentication," i.e., "Are you who you say you are?" However, what about the trust question on "authorization," i.e., "Do you have the right to do what you want to do?" The authorization question is addressed by the distributed consensus on a peer-to-peer (P2P) network of blockchain. Unlike the centralized system of today where there is a central body that defines the rules and has full control over the network data and users, in blockchain all the nodes or devices (computers, mobiles, etc.) in the network share the responsibility of ensuring the right rules or blockchain protocols are applied. There is no hierarchy within the network. All nodes are peers, each having an identical copy of the database or ledger. Hence the Blockchain network is called a peer-to-peer network since the tasks of maintaining the network are continuously shared from peer to peer. This spreads out control and responsibility among lots of different peers, thus ensuring no single point of failure and reduces the risk of corruption.

Blockchain leverages Distributed Ledger Technology to support immutable transactions on a P2P network without any centralized coordinating entity. Authentication and Authorization in blockchain systems are maintained using cryptographic signatures and consensus-based validation procedures. More details on Distributed Ledger Technology (DLT) and Consensus Mechanisms are explained in Chapter 2.

1.4 Components of Blockchain

So far, we learned about the origin of Blockchain, its inherent concepts and the technologies that make blockchain transaction secure, immutable, and indisputable. But from where does blockchain get its name, and what makes it unique? To understand the block in Blockchain, we need first to understand some of the concepts, terms, and components used in Blockchain technology.

1.4.1 Node

A node is an electronic device (computers, mobile devices, servers, etc.) that is connected to the internet. In blockchain parlance, any computer or hardware device that is connected to the blockchain network is a node. All the nodes in the network have a copy of the blockchain ledger and are interconnected. So theoretically, we can say that the Blockchain exists on nodes whose primary purpose is to preserve the integrity of the blockchain. The node supports the network by maintaining a copy of the blockchain. A node can be:

a) A Full Node

The node maintains a full copy of the transaction history of the blockchain. Computers run full nodes to help synch the blockchain. They also help the network by processing and accepting transactions/blocks, validating those transactions/blocks, and then broadcasting them to the network. This is called mining or forging, explained later in this section.

b) A Partial or Lightweight or Light Node

Nodes maintain only a partial copy of the ledger as they could be early users or those who do not have sufficient disk space for the full blockchain. Light nodes download only the block headers to validate the authenticity of transactions using a method called SPV or Simplified Payment Verification. They rely on the full nodes for the latest headers, account balance, and any transaction that affects their wallet.

However, a blockchain network needs to have more full nodes operating within the network to make it a truly trustless and decentralized system.

As a principle, blockchain is open to all. Anyone can join the network, i.e., be a node and participate in the blockchain network to validate and create blocks. This is called a **public permissionless blockchain**, e.g., Bitcoin and Ethereum. However, based on business needs, other types of blockchain are evolving – such as the private permissioned blockchain and consortium blockchain, where the blockchain is not open to everyone. Nodes need to fulfill pre-requisite criteria to join and/or to participate in the blockchain. Examples include Ripple, Hyperledger Fabric, and Monero. Different types of blockchain will be explored in Chapter 2.

1.4.2 Ledger

A ledger, in blockchain technology, refers to a digital database of information that is immutable. Blockchain is commonly referred to as a public distributed decentralized ledger. Let us see why.

a) Ledger Is Public

Anyone in the blockchain has access to the ledger and can read or verify the transactions therein.

b) Ledger Is Distributed

All the nodes in the blockchain network have a copy of the blockchain ledger. The traditional database works in a client–server environment, while the blockchain works on the principle of replication in every node.

c) Ledger Is Decentralized

Blockchain protocols are built such that no one node or group of nodes has excessive control over the ledger. There is no central control; hence it is decentralized with no single point of failure. So, while the traditional database works on central principle integrity (only the central body can validate the record), the blockchain works on the principle that anybody can validate the records.

Also, the traditional database works on the CURD (Create, Read, Update, Delete) principle. In contrast, the blockchain works in the principle of Append-only, i.e., blocks are only added to the existing blockchain (Section 1.5).

1.4.3 Wallet

A Wallet in the blockchain world is a digital wallet that allows users to manage cryptocurrency like bitcoin, litecoin, ether, etc. With a blockchain wallet, one can receive and send cryptocurrency. The term "wallet" is a misnomer, as real money is not stored. When Person A sends cryptocurrency coins to Person B's wallet, Person A is, in effect, signing off ownership of the coins to Person B. However, Person B can spend the coins only if the private key stored in Person B's wallet matches the public address the cryptocurrency is assigned to. If the keys match, the balance in Person B's wallet increases while that in Person A's wallet decreases accordingly. It should be noted that there are no real coins at play here. The said transaction is a new record added on the blockchain along with a change in the balance in the cryptocurrency wallet.

The wallet provides all the features that are needed for a safe, easy and secure transfer of funds between two parties, namely,

a) Privacy Is Maintained

Whenever a user creates a wallet, the public and private key associated with the wallet is also generated. It can be compared to how your email account works. The public key is like email id, and the private key is your email id password. Just like you share your email id to receive an email, you share your public key to receive funds (Section 1.3.3). However, your identity and personal details are kept private.

b) Transactions Are Secure

The private key is used to send funds as well as to open encrypted messages. This keeps the transactions secure.

c) Ease of Usage

Wallets can be installed and accessed from the web, desktop, or any mobile device. Transfer of funds is relatively instantaneous, without any geographical constraints or intermediaries like banks.

d) Currency Conversion

Wallets help you to transact across various types of cryptocurrencies like BTC, ETH, XMR, LTC, and others, without worrying about the currency conversion.

Did you know?

According to Statista, one of the world's largest consumer and market data providers, the number of blockchain wallets has reached nearly 35 million users at the end of March 2019 since its inception in 2009 with the bitcoin wallet.

There are around 120 wallets that are used for cryptocurrency transactions today. Blockchain.info, Coinbase, Mycelium and Electrum are some examples of blockchain wallets.

1.4.4 Nonce

A nonce is a number generated randomly that can be used just once in the cryptographic communication. Adding a nonce to a transaction's identifier makes it additionally unique, thus reducing the chance of duplicate transactions. Nonce is the key to creating a block in a blockchain database.

1.4.5 Hash

A hash function can take data of any size, perform an operation on it, and return a "hash" that is a data of a fixed size. Whether it is a single sentence or the Oxford Dictionary, the resulting hash will always be of the same size.

Critical characteristics of hashing are:

a) It creates an almost unique identifier. In the blockchain, hashes are used as identifiers for blocks, transactions and addresses. In the Bitcoin, blockchain hashes are 256 bits or 64 characters. The hashing algorithm used in the blockchain is called SHA-256. SHA stands for Secure Hash Algorithm generating a 256-bit hash.

Input (A Text String)	Hash Result (SHA-256)
OK	565339bc4d33d72817b583024112eb7f5cdf3e5eef0252d6ec1b9c9a94e12bb3
The world is a balloon!	7e4a05ad886f9dbf1f0a167c2a5d5dd5e41b853ad5e0c331e065c2fb2c85b3da
The world is a balloon	6978378fe2785a3159b6a5e5284abc7e4140ad829a949799b7419fd72ac74813
the world is a balloon	84261d4cc1eb9e0571fb5c247f434104e10b8268846c33c2bb63322d8309e8c9

- No matter the size of the input string, the output is always 64 characters
- A completely different hash result though the change in text string "The world is a balloon!" is minute, i.e., removal of "!" or replacing "The" with "the". Provides security from tampering during transmission.

Figure 1.8: Hash output example

b) It is one-directional and hence a right candidate for encryption (Section 1.3.3). A hash function can take a string or input of any length to create a fixed-length data output. However, we cannot take the data output and recreate the string/input.

c) A very minute change gives a different hash making it one of the most secure functions (refer Fig. 1.8).

d) It keeps the database small. As mentioned earlier, the data output is always a fixed size. Hence, storing the hash of a file instead of the original file considerably reduces the size of the database. For example, you can hash the photo of the painting, including details of the painter, when and where it was painted, etc. and create a hash output. Only the hash needs to be stored in the database since the hash is mathematically associated with the original painting and data.

A hash with 64 alphanumeric characters creates an astronomical number of combinations, and there is little chance of duplication. More on hashing and SHA-256 is explored in Chapter 4.

1.4.6 Mining

Mining is the mechanism whereby nodes called "miners" in the Bitcoin world or "forgers" in the Ethereum world validate new transactions and add them to the blockchain ledger. Miners/forgers compete to solve a complex mathematical problem based on a cryptographic hash algorithm referred to earlier, basically the nonce and hashing. Mining comprises hashing a block and then introducing a nonce to the hashing function and running the hash all over again.

However, this is where the complexity occurs: The resulting hash value should be less than the "target value." The target value is a number that a hashed block header must be less than, for a new block to be awarded. If the hash is not less than or equal to the target value, the miner has to increment the nonce and rerun the hash (refer Fig. 1.9). This process is run multiple times until the required hash is reached. Once the solution is found, the new block is added to the network and propagated.

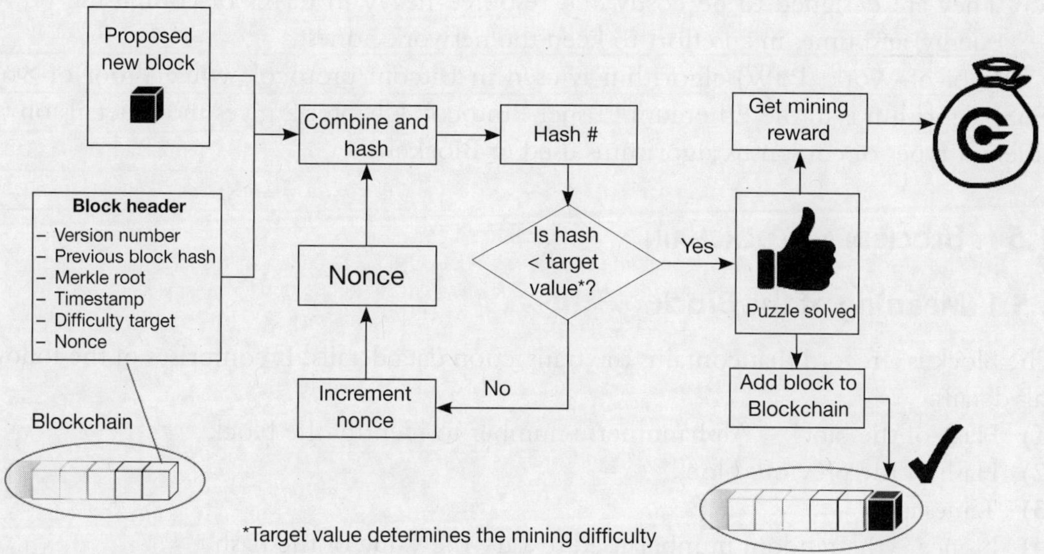

Figure 1.9: Mining process

Miner nodes compete to solve a complex mathematical problem using massive computing power, energy, and time. The first miner to create the winning hash will receive rewards in the form of transaction fees or new bitcoins. This process of computing the hash is called proof-of-work or consensus mechanism and provides **integrity** to the blockchain. Adding blocks to a blockchain without the consensus mechanism makes it highly vulnerable to either accidental faults or malicious attacks from hackers.

More details on consensus mechanism and mining can be found in the forthcoming chapters.

1.4.7 Consensus Protocol

Consensus protocols are a set of rules whereby nodes in a network can achieve agreement on the data value or state of the network such that it benefits the network as a whole and does not focus on individual interests. In the decentralized world of Blockchain technology, all the participating nodes must agree on a single source of truth, for example, whether Joe has enough money in his wallet or he is double-spending. Consensus protocols are used in blockchain to ensure that all transactions are validated before being added to the blockchain, i.e., there should be a 'consensus' or agreement between the nodes on the network on the state of a blockchain. This allows for three critical functions of the blockchain:

a) Consensus protocols
b) Spread control between nodes, thus preventing any single entity from taking control or disrupting the blockchain system. Consensus rules aim to guarantee that a single chain is used and followed.

c) They are designed to be costly and resource-heavy in terms of computing power, energy, and time, in an effort to keep the network honest.

Proof-of-Work (PoW) algorithm is used in Bitcoin protocol, while Proof-of-Stake (PoS) algorithm is in the Ethereum Casper Protocol. Chapter 2 gives more details on the different types of consensus algorithms used in Blockchain.

1.5 Block in a Blockchain

1.5.1 Meaning of the Block

The block is a record that contains the transaction data details. It comprises of the following details:

1) Hash of the block – Alphanumeric number to identify the block
2) Hash of the previous block
3) Timestamp
4) Nonce – the random number used to vary the value of the hash
5) Merkle root – hash of all the hashes of all the transactions in the block
6) Transaction data. This contains details of several transactions

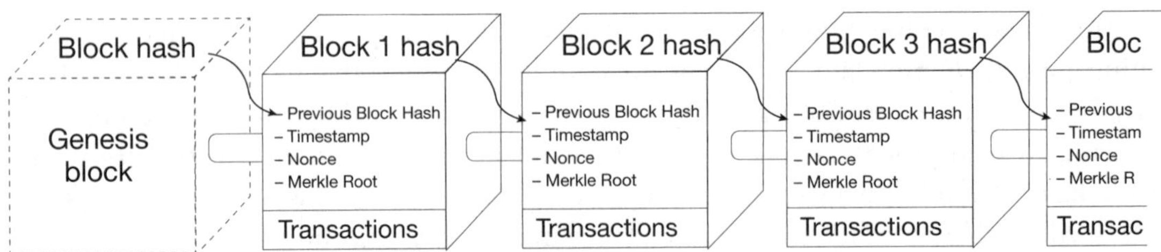

Figure 1.10: Block representation

The genesis block is the first block. It is the only block with no data hash of the previous block because no block precedes it. The genesis block contains transactions that are combined and validated to produce a unique hash.

The genesis block hash, along with all the new transactions, are processed and used as input to create a new hash that is used in the next block, i.e., say Block 1 (refer Fig. 1.10). Block 2 hash is created with a hash of Block 1 and another set of new transactions. This goes on with each block linking back to its previous block via its hash, thus forming a chain leading back to the genesis block. Hence the name, blockchain.

This continuous linkage makes it impossible for any malicious actor to alter the information or to insert a block between two existing blocks. To do so, all connected blocks would need to be altered as well. As a result, each block strengthens the previous block, including the security of the entire blockchain, because a bigger chain means more blocks would need to be changed if one wants to tamper with any information.

1.5.2 Blockchain Transaction in a Nutshell

Let us look at all the high-level steps involved in a blockchain transaction.

Remember:

- ✓ Blockchain is a digital ledger or digital database
- ✓ The blockchain ledger is distributed to all the nodes in the network, i.e., all the nodes have the same copy of the ledger
- ✓ Blockchain is decentralized, i.e., there is no central control. All the nodes in the network can participate in the processing and creation of a block
- ✓ A unique cryptographic key secures every record on the blockchain

Let us take our example of Joe's plan to send 0.5 BTC to Ann through the blockchain.

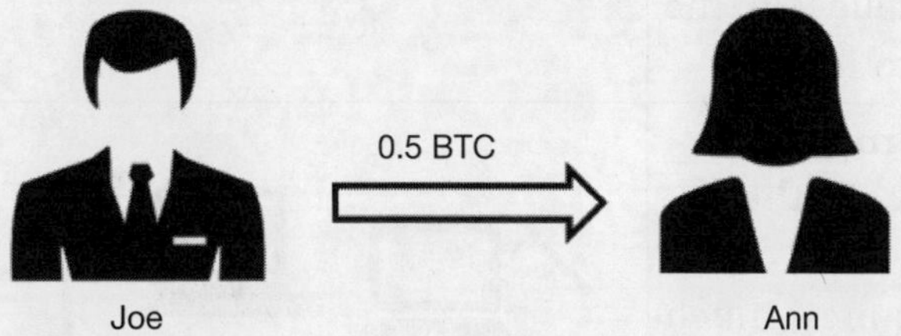

The following table gives a step-by-step representation of the transaction between Joe and Ann in the blockchain.

Table 1.2 Step-by-step representation of a blockchain transaction

Step 1: Joe requests the proposed transaction Joe sends 0.5 BTC from his Wallet app.	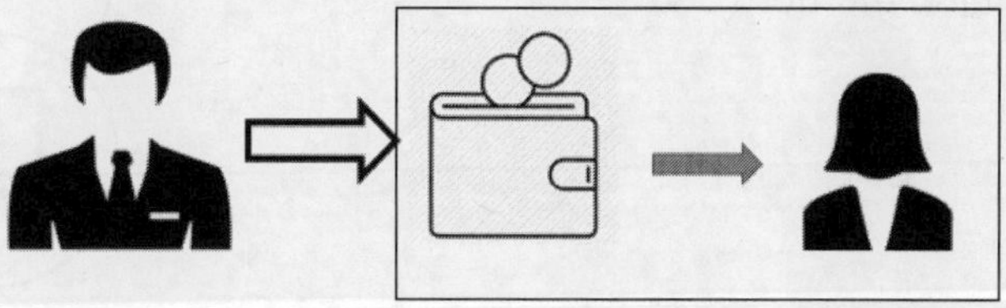
Step 2: The proposed transaction is broadcast to the network	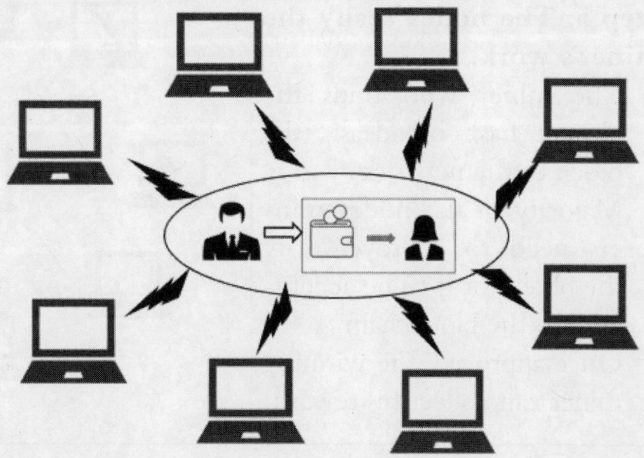

(*Continued*)

Table 1.2 (*Continued*)

Step 3: Miners verify the transaction and bundle it into a block along with other transactions. – The miner will validate the authenticity of the transaction, i.e., the status of Joe, his balance, etc. **Note**: Miners validate all the transactions they wish to include in the block they plan to mine.	
Step 4: Miners compete to solve the complex mathematical puzzle. – The puzzle requires much computational power to solve. – This protects the blockchain against hackers as it would be difficult and expensive to attack the network.	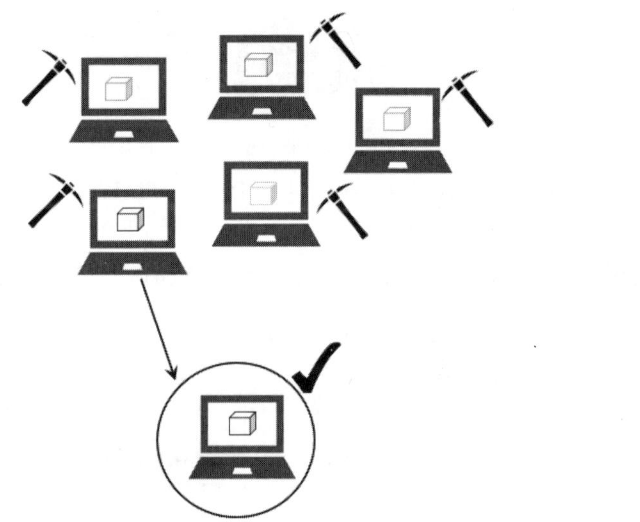
Step 5: The nodes verify the miner's work. – The miner who finds the correct hash broadcasts the block to the network – Majority of the nodes/miners need to approve/verify the block for it to be accepted into the blockchain – Once approved, the winning miner can collect his reward.	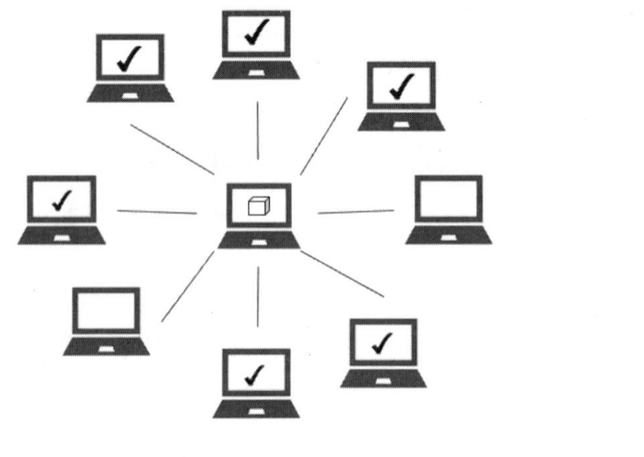

(*Continued*)

Table 1.2 (*Continued*)

Step 6: Block is added to the blockchain. – Once the block is verified, the winning miner adds his block to the existing blockchain. **Note**: Joe's transaction is added to the blockchain along with the other transactions	
Step 7: The updated copy of the blockchain is circulated throughout the network.	
Step 8: Transaction completion Ann receives 0.5 BTC in her wallet. The transaction is complete.	

So in a nutshell, the process of blockchain transaction consists of:

- A node in the blockchain (P2P network) requests a transaction via a wallet.
- The transaction is broadcasted to all the nodes in the network.
- The transaction is validated/verified by the network using consensus algorithms, i.e., preset rules set by the specific blockchain.
- The transaction is either accepted or rejected. If accepted, the transaction is added in a chronological order along with other transactions to create a new block of data that is sealed (hash).
- The transaction is now part of the blockchain and is permanent and immutable.

1.5.3 Double-spending

With the ever-growing popularity of blockchain, more and more people are moving towards transacting in digital currency. This brings about the double-spend problem, a flaw that is unique to digital currencies. Double-spending, as the name suggests, is spending the money more than once. Just as one can copy a digital file and send it to several people, it is possible to duplicate crypto-coin or token and reuse it. If this occurs in the blockchain network, it could not only breakdown the concept of trusted distributed ledger but also lead to inflation with fraudulent, duplicate currencies in the network. Remember, there is no trusted third-party to validate that the transaction is not a double-spend.

The double-spend problem is circumvented in blockchain through its consensus mechanism and the basic chronological structure of how the blocks are chained together.

A transaction is verified and added to a block via the consensus mechanism after a considerable amount of computational power and resources are spent. Going back and attempting to double-spend that transaction would require the same if not more computational power, especially with more blocks added to the chain in the interim; it is practically impossible to modify all the blocks. No one will expend enormous amount of resources if the payback is not worth the effort.

Also, the first transaction would be time-stamped and cryptographically linked to previous blocks and broadcasted to all nodes in the network. When the second (fraudulent) transaction is proposed, it will fail at the verification process and be rejected.

Once a transaction is confirmed, it is nearly impossible to double-spend it. The more confirmed blocks in the chain, the harder it is to double-spend the crypto.

However, it is theoretically possible to double-spend a cryptocurrency. Though rare, this can be done by the 51% attack (where there is control of more than half the network's hash rate), Finney attack, or Race attack. More on blockchain attacks are explained in Chapter 11.

1.6 THE TECHNOLOGY AND THE FUTURE

1.6.1 Blockchain Layers

A simple structure of the blockchain ecosystem is represented in Fig. 1.11.

It typically consists of:

a) The Presentation Layer

This layer is a door to users that gives access to the whole network and blockchain applications. The UI (User Interface) and UX (User Experience) of the blockchain app come into play here. In today's competitive market, mobility and ease of usage are paramount to customer satisfaction. It is no different for blockchain where UI and UX design could make or break a blockchain adoption. A good UX design will enhance customer satisfaction and loyalty by improving the product usability and interaction between the user and the rest of the blockchain ecosystem or company. This could relate to faster speeds, relevant information, etc. UI design produces blockchain visuals or graphical presentations. It enhances the user's experience with aesthetics, visual hints, and icons leading to an intuitive interface.

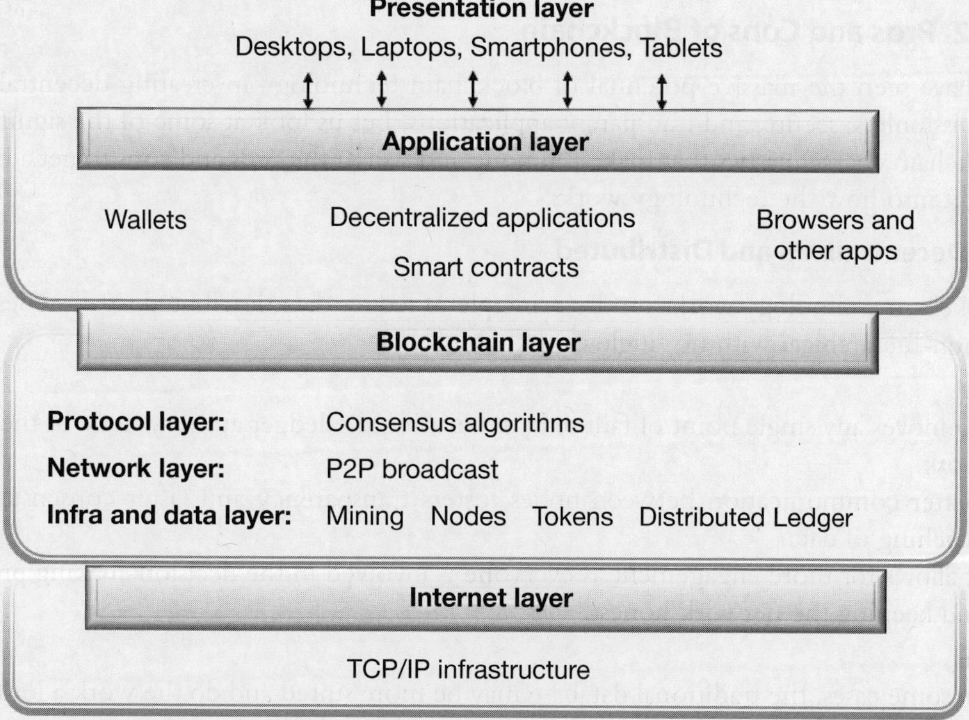

Figure 1.11: Blockchain ecosystem

b) The Application Layer

This is the layer than combines the business logic with user interactions. It consists of the Decentralized Application, better known as DApps running on a P2P network. It is the DApps that sets the communication between the Presentation and Blockchain layers. DApps is similar to a web application except that while APIs are used to connect a website to a database, DApps connects to the Blockchain via Smart Contracts. The Smart Contract is a program containing a set of rules and criteria that automatically execute if/when the pre-defined conditions are met. More on Smart Contracts and DApps are explained in other chapters within this book.

c) The Blockchain Layer

This is where the core of the blockchain subsists. It consists of the consensus algorithms (PoW, PoS, pBFT, etc.), the medium, and interface for the P2P network that decide how data is packetized and transmitted between peers. It controls the mining layer, protocols that decide the consensus methods and participation, nodes that execute protocols, and the distributed ledger.

d) The Internet Layer

Blockchain works over the internet. This layer ties all the networks together, i.e., the computers, IoT devices, smartphones, etc.

1.6.2 Pros and Cons of Blockchain

We have seen the massive potential of blockchain technology in creating decentralized, permissionless, secure, and transparent applications. Let us look at some of the significant blockchain characteristics that make it unique and weigh the pros and cons to get a better insight into how the technology works.

a) Decentralized and Distributed

Blockchain technology works on the principle of ledger data distributed to all nodes that are non-hierarchical with no single central control.

Pros

- Removes any single point of failure by replicating the ledger at every node in the network.
- Better communication between nodes fosters transparency and faster consensus and synching of data.
- It allows for more engagement as everyone is involved in the decision-making process and keeping the network honest.

Cons

- In some cases, the traditional database may be more suited and do the work a lot faster and cheaper
- Specific and trusted third parties exist in some domains that may guarantee more efficient and specialized services using other technologies
- If a time-tested and fully functional database and the operational network are already in place, the benefits of replacing or introducing blockchain may not produce the required return on investment.
- Stronger players (nodes with higher computing power or with pooling) can take control of the network, impacting decentralization.

b) Trustlessness

In the blockchain, cryptography completely replaces the need for third parties to ensure trust. Also, the complex consensus protocols that are run within the blockchain network to unanimously and securely agree on what should be added and should not be added to the ledger secures it further, thus ensuring its integrity at all times.

Pros

- Allows for multiple entities or key players who do not trust each other (i.e., unknown to each other or across borders) to transact directly with one another.
- Ensures valid and accurate data.

– Disintermediation (removal of the middleman) reduces the overall cost of transacting.

Cons

– The integrity of data is obtained at the expense of time. Every node needs to run the blockchain to verify transactions and maintain consensus. Currently, blockchain can, on an average, process only 5 transactions/sec.
– Significant computing power is expended by miners leading to substantial energy consumption and wastage. Hence, it is not suitable for organizations that require instant transaction results within milliseconds.
– Nodes may prioritize transactions with higher rewards.

c) Immutability

In blockchain technology, one cannot modify data or transactions once they are recorded in the blockchain database. It becomes a permanent record that is close-to-impossible to undo. Any change required can be addressed only by adding a new block of data to the existing chain of blocks in chronological order, ensuring that the database is complete and consistent.

Pros

– Contains a verifiable record of all transactions made that is auditable
– Consensus algorithms and block propagation mitigate the risk of double-spending, fraud, and manipulation of data.
– There is provenance, i.e., ability to track transaction or product movement across accounts.

Cons

– Not every node has the capacity to maintain and run a full copy of the blockchain. This can potentially affect consensus and immutability.
– In smaller blockchains, there is a risk of a 51% attack. If one or group of malicious nodes can get 51% of the mining hash rate, they can manipulate the transactions.
– Quantum computing can potentially break the cryptographic algorithm to reverse engineer public keys of blockchain networks to obtain the private keys.

With FinTech investing in blockchain technology and innovations, there are resolutions to many of the blockchain issues faced. It is the onus of the organization to evaluate the pros and cons of implementing blockchain technology solutions. We shall deal with this in more detail in Chapter 11.

1.6.3 Potential Applications in the Industry

The issues relating to blockchain technology may be significant, but its inherent benefits of immutability, openly shared ledger, security and non-dependence of intermediaries means that blockchain cannot be ignored. All agree that blockchain is way faster than the manual validation and audit process. Following are some of the potential applications that could cut across any industry and revolutionize the way we do business:

a) Digital Identity

In today's world, proof of identity is a must for verification; be it your security number, passport, birth certificate, driver's license, etc. Without a valid form of id, one cannot open a bank account, own property, get government services, or employment. Digital identities on the blockchain can reduce the burden of maintaining physical copies or ID cards and protect from identity theft, providing control over your own identity (to share only specific information that is relevant for the purpose). Blockchain technology can provide a breakthrough for helping people who do not have legal identities such as refugees and people who live way below the poverty line and do not have ids.

Elections are another area that could benefit from digital identity. There will be greater voting participation if the voting system has a verifiable audit trail with no fraudulent or illegitimate votes, while still keeping the votes confidential.

Adoption of digital identity under blockchain technology would substantially eliminate commonly occurring issues such as impersonation and fraud.

b) Payments and Settlement

Swiss bank UBS and UK-based Barclays are exploring blockchain solutions as a means to expedite back-office functions and settlement. Some in the banking industry say this could cut up to $20B in middleman costs. Banks and other financial sectors are exploring various cases of possible blockchain use in areas such as payments and settlement of currencies, asset registries, enforcement and clearing derivative contracts, regulatory reporting, KYC, AML registries, improving post-trade processing services, etc.

c) Proof of Ownership

Blockchain can be used to track any asset, be it jewelry, property deeds, vehicle, paintings, or any artefact, for proof of ownership. Fake and stolen goods are a significant issue impacting global commerce. Currently, proof of ownership is mostly paper-based that can be stolen or faked. With blockchain, all the properties of the asset can be digitized and stored in the blockchain. As the blockchain ledger is public and immutable, the ownership of the asset can be tracked to ensure its authenticity. This aids consumers and insurance companies in terms of time and money for evaluations, legal purchases, and arbitration.

d) Records Management

Industries that rely on the manual process of verification that is heavy on paper documentation and case-by-case checking can leverage blockchain technology. The education industry can benefit from blockchain solutions for verifying academic credentials, thereby reducing fraudulent claims of unearned educational qualifications. In healthcare, vast volumes of patient data can be stored securely and made accessible only to verified authorities, thus ensuring the privacy of patient data. The blockchain can securely store intellectual property and creative digital products like music, photos, software apps, etc.

e) Supply Chain Management

The undisputable nature of blockchain data makes it possible to track the journey of a product/product starting from its origin to its destination. There is hardly any industry without a supply chain, be it pharmaceuticals, food & beverages, automotive, construction, etc. When all parties see the same data and can verify shipment, receipt, and customs clearance, this cuts cost and streamlines the whole system. By using blockchain technology and removing paper-based trails and intermediaries, businesses can quickly pinpoint inefficiencies within their supply chains is nearly real-time. Contaminated products can be tracked and traced back to the source for containment and prevention. Some say that supply chain and logistics companies can benefit the most from adopting blockchain solutions.

f) Loyalty Programs and Rewards

The application of blockchain in Retail can step up the process of customer loyalty and reward programs. Blockchain technology-based token system that rewards customers and stores can be used to incentivize customers to return to a specific store or chain for their shopping. The very nature of blockchain will eliminate the fraud and waste commonly associated with paper and card-based loyalty programs with real-time updating of points balance and improving points management across franchised operations.

g) Decentralized IoT

Blockchain can support the Internet of Things (IoT) applications by supporting transaction processing devices. The distributed nature of the ledger can foster coordination among multiple devices. Smart Energy, Smart City, Smart Building and Smart Health can all leverage blockchain for cross-communication, data accuracies, permanence and privacy, all the while circumventing single points of failure.

h) Charity

Blockchain can address the issue of accountability and transparency in transactions related to charitable donations, thereby checking organizational inefficiency and financial misconduct that can prevent money from reaching those it was meant for. Blockchain provides the donors with the ability to precisely track their donations at all points until it reaches the right hands. It provides a permanent record of charitable financial transactions across the globe, thus driving stronger trust with donors.

Blockchain technology can be adapted to different business domains and industries according to standards and needs that each company requires. A more detailed look at current blockchain implementations and projects can be found in Chapter 10.

Summary

In this chapter, we have seen the history of Blockchain and how it got its name. Blockchain is a shared distributed digital ledger that maintains a permanent, transparent, immutable record of transactional data that is tamperproof and auditable. Data is organized into blocks that are cryptographically chained together in an "append-only" mode, thus forming a chain of blocks. Hence the name Blockchain.

Blockchain technology is what makes the blockchain transactions unique. Unlike traditional databases that work on client-server network architecture with centralized control, the blockchain is a type of digital ledger technology (DTL). The ledger exists on multiple devices called nodes that communicate with each other on the peer-to-peer (P2P) network. Transactions are verified, blocks validated, and created through cryptography and consensus protocols. Figure 1.12 sums up the characteristics of blockchain.

The central vision of the creator of blockchain-enabled bitcoin Satoshi Nakamoto was an electronic payment system that could remove the need for trusted third-parties and protect the digital currency from the double-spending problem. In the blockchain, asymmetrical cryptography and consensus algorithms completely replace third-parties as the governor of trust. Network nodes run complex consensus protocols to unanimously and securely agree on what should be added to the blockchain ledger. With transactional blocks being time-stamped and cryptographically linked to previous blocks and broadcasted to all nodes in the network, it is near- to-impossible to double-spend.

Figure 1.12: Characteristics of blockchain

Blockchain guarantees the integrity of transactions through digital signatures. Digital signatures secure not only the data but also the identity of the individual sending it. It ensures the authenticity of the data being sent and the person who is sending it.

Companies can leverage the benefits of deploying blockchain applications. The blockchain public ledger is secure, reliable, and transparent. Its distributed nature reduces the risk of a single point of failure. The complex consensus protocols that are run by miners authorize transactions to ensure an immutable, hacking resistant ledger. The trustless nature of blockchain ensures fast settlements, reduced operating costs, and savings.

Blockchain Technology has found user cases in several domains including but not limited to Banking, Healthcare, Automotive, Retail, Supply chain, and others. In the following chapters, we will delve deeper into the various concepts, components, various blockchain systems and platforms, challenges, and use cases.

EXERCISES

Multiple Choice Questions

1. **Blockchain is a type of:**

   A. Distributed ledger technology

   B. Client server

   C. Centralized ledger technology

   D. Physical ledger

   Answer: A

   Explanation: Technically, Blockchain is defined as a distributed, replicated peer-to-peer network of databases that allows multiple non-trusting parties to transact without a trusted intermediary and maintains an ever-growing, append-only, tamper-resistant list of time-sequenced records.

   In short, Blockchain is a type of a distributed ledger that sits on the internet for recording transactions and maintaining a permanent and verifiable record-set of information.

2. **Technically, the Blockchain and Bitcoin are not the same.**

   A. False

   B. True

   Answer: B

   Explanation: Though bitcoin and blockchain are often referred to interchangeably, they are not the same. Blockchain is the underpinning technology that the Bitcoin was built on.

3. **Bitcoin is a cryptocurrency, while blockchain is a ledger.**

   A. False

   B. True

Answer: B

Explanation: Bitcoin is a cryptocurrency, while Blockchain is a ledger or database of information.

4. **Bitcoin is limited to currency transactions, while the blockchain has numerous applications.**
   A. True
   B. False
   Answer: A
   Explanation: Bitcoin is limited to currency transactions, while the blockchain has numerous applications. It can not only trade currencies, property rights of stock but also be used for identity management, records management, research management, as well as many other areas that cover all aspects of the business.

5. **Bitcoin has high degree of**
   A. Transparency of users
   B. Access control
   C. Anonymity of users
   D. Identity control
   Answer: C
   Explanation: Bitcoin has a high degree of anonymity. Though the transactions are visible, it is close-to-impossible to identify the user. Until today, the identity of Satoshi Nakamoto is a mystery even though Satoshi has roughly one million bitcoins (BTC) worth around 8 billion dollars as of 2019.

6. **The first Bitcoin was mined in the year:**
   A. 2009
   B. 1999
   C. 2019
   D. 2015
   Answer: A
   Explanation: The first Bitcoin was mined on January 03, 2009, marking the genesis of Bitcoin.
   On January 12, 2009, Hal Finney received first Bitcoin transaction, thus launching the first application of a public blockchain.

7. **____________ was released as an alternative to Bitcoin with different mining algorithm and faster transaction speed.**
   A. Namecoin
   B. Heavycoin
   C. Altercoin
   D. Litecoin
   Answer: D

Explanation: Namecoin is the first Bitcoin fork. Litecoin was released as an alternative to Bitcoin with different mining algorithm and faster transaction speed.

8. **To date, ____________ is one of the slowest CryptoCurrency.**
   A. CORDA
   B. Ethereum
   C. Ripple
   D. Bitcoin
   Answer: D
   Explanation: The first generation of blockchain promised transparency, immutability, accountability, and security in transactions. However, the protocols used in the blockchain (proof-of-work) necessitated the use of heavy mining hardware and significant resources leading to problems in scalability, interoperability and speed. To date, Bitcoin is one of the slowest cryptocurrency, taking about 10 minutes to confirm a transaction.

9. **Ethereum requires ____________ energy for maintenance than Bitcoin.**
   A. lesser
   B. more
   C. exactly same
   D. almost same
   Answer: A
   Explanation: Though Ethereum required lesser energy to be maintained, there were still concerns about future scalability. However, with the hosting of Smart Contracts that made available new functionalities, Ethereum was the go-to blockchain for Enterprise use.

10. **Smart Contracts of Ethereum opened up the possibility of a ____________ consensus mechanism.**
    A. Proof-of-Work (PoW)
    B. Proof-of-Stake (PoS)
    C. Proof-of-Authority (PoA)
    D. RAFT
    Answer: B
    Explanation: The third generation saw the arrival of Hyperledger from the Linux Foundation and Decentralized Applications (DApps) of Ethereum. Though the Hyperledger is a platform that can plug in any consensus mechanism, Smart Contracts of Ethereum opened up the possibility of a Proof-of-Stake consensus mechanism.

11. **In the fourth generation of Blockchain, ____________ is expected to accelerate the processing speeds, without sacrificing security significantly.**
    A. Blockchain Scaling
    B. Blockchain Mining

C. Blockchain Propagation
D. Blockchain Difficulty
Answer: A

Explanation: Fourth Generation Blockchain (2018–future): Blockchain technology's future appears optimistic as many governments and organizations are investing heavily in innovations and applications. A significant innovation on the horizon is called blockchain scaling. A scaled blockchain is expected to accelerate the processing speeds, without sacrificing security significantly. This is done by assessing the computing power required to validate each transaction and then dividing the work efficiently. Progress in this front is yet to be seen, but the outlook has been positive and more is expected on this front.

12. **Satoshi Nakamoto was proposing a digital currency trading or payment solution that addressed the ____________ problem that is unique to digital currency transactions.**
A. liquidity
B. security
C. double-spending
D. transparency
Answer: C

Explanation: As part of the abstract from the paper *A Peer-to-Peer Electronic Cash System*, it is evident that Satoshi Nakamoto was proposing a digital currency trading or payment solution that addressed the double-spending problem that is unique to digital currency transactions.

13. **Business transactions are recorded in ledgers since time immemorial. History credits ____________ as the first record keepers around 7000 years ago.**
A. Iran
B. India
C. Australia
D. Iraq
Answer: D

Explanation: Business transactions are recorded in ledgers since time immemorial. History credits the Mesopotamians (now Iraq) as the first record keepers with evidence dating back to about 7000 years. Clay tablets were used to record the expenditures of goods received and traded.

14. **The P2P network was first introduced in 1969 with ____________, a precursor to the Internet where every participating node (computer) could request and serve content.**
A. Betanet
B. Arpanet
C. LAN
D. Intranet

Answer: B

Explanation: The peer-to-peer (P2P) architecture is the cornerstone of the blockchain network that makes the deed of trust from a mediator or third parties like banks, lawyers, etc. redundant. The P2P network was first introduced in 1969 with Arpanet, a precursor to the Internet where every participating node (computer) could request and serve content. Arpanet (Advanced Research Projects Agency Network) was the first packet-switching network and the first network to implement the TCP/IP protocol suite. Both technologies became the technical foundation of the Internet. The Arpanet was initially founded by the Advanced Research Projects Agency (ARPA) of the United States Department of Defense.

15. In a ___________ network, peers share resources among each other.

A. client–server network
B. centralized administrative system
C. thin client server network
D. peer-to-peer (P2P)

Answer: D

Explanation: A client–server network is where individual clients request services from centralized servers. In a peer-to-peer (P2P) network, interconnected nodes or peers share resources among each other without any centralized administrative system.

16. Public key cryptography enables two essential cryptographic capabilities, namely,

A. Confidentiality and Integrity
B. Anonymity and Integrity
C. Confidentiality and Availability
D. Manageability and Controllability

Answer: A

Explanation: Public key cryptography, also called asymmetric cryptography, was discovered in 1976 by two Stanford mathematicians, Whitfield Diffie and Martin Hellman. Used in PKI (Public Key Infrastructure), two keys, called the private key and the public key, are generated, which can be used to encrypt/decrypt a message. They need to be used in combination. The public key cannot decrypt the message encrypted by the public key. Similarly, the private key cannot decrypt the message encrypted by the private key. This enables two essential cryptographic capabilities, i.e., confidentiality and integrity.

17. If you relate the public key to your bank account number that you would need to share to receive money, private key is similar to your ___________

A. Bank Balance
B. Bank Account Password
C. Bank Transaction
D. Bank Name

Answer: B

Explanation: While the private key must be maintained confidential, the public key might be distributed to anyone. You can relate the public key to your bank account number that would need to be shared to receive money while your private key is similar to your bank account password or PIN that only you can access. So if anyone sends you a message with your public key, only you can open it with your private key provided no one else has access to your private key. Thus it goes without saying that the private key should not be shared with anyone.

18. In blockchain concept, a ____________ is an electronic device (computers, mobile devices, servers, etc.) that is connected to the internet.

A. Blockchain
B. Wallet
C. Ledger
D. Node

Answer: D

Explanation: A node is an electronic device (computers, mobile devices, servers, etc.) that is connected to the internet. In blockchain parlance, any computer or hardware device that is connected to the blockchain network is a node. All the nodes in the network have a copy of the blockchain ledger and are interconnected.

19. A ____________ in blockchain technology refers to a digital database of information that is immutable.

A. Blockchain
B. Wallet
C. Ledger
D. Node

Answer: C

Explanation: A ledger in blockchain technology refers to a digital database of information that is immutable. Blockchain is commonly referred to as a public distributed decentralized ledger.

20. In general, which of the following is not a characteristic of blockchain ledger?

A. Centralized
B. Public
C. Distributed
D. Decentralized

Answer: A

Explanation: The ledger is in public domain. Anyone in the blockchain has access to the ledger and can read or verify the transactions therein. It is also distributed since all the nodes in the blockchain network have a copy of the blockchain ledger. The traditional database works in a client–server environment, while the blockchain works on the principle of replication in every node. The ledger is decentralized as blockchain protocols are built such that no one node or group of nodes has excessive control over it. There is no central control; hence, there will be no single point of failure.

21. A ___________ is a number generated randomly, that can be used just once in the cryptographic communication.

A. Private number
B. Hash
C. Difficulty
D. Nonce
Answer: D

Explanation: A nonce is a number generated randomly, that can be used just once in the cryptographic communication. Adding a nonce to a transaction's identifier makes it additionally unique, thus reducing the chance of duplicate transactions. The nonce is the key to creating a block to a blockchain database.

22. A ___________ function can take data of any size, perform an operation on it, and return data of a fixed-size.

A. Private number
B. Hash
C. Difficulty
D. Nonce
Answer: B

Explanation: A hash function can take data of any size, perform an operation on it, and return a "hash" that is a data of fixed size. Whether it is a single sentence or the Oxford Dictionary, the resulting hash will always be of the same size.

23. Which of the following are the characteristics of a hash function?

A. One-directional
B. No direction
C. Bi-directional
D. Multi-directional
Answer: A

Explanation: Critical characteristics of hashing are: It is one-directional and hence a right candidate for encryption. A hash function can take an input or string of any length to create a fixed-length data output. However, we cannot take the data output and recreate the string/input.

24. ___________ are a set of rules whereby nodes in a network can achieve agreement on the data value or state of the network.

A. Hashing
B. Consensus
C. Mining

D. Nonce

Answer: B

Explanation: Consensus protocols are a set of rules by which nodes in a network can achieve agreement on the data value or state of the network such that it benefits the network as a whole and not focus on individual interests. In the decentralized world of Blockchain technology, all the participating nodes must agree on a single source of truth.

25. ____________ is a hash of all hashes of all the transactions in the block.

A. Merkle Root

B. Hash

C. Transaction data

D. Timestamp

Answer: A

Explanation: The block is a record that contains the transaction data. It comprises of following details:

1) Hash of the block – Alphanumeric number to identify the block

2) Hash of the previous block

3) Timestamp

4) Nonce – the random number used to vary the value of the hash

5) Merkle Root – hash of all the hashes of all the transactions in the block

6) Transaction data. (**Note**: This contains details of several transactions.)

26. The first block in a blockchain is called as:

A. First block

B. Front block

C. Genesis block

D. Parent block

Answer: C

Explanation: The Genesis block is the first block. It is the only block with no data hash of the previous block because no block precedes it. The Genesis block contains transactions that are combined and validated to produce a unique hash.

27. This is the only block with no data has of the previous block:

A. First block

B. Front block

C. Genesis block

D. Parent block

Answer: C

Explanation: Same as for Question 27.

28. This layer is a door that gives users access to the whole network and blockchain applications.

A. Presentation layer

B. Application layer
C. Blockchain layer
D. Internet layer
Answer: A

Explanation: The Presentation Layer is a door that gives users access to the whole network and blockchain applications. The UI (User Interface) and UX (User Experience) of the blockchain app come into play here. In today's competitive market, mobility and ease of usage are paramount to customer satisfaction. It is no different for a blockchain where UI and UX design could make or break its extent of adoption.

29. This is the layer that combines business logic with user interactions.

A. Presentation layer
B. Application layer
C. Blockchain layer
D. Internet layer
Answer: B

Explanation: The Application Layer is the layer that combines business logic with user interactions. It consists of the Decentralized Application, better known as DApps, running on a P2P network. It is the DApps that sets the communication between the Presentation and Blockchain layers.

30. This is similar to a web application except that while APIs are used to connect a website to a database, here, this connects to the Blockchain via Smart Contracts.

A. Connector
B. Blockchain Engine
C. Process Integrator
D. DApps
Answer: D

Explanation: DApps is similar to a web application except that while APIs are used to connect a website to a database; here, DApps connects to the Blockchain via Smart Contracts. Smart Contract is a program containing a set of rules and criteria that automatically execute if/when the pre-defined conditions are met.

31. This layer consists of the consensus algorithms (PoW, PoS, pBFT, etc.), the medium, and interface for the P2P network and decides how data is packetized and transmitted between peers.

A. Presentation Layer
B. Application Layer
C. Blockchain Layer
D. Internet Layer
Answer: C

Explanation: The Blockchain Layer is where the core of the blockchain subsists. It consists of the consensus algorithms (PoW, PoS, pBFT, etc.), the medium, and interface for the P2P network to decide how data is packetized and transmitted between peers. It controls the mining layer, protocols that decide the consensus methods and participation, nodes that execute protocols, and the distributed ledger.

32. **Which of the following is not a pro-blockchain that is Decentralized and Distributed?**
    A. It removes any single point of failure by replicating the ledger on every node in the network.
    B. Better communication between nodes that fosters transparency and faster consensus and synching of data.
    C. It allows for more engagement as everyone is involved in the decision-making process and keeping the network honest.
    D. Specific and trusted third parties exist in some domains that may guarantee more efficient and specialized services using other technologies.

    Answer: D

    Explanation: Option D is one of the disadvantages of Blockchain.

33. **In blockchain technology, one cannot modify data or transactions, once they are recorded in the blockchain database. This is called as:**
    A. Immutability
    B. Changeability
    C. Scalability
    D. Trustlessness

    Answer: A

    Explanation: In blockchain technology, one cannot modify data or transactions once they are recorded in the blockchain database. It becomes a permanent record that is close-to-impossible to undo. Any change required can be addressed only by adding a new block of data to the existing chain of blocks in chronological order. Ensuring this guarantees that the database is complete and consistent.

34. **Which of the following is not a pro-statement for a Blockchain which is Trustless?**
    A. Allows for multiple entities or key players who do not trust each other (i.e., unknown to each other or across borders) to transact directly with one another.
    B. Ensures valid and accurate data.
    C. Disintermediation (removal of the middleman) reduces the overall cost of transacting.
    D. The integrity of data is obtained at the expense of time. Every node needs to run the blockchain to verify transactions and maintain consensus.

    Answer: D

    Explanation: Option D is one of the disadvantages of Blockchain.

Short-answer Questions

1. **Define blockchain.**

   Technically, blockchain is defined as a distributed, replicated peer-to-peer network of databases that allows multiple non-trusting parties to transact without a trusted intermediary and maintains an ever-growing, append-only, tamper-resistant list of time-sequenced records. In short, Blockchain is a type of distributed ledger that sits on the internet for recording transactions and maintaining a permanent and verifiable record-set of information.

2. **What is the main difference between blockchain and bitcoin?**

   Though bitcoin and blockchain are often referred to interchangeably, they are not one and the same. Blockchain is the underpinning technology that the Bitcoin was built on. In addition to Bitcoin blockchain, we now also have the Ethereum Blockchain and other private blockchains that are adapted to cater to various industry applications such as supply chain, cross-border payments and many others.

3. **What is the scope of usage of bitcoin and blockchain?**

   Bitcoin is limited to currency transactions while the blockchain has numerous applications. It can not only trade currencies and property rights of stock but also be used for identity management, records management, research management as well as many other areas that cover all aspects of business.

4. **What is the difference between bitcoin and blockchain with respect to Transparency?**

   Bitcoin has a high degree of anonymity. Though the transactions are visible, it is nearly impossible to identify the user. For example, the identity of Satoshi Nakamoto, the original creator of bitcoin, is a mystery even though Satoshi has roughly one million bitcoins (BTC) worth around 8 billion dollars as of 2019. On the other hand, Blockchain is quite transparent as it is expected to work across multiple industry applications. However with Smart contracts and various consensus mechanisms, Blockchain can assure compliance to KYC and other industry standards.

5. **Write notes on first generation of blockchain.**

   In 2009, Bitcoin became the first application of the Blockchain technology. Satoshi Nakamoto formed the Genesis block. On 12 Jan 2009, the first successful bitcoin transaction on blockchain took place between Nakamoto and Hal Finney.

   The first generation of blockchain promised transparency, immutability, accountability and security in transactions. However the protocols used (Proof-of-work consensus) necessitated the use of heavy mining hardware and significant resources leading to problems in scalability, interoperability and speed. To date, bitcoin is one of the slowest cryptocurrency, taking about 10 minutes to confirm a transaction.

6. **Write notes on second generation of blockchain.**

   The second generation sought to overcome the limitations of the Bitcoin. Thus emerged Ethereum in 2013, when it was realized that the underlying technology of Bitcoin could be used for all kinds of B2C and other general applications. It still used Proof-of-work algorithm and had less-than-optimal speed. Though Ethereum required lesser energy to maintain, there were still concerns on future scalability. However, with the hosting of Smart Contracts that made available new functionalities, Ethereum was the go-to blockchain for Enterprise use.

7. Write notes on third generation of blockchain.

This generation saw the arrival of Hyperledger from the Linux Foundation and Decentralized Applications (DApps) of Ethereum. Though the Hyperledger is a platform that can plug in any consensus mechanism, Smart Contracts of Ethereum opened up the possibility of a Proof-of-Stake consensus mechanism. The focus of the generation was consensus mechanisms that can bring greater interoperability and boost network speeds. Promoting cross-chain transactions, using sharding (a type of database partitioning that separates huge databases into smaller, faster and more easily managed parts) and establishment of parallel chains are only some of the approaches being taken by the third generation blockchain solutions. However, EOS and TRON have taken over the DApps market due to the scaling issues still inherent in Ethereum platform.

8. Write notes on fourth generation of blockchain.

The future of Blockchain technology appears optimistic as many governments and organizations are investing heavily into innovations and applications. It does not seem far-fetched to suppose that one day there will be a public blockchain that anyone can use.

Advocates expect the technology to help in the automation of most tasks handled by professionals in all sectors by integrating AI, Big Data, IoT and others. A major innovation on the horizon is called blockchain scaling. A scaled blockchain is expected to significantly accelerate the processing speeds, without sacrificing security. This is done by assessing the computing power required to validate each transaction and then dividing the work efficiently. Progress in this front is yet to be seen but the outlook is positive.

9. Write notes on distributed consensus.

The distributed consensus protocol is fundamental in avoiding double-spending and other internet attacks. In 1992, researchers Cynthia Dwork and Moni Naor presented an approach to combat junk emails through a protocol known as the pricing function. Users would be able to assess their email service once they are able compute a function or puzzle by engaging their processing power. In 1997, cryptographer Adam Back proposed a similar function called Hashcash. Hashcash utilized the cryptographic hash function SHA-1 (Secure Hash Algorithm 1) that would help email recipients to identify spam. Hashcash served as the inspiration behind the PoW consensus mechanism used within the Bitcoin distributed ledger. Once a transaction is verified, it is broadcasted over the network for a unified agreement or consensus.

10. Write notes on public key cryptography.

Public key cryptography is also called as asymmetric cryptography. It was discovered in 1976 by two Stanford mathematicians, Whitfield Diffie and Martin Hellman. Used in Public Key Infrastructure (PKI), two keys called the private key and the public key are generated. These keys are used in combination to encrypt/decrypt a message. A message encrypted by the public key can only be decrypted by its matching private key and vice versa. In other words, the public key cannot decrypt the message encrypted by the public key nor can the private key decrypt message encrypted by the private key. This enables two essential cryptographic capabilities, i.e., confidentiality and integrity.

11. What is node in a blockchain?

A node is any electronic device (computer, mobile device, server, etc.) that is connected to the internet. In blockchain parlance, any computer or hardware device that is connected to the blockchain network is a node. All the nodes in the network have a copy of the blockchain ledger and are interconnected. So theoretically, we can say that the Blockchain exists on nodes whose primary purpose is to preserve the integrity of the blockchain. The node supports the network by maintaining a copy of the blockchain.

12. What is full node in a blockchain?

Node maintains a full copy of the transaction history of the blockchain. Computers run full nodes to help synch the blockchain. They also help the network by processing and accepting transactions/blocks, validating those transactions/blocks, and then broadcasting them to the network. This is called mining or forging.

13. What is light node in a blockchain?

Nodes maintain only a partial copy of the ledger as they could be early users or may not have sufficient disk space for the full blockchain. Light nodes download only the block headers to validate authenticity of transactions using a method call SPV or Simplified Payment Verification. They rely on the full nodes for the latest headers, account balance and any transaction that affects their wallet.

14. What is public permissionless blockchain?

As a principle, blockchain is open to all. Anyone can join the network, i.e., be a node and participate in the blockchain network to validate and create blocks. This it is called a public permissionless blockchain, e.g., Bitcoin and Ethereum. However based on business needs, other types of blockchain like the private permissioned blockchain and consortium blockchain are evolving, where the blockchain is not open to everyone. Nodes need to fulfill pre-requisite criteria to join and/or to participate in the blockchain. Examples include Ripple, Hyperledger Fabric, and Monero.

15. What is wallet in blockchain?

A Wallet in the blockchain world is a digital wallet that allows users to manage cryptocurrency like bitcoin, litecoin, ether, etc. With a blockchain wallet, one can receive and send cryptocurrency.

16. What is nonce in blockchain?

A nonce is a random number that can be used just once in cryptographic communication. Adding a nonce to a transaction's identifier makes it additionally unique, thus reducing the chance of duplicate transactions. The nonce is a key to creating a block to a blockchain database.

17. What is mining in blockchain?

Mining is the mechanism whereby nodes called "miners" in the Bitcoin world or "forgers" in the Ethereum world validate new transactions and add them to the blockchain ledger. Miners/forgers compete to solve a complex mathematical problem based on the nonce and cryptographic hash algorithms. Mining comprises of hashing a block and then introducing a nonce to the hashing function and running the hash all over again.

18. What is consensus protocol in blockchain?

Consensus protocols are a set of rules whereby nodes in a network can achieve agreement on the data value or state of the network such that it benefits the network as a whole without focusing on individual interests. In the decentralized world of Blockchain technology, all the participating nodes must agree on a single source of truth such as whether the user has enough money in his wallet or if he is double-spending. Consensus protocols are used in Blockchain to ensure that all transactions are validated before they are added to the blockchain, i.e., there should be a 'consensus' or agreement between the nodes on the network on the state of a blockchain.

Essay-type Questions

1. Write an essay on the evolution of blockchain.
2. Differentiate between traditional and blockchain transactions using an example.
3. Explain Peer-to-Peer Network, Public Key Cryptography and Distributed Consensus.
4. In the digital world, trust relates to two questions: "Are you who you say you are?" and "Do you have the rights to do what you want to do?" Explain blockchain with respect to the above two aspects.
5. Explain ledger and wallet in detail.
6. What is hash in a blockchain? What are the key characteristics of Hash?
7. Discuss the concept of mining in blockchain in detail.
8. Discuss the concept of block in blockchain in detail.
9. Discuss the concept of double-spending.
10. Explain blockchain ecosystem in detail.
11. Discuss the pros and cons of blockchain in detail.
12. Discuss the potential application of blockchains in the industry.

CHAPTER 2

Blockchain Types and Consensus Mechanism

LEARNING OBJECTIVES

This chapter covers the different types of Blockchain and provides a brief on the consensus algorithms in a decentralized network architecture where anyone can be a node and decisions are collectively reached. The consensus mechanism enables the trustless nature of Blockchain. Consensus algorithms form the governance mechanism that prevents malicious actors from tampering with the data or record-sets. Transactions are entered only with the consensus of all the relevant parties after they validate all relevant data.

2.1 INTRODUCTION

As described in the previous chapter on fundamentals, blockchain is a **Digital Ledger** of information that is:

1) **Distributed** – Every participating node has a digital copy of the blockchain database, whereby all can contribute to processing the blockchain.
2) Maintained by **Consensus** – The consensus algorithms are the governance mechanisms to guarantee that the data/records are legitimate and not tampered with. Transactions can be entered only with the agreement of all the relevant parties, on validation of the data.
   Hence, it is ensured that the records are
3) **Immutable** – Once consensus is reached on the validity of a transaction/data and recorded on the blockchain, it cannot be changed or deleted. Any subsequent changes are recorded in a new transaction block (refer Fig. 2.1);
   and
4) **Auditable** – The immutable tracking of a record with timestamp allows for the provenance of the asset at every step.

Figure 2.1: Characteristics of blockchain (Immutable)

With blockchain, the dependency of "trust" on third-party or intermediaries (like banks, brokers, lawyers) is made redundant by the direct peer-to-peer handshake within the network of nodes. The following characteristics enable the **Trustless** nature of the Blockchain:

a) **Decentralization** – The data is digital and decentralized, i.e., it can be shared across a network of computers or servers without the need for a central authority for making decisions.
b) **Transparency** – Data is transparent and open to everyone in the P2P network, i.e., anyone can view the transactions.
c) **Privacy** – User identity is kept private by robust cryptography.
d) **Security** – Transactions are cryptographically secure using hash algorithms.

However, even with its rich set of features, it cannot be plugged into any and every industry in its vanilla form. Some industries may consider the full transparency of data to be a security risk, especially for government and research organizations. Other industries find the accessibility to the network a source of concern.

To circumvent these business concerns, technologists have attempted to find ways to strike a balance between the organizational risk appetite and the blockchain's capability. The two main blockchain capabilities taken into consideration to enable fit for service are its decentralization and consensus mechanism. This chapter will explore the different types of Blockchain based on the kind of access and the various consensus mechanisms.

2.2 Decentralization and Distribution

2.2.1 Decentralization

One of the main characteristics of blockchain technology is its decentralization, where transactions are not under the control of any single party.

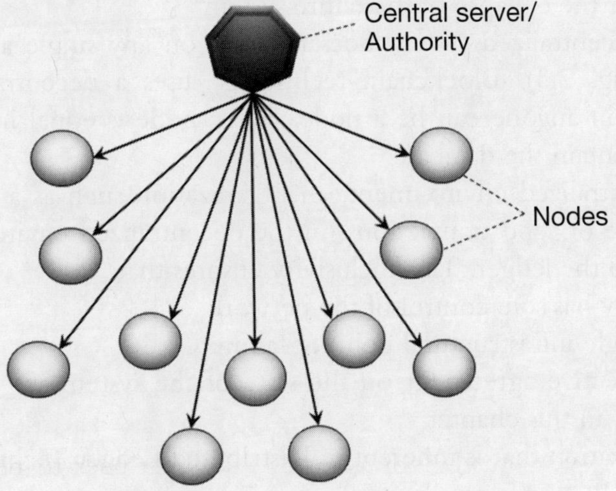

Figure 2.2: Centralized network

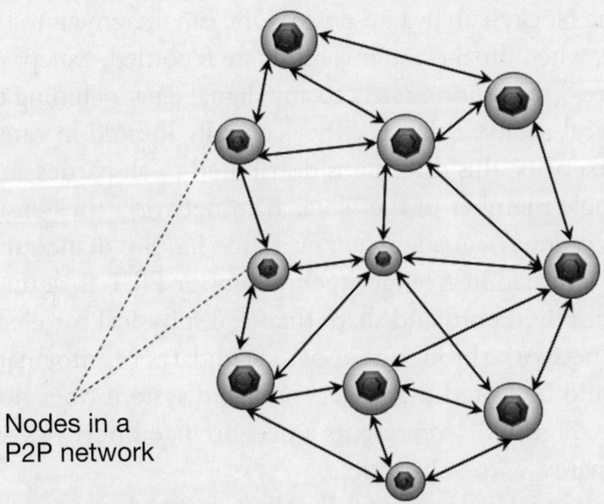

Figure 2.3: Decentralized network

In centralized systems (refer Fig. 2.2), only a central authority or administrator has the power to maintain and update the database. Thus, data flows are controlled and managed by the central authority. The central authority maintains the database by defining the rules and procedures users can have on adding, deleting, or updating the data. All the nodes (computers and devices) connected to the network are subject to access granted by the central authority. In simple words, the central authority makes all the decisions. A failure at the centre means the collapse of the entire system.

However, a decentralized system does not rely on any single authority and is self-regulated (refer Fig. 2.3). Blockchain technology uses a decentralized P2P network architecture wherein anyone can be a node. Every node is equal in the hierarchy with equal access to maintain the database.

While in a centralized environment, an organization such as a bank holds the sole right to read, write or send transactions, in the decentralized environment, anyone can access or write into the ledger. This inclusivity ensures that

a) no single entity has sole control of the network
b) there is no single infrastructural point of failure,
c) there is a collective agreement on the state of the system via consensus, which is explained later in this chapter.

The decentralization that is inherent in Distributed Ledger Technology is the core of Blockchain Technology.

2.2.2 Distributed Ledger Technology

In blockchain technology, a distributed ledger is a decentralized ledger of all the information that is recorded on the blockchain by consensus. One can imagine it to be a database similar to an accounting ledger where financial transactions are recorded, except that it is not restricted to financial data. Here, a transaction refers to any digital data including text, picture, or audio files. Unlike the typical database that is either centrally located in one server or spread out among several select servers, this database is distributed to all parties and locations. It can be accessed by every single member of the blockchain network, thus ensuring incorruptibility as malicious changes cannot be made when everyone has simultaneous access to all records.

In other words, a distributed ledger technology or DLT is defined as a decentralized database that can securely record and share financial, physical, or electronic assets across a geography agnostic network through transparent updates of information.

However, it should be noted that a decentralized system does not necessarily mean a fully distributed one. Figure 2.4 represents a decentralized network where the processing work or control is shared with sub-nodes.

This decentralization is different from the decentralized network of blockchain, where the work is shared between all the nodes (refer Fig. 2.3). In DLT of blockchain, a copy of the database is available on every computer or device of the users in the network as depicted in Fig. 2.5. The database is independent of a central authority and any changes or entries to the ledger have to be "agreed" upon by all users/nodes/parties. This agreement mechanism is referred to as the consensus mechanism.

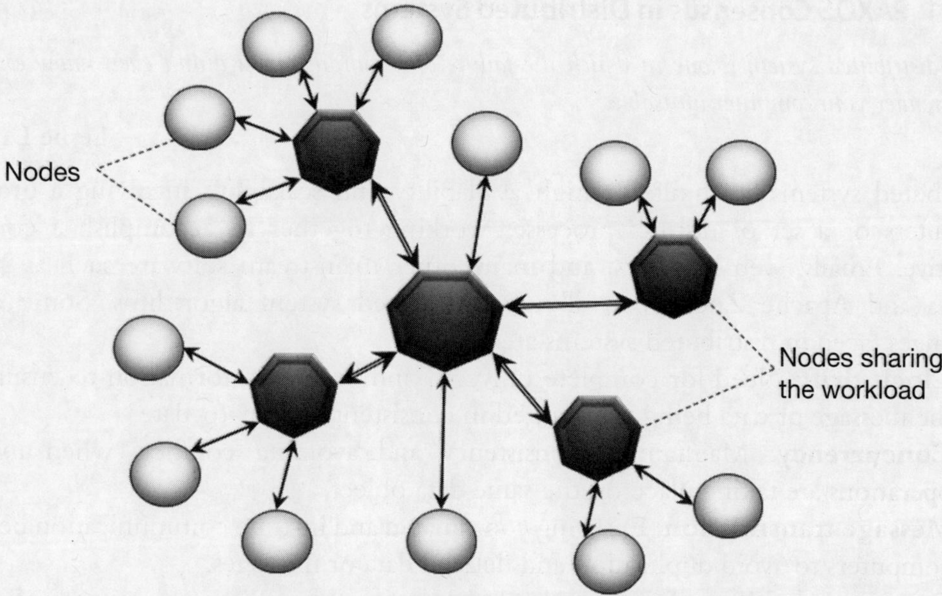

Figure 2.4: Decentralized network

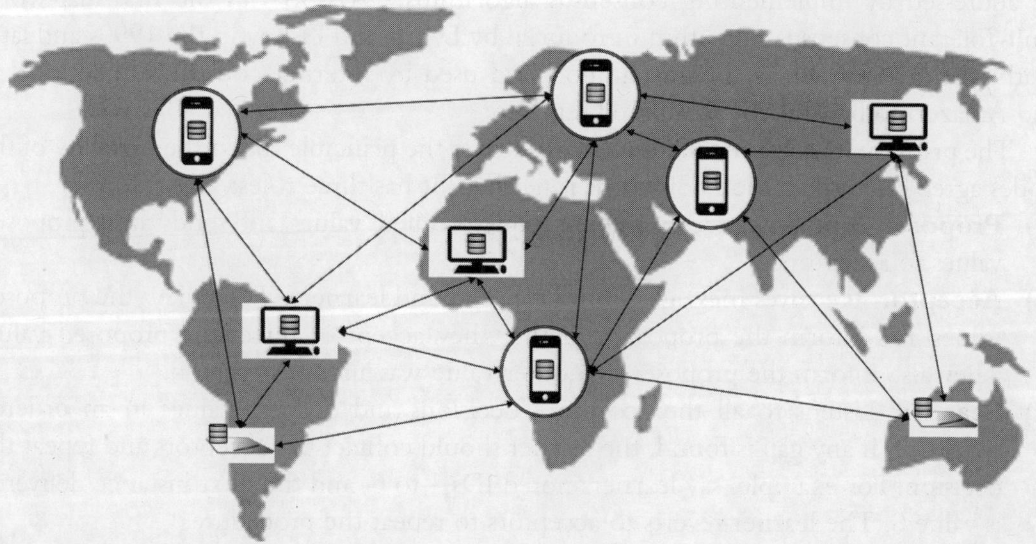

Figure 2.5: Decentralized distributed network (Blockchain)

Once consensus is reached, the database is updated to all the nodes in the network. Thus, at any given time, information is synchronized across all the nodes. Hence, it is also referred to as distributed consensus.

2.2.2.1 PAXOS Consensus in Distributed Systems

"A distributed system is one in which the failure of a computer you didn't even know existed can render your computer unusable."

– Leslie Lamport

Distributed systems are built for high availability and scalability involving a group of computers or a set of distinct processes working together to accomplish a common objective. Emails, web browsers, and many other mainstream software such as Netflix Eureka and Apache Zookeeper, all use distributed system algorithms. Some of the challenges faced in distributed systems are

a) **Clock drift**: Need for complete universal and ordered information to ensure that the message or data being transmitted in consistent and up-to-date.
b) **Concurrency**: Maintaining consistency and avoiding conflicts when multiple operations are taking place on the same data object.
c) **Message transmission**: Ensuring coordinated and in-time communication between computers to avoid duplication and delayed data or messages.
d) **Component failure**: Ensuring that breakage in one node/machine does not impact the function of another node or network.

The above challenges in making the network fault-tolerant can, in most instances, be addressed by implementing consensus algorithms. PAXOS was the first real-world fault-tolerant consensus algorithm introduced by Lynch and Liskov in the 1990s and later mathematically proven by Leslie Lamport and used by internet companies like Google and Amazon to build their distributed services.

The primary PAXOS mechanism works under the principle that if the majority of the nodes agree on a value, then consensus is reached. It has three roles:

1) **Proposer**: A proposer receives client requests called 'values' and sends these proposed values to acceptors.
2) **Acceptor**: Receives messages from proposers and learners. They view the proposed values and inform the proposer whether they accept or reject the proposed value. They also inform the proposer if another value was already accepted.
3) **Learner**: Listens to all the acceptor's decisions and delivers values in an ordered sequence. If any gap is found, the learner should contact the acceptors and repeat the decision. For example, say learners noted IDs 1 to 6, and the next instance delivered is value 8. The learner reverts to acceptors to repeat the procedure.

In practice, a node or server can function in all three roles. There are two phases (refer Fig. 2.6) in the underlying PAXOS protocol.

Phase I: A proposer prepares a unique number **N** and gets acceptors to accept the proposed number **N,** i.e., get their promise to accept the values within a set timeout period. If any acceptor has previously accepted a value, he or she should inform the proposer of the already accepted proposal number and value.

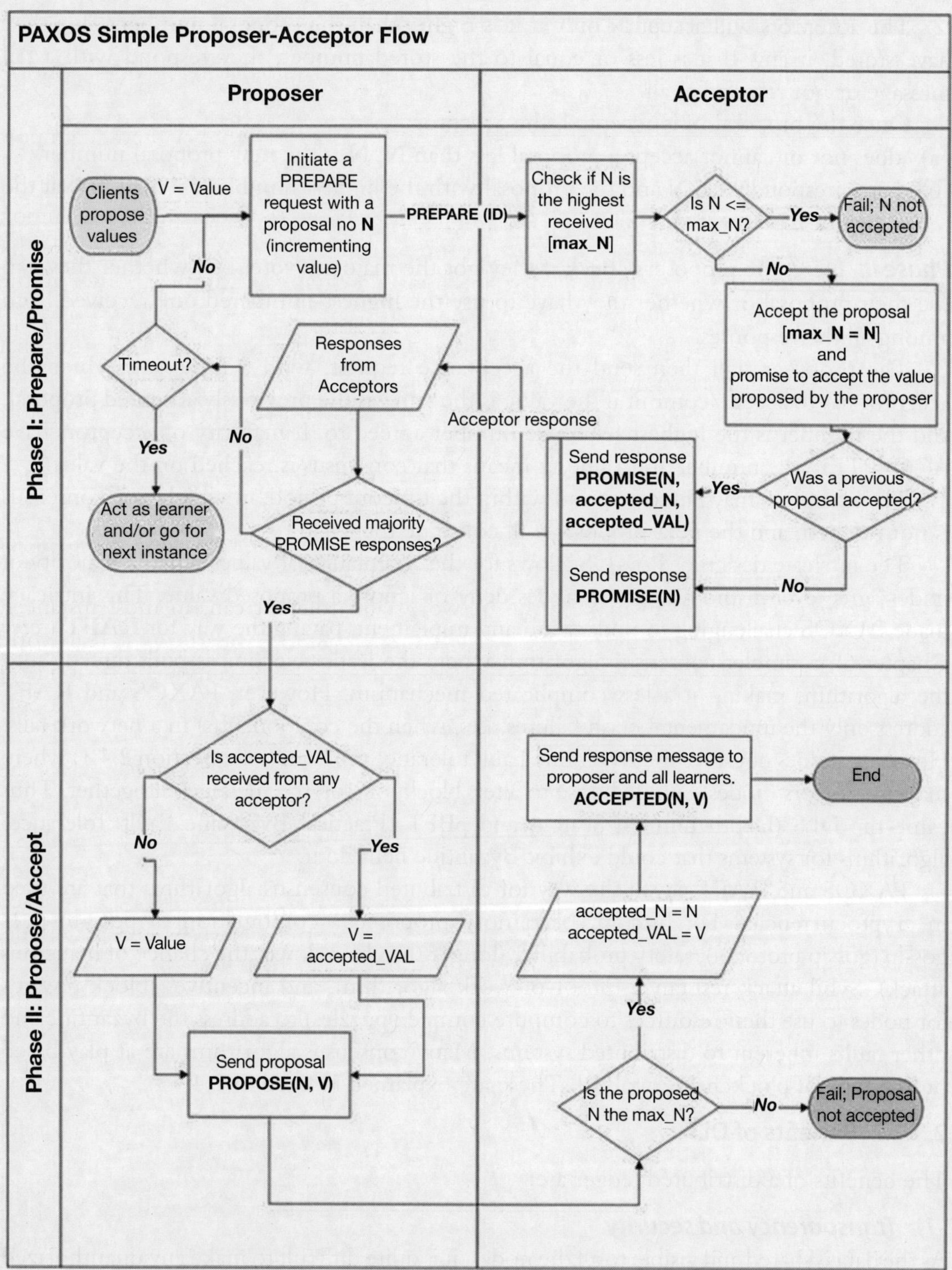

Figure 2.6: PAXOS simple proposer–acceptor flow

The acceptors will accept **N** only if it is higher than the proposal number value they have stored, if any. If it is less or equal to the stored number, they respond with a fail message or not respond at all.

Once the proposal **N** is accepted, the acceptor:

a) does not or cannot accept a proposal less than **N**; **N** is the new proposal number.
b) has to respond with **N** and the proposal with the highest number less than **N** that the acceptor has accepted.

Phase II: Here, the proposers check if they got the majority vote, i.e., whether they can use their proposal or whether they have to use the highest-numbered one received from among all the responses.

The proposer will then send the acceptance request, with a time-out, which the acceptor has to accept/commit if the value is the same as the previously accepted proposal, and the number is the highest sequence number agreed to. If majority of acceptors have ACCEPTED the number and value, it means that consensus is reached on the value.

The acceptors may fail to respond within the timeout period, in which case consensus is not reached, and the next proposer will continue the cycle.

The intricate design of PAXOS allows for the acceptance of values when a majority of nodes agree even if the rest of the nodes deny or ignore a proposed value. This intricacy made PAXOS challenging to understand and implement, paving the way for **RAFT** (refer Chapter 6), a simpler consensus algorithm where the leader election is built directly into the algorithm, making it a less complicated mechanism. However, PAXOS and RAFT address only the fundamental crash failures, i.e., when the component(s) in a network fails. However, it does not solve the Byzantine Fault tolerance problem (refer Section 2.4.1) where malicious actors (nodes) could choose to alter, block, or stop the messages altogether. Thus came the DLS (Depth Limited Search) and pBFT (Practical Byzantine Fault Tolerance) algorithms for systems that could exhibit Byzantine behaviour.

PAXOS and RAFT paved the way for distributed consensus algorithms that are used in cryptocurrencies. For bitcoin, Nakamoto proposed a combination of peer-to-peer gossip (gossip protocol), safety probability (longer the chain, lower the chance of malicious attack), Sybil attack resistance (proof-of-work algorithm), and incentives (block rewards for nodes to use their resources to compute complex puzzles) to address the byzantine and other faults inherent to distributed systems. Many consensus algorithms are at play based on the type of blockchain involved. These are explained in Section 2.4.

2.2.2.2 Benefits of DLT

The benefits of a distributed ledger are:

1) Transparency and security

As the data is shared and visible to all the nodes, it is quite difficult to make any unauthorized changes. Every participating node maintains a copy of the ledger, thus preventing a single point of failure. Any entry has to be consensually agreed upon by all parties making the distributed ledger secure and almost hack-proof.

2) Decentralization

Every owner or node has control over his/her data, unlike a centrally managed system, where a corporate or central authority has sole control. A centrally managed system can lead to a single point of failure. Decentralization gives the users more decision-making power over what goes into their data record. The consensus protocols help to unanimously and securely agree on what should or should not be added to the distributed ledger and bring about an inherent trust within the system.

3) Speed and efficiency

In today's world, the validation and reconciliation of information between various disparate systems are mostly manual, involving time-consuming procedures and prone to errors. With the trustless and distributed nature of blockchain's DLT, the administrative effort of capturing, validating, and synchronizing individual sets of information by staff and intermediaries can be eliminated. This practically diminishes the chances of human error and improves operational efficiencies.

4) Cost savings

Intermediaries such as banks, brokers, lawyers, and administrative staff are required to foster trust between parties. However, this also leads to delays and costs. It goes without saying that with disintermediated systems, companies can save on bottom-line costs while also realizing near-time transactions and efficiencies.

Blockchain uses the underlying technology of DLT that supports immutable transactions on a peer-to-peer (P2P) network without any centralized coordinating entity. Blockchain systems maintain a distributed database in a decentralized manner using cryptographic signatures and consensus-based validation procedures, making it secure, frictionless, and indisputable.

2.3 Types of Blockchain

2.3.1 Accessibility and Permissions

Organizations, including blockchain protagonists, ask the quintessential question – "Are the people really whom they say they are AND is there a risk in granting them rights to the system?"

A node is any computer or hardware device connected to the blockchain network via the internet. There are people behind these devices. Hence a node is also referred to as a member or actor or participant or party or user in the blockchain network. Blockchains are categorized based on user identity/authentication and user rights/authorization.

1) Public and Private Blockchain

A blockchain is considered to be either public or private, based on whether the network is open or accessible to anyone with an internet connection. In a **public** blockchain, anyone can join the network. They can download a copy of the ledger and initiate, broadcast, or mine blocks. Users are anonymous (refer Fig. 2.7).

In a **private** blockchain, membership or association with the blockchain is restricted. One has to meet certain pre-requisite conditions to be a part of the blockchain network. Users are not anonymous.

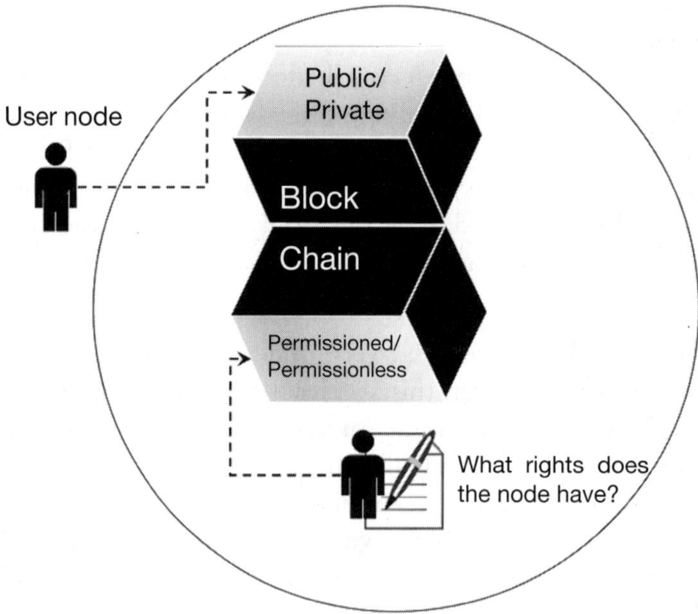

Figure 2.7 Blockchain user anonymity

2) Permissionless and Permissioned Blockchain

This category is based on the type of rights the user or node has within the blockchain network. The rights, if any, are defined by a central entity or group of entities. A Blockchain is considered **permissionless** if no such control entity exists, and all the nodes have equal rights to the network, i.e., they can all read, receive and send transactions and participate in the consensus mechanism.

In a **permissioned** blockchain, the central entity or group restricts the roles that the nodes can play. It can vary from nodes having rights to only initiate transactions to those who validate transactions and to still others that deploy or execute smart contracts. In other words, only selected nodes will participate in the consensus mechanism for permissioned blockchain. In contrast, in a permissionless blockchain, all or majority nodes in the network need to agree on the validity of a record collectively.

Based on the authentication and authorization privileges, Blockchain can be classified as below:

a) Public Blockchain (or Public Permissionless Blockchain)
b) Private Blockchain (or Private Permissioned Blockchain)
c) Consortium Blockchain (or Public/Private Permissioned Blockchain)
d) Hybrid Blockchain (interconnected Public-Private Blockchain)

2.3.2 Public Blockchain

The **public permissionless** blockchain is synonymous with a **public** blockchain. Bitcoin, the first blockchain created, is a public permissionless blockchain.

In a public blockchain, anyone in the world can access the blockchain, download a copy of the code, and run a node. It is a fully decentralized distributed network. One does not need any permission to read/access a transaction, initiate a transaction, or participate in the consensus process (PoW) to create a block. Participants or nodes remain anonymous through high cryptographic protocols. Anonymity, transparency, and immutability are valued over efficiency.

The distinct features of a public blockchain (refer Fig. 2.8) are:

- It is open to the public, as the name suggests. Anyone can join the network and be a participant/node.
- No permissions are required for anyone to read/send transactions
- The standard consensus algorithm used is Proof-of-Work (PoW), where nodes (miners) solve the hash puzzle and submit their resultant block to the rest of the network participants for consensus.

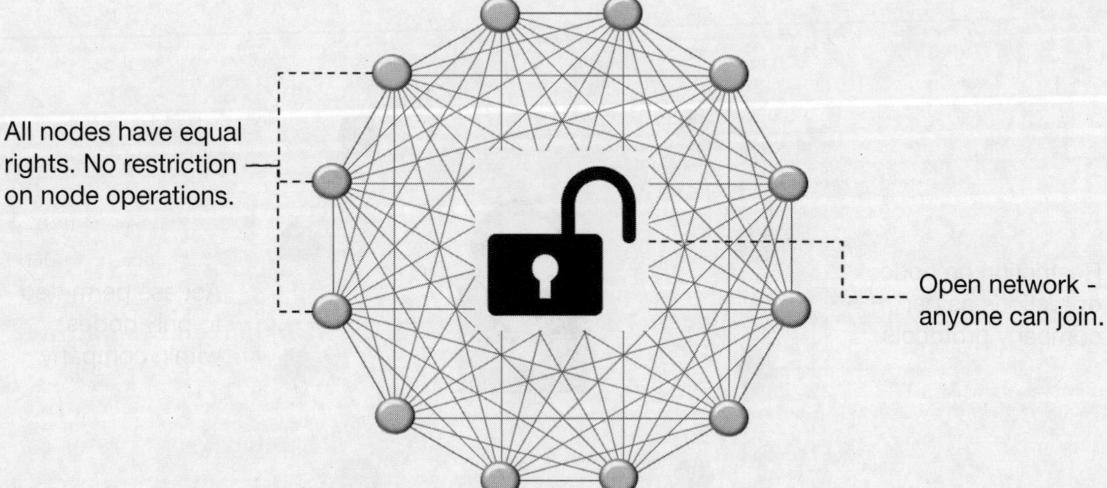

Figure 2.8: Public blockchain

- There is no single point of failure (SPOF) as validation (consensus) is done by all the nodes.
- High cryptographic methods are used to secure data
- It establishes a process of trust.

The downside of a public blockchain is its poor scalability:

- It has low transaction processing speed – ten minutes to create a block.
- Consensus mechanism requires an immense amount of energy and computational power.
- Participants with supercomputers or more powerful ASICs have a better chance of mining than the others, hence the risk of decentralization with mining pools (refer 51% risk in Section 2.4.3.1).

Bitcoin, Litecoin, Ethereum are the most common examples of a public blockchain.

2.3.3 Private Blockchain

The **private permissioned** blockchain, also known as the **private** blockchain (refer Fig. 2.9), differs from the public blockchain in its accessibility and permission. The network is not open to everyone. It leverages the blockchain features of distributed database, immutability, and security. However, the innovative blockchain feature of decentralization and openness is lost as all the permissions are controlled by a few nodes in the organization. Blockchain technology was invented to remove the power of central authority. So in the case of a private blockchain network or organization where the owner has sole control over who can read, write, and validate data, it stands to reason why many may not consider it to be a real blockchain. In public blockchain, efficiency and immutability take precedence over anonymity and transparency.

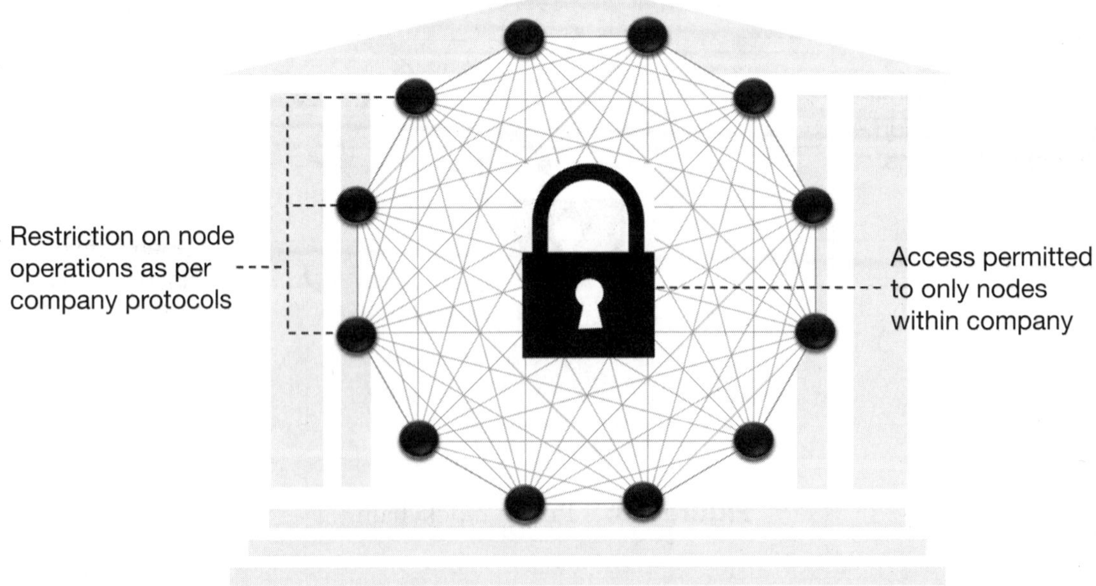

Figure 2.9: Private blockchain

The key features of a private blockchain are:

- It is not open to the public. However, participants are known to each other and hence, trust is assured.
- All participants are pre-approved by the organization. The ledger and access therein are distributed within the network of participants.
- The owner or central authority controls the permission to read, write, or audit the ledger.
- The central authority controls the consensus process.
- High cryptographic methods secure the data.
- It has high transaction processing speed, taking only seconds to create a block.
- Very low energy consumption as supercomputers are not required for processing.

Challenges of a private blockchain:

- They are not decentralized. The trade-off is better scalability and security.
- Blockchain is supposed to function in a trustless environment. If nodes are trusted, it may be cheaper to go with a traditional database
- A central authority means a single point or point of failure, unlike the public blockchain, where there is zero downtime.
- Not considered a legitimate blockchain as it is permissioned, and there is the inherent skepticism on the immutability and trust of transactions, if controlled by a singular authority.
- The organization must agree on who has the highest power to be the central authority.

The private blockchain is scalable and cryptographically secured from the organization's point of view and hence more cost-effective. It is mostly used by organizations that have strict privacy and compliance requirements.

Examples of private blockchain are Multichain and Monax.

2.3.4 Consortium Blockchain

The **consortium** blockchain (refer Fig. 2.10), also known as the **federated** blockchain, is a permissioned blockchain and considered to be a hybrid between public and private blockchain. It is a distributed ledger that anyone can download and access. It has the security features inherent to public blockchains while also maintaining a fair amount of control over the network. Unlike private blockchain, the consensus process is not controlled by one company but by a predetermined consortium of companies or representative individuals. Only the predetermined group has the right to take part in the validation process to create a block. For example, in a supply chain user case, the consortium may be the importer, the exporter, the shipping company, the customs, the participating banks, and inspectors.

Based on the type of enterprise, there may be an overlay of smart contracts and/or other protocols in place that restrict user access. This type of blockchain is best used in government applications.

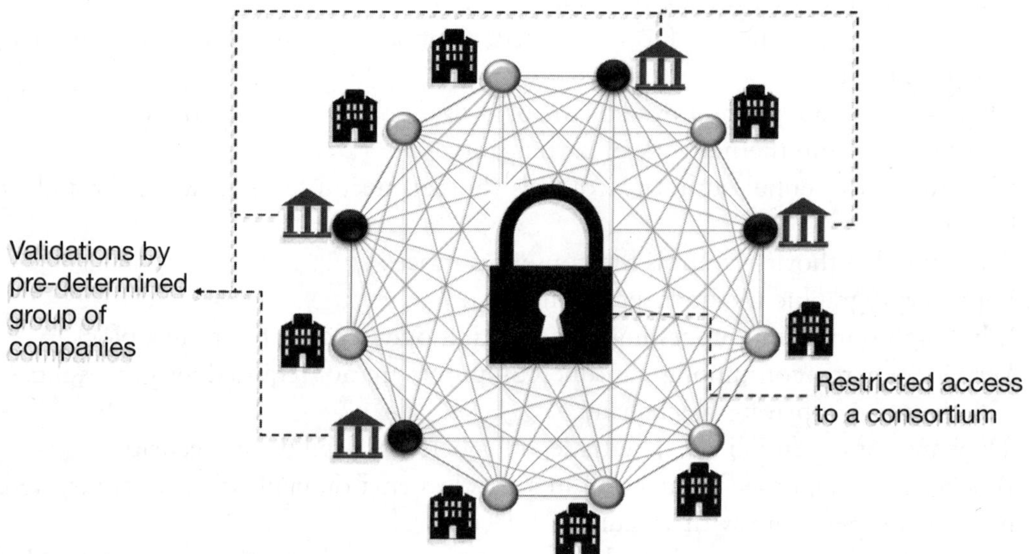

Figure 2.10: Consortium blockchain

> *"So far there has been little emphasis on the distinction between consortium blockchains and fully private blockchains, although it is essential. The former provides a hybrid between the 'low-trust' provided by public blockchains and the 'single highly-trusted entity' model of private blockchains, whereas the latter can be more accurately described as a traditional centralized system with a degree of cryptographic auditability attached."*
>
> – Vitalik Buterin on the consortium blockchain

The key features of a consortium blockchain are:

- Any member node can initiate and receive transactions. However, permission to write or audit the ledger is determined by a group of pre-approved individuals or organizations (consortium).
- The consensus process is done by a group of pre-approved nodes that have full access to the ledger.
- High cryptographic methods secure data.
- They are faster with higher scalability as compared to a public blockchain.
- They have better transaction privacy and traceability.

Challenges of a consortium blockchain:

- They are not fully decentralized. The trade-off is better scalability and security.
- Different organizations have different requirements. Agreement on a standard set of rules may get challenging.

Examples of consortium blockchain platforms are R3 Corda and Hyperledger Fabric.

2.3.5 Hybrid Blockchain

The public blockchain is fully decentralized, tamper-proof, anonymous, transparent, immutable, and open to the public. These features are achieved at the cost of low throughput, poor scalability, expensive hardware, and significant energy consumption. The private blockchain boasts of higher speeds, lower costs, and better scalability while being criticized for its centralization and restricted access. The **hybrid blockchain** (refer Fig. 2.11), as the name suggests, incorporates the best practices of both models. It takes the benefits of both the public and private blockchain, thus attempting to neutralize the negatives.

Hybrid blockchain is best suited for highly regulated enterprises or government organizations that require control over what data is kept private and what can be shared with the public. The hybrid blockchain consists of a public blockchain and a private network that is restricted to only those nodes that are invited by a centralized body. The private network(s) generate the hashed data blocks that are then shared with the public network without compromising the privacy of data. The public blockchain does the verification and time-stamping. Hence, it can be considered as a private network(s) sitting within the main public blockchain. The private network does the creation of transactions, thus maintaining the privacy of data while the public blockchain stores and verifies the blocks, ensuring disintermediation.

Privacy is a concern in public blockchain as all the data are visible to everyone. This risk is mitigated in the hybrid blockchain. If members do not want their transaction data to be accessible without their permission, they can assign exclusive rights for view or modification, in which case it goes to the different members for acceptance and consensus.

The public network uses Delegated Proof-of-Stake (refer Section 2.4.3.4) as its consensus protocol. However, the private network can use any consensus protocol of its choice.

Figure 2.11: Hybrid blockchain

The key features of a hybrid blockchain are:

- It is open to the public; hence the public blockchain is an element where anyone can participate. It also has the private network element that consists of participants invited by a central authority.
- The ledger is distributed within the network of participants. Either the central body or the network members (based on the application need) can decide which transaction data can be public and which needs to be confined to specific members.
- The private network creates the hash of transactions and passes it on to the public network for validation and approval, thus maintaining the trustless nature of blockchain.
- As a combination of a public and private blockchain, some processes are public while others are private. The processes can be changed by the private or central authority to fit the purpose, with the consensus of all the nodes in the network.
- Consensus protocols are available in both public and private networks. The main public network uses the DPoS consensus while the private network can have its own.
- Data is immutable and secured by high cryptographic methods.
- It has the highest transaction processing speed.
- Practically hack-proof as no malicious actor can enter either the private network or the robust consensus mechanism of the public network.
- Data is auditable. Though the privacy of transactions is maintained, it is open for verifiability as and when required.

Challenges of a hybrid blockchain:

- It is a relatively new ecosystem, with XinFin being the only genuinely functional hybrid blockchain protocol/platform currently available for highly regulated markets.

XinFin, a non-profit organization based in Singapore, with a focus on cross-border trade and finance, has built the first hybrid blockchain platform, TradeFinex, combining Ethereum (for the public blockchain state) and Quorum (for the private blockchain state). Organizations like Ripple, IBM, and other technology companies are exploring hybrid blockchains. It is hoped that we shall see more applications of the technology in the coming years.

Table 2.1 Comparison: Public, Private, Consortium and Hybrid Blockchain

	Public blockchain	**Private blockchain**	**Consortium blockchain**	**Hybrid blockchain**
Organization type				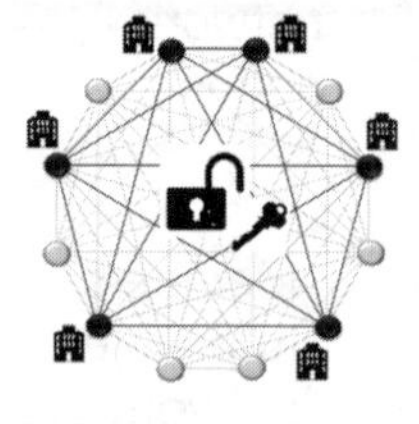
	Public	Single entity or organization	Multiple organizations or enterprises	Highly regulated enterprise

(Continued)

Table 2.1 (*Continued*)

	Public blockchain	**Private blockchain**	**Consortium blockchain**	**Hybrid blockchain**
Common features	- Chain of blocks - Peer-to-peer architecture - Public-key cryptography – Immutable - Byzantine fault tolerance - Auditable -			
Users	Anonymous, but web tracking and cookies pose a risk to privacy	Known and trusted participants	Known and trusted participants	Anonymity for public network members; Private network members are known within the private network.
Access	Open and transparent to all	Access fully restricted	Selectively open; relevant transparency provided	Centralized control of providing access, hence privacy and confidentiality maintained
Network type	Decentralized; zero points of failure	Centralized; single point of failure	Partially decentralized; multiple points of failure	Zero points of failure
Operation	Anyone can read or initiate or receive transactions	Pre-approved participants can read and/ or initiate transactions	Pre-approved participants can read and/ or initiate transactions	Any combination is possible; Operations are customizable. Central authority decides which transactions can be made public and which are private

(*Continued*)

Table 2.1 (*Continued*)

	Public blockchain	Private blockchain	Consortium blockchain	Hybrid blockchain
Verification	Anyone can be a node and take part in the consensus process to validate transactions and create a block	Single validator node or central authority to create a block	Only privileged members of the consortium can validate and create a block	The public network verifies the block
Immutability	Secured by hashing	Secured by distributed consensus	Secured by distributed consensus	Secured by hashing at the private network and secured by distributed consensus by the public blockchain
Consensus mechanism	PoW, PoS, etc.	Voting or variations of PoW/PoS consensus algorithms	Voting or variations of PoW/PoS consensus algorithms	DPoS in public and variations in private
Incentivization	Incentivizes miners to grow the network	Users limited to within a company; hence incentivization is not relevant	Limited incentivization	Can incentivize users in the main public network
Security	Security based on consensus protocols and hash functions. Higher the security, lower the performance	Security is dependent on the blockchain architecture adopted	Security is dependent on the blockchain architecture adopted	Very high as hackers or unknown parties cannot access the system

(*Continued*)

Table 2.1 *(Continued)*

	Public blockchain	Private blockchain	Consortium blockchain	Hybrid blockchain
Trust	Trust-free system; trust is enforced via cryptographic proof	Trusted; central control	Trusted; need to trust the majority	Trust-free system; consensus by public blockchain
Transaction speed	Slow; takes more than 10 minutes for creating a block	High; takes seconds to create a block	Very high; takes seconds to create a block	Highest
Energy consumption	Very high	Low	Low	Low
Scalability	Limited; as the network grows, the node requirements of bandwidth, storage, and computational power exponentially increases.	Better scalability as high storage and computational power is not required.	Better scalability as high storage and computational power is not required.	Highly scalable

2.3.6 Blockchain-as-a-Service

Blockchain-as-a-Service (BaaS) is a concept similar to Software-as-a-Service (SaaS) model. The complexity of the infrastructure and cost of setting up and operating the blockchain may discourage many start-ups and SMEs from adopting blockchain. By adopting the BaaS model, businesses can leverage the cloud-based solution to build their blockchain apps, smart contracts, and other blockchain functions. In contrast, the cloud-based service provider maintains the infrastructure and other back-end operations, including storage, bandwidth management, security, and resource allocation.

For a fee, the service provider takes care of setting up and keeping the blockchain infrastructure up and running on behalf of the company. This will enable the company to focus on its core business areas without worrying about infrastructure and performance-related issues.

It is a lot cheaper to host and run the blockchain on a BaaS solution rather than developing a blockchain in-house. However, security must be assessed thoroughly as mistakes in set-up or configuration/code errors can severely disrupt the blockchain.

All the major technology companies like Amazon (on AWS-Amazon Web Services), Microsoft (by MS Azure), IBM, and Oracle have launched blockchain-as-a-service (BaaS) offerings.

MTBC Inc., a healthcare information technology company based in the US, is the industry's first decentralized integrated suite of web-based solutions, including AI that operates on the BaaS platform. The solution offers interoperability between highly secure blockchain networks for better business/clinical decisions, seamless interoperability between electronic health records (EHRs) giving patients full control over their health records, faster billing cycles, reduced administrative overheads and operating costs.

2.4 Consensus Protocol

As per Webster dictionary, a consensus is a general agreement or opinion shared by all the people in a group. A protocol is a system of standard rules that are acceptable by all parties to control the exchange of information in a network. Thus, a **consensus protocol** in blockchain can be defined as a set of rules and procedures for attaining a unified agreement (consensus) between the participating nodes on the status of the network.

Blockchain technology uses a decentralized network architecture where anyone can be a node. All nodes are equal in the hierarchy, with no individual node having more access or advantage over the other. Nodes accept decisions collectively for the good of the whole network. The consensus protocol enables this trustless nature of blockchain.

Consensus protocols are the rules that define how the different actors in a distributed ledger authenticate and validate the transactions added to it to prevent different versions of the ledger from being created or previous transactions from being edited.

The consensus protocol aims to overcome the classic problem of a distributed computing system known as the Byzantine Generals Problem.

2.4.1 Byzantine Generals Problem

Originally coined as the "Two-generals Problem," the Byzantine Generals Problem is a thought experiment introduced in computer networking classes, especially concerning TCP (Transmission Control Protocol), to highlight that there is no guarantee that failure will not occur when it applies to a two-party communication.

The Problem: Two Byzantine (Roman) armies led by different generals are preparing to attack a fortified city. They are based on either side of the city. The city is strong enough to withstand an individual attack of either army, but not strong enough to defend itself from a coordinated attack by the two armies at the same time. In other words, the two armies must attack the city at the **same time** to win the battle (refer Fig. 2.12).

The two generals, say General A and General B need to agree on the time of the attack. The only way of communication is by sending messengers through the city. The simplest way to send the message will be for one general to take the leadership role and send a messenger through the enemy lines with a proposed day and time. The second general, on receiving the message, sends back the acknowledgment or agreement message back to the first general.

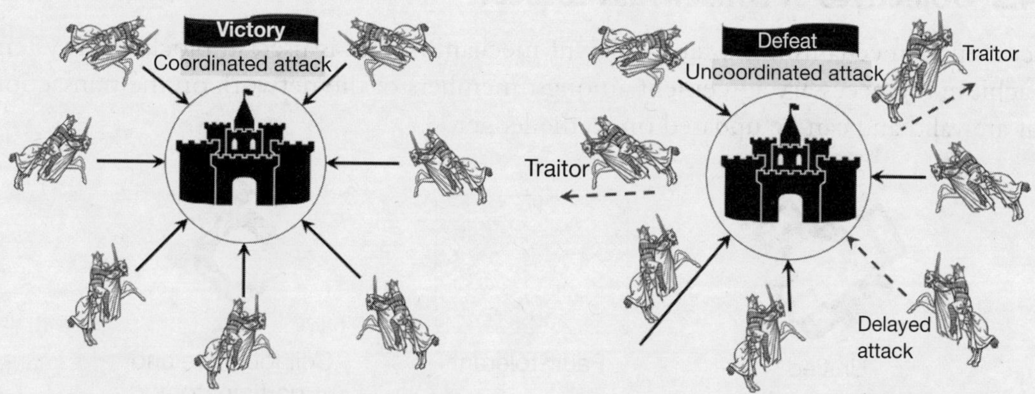

Coordinated attack leading to victory **Uncoordinated attack leading to defeat**

Figure 2.12: Byzantine generals problem

Here are the issues that may hamper victory:

- General A will hesitate to attack at the appointed time if he does not get the acknowledgment from General B.
- The enemies could capture General A's messenger and the intercept the message.
- General A's message could be intercepted and replaced with a fake message. General B may hesitate to attack as he cannot verify the authenticity of the message.
- General B may send an acknowledgment message, but there is no assurance that General B's messenger will not be caught by the enemies and the message intercepted and/or replaced.
- One of the Generals could potentially be a traitor.
- The lack of confidence or doubt between the Generals may trickle down to the soldiers in the army leading to some deserters, thus compromising the strength of the army. The desertion can lead to defeat even if there is a coordinated attack.

Hence, there is no way for either of the generals to guarantee that their counterparts have received their message. Here, the dilemma is between two generals or two participants. In a distributed network, the dilemma is between all the participants or nodes. All the participants need to verify and reach agreements neutralizing corrupt parties and disseminating false and unreliable information.

Blockchain's key feature of consensus mechanism or consensus algorithms is seen as a solution to the Byzantine Generals Problem. The consensus mechanism of blockchain aims to overcome the trust risks attributed to a distributed network system, namely,

a) *Authenticity*: The message should be easily verifiable to guarantee that it is genuine and not tampered with.
b) *Unity*: There should be a collective agreement by all parties (nodes) on action to be taken
c) *Fault-tolerance*: A few traitors or hackers should not be able to compromise the process.

2.4.2 Objectives of Consensus Protocol

A consensus mechanism is a fault-tolerant mechanism that is used in blockchain systems to achieve the necessary agreement amongst members of the network on the transactions that are valid and can be updated on to the ledger.

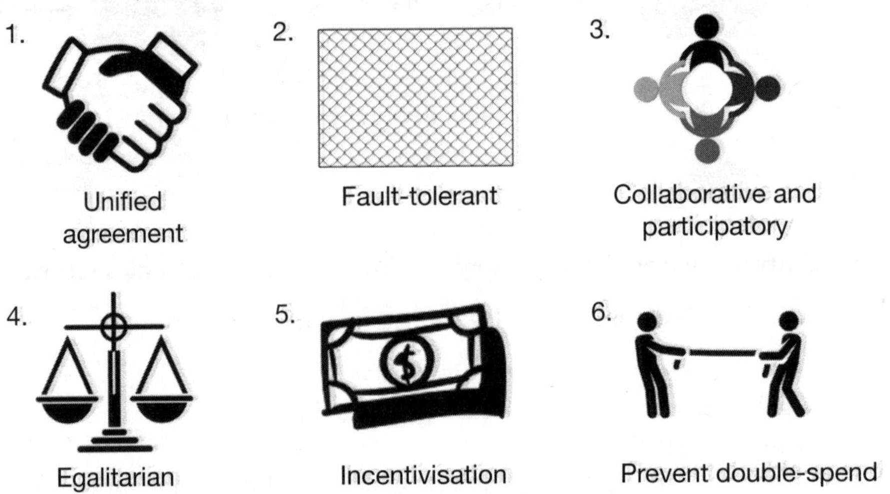

Figure 2.13: Objectives of consensus mechanism

The main objectives of a consensus mechanism (refer Fig. 2.13) are

a) Unified Agreement

There should be a unified agreement on which data is valid and accurate. Protocol rules embedded in the network ensure this.

b) Fault-tolerant

In a blockchain, every node acts as both a client and a server. In a blockchain network, there are thousands of such nodes that lead to high fault tolerance. Even if some of the nodes are unresponsive, there are still a considerable number of nodes communicating to keep the system functional.

c) Collaborative and Participatory

The consensus mechanism should ensure that all nodes participate in the overall process, in the best interests of the group as opposed to the interest of a single or a few nodes. The features of blockchain P2P network architecture enable constant communication between nodes. It also allows for any node to inspect and verify that the underlying process is fair to all participants in the network.

d) Egalitarian

There should be no discrimination between nodes. Blockchain runs on a peer-to-peer network architecture where every node should be considered equal to every other node. Every node should have equal weightage.

e) Incentivization

Validating new transactions and securing them on the blockchain requires "miners" to solve complex mathematical problems based on cryptographic hash algorithms requiring vast amounts of computational resources, including electricity. Rules are built into the consensus mechanism to incentivize the miners to work for the system, making it more secure.

f) Prevent Double-spend

Double-spending refers to the possibility of digital currency or token being spent more than once by falsification or duplication. Protocol rules embedded in the blockchain consensus mechanism ensure valid and authentic transactions. Additionally, substantial computational resources are extended by miners to secure transactions, making it difficult to double-spend or alter transactions.

There are different kinds of consensus algorithms that work on different principles. Proof-of-Work (PoW) and Proof-of-Stake (PoS) are the most commonly used consensus mechanisms in blockchain for public distributed ledgers. In contrast, Proof-of-Authority (PoA) and Proof-of-Elapsed Time (PoET) algorithms are applied in private blockchains.

2.4.3 Consensus Algorithms

2.4.3.1 Proof of Work

Founded by Satoshi Nakamoto, Proof-of-Work is the most well-known consensus mechanism used by the first blockchain **Bitcoin** in 2009. Here, several nodes of the distributed ledger called **miners** compete to solve a complicated mathematical problem based on a cryptographic hash algorithm. The solution found is called Proof of Work or **PoW**. Without proof of work, adding blocks to the blockchain would be too easy and could make it vulnerable to hackers.

The mining node releases the proof of work to the other nodes for verification to reach consensus.

The solution to the problem is difficult to produce but easy for the network to verify. The process of mining is extremely computation-intensive. So the first miner who manages to produce the PoW will be rewarded either in the form of bitcoins or digital currency.

Disadvantages of the PoW consensus mechanism are

1) **Time-consuming**: Miners have to iterate over many nonces before finding the right solution, which is a time-consuming process (refer Section 1.4.6).
2) **High energy consumption**: Miners conduct significant work in terms of processing power and electricity to find the nonce for creating the winning hash. As only one miner can be successful, for all other miners who competed, it is wasted energy.
3) **51% risk**: To counteract the high time and energy consumption in transaction validation, some miners group the mining pools together to combine their mining resources for more efficiency and savings. Mining pool goes against the basic principle of distributed ledgers as a person or group gaining control of over 50% of the network's computing power can control the validation process. This is usually referred to as a 51% attack.

Bitcoin, Litecoin, Dash, Monero, and Ethereum use PoW as the underlying consensus mechanism.

2.4.3.2 Proof of Elapsed Time

Proof of elapsed time (**PoET**) was conceived in 2016 by Intel. It is commonly used in permissioned blockchain networks to decide on the mining rights or the block winners on the network.

PoET mechanism is based on the principle of a fair lottery system where every single node is equally likely to be a winner. Each miner node in the blockchain network is provided with a randomized timer object from a trusted code that generates a random wait time. This method of randomization aims to circumvent any attempt by a miner to get a timer with a shorter period. The miner who completes the designated waiting time commits a new block to the blockchain and broadcasts the relevant information across the blockchain network. The process is then repeated for the discovery of the next block.

The PoET mechanism is similar to the PoW consensus mechanism except that instead of being resource-intensive, it allows a miner's processor to sleep and switch to other tasks for the specified time, thereby increasing its efficiency and reducing power consumption. Also, in PoET, the identity of the miners is known unlike in PoW, where it remains anonymous.

Disadvantage attributed to PoET consensus mechanism:

1) **Vulnerability** – It relies heavily on the use of a Trusted Execution Environment (TEEs), i.e., Intel SGX-enabled CPUs. Though the protocol prevents nodes from running multiple instances of "wait time" to boost their chances of success, it is vulnerable to various other security attacks such as "Foreshadow", which attacks the secure enclave of SGX.

Hyperledger Sawtooth architecture, developed by Intel, uses PoET consensus.

2.4.3.3 Proof of Stake

The Proof of Stake (**PoS**) was implemented as a consensus algorithm for Peercoin in 2012. The more stake one has in the validating node, the less chance one will be tempted to corrupt the validating process. In other words, the users with the highest stake in a cryptocurrency will have the most interest in maintaining and securing the network because any attacks would diminish the reputation and price of the cryptocurrency that they hold.

In PoS, the mining nodes are called **validators** or **forgers** or **delegates**. A forger has to commit some of his/her stake (cryptocurrency) in the network as collateral to be in the running for a chance to validate the transaction. An algorithm will randomly select a forger based on the percentage stake or collateral he or she has put forward. Validating nodes can forge or create new blocks proportional to the amount they have staked; i.e., a node with a 10% stake in the network can validate 10% of transactions.

Energy consumption is less here as compared to PoW consensus. Also, the forgers are paid a transaction fee as against the block reward of PoW consensus.

PoS addresses all the disadvantages of PoW with low time and energy consumption and a reduced threat of 51% attack. It also incentivizes forgers to validate legitimately as their staked amount will be forfeited in case of fraudulent behaviour.

Disadvantages of the PoS consensus mechanism are:

1) ***Cheaper to attack:*** A PoS based network is cheaper to attack as the perpetrator would just need to spend some money and not invest in the combined set of money, time, hardware, electricity, and other resources.
2) ***Centralization risk:*** The richest forger can control the consensus mechanism and get even richer.

Examples of Blockchain using the PoS mechanism are Peercoin, NXT, and BlackCoin. Ethereum uses PoS over an existing PoW blockchain, resulting in a hybrid PoW/PoS system called Casper Friendly Finality Gadget (FFG).

2.4.3.4 Delegated Proof of Stake

Delegated Proof of Stake (**DPoS**) is a variation of the PoS consensus mechanism. Here, the network participants or nodes use their cryptocurrency or tokens to vote for the delegates. Just as in PoS, the delegates are responsible for validating transactions and maintaining the blockchain ledger. These elected delegates are called **witnesses**. The more the crypto-coins or tokens, the more the voting power.

In addition to the PoS benefits, DPoS enables better security and even distribution of wealth. Any fraudulent activity by the witnesses can be easily detected by the voters and penalized. As it is a democratic system, it is not only the rich, but all users have a chance to be elected as witnesses and earn rewards. This makes DPoS more decentralized than either PoS or PoW.

Disadvantages of DPoS consensus mechanism are:

1) **51% risk**: Since fewer people are in charge of maintaining the network, it is easier to organize a 51% attack
2) **Potential centralized power**: Sufficient decentralization cannot be achieved without compromising the scalability of the network. The more the validators, the

more is the risk of slowing the network down. Hence there is a risk of power getting concentrated in the hands of a few.

Bitshare, Lisk and EOS are examples of blockchains that use the DPoS consensus mechanism.

2.4.3.5 Proof of Authority

Proof of Authority (**PoA**) consensus mechanism proposed in 2017 is used in private blockchains. It is similar to PoS and DPoS in the sense that only a group of pre-selected authorities called **validators** secure the blockchain and can produce new blocks. However, instead of staking coins or tokens, the validators stake their identity.

The identities of the validators are public and verifiable by a reliable third party, such as a public notary database. This incentivizes the validators to act in the best interest of the network, for otherwise, their reputation is ruined. The validators are ideally limited to 25 or less to ensure the efficiency and security of the network. In addition to low energy consumption, PoA also benefits from zero node-to-node data transfer requirements. Once the nodes are verified and approved as validators, re-verification is not done unless required.

The following conditions must be met to identify validators:

- Validators must have a valid identity in the public domain that must match the records found in the public notary database.
- The authority needs to be uniform and unbiased for all validators.
- Eligibility criteria for staking identity must be stringent to ensure the trustworthiness of the validator.

Some of the issues attributed to PoA are:

1) **Semi-centralized**: Blockchains with PoA consensus mechanism lean more towards a centralized system in the form of an authority node as the validators are predetermined. However, this mechanism works well with private or consortium blockchain, enabling better scalability, such as a network of banks where each bank acts as a validator for the others.
2) **Reputational indifference**: If the payoff is strong enough, validator(s) may sacrifice their reputation. However, this issue is a high risk only if the validators are limited in number. If they fall to the influence of third parties with malicious interests, the network could fail.

VChainThor blockchain, Ethereums' Kovan and Rinkeby testnets use the PoA consensus mechanism algorithm. Hyperledger and Ripple also use optimized versions of PoA.

2.4.3.6 Practical Byzantine Fault Tolerance

Practical Byzantine Fault Tolerance (**pBFT**) was introduced by Miguel Castro and Barbara Liskov at the MIT Laboratory for Computer Science in 1999. It is considered as one of the potential solutions to the Byzantine Generals problem. Here, the goal is to decide whether to accept a piece of information submitted to the blockchain or not. It tolerates "Byzantine faults" based on the assumption that the number of malicious nodes in the network cannot simultaneously be equal to or exceed one-third of the overall nodes in the system in a given window of vulnerability.

It works on the format of the Byzantine Generals Problem, where all "generals" (nodes) are considered equal and take their work instruction from the **leader** node. The leader node is the primary node, and all other nodes are called secondary or **backup** nodes. The leader is selected at random in a round-robin fashion. A node **client** sends a transaction request to the leader who then broadcasts it to all the backup nodes. The leader and backup nodes will use the message with their internal state to run computation and transmit the decision result to all the client nodes. The final decision is arrived at based on the agreement of the majority.

A high hash rate is not required as pBFT relies on the minimum number of backup nodes to confirm trust, namely (f+1), where f represents the maximum number of faulty nodes. Hence, it is not computationally intensive and as a result, there is substantial energy saving.

The disadvantages of the pBFT protocol are:

1) **Small group sizes** – Due to the amount of communication required between all the nodes, this model works best with small-group networks for better response times.
2) **Sybil Attacks** – A single party can assume several identities or nodes and manipulate the network. This is mitigated with larger network sizes, but scalability and throughput will be compromised.

Stellar, Ripple, and Hyperledger Iroha are some blockchains that use variants of the pBFT consensus mechanism algorithm.

2.4.3.7 RAFT Algorithm

RAFT was built as a simpler version of PAXOS consensus. The principle of the RAFT consensus mechanism is similar to that of pBFT consensus except that only the leader node can communicate with the other nodes and decide on the state of the transaction. Nodes can have three states: leader, follower, and candidate. RAFT uses randomized timers to elect the leader for each term. If a leader is not elected in a term, candidates will time out and start the election for the next term. The leader candidates log must be more up-to-date than the follower logs. If a candidate's log is less up-to-date than a potential follower, then the candidate is rejected by the follower.

A node starts as a **follower** expecting a "heartbeat" from a leader. If it does not receive it within the "election time," it assumes the leader is dead and takes the **candidate** state to send out a "RequestVote." If the candidate node receives majority approvals from follower nodes, it transitions to a **leader** state.

Only the leader can append log entries based on client requests. When the leader node receives a request, it appends the entry to its log as a new entry and sends it to all the follower nodes. After receiving the confirmation from the majority of the followers, the leader, in turn, commits the message and sends a confirmation (heartbeat) message to the client and followers.

RAFT based consensus is used in Quorum for consortium settings.

2.4.4 Other Consensus Mechanisms

Some other variations and evolving consensus mechanisms are listed below:

a. Proof of Stake Anonymous (PoSA)

A variation of PoS consensus mechanism, PoSA was first introduced in Cloakcoin in 2014. Here, nodes are incentivized for "cloaking" the transaction. There are no master nodes, making it a truly decentralized and secure network.

The cloaking nodes provide the transaction with inputs and outputs, rendering it close-to-impossible to establish the identity of the receiver or the sender of a transaction and ensuring anonymity.

b. Leased Proof of Stake (LPoS)

LPoS is another variation of the PoS mechanism. In PoS, one needs a large stake to get a chance to validate a block. Hence many users with low balances do not get a chance to generate a block. The LPoS mechanism enables users to sublet their balances to staking nodes. This allows for small holders also to forge a block of transaction in the blockchain. Any reward received is shared proportionally.

c. Proof of Importance (PoI)

First established with the NEM cryptocurrency platform, the PoI consensus mechanism works on the principle that users with the highest balance as also users who provide maximum value to the network should be incentivized. Thus, the chance of forging a block depends on many factors, including coin balance and authority.

d. Proof of Storage

Proof-of-Storage consensus mechanism is implemented in **the Storj** system. Here, the network uses a block tree. Instead of going through every single transaction listed on the blockchain, the user can only see the transactions that are of particular importance to him.

e. Proof of Burn

Iain Stewart created a Proof-of-Burn consensus. When coins are destroyed on the blockchain, it is referred to as being burned. Technically, the coins in circulation are sent to an unspendable address, known as an **eater address**. Just like in PoW consensus where the more that is invested in supercomputers and electricity, the more the chances of mining, in Proof of Burn, more the coins one burns, the more chance one gets to mine blocks. Proof of Burn is used in Counterparty and Slimcoin.

f. Proof of Activity

It is a hybrid of PoW and PoS consensus mechanisms. It starts with miners vying to be the first to solve the cryptographic puzzle and claim their reward. However, the blocks being mined are not transactions but templates with header information and the mining reward address. Once the template block is mined, the PoS selects a random group of validators to sign the block. Once all validators sign the block, it becomes part of the blockchain. If the block remains unsigned by a few, it is discarded, and the next winning block template is used. Proof of Activity reduces the risk of a 51% attack to zero. However, the energy consumption issue is not eased.

g. Proof of Capacity (PoC)

PoC consensus algorithm is currently used only in **Burstcoin**. It was built to circumvent the high energy consumption of PoW and coin hoarding risks of PoS. Here, mining nodes can use the space available on their hard drive to mine crypto-coins instead of using the mining device's computing power.

In the PoC mechanism, the miners will first "plot" their hard drives, i.e., they will create a list of all possible nonce values through repeated hashing of data, including a miner's account. In other words, the miners will compute the solutions and store them ahead of time. Once the actual mining starts, the miner with the fastest solution wins the block.

Using hard drives is said to be 30 times more energy-efficient than ASIC (application-specific integrated circuit)-based mining. It is also more decentralized as anyone can own a basic hard drive.

h. Directed Acyclic Graph (DAG)

DAG was created to circumvent the inefficiencies of PoW. In PoW consensus, it takes around 10 minutes or more to create a block, and blocks cannot be created simultaneously. With DAG, transactions can run on different chains simultaneously. ITC (IoT Chain), built on DAG consensus protocol, is said to process over 10,000 transactions per second.

Consensus protocol algorithms are continuously evolving with multiple variants of PoW and PoS and hybrids.

Summary

Though the distributed ledger technology (DLT) is not complex, it enables complex solutions with blockchain, thus enabling the transformation from "the internet of information" to "the internet of value."

Blockchain is still an evolving technology (refer Fig. 2.14). Though Bitcoin was introduced to replace government-controlled currency with a digital decentralized and distributed alternative, it is yet to be seen whether cryptocurrency is a viable and lasting option.

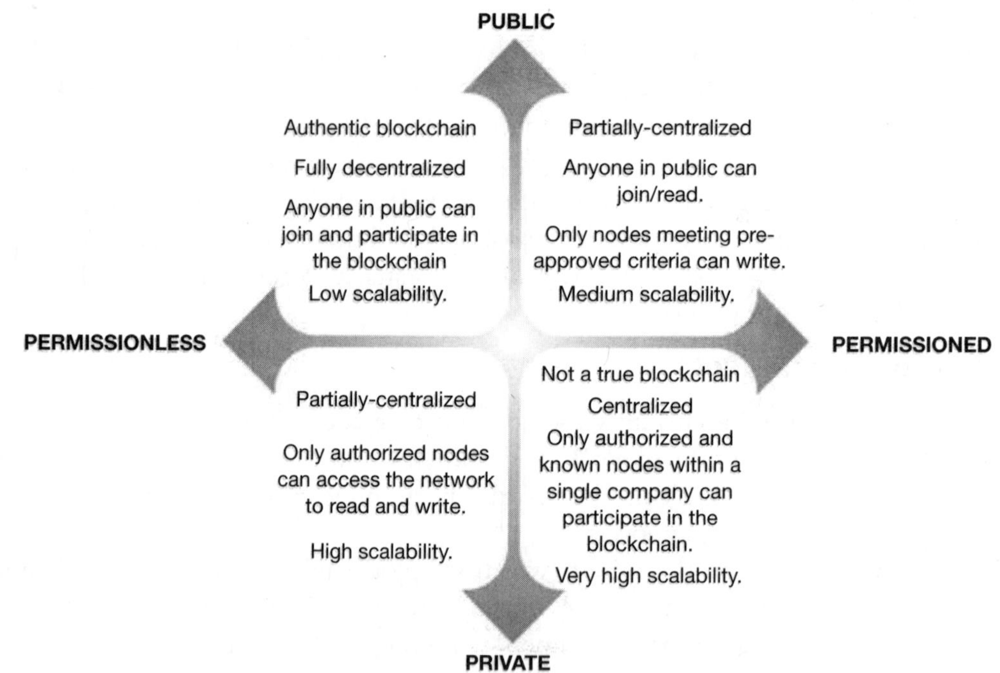

Figure 2.14 Evolution of blockchain

However, one cannot ignore the revolutionary features of blockchain and the opportunities that it provides to financial and various other industrial sectors. With consortium and hybrid blockchain, FinTech and other large technology companies have gone beyond the use-cases of the public and private blockchain.

The consensus mechanisms are continually transforming to cater to various industry needs and mitigating the risks identified with each blockchain type. PoW and PoS were the most popular consensus mechanisms initially used for public blockchains. However, many variants and hybrids like DPoS, PoET, PoA, and others have since been introduced to either circumvent the limitations or improve upon the existing consensus algorithms.

This chapter skims the surface of blockchain types and their underlying protocols.

EXERCISES

Multiple Choice Questions

1. **One of the main characteristics of blockchain technology is its ____________, where transactions are not under the control of any single party.**
   A. Decentralization
   B. Centralization
   C. Privatization
   D. Immutability

Answer: A

Explanation: One of the main characteristics of blockchain technology is its decentralization, where transactions are not under the control of any single party.

2. **In this case, a failure at the centre means the collapse of the entire system:**
   A. Decentralized System
   B. Centralized System
   C. P2P System
   D. Immutable System
   Answer: B
   Explanation: In centralized systems, only a central authority or administrator has the power to maintain and update the database. All the data flows are controlled and managed by the central authority. A failure at the centre means the collapse of the entire system.
3. **In this type of blockchain, anyone can join the network. They can download a copy of the ledger and initiate, broadcast, or mine blocks.**
   A. Private Blockchain
   B. Hybrid Blockchain
   C. Consortium Blockchain
   D. Public Blockchain
   Answer: D
   Explanation: A blockchain is considered to be either public or private, based on whether the network is open or accessible to anyone with an internet connection. In public blockchain, anyone can join the network. They can download a copy of the ledger and initiate, broadcast, or mine blocks. The users are anonymous.
4. **These systems are built for high availability and scalability involving a group of computers or a set of distinct processes working together to accomplish a common objective.**
   A. Distributed Systems
   B. Centralized Systems
   C. Scalable Systems
   D. Immutable Systems
   Answer: A
   Explanation: Distributed systems are built for high availability and scalability involving a group of computers or a set of distinct processes working together to accomplish a common objective.
5. **Emails, web browsers, and many other mainstream software like Netflix Eureka and Apache Zookeeper, all use ____________ algorithms.**
   A. Distributed Systems
   B. Centralized Systems

C. Scalable Systems
D. Immutable Systems
Answer: A
Explanation: Emails, web browsers, and many other mainstream software like Netflix Eureka, Apache Zookeeper, all use distributed system algorithms.

6. **The challenge faced in distributed systems, where there is a need for complete universal and ordered information to ensure that the message or data being transmitted in consistent and up-to-date, is called**
A. Concurrency
B. Message transmission
C. Component failure
D. Clock drift
Answer: D
Explanation: Some of the challenges faced in distributed systems are
a) *Clock drift*: Need for complete universal and ordered information to ensure that the message or data being transmitted in consistent and up-to-date.
b) *Concurrency*: Maintaining consistency and avoiding conflicts when multiple operations are taking place on the same data object.
c) *Message transmission*: Ensuring coordinated and in-time communication between computers to avoid duplication and delayed data or messages.
d) *Component failure*: Ensuring that breakage in one node/machine does not impact the function of another node or network.

7. **The challenge faced in distributed systems for maintaining consistency and avoiding conflicts when multiple operations are taking place on the same data object is known as**
A. Concurrency
B. Message transmission
C. Component failure
D. Clock drift
Answer: A
Explanation: Same as for Question 6.

8. **The challenge faced by distributed systems while ensuring coordinated and in-time communication between computers to avoid duplication and delayed data or messages occurs in the area of**
A. Concurrency
B. Message Transmission
C. Component Failure
D. Clock Drift
Answer: B
Explanation: Same as for Question 6.

9. The challenge faced in distributed systems while ensuring that breakage in one node/machine does not impact the function of another node or network is described as

A. Concurrency

B. Message Transmission

C. Component Failure

D. Clock Drift

Answer: C

Explanation: Same as for Question 6.

10. The primary PAXOS mechanism works under the principle that if the majority of the nodes agree on a value, then consensus is reached. It has three roles, namely,

A. Proposer, Acceptor and Miner

B. Voter, Acceptor and Learner

C. Proposer, Acceptor and Learner

D. Receiver, Sender and Observer

Answer: C

Explanation: The primary PAXOS mechanism works under the principle that if the majority of the nodes agree on a value, then consensus is reached. It has three roles:

1) Proposer: A proposer receive client requests called 'values' and sends them to acceptors who may accept the proposed values.
2) Acceptor: Receives messages from proposers and learners. They view the proposed values and inform the proposer whether they accept or reject the proposed value. They also inform the proposer if another value was already accepted.
3) Learner: Listens to all the acceptor's decisions and delivers values in an ordered sequence. If any gap is found, the learner should contact the acceptors and repeat the decision. For example, say the learners note IDs 1 to 6, and the next instance delivered is value 8. The learner reverts to the acceptors to get the procedure repeated.

11. In the primary PAXOS mechanism, ____________ receive client requests called ____________.

A. Proposer, Values

B. Acceptor, Values

C. Learner, Values

D. Proposer, Access

Answer: A

Explanation: Same as for Question 10.

12. In the primary PAXOS mechanism, ____________ also informs the proposer if another value was already accepted.

A. Proposer
B. Learner
C. Informer
D. Acceptor

Answer: D

Explanation: Same as for Question 10.

13. In the primary PAXOS mechanism, ___________ listens to all the acceptor's decisions and delivers values in an ordered sequence.

A. Proposer
B. Learner
C. Informer
D. Listener

Answer: B

Explanation: Same as for Question 10.

14. There are ___________ phases in the underlying PAXOS protocol.

A. 3
B. 2
C. 4
D. 6

Answer: B

Explanation: In practice, a node or server can function in all three roles in PAXOS Algorithm. There are two phases (refer Fig. 2.6) in the underlying PAXOS protocol.

15. There are ___________ roles in the underlying PAXOS protocol.

A. 3
B. 2
C. 4
D. 6

Answer: A

Explanation: Same as for Question 10.

16. In Distributed Ledger Technology (DLT), as the data is shared and visible to all the nodes, it is quite difficult to make any unauthorized changes. This leads to:

A. Decentralization
B. Speed and efficiency
C. Transparency and security
D. Cost savings

Answer: C

Explanation: Transparency and Security: As the data is shared and visible to all the nodes, it is quite difficult to make any unauthorized changes. Every participating node maintains a copy of the ledger, thus preventing a single point of failure.

17. In Distributed Ledger Technology (DLT), intermediaries like banks, brokers, lawyers, and administrative staff are required to foster trust between parties. This leads to:

A. Decentralization
B. Speed and efficiency
C. Transparency and security
D. Cost savings

Answer: D

Explanation: Cost savings: Intermediaries like banks, brokers, lawyers and administrative staff are required to foster trust between parties. However, this also leads to delays and costs. It goes without saying that with dis-intermediated systems and near-time transactions and efficiencies, companies can save on bottom-line costs.

18. In Distributed Ledger Technology (DLT), the administrative effort of capturing, validating and synchronizing individual record sets of information by staff and intermediaries can be eliminated. This leads to:

A. Decentralization
B. Speed and efficiency
C. Transparency and security
D. Cost savings

Answer: B

Explanation: Speed and Efficiency: In today's world, the validation and reconciliation of information between various disparate systems are mostly manual, involving time-consuming procedures and prone to errors. With the trustless and distributed nature of Blockchain's DLT, the administrative effort of capturing, validating, and synchronizing individual record sets of information by staff and intermediaries can be eliminated. This practically diminishes the chances of human error and improves operational efficiencies.

19. Distributed Ledger Technology (DLT), gives the users more decisioning power over what goes into their data record. This leads to:

A. Decentralization
B. Speed and efficiency
C. Transparency and security
D. Cost savings

Answer: A

Explanation: Decentralization: Every owner or node has control over his/her data, unlike a centrally managed system, where a corporate or central authority has sole control. A centrally managed system can lead to a single point of failure. Decentralization gives the users more decisioning power over what goes into their data record. The consensus protocol is used to unanimously and securely agree on what should or should not be added to the distributed ledger, bringing about an inherent trust within the system.

20. In a private blockchain, just anyone cannot join the blockchain. One has to meet certain pre-requisite conditions to be a member of the blockchain network. Users are not anonymous.

A. False

B. True

Answer: B

Explanation: In a private blockchain, just anyone cannot join the blockchain. One has to meet certain pre-requisite conditions to be a member of the blockchain network. Users are not anonymous.

21. The first blockchain created, is a public permissionless blockchain, called

A. Bitcoin

C. Ethereum

D. MultiChain

E. Hyperledger

Answer: A

Explanation: The public permissionless blockchain is synonymous with a public blockchain. Bitcoin, the first blockchain created, is a public permissionless blockchain.

22. The downside of a public blockchain is its poor:

A. transparency

B. trust

C. single point of failure

D. scalability

Answer: D

Explanation: The downside of a public blockchain is its poor scalability. It has low transaction processing speed – ten minutes to create a block. Consensus mechanism requires an immense amount of energy and computational power. Participants with supercomputers or more powerful ASICs have a better chance of mining than the others, hence the risk of decentralization with mining pools.

23. In a private blockchain, the owner or central authority controls the permission to read, write, or audit the ledger.

A. True

B. False

Answer: A

Explanation: The key features of a private blockchain are: It is not open to the public. However, participants are known to each other and hence, trust is assured. All participants are pre-approved by the organization. The ledger and access thereof are distributed within the network of participants. The owner or central authority controls the permission to read, write, or audit the ledger. The central authority controls the consensus process. High cryptographic methods secure the data. It has high transaction processing speed taking only seconds to create a block. Very low energy is consumed as supercomputers are not required for processing.

24. Only the predetermined group has the right to take part in the validation process to create a block in:

A. Consortium Blockchain
B. Private Blockchain
C. Public Blockchain
D. Bitcoin

Answer: A

Explanation: Unlike private blockchain, in the consortium blockchain, the consensus process is not controlled by one company but by a predetermined consortium of companies or representative individuals. Only a predetermined group has the right to take part in the validation process to create a block. For example, in a supply chain user case, the consortium may be any party such as the importer, the exporter, the shipping company, the customs, the participating banks, or inspectors.

25. There should be no discrimination between nodes. Blockchain runs on a peer-to-peer network architecture where every node should be considered equal to every other node. Every node should have equal weightage. This is referred to as:

A. Egalitarian
B. Collaborative and Participatory
C. Incentivization
D. Fault-tolerant

Answer: A

Explanation: Egalitarian: There should be no discrimination between nodes. Blockchain runs on a peer-to-peer network architecture where every node should be considered equal to every other node. Every node should have equal weightage.

26. The consensus mechanism should ensure that all nodes participate in the overall process, putting the best interests of the group as against the interest of a single or few nodes. This is known as:

A. Egalitarian
B. Collaborative and Participatory
C. Incentivization
D. Fault-tolerant

Answer: B

Explanation: The consensus mechanism should ensure that all nodes participate in the overall process, in the best interests of the group as against the interest of a single or few nodes. The blockchain's P2P network architecture and features enable constant communication between nodes. It also allows for any node to inspect and verify that the underlying process is fair to all participants in the network.

27. **In a blockchain, every node acts as both a client and a server. Even if some of the nodes are unresponsive, there are still a considerable number of nodes communicating to keep the system functional. In a blockchain network, there are thousands of such nodes that lead to:**
    A. Egalitarianism
    B. Collaborative and participatory approach
    C. Incentivization
    D. Fault-tolerance
    Answer: D
    Explanation: Fault-tolerant: In a blockchain, every node acts as both a client and a server. In a blockchain network, there are thousands of such nodes that lead to high fault tolerance. Even if some of the nodes are unresponsive, there are still a considerable number of nodes communicating to keep the system functional.

28. **Validating new transactions and securing them on the blockchain requires "miners" to solve complex mathematical problems based on cryptographic hash algorithms requiring vast amounts of computational resources, including electricity. Rules are built into the consensus mechanism to compensate the miners to work for the system making it more secure. This is known as:**
    A. Egalitarianism
    B. Collaborative and participatory approach
    C. Incentivization
    D. Fault-tolerance
    Answer: C
    Explanation: Incentivization: Validating new transactions and securing them on the blockchain requires "miners" to solve complex mathematical problems based on cryptographic hash algorithms requiring vast amounts of computational resources, including electricity. Rules are built into the consensus mechanism to incentivize the miners to work for the system making it more secure.

29. **This is the most well-known consensus mechanism used by the first blockchain, Bitcoin, in 2009.**
    A. Proof of authority
    B. Proof of work
    C. Proof of stake
    D. RAFT
    Answer: B
    Explanation: Founded by Satoshi Nakamoto, Proof-of-Work is the most well-known consensus mechanism used by the first blockchain Bitcoin in 2009. Here, several nodes of the distributed ledger called miners compete to solve a complicated mathematical problem based on a cryptographic hash algorithm.

30. In PoW, the solution to the problem is __________ to produce and __________for the network to verify.

A. Easy, Easy

B. Easy, Difficult

C. Difficult, Easy

D. Difficult, Difficult

Answer: C

Explanation: The solution to the problem is difficult to produce but easy for the network to verify. The process of mining is extremely computation-intensive. So the first miner who manages to produce the PoW will be rewarded either in the form of bitcoins or digital currency.

31. In PoW consensus, it takes around 10 minutes or more to create a block and blocks cannot be created simultaneously. This is created to circumvent the inefficiencies of PoW:

A. DAG

B. POC

C. POI

D. DPOW

Answer: A

Explanation: Directed Acyclic Graph (DAG) was created to circumvent the inefficiencies of PoW. In PoW consensus, it takes around 10 minutes or more to create a block and blocks cannot be created simultaneously. With DAG, transactions can run on different chains simultaneously. ITC (IoT Chain) is built on DAG consensus protocol and is said to process over 10,000 transactions per second.

32. PoC consensus algorithm is currently used only in:

A. Bitcoin

B. Burstcoin

C. Ethereum

D. Bitcoin

Answer: B

Explanation: PoC consensus algorithm is currently used only in Burstcoin. It was built to circumvent the high energy consumption of PoW and coin hoarding risks of PoS. Here, mining nodes can use the space available on their hard drive to mine crypto-coins instead of using the mining device's computing power.

33. __________ is a hybrid of PoW and PoS consensus mechanisms.

A. Hybrid PoW

B. RAFT

C. PAXOS

D. PoA

Answer: D

Explanation: Proof of Activity (PoA) is a hybrid of PoW and PoS consensus mechanisms. It starts with miners vying to be the first to solve the cryptographic puzzle and claim their reward. However, the blocks being mined are not transactions but templates with header information and the mining reward address. Once the template block is mined, the PoS selects a random group of validators to sign the block. Once all validators sign the block, it becomes part of the blockchain. If the block remains unsigned by a few, it is discarded and the next winning block template is used. PoA reduces the risk of 51% attack to zero. However, the energy consumption issue is not eased.

34. **This consensus mechanism is implemented in Storj system. Here, the network uses a block tree. An individual node on the block tree consists of a blockchain. So instead of going through every single transaction listed on the blockchain, the user can only see the transactions that are of particular importance to him.**
    A. PoA
    B. PoS
    C. PoB
    D. PAXOS
    Answer: B
    Explanation: Proof of Storage (PoS) consensus mechanism is implemented in Storj system. Here, the network uses a block tree. An individual node on the block tree consists of a blockchain. So instead of going through every single transaction listed on the blockchain, the user can only see the transactions that are of particular importance to him.

35. **In this consensus mechanism, technically, the coins in circulation are sent to an unspendable address, known as an eater address.**
    A. PoA
    B. PoS
    C. PoB
    D. PAXOS
    Answer: C
    Explanation: Proof of Burn consensus was created by Iain Stewart. When coins are destroyed on the blockchain, it is referred to as being burned. Technically, the coins in circulation are sent to an unspendable address, known as an eater address. Just like in PoW consensus where the more that is invested in supercomputers and electricity the more the chances of mining, in Proof of Burn, more the coins one burns, the more chance one gets to mine blocks. PoB is used in Counterparty and Slimcoin.

36. **In PoS, one needs a large stake to get a chance to validate a block. Hence, many users with low balances do not get a chance to generate a block. This mechanism enables users to sublet their balances to staking nodes.**

A. PoA
B. LPoS
C. PoB
D. PAXOS
Answer: B

Explanation: Leased Proof of Stake (LPoS) is another variation of the PoS mechanism. In PoS, one needs a large stake to get a chance to validate a block. Hence, many users with low balances do not get a chance to generate a block. The LPoS mechanism enables users to sublet their balances to staking nodes. This allows for smallholders also to forge a block of transaction in the blockchain. Any reward received is shared proportionally.

Short-answer Questions

1. **Define the concept of Centralization.**

   In centralized systems, only a central authority or administrator has the power to maintain and update the database. All the data flows are controlled and managed by the central authority. A failure at the centre means the collapse of the entire system.

2. **Define decentralization concept of blockchain.**

   One of the main characteristics of blockchain technology is its decentralization, where transactions are not under the control of any single party. A decentralized system does not rely on any single authority and is self-regulated. Blockchain technology uses a decentralized P2P network architecture wherein anyone can be a node. Every node is equal in the hierarchy with equal access to maintain the database.

3. **Define Distributed Ledger Technology (DLT).**

   A distributed ledger technology or DLT is defined as a decentralized database that can securely record and share financial, physical or electronic assets across a geography agnostic network through transparent updates of information.

4. **Define public blockchain.**

   A blockchain is considered to be either public or private based on whether the network is open or accessible to anyone with an internet connection. In public blockchain, anyone can join the network. They can download a copy of the ledger and initiate, broadcast, or mine blocks. Users are anonymous.

5. **Define private blockchain.**

   A blockchain is considered to be either public or private based on whether the network is open or accessible to anyone with an internet connection. In the case of a private blockchain, access to the network is not open to everyone. One has to meet certain prerequisites to be a member of the blockchain network. Users are not anonymous.

6. Define permissionless blockchain.

This category of blockchain is based on the type of rights the user or node has within the blockchain network. The rights, if any, are defined by a central entity or group of entities. A blockchain is considered permissionless if no such control entity exists, and all the nodes have equal rights to the network, i.e., they can all read, receive and send transactions and participate in the consensus mechanism.

7. Define permissioned blockchain.

In a permissioned blockchain, the central entity or group restricts the roles that the nodes can play. It can vary from nodes having rights to only initiate transactions to those who validate transactions and to still others that deploy or execute smart contracts. In other words, only selected nodes will participate in the consensus mechanism for permissioned blockchain, while in a permissionless blockchain, all or majority nodes in the network need to agree on the validity of a record collectively.

8. What are the classifications of blockchain based on the authentication and authorization privileges?

Blockchain can be classified as below:

a) Public Blockchain (or Public Permissionless Blockchain)
b) Private Blockchain (or Private Permissioned Blockchain)
c) Consortium Blockchain (or Public/Private Permissioned Blockchain)
d) Hybrid Blockchain (interconnected Public–Private Blockchain)

9. Give any two examples for public blockchain.

Bitcoin, Litecoin and Ethereum are the most common examples of a public blockchain.

10. Give any two examples for private blockchain.

Examples of private blockchain are Multichain and Monax.

11. What is consortium blockchain?

The consortium blockchain, also known as the federated blockchain, is a permissioned blockchain and considered to be a hybrid between public and private blockchain. It is a distributed ledger that anyone can download and access. It has the security features inherent to public blockchains, while also maintaining a fair amount of control over the network.

12. List any two challenges of consortium blockchain.

Challenges of a consortium blockchain:

- They are not fully decentralized. The trade-off is better scalability and security.
- Different organizations have different requirements. Agreement on a standard set of rules may become challenging.

13. Give two examples for Consortium blockchain.

Example of consortium blockchain platforms is R3 Corda and Hyperledger Fabric.

14. What are the challenges of hybrid blockchain?

Hybrid blockchain is a relatively new ecosystem, with XinFin being the only genuinely functional hybrid blockchain protocol/platform currently available for highly regulated markets.

15. What is Proof-of-Work algorithm?

Founded by Satoshi Nakamoto, Proof-of-Work is the most well-known consensus mechanism used by the first blockchain Bitcoin in 2009. Here, several nodes of the distributed ledger called miners compete to solve a complicated mathematical problem based on a cryptographic hash algorithm. The solution found is called Proof of Work or PoW. Without proof of work, adding blocks to the blockchain would be too easy and could make it vulnerable to hackers.

16. What are the trust risks attributed to a distributed network system, overcome by the consensus mechanism of blockchain?

a) Authenticity: The message should be easily verifiable to guarantee that the message is genuine and not tampered with.

b) Unity: There should be a collective agreement by all parties (nodes) on action to be taken

c) Fault-tolerance: Miscreants or hackers should not be able to compromise the process.

17. Define Proof of Stake Anonymous (PoSA) algorithm.

PoSA is another variation of a PoS consensus mechanism. It was first introduced in Cloakcoin in 2014. Here, nodes are incentivized for "cloaking" the transaction. There are no master nodes, making it a truly decentralized and secure network. The cloaking nodes provide inputs and outputs to the transaction, making it near-to-impossible to identify the sender and recipient of a transaction, thus ensuring anonymity.

18. Define Leased Proof of Stake (LPoS) alogorithm.

LPoS is another variation of PoS mechanism. In PoS, one needs a large stake to get a chance to validate a block. Hence, many users with low balances do not get a chance to generate a block. The LPoS mechanism enables users to sublet their balances to staking nodes. This allows for small holders also to forge a block of transaction in the blockchain. Any reward received is shared proportionally.

19. Define Proof of Importance (PoI)

First established with NEM cryptocurrency platform, PoI consensus mechanism works on the principle that not only users with the highest balance but also users who provide a lot of value on the network should be incentivized. Thus, the chance of forging a block depends on many factors including coin balance, authority and the number of transactions made to and from a particular node address.

20. Define Proof of Storage.

Proof of Storage consensus mechanism is implemented in Storj system. Here, the network uses a block tree. An individual node on the block tree consists of a blockchain. So instead of going through every single transaction listed on the blockchain, users can only see the transactions that are of particular importance to them.

21. Define Proof of Burn.

Proof of Burn consensus was created by Iain Stewart. When coins are destroyed on the blockchain, it is referred to as being burned. Technically, the coins in circulation are sent to an unspendable address, known as an eater address. Just like in PoW consensus, where the more that is invested in supercomputers and electricity the more the chances of mining, in Proof of Burn, more the coins one burns, the more chance one gets to mine blocks. Proof of Burn is used in Counterparty and Slimcoin.

22. Define Proof of Activity.

Proof of Activity is a hybrid of PoW and PoS consensus mechanisms. It starts with miners vying to be the first to solve the cryptographic puzzle and claim their reward. However, the blocks being mined are not transactions but templates with header information and the mining reward address. Once the template block is mined, the PoS takes over to select a random group of validators to sign the block. Once all validators sign the block, it becomes part of the blockchain. If the block remains unsigned by a few, it is discarded and the next winning block template is used. Proof of Activity reduces the risk of 51% attack to zero. However, the energy consumption issue is not eased.

23. Define Proof of Capacity (PoC).

PoC consensus algorithm is currently used only in Burstcoin. It was built to circumvent the high energy consumption of PoW and coin hoarding risks of PoS. Here, mining nodes can use the space available on their hard drive to mine crypto-coins instead of using the mining device's computing power.

24. Define Directed Acyclic Graph (DAG).

DAG was created to circumvent the inefficiencies of PoW. In PoW consensus, it takes about 10 minutes or more to create a block and blocks cannot be created simultaneously. With DAG, transactions can run on different chains simultaneously. ITC (IoT Chain) is built on DAG consensus protocol and is said to process over 10,000 transactions per second.

25. Write short notes on Vulnerability problem of PoET consensus mechanism.

PoET consensus mechanism relies heavily on the use of a Trusted Execution Environment (TEEs) i.e., Intel SGX-enabled CPUs. Though the protocol prevents nodes from running multiple instances of "wait time" to boost their chances of success, it is vulnerable to various other security attacks.

Essay-type Questions

1. Write an essay on the concept of Centralization and Decentralization in detail.
2. Discuss DLT concepts of Blockchain in detail along with the benefits.
3. What is public blockchain? What are the features and drawbacks of public blockchain?
4. What is private blockchain? What are the features and drawbacks of private blockchain?
5. What are the distinct features of a public blockchain?
6. What are the distinct features of a private blockchain?
7. List the challenges of a private blockchain.
8. What is consortium blockchain? What are the features and drawbacks of consortium blockchain?
9. What is hybrid blockchain? What are the features and drawbacks of hybrid blockchain?
10. Compare Public, Private, Consortium and Hybrid blockchain.
11. Write notes on consensus protocol.
12. Write an essay on Byzantine Generals problem.
13. What are the main objectives of a consensus mechanism?
14. Discuss Proof-of-Work algorithm in detail.
15. Discuss Proof-of-Elapsed Time algorithm in detail.
16. Discuss Proof-of-Stake algorithm in detail.
17. Discuss Delegated Proof-of-Stake algorithm.
18. Discuss Proof-of-Authority algorithm.
19. Discuss RAFT algorithm.

CHAPTER 3

Cryptocurrency – Bitcoin, Altcoin and Token

LEARNING OBJECTIVES

The previous chapters expounded on what is meant by a blockchain transaction, its underlying technology, the various kinds of consensus algorithms, and introduction to blockchain terminologies. We have seen that bitcoin is a cryptocurrency or digital currency, while blockchain is the underlying technology that supports it. The birth of bitcoin in 2009 has resulted in an explosion of different kinds of digital coins having their unique features and functional purpose.

This chapter briefly expands on the reasons why bitcoin was conceived and delves deeper into the meaning of cryptocurrency, its characteristics, the different types of cryptocurrency, and its general usage in the blockchain ecosystem.

3.1 Introduction

The term 'cryptocurrency' evolved from the words 'crypto' meaning concealed or secret and 'currency' indicating a system of money. It refers to all encrypted decentralized digital currencies. Bitcoin was the first cryptocurrency. Cryptocurrencies are digital currencies designed to be faster, cheaper, and more reliable than money.

Cryptocurrency is an open-source encrypted ledger of transactions built on blockchain technology. On the blockchain network, the cryptocurrency is created via mining, and users can transact directly with each other over the internet/online platforms. Thus in the Cryptocurrency world, the need for government to create money and banks to store and facilitate the money transactions is practically redundant. Therefore, the removal of the middleman makes for quicker and affordable transactions.

The system is also highly resilient against fraud as the cryptocurrency users can record and verify each other's transactions, and the acceptance of the network is mandatory before it can be committed on to the blockchain ledger. This ledger of digital transactions is public and secure; thus it does not require any entity like banks or financial institutions to provide trust. Hence, it is referred to as a 'trustless' system.

Cryptocurrency is still in its infancy and considered more as an investment or commodity rather than as a currency. Since bitcoin, the number of cryptocurrencies has increased exponentially. Trading on cryptocurrency exchanges is gaining popularity despite its volatile prices. Ether, litecoin and tether are some popular cryptocurrencies other than bitcoin.

3.2 Bitcoin and the Cryptocurrency

3.2.1 Evolution of Currency

The terms 'currency' and 'money' are often used interchangeably. However, there is a difference between the two terms. Economists define money as a medium of exchange, a unit of account, and a store of value, i.e., they define money by the functions that it serves. Currency is the physical or tangible representation of money in the form of paper notes and coins. Figure 3.1 represents the evolution of physical currency/commodity to the virtual currencies we see now.

The concept of currency evolved out of trading economies. Bartering was the earliest form of trading where people exchanged goods and services for other goods and services according to their needs. For instance, animal skins and tools were traded for barley, salt, spices, or other commodities/services. Historians believe that the Mesopotamians used the earliest barter system at around 6000 BC.

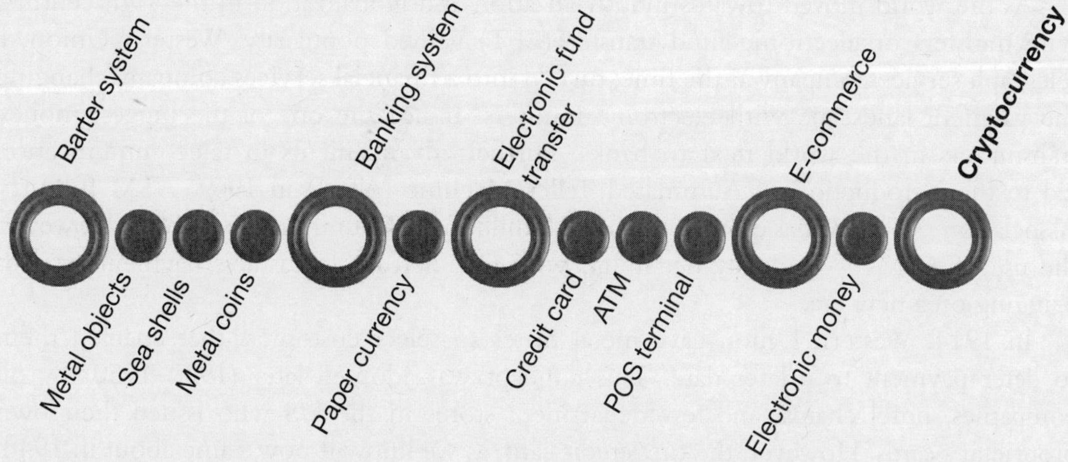

Figure 3.1: Evolution of currency

The difficulty in agreeing to a standard deal as to the fair value of the commodities or services being exchanged, as well as the issues in storage and transportation, made the bartering system impractical, long-term. Hence for practical purposes, around 1000 BC, China, India, and Africa used cowry shells as money. The Chinese also used miniature replicas of the commodity being traded that later evolved into circular coins made of bronze. Thus shell money, coin money, and other forms of commodity money were used as a medium of exchange for goods and services across all trade continents.

The first known currency coin was minted in 600BC in Lydia, which is now part of Western Turkey. The Lydian coin was made of electrum, a naturally occurring alloy of gold and silver with the image of a lion head imprinted on it. Since then, coins became associated with money and commerce. Countries started minting their specific brand of metal coins and designating a value to it. The florin gold coin that was first introduced in the Republic of Florence in 1252 was widely adopted across Europe, encouraging international commerce. This brought in the first banking system gaining dominance in the monetary economy of Europe. These banks became the trusted intermediaries in money exchange by taking the excess money of the 'savers' and lending it out to the 'borrowers' for a fee.

Paper money was first introduced in China in around 960 AD in the form of a credit note or promissory note that could be traded for goods, and the seller could redeem the note for coins. The concept of paper money was introduced in Europe by Marco Polo and other travelers in the 13th century. However, the permanent issue of banknotes did not take place until the late 17th century. The Bank of England began to issue paper notes in 1695. It was initially handwritten with the precise amount deposited or loaned. Standardized printed notes were established in 1745. The 18th century saw governments across countries moving towards standardized paper currency.

As the world moved towards industrialization and globalization in the 19th century, wire transfers or electronic fund transfer (EFT) gained popularity. Western Union, a telegraph service company at the time, turned into a financial services company changing the payment landscape with electronic transfers. It became one of the biggest money transmitters in the world next to banks. Further advancements in telecommunication led to the introduction of Automated Teller Machine (ATM) in 1969. ATM Industry Association (ATMIA) has estimated around 3 million ATM units in use globally. However, the use of ATMs is gradually declining, with the increase in cashless payment systems bringing on a new era.

In 1914, Western Union gave metal plates to select customers that enabled them to defer payment to a later date. This concept was adopted later (1930–1950) by oil companies, hotel chains, and few department stores in the US who issued their own proprietary cards. However, the first credit card, as we know it now, came about in 1946. It was called 'Charge-it' and introduced by New York banker John Biggins. Other credit card types that could be used in a variety of establishments soon followed like the Diners'

Club card in 1950 and the travel/entertainment card established by the American Express Company in 1958. Visa credit cards came out in 1958 and Mastercard in 1966.

With the advent of the internet, people started experiencing a virtual world, bringing about a gradual rise in e-commerce, creating a demand for money transfer solutions in the online world. In 1999, mobile banking was introduced by European banks using SMS and hence called SMS banking. The first fully functional banking app was launched by RBS in 2011 on Apple devices but was soon available on Blackberry and Android devices.

Next came the contactless payment cards in 2007, wherein chips in the credit/debit/smart card allowed for a secure means of payments by using RFID technology or near-field communication (NFC). Mobile contactless payment apps like ApplePay, GooglePay, SamsungPay, and others were introduced in 2014–2018.

Though the use of currency notes is gradually being displaced by credit and debit cards, electronic money transfers and mobile payments, it remains the primary means of payment or store of value for the unbanked people. Hence, the concept of a secure decentralized digital currency that can be operated by anyone who has an internet facility is truly revolutionary. This concept became a reality with the introduction of Bitcoin in 2008 as a fully decentralized cryptocurrency. As of 2018, the number of cryptocurrencies in the world is 1600 and growing. With the growing acceptance of cryptocurrencies and virtual money, Fintechs are focusing on disruptive technologies like blockchain, IoT, and AI. The combined effort of Fintechs, banks, and governments could bring about a new era of currency economics.

3.2.2 Birth of Bitcoin

Money facilitates economic exchange, and hence the saying 'Money makes the world go round.' To comprehend the motivation behind the creation of the Bitcoin, one must understand the basic financial circumstances at the time.

The banking system is the backbone of any economy. With the world dominated by monetary economies, banks acted as the ultimate gatekeepers of the financial world and charged fees for the services that they provided. Eventually, banks became the mandatory middlemen providing intermediary 'trust' for a person to confidently avail the specialized financial services provided by payment processors, credit card companies, insurance companies, bonds, and security brokers, etc. This promoted a heavy dependence on banks. This monopoly, however, had its disadvantages, especially for people in lower-income groups who did not have accounts or IDs or for whom the transaction fees took a toll on their earnings.

Financial institutions incur significant costs related to back-office expenses, reconciliations, legalities, secure-data storage, and prevention measures for security breaches and potentially fraudulent activities. These costs are passed down to the end-users as fixed transaction fees irrespective of the size of the transaction. While this ensured profitability for the financial institution, it did not justify the hit on the customer base irrespective of the transaction size. Also, the institutions could raise the fees at any time.

Paying fees for trusted transactions was not the only factor. There was also the question of transparency. People deposit money, trusting the banks to keep them safe. However, these deposits are used by banks to find opportunities for additional financial returns like extending mortgage and other loans, and investments. The anticipated returns on loans granted are tied down to various other investments. Financial institutions are incentivized to take on risky behaviour as they do not have to face the brunt of a system-wide failure alone. It is a complex system including investment banks, insurance companies, and others, and in dire circumstances, they are protected by the Government. When people defaulted on loan payments and the investments the banks made did not pay off, the banks declared bankruptcy. The result was that while the Government bailed out many of the financial institutions, the depositors lost all the money that they trusted the banks to keep safe, as was seen during the financial crisis of 2008. It should also be noted that the money offered by the government was the taxpayers' money. Countries printed more currency notes to boost circulation. However, that brought down the value of the country's currency, thus impacting the economy of the country.

So the three key requirements that eventually brought about the birth of the Bitcoin was the need for a monetary system that could:

1) Directly transact with another person without involving a third party, like a bank, to verify and validate the transaction and thus avoid the cost of mediation
2) Operate without being backed and controlled by a central authority and assure the value of the money is maintained
3) Preserve transparency in transactions, while still maintaining the users' privacy

The above requirements were made a reality when Satoshi Nakamoto (alias adopted by a programmer or group of programmers), conceptualized a peer-to-peer electronic cash system in 2008 and brought to life the first cryptocurrency – the Bitcoin in 2009 as open-source software. The Bitcoin system or protocol solved the **problem of trust** by allowing for willing parties to transact with one another using cryptography directly. The hashing algorithm used in the Bitcoin protocol is the hash function SHA256 (Secure Hash Algorithm 256-bit). The term 'Bitcoin' is used intermittently as both a currency and a technology, but basically, the Bitcoin is powered by the blockchain technology. As a digital currency, it lives on the internet and does not have any physical form. Hence, it is transferable anywhere and not restricted by country borders or tax requirements.

Bitcoin transactions are done via the digital wallets where one can store the bitcoins. With the bitcoin wallet, you are your banker and have sole control of your money. Unlike the typical currency where the Government can print more and thereby reduce the value of the money that is already in circulation, the total number of bitcoins that can ever be in circulation is fixed at 21 million. Bitcoins are released at a decreasing rate, i.e., as the number of bitcoins in circulation increases, fewer the number of bitcoins that can be created, thus **appreciating its existing value**. The value of the bitcoin is dependent only on the market supply and demand and **free of intervention** from any central authority.

While a central authority backs fiat currencies, the Bitcoin is backed by a widely distributed network of nodes or computers. Transactions are made in chronological order and non-reversible. The network does the verification and validation of the transaction. Whenever a transaction is made, it is broadcasted to the whole network and accepted only if it accepted by the majority of the network. This ensures security and prevents duplication of bitcoins, i.e., double-spending and fraud. Also, **user anonymity** is maintained as a persons' transaction is not associated with his or her identity, unlike in traditional databases where the transaction is associated with the customer id.

Ever since Satoshi Nakamoto mined the first Bitcoin block known as the Genesis Block in 2009, many developers have used the open-source code and developed their variation of the cryptocurrency, by a method called forking, to improve upon it, e.g., Bitcoin XT, Segwit, Bitcoin Cash, etc. However, the original Bitcoin is still the most popular cryptocurrency is usage and value.

As of January 2020, there are over 2.8 million bitcoins left to mine. The initial value of the Bitcoin was 0.0001 USD, and as of January 2020 stands at over 8,000 USD. As bitcoins and other cryptocurrencies gain popularity among the merchants, Governments are conscious of the need for forming regulations around cryptocurrency trading.

3.3 Cryptocurrency Basics

3.3.1 What Is Cryptocurrency

A cryptocurrency is a digital asset that is used as a medium of exchange on the blockchain. The most popular cryptocurrency is Bitcoin, followed by Ether of Ethereum and Ripples XRP tokens. Since the creation of the first decentralized cryptocurrency Bitcoin, thousands of alternative digital currencies, referred to as altcoins or coins (used for buying or selling products or services) and tokens (used as a utility or security) have emerged. According to CoinMarketCap, there are over 3000 active cryptocurrencies as of October 2019.

Though the concept of cryptocurrency first started as a simple peer-to-peer payment system focusing more on replacing the traditional cash system, it has now found unique user cases in almost every domain spanning finance, identity management, supply chain management, security, health management, ownership, records management and even powering Internet of Things. Cryptocurrency has also given rise to a new form of raising capital for start-up ventures called ICO or Initial Coin Offering that is much faster and simpler than the traditional fund-raising models.

Cryptocurrency and blockchain projects all have open-source code to foster trust and safety. Developers can copy the open-source software and modify it to either build compatible applications, e.g., wallets on the existing blockchain or build a completely new cryptocurrency network, e.g., Litecoin. This process of duplication from an existing blockchain is called **forking**.

A fork creates an alternative version of a blockchain. It is typically done to apply upgrades or new governance rules to a network. A fork can be of two types:

- ***Hard fork:*** A hard fork occurs when the protocol upgrade results in a split in the blockchain that is not backward compatible, i.e., the software validating according to the old protocol will see the new protocol as invalid and vice versa. Hence, all clients need to upgrade to the new version if they want to continue participating in the network. The miners/nodes need to decide on the version of blockchain they want to run with, resulting in two versions (original version and new version) running independently of each other. In some instances, the old chain is eventually abandoned while, in other instances, both chains remain active, e.g., Bitcoin–Bitcoin Cash fork, Ethereum Classic–Ethereum fork.
- ***Soft fork:*** A soft fork is a protocol upgrade that is backward compatible. Here the old software will recognize the blocks made by the new protocol as valid. Hence it does not require all the nodes in the network to upgrade to maintain consensus as the soft-forked chain follows both the old as well as the new set of protocol rules. However, the majority of the miners need to agree on the upgrade so that any blocks made with the old set of rules are rejected, and the old network is eventually abandoned. For example, Bitcoin protocol upgraded to SegWit.

However, there exist original cryptocurrency blockchains such as Ethereum, Ripple, Lisk, etc. which have their codebase. Ethereum has brought in a new category of software applications called Decentralized Applications (DApps) and smart contract protocols.

The source code of cryptocurrency can be found in software repositories like GitHub.com. Creating a cryptocurrency may be easy, but sustaining it is a long-term endeavour. Some of the aspects of the cryptocurrency that must be decided on while creating it is the type of usage, the total number of coins that will be available, the consensus mechanism that will be used, the privacy of transactions, i.e., whether public or private or hybrid and other technical and maintenance aspects related to security, scalability and other community needs.

For the cryptocurrency to succeed and sustain in the market, it needs to foster a sturdy community of users, miners, merchants, and other players to build a strong blockchain ecosystem.

3.3.2 Characteristics of Cryptocurrency

Both fiat currency and cryptocurrency can be considered as money as they are both used as mediums of exchange, i.e., used to buy and sell goods/services. They are a means to store and transfer value and can be traded on exchanges. Both forms of currency are governed by supply and demand and other economic factors.

However, cryptocurrencies have distinct characteristics that set them apart from the traditional fiat currency.

1) Decentralized

Fiat currencies are issued by the government and regulated by the Central Bank. Thus, government-issued policies and control of supply can affect the value of the currency. Cryptocurrency, on the other hand, is not backed or controlled by any bank or central government. All the nodes/computers in the network work together in mining or processing a transaction. This decentralized feature means that no central body can control or influence the value of the cryptocurrency.

2) Form of Existence

While fiat can exist in both physical (coins and paper notes) and digital forms that allow for electronic transfer of money, cryptocurrency can only exist in digital form. They are not tangible and are essentially developed by software code and cryptographic algorithms.

3) Limited Supply

Another key feature of cryptocurrency is that the supply is limited. The maximum supply of cryptos that can ever be generated or mined is defined when the genesis block is created.

4) Global Access

The decentralized nature of blockchain allows anyone to access and transact in cryptocurrency irrespective of their geographical location so long as they have access to the internet. There are no cross-border restrictions and their use is not limited to only those having access to banks. This allows for faster processing times and very low transaction fees as third-party intervention is removed.

5) Anonymity and Transparency

In standard transactions, the personal details are stored in traditional databases, and these transactions can be monitored, tracked, and retrieved by the central authority. Privacy and anonymity is a major concern. In cryptocurrency, a transaction is linked to the person's cryptocurrency addresses and not the person's name, address, or any other personal details. Thus anonymity is maintained. At the same time, every transaction record is stored in the blockchain ledger as a list of open, yet encrypted records making it transparent, verifiable, and honest.

6) Impossible to Duplicate

The underlying blockchain technology of the cryptocurrency is that of decentralization. Cryptographic encryption and consensus protocols make counterfeiting cryptocurrency impossible.

7) Irreversible

Cryptocurrency transactions are irreversible, unlike traditional currency transactions. This feature has its pros and cons. Once a transaction is recorded and verified, it is irreversible and impossible to change. While this fosters trust, integrity, and auditability, it also means that if a cryptocurrency is sent to the wrong address, the coin is lost forever. Hence, users must we vigilant not to fall for cryptocurrency scams or frauds.

3.3.3 Cryptocurrency Wallets

Bitcoin or any other cryptocurrency transactions can be done only using a digital wallet. A digital wallet or **cryptocurrency wallet** is a software program that stores the user's private and public keys enabling the user to transact crypto assets. It is a management system that interacts with various blockchains to enable users to send and receive digital currency and monitor their balance. A basic flow of cryptocurrency transactions within the blockchain by using a pair of public and private cryptographic keys is described in Chapter 1. Cryptocurrencies are stored immutably on the blockchain using the user's public key. This public key is used by other wallets to send funds to the user's wallet address. However, the private key is required if users want to spend cryptocurrency from their address.

There are many classifications of cryptocurrency wallets. But typically, they are classified based on their storage, as shown in Fig. 3.2.

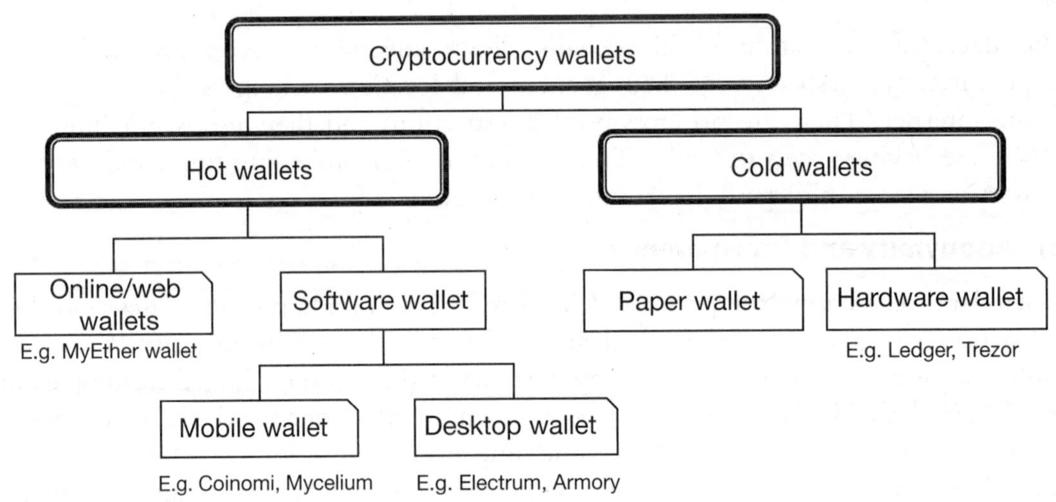

Figure 3.2: Types of wallet

Hot Wallet

A hot wallet is designed for online day-to-day transactions. It is connected to the internet at all times and hence a strong candidate for hackers. For this very reason, it is not advisable to use a hot wallet for long-term storage. Hot wallets are typically free, easy to set up, and convenient to use.

Based on their installation characteristics, they are differentiated into three types as follows.

Desktop wallets: Desktop wallets, as the name suggests, can be downloaded and installed on your desktop or laptop. As they are locally stored, the user has complete control of the wallet. They come with a variety of features, including build-in exchange platforms, multi-currency support, etc. The downside, however, is that the wallet is dependent on the security features installed on your computer. Hence, if the computer gets hacked or is virus-infected, there is the risk of losing all your funds.

Bitcoin Core, Electrum, and Exodus are examples of desktop wallets.

Mobile wallets: Mobile wallets are designed to operate on smartphone devices. Once installed, the smartphone application operates just like its desktop counterpart, but it may not have all the savvy features. However, it circumvents the constraints of the desktop wallet with its ease of accessibility, as you can easily carry it with you wherever you go and make purchases from merchants who accept cryptocurrency payments using QR codes. However, they are more vulnerable to malicious apps and viruses than desktop wallets. Hence, it is recommended that security features like wallet encryption are installed on the mobile devices and adequate backup is provided for private keys in case the smartphone is lost or broken.

Well-known examples of mobile wallets are Coinbase Wallet, Mycelium, and TrustWallet.

Online wallets: Online wallets are also called web wallets. They run on the cloud, and hence the user does not need to download or install any application. They can be accessed from any computing device via a web browser. They are the most convenient of the hot wallets and preferred by fresh cryptocurrency users. Online wallets are hosted and controlled by a third party and hence are the most vulnerable. Though some service providers offer to manage your private keys on your behalf, many web wallets have come with the option where the user can take full control of the keys or share control via multiple signatures.

MyEtherWallet, GreenAddress, and MetaMask are examples of online wallets.

Cold Wallet

A cold wallet is a digital wallet that is not connected to the internet. Unlike hot wallets, they are not free. Being offline, they are more secure and used for long-term storage of cryptocurrencies. Hardware wallets and paper wallets are the two types of cold wallets.

Hardware wallets: Hardware wallets are physical, electronic devices that use Random Number Generator (RNG) to generate the public/private key that is stored in the device. It connects to the internet whenever the user needs to send or receive payments and disconnects once the transaction is executed. Transactions are confirmed through the private keys that are saved offline. Hardware wallets have the facility to generate a PIN to protect the device as well as a recovery phrase in case the wallet is lost. Though more secure than hot wallets, they are less user-friendly and difficult to access. Trezor and Ledger wallets are popular hardware wallets.

Paper wallets: Paper wallets are offline and, as the name suggests, it is a piece of paper with the crypto address and its private key is physically printed out in the form of QR codes. These codes can then be scanned to execute cryptocurrency transactions.

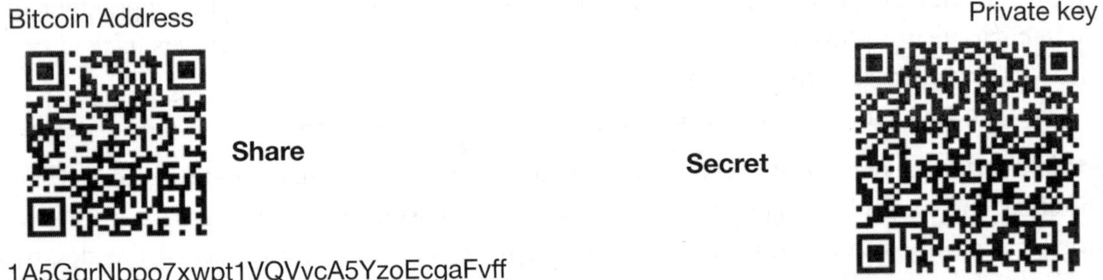

Reference: FlippyFlink [CC BY-SA (https://creativecommons.org/licenses/by-sa/4.0)]

Figure 3.3: A paper bitcoin wallet

Figure 3.3 shows a printable bitcoin wallet with the public bitcoin address for receiving and the corresponding private key for spending. Paper wallets are completely secure as the question of hacking does not arise with paper wallets. However, a big disadvantage is that the paper wallet contains only a single public/private pair, and hence they can be used only once for the whole amount in one transaction. If you spend a partial amount, you will lose the remaining balance. Also, they need to be kept safe from water, fire, wear and tear, or anyone photographing the QR code.

3.4 Types of Cryptocurrency

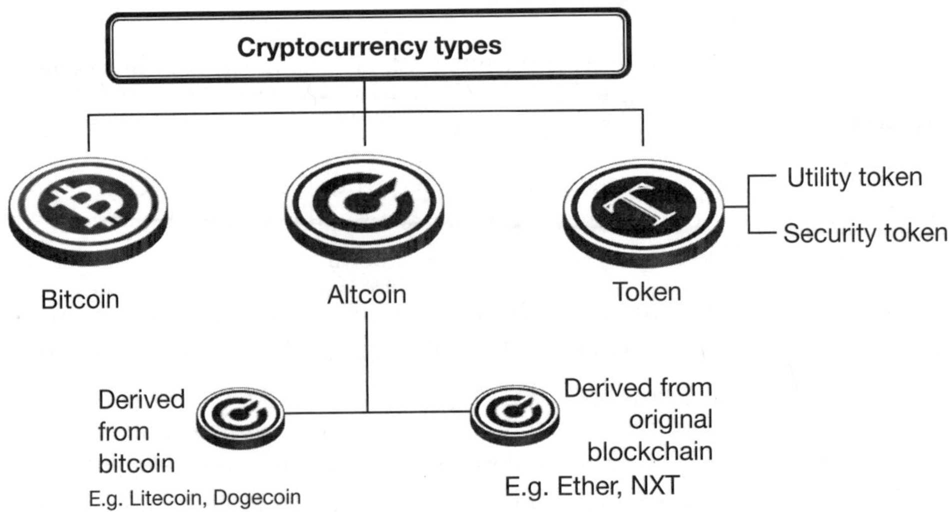

Figure 3.4: Cryptocurrency types

3.4.1 Altcoins

Bitcoin is the original blockchain-based cryptocurrency (refer Fig. 3.4). It is the frontrunner in the crypto world, being the best known, widely used, and valuable of the cryptocurrencies. Since its induction, many variants of the bitcoin have arisen using its open-source protocol, creating a new currency coin with a different set of features that operates on its blockchain. Namecoin, Peercoin, and Litecoin are examples of Bitcoin-derived blockchains.

Some coins are not derived from Bitcoin but created as a native currency of an original blockchain and protocol, namely Monero, XRP (Ripple), and NxtCoin.

These cryptocurrency alternatives to the Bitcoin are referred to as 'Alternative Cryptocurrency Coins,' abbreviated as **Altcoins** or simply **Coins**. Many of the altcoins come from a fork of famous and durable cryptocurrencies like Bitcoin, Litecoin, and Ethereum. While some altcoins are similar to their predecessors, most attempt to improve or set themselves apart by bringing in additional features or security like improved block times, different parameters, transaction management, scripting language, consensus mechanism, etc.

However, the main features that are common to all altcoins are:

1) They are peer-to-peer digital currencies that involve a mining process.
2) They possess their independent blockchain.
3) They possess the characteristics of money, i.e., they are fungible, divisible and have limited supply, and typically meant to operate only as a means of payment.

Altcoins or coins are, in some instances, also referred to as 'currency tokens' that should not be confused with the broader term of tokens.

3.4.2 Tokens

Tokens are another category of cryptocurrency that reside on top of an existing blockchain. It facilitates the creation of decentralized applications (DApps). Tokens and their underlying applications benefit from the technology and security of the native blockchain as they can leverage their existing mining strength and inherent resistance to attacks. The basic principle is that running several applications on a shared sturdy blockchain is more practical and business savvy than executing a single application on a blockchain that needs years to build up its network and reputation.

Unlike coins that need to be built from scratch, tokens are created using standard templates from blockchain platforms like Ethereum, Lisk, or Waves platform. The functional variances are possible through the use of smart contracts. Developers widely use smart contracts compliant to ERC20 standard due to its ease of implementation and wide adoption in the Ethereum platform. Status (SNT) and OmiseGO (OMG) are well known ERC20 tokens. Aside from ERC20, few other Ethereum token standards are ERC223, ERC 721, and ERC 777. Compliance to standard enables the storage of different types of tokens in a single wallet.

Though a token can be used as a method of payment within the project ecosystem, its primary purpose is its usage as a utility or asset to give the holder specific rights to participate in the blockchain network or even represent tangible objects like a house or painting. For example, the WPR (WePower) token represents electricity and allows the public to buy and sell electricity on the blockchain using smart contracts.

Tokens differ from altcoins in following respects:

1) They are digital assets that are built to perform a specific function(s) within the project ecosystem and not operational as a typical cryptocurrency. For example, a dinner voucher will entitle you to dinner at a restaurant and a cinema ticket to a movie. However, you cannot use the dinner voucher to watch a movie and vice versa. The same principle stands true for tokens, where the token can be used to provide a service or right of product used only for the specific project/blockchain.
2) They do not have purchasing power, i.e., tokens can be bought with coins, but coins cannot be bought with tokens
3) They are built to interact with decentralized applications (DApps) that sit on an existing blockchain network like Ethereum, NEO, etc. Hence, they are easier to create than coins that need new coding or code modification, thus saving the application developer time and resources.

Tokens are created and distributed to the public through a crowd-funding method called Initial Coin Offering (refer to Chapter 8 on ICOs), where tokens are issued in exchange for funding a potential blockchain project. Based on the function of the token, they are typically divided into two types as below.

3.4.2.1 Utility Tokens

Utility tokens offer the holder the right or license to use a product or service, e.g., ETH (Ethereum). Issued during an ICO, these tokens can, in the future, be redeemed to access the product or services of the company or project. They are considered free of regulatory restrictions as they are not built to function as an investment instrument, but to facilitate the funding of the ICOs. The token holder may have some voting rights within the ecosystem. However, they do not have any decisioning power in the direction that the company or project can move.

The most common type of utility token is the ERC20 token, which is the technical standard used for all smart contracts on the Ethereum blockchain for token implementation. According to Investopedia, more than 181,000 ERC20 compatible tokens exist on Ethereum's main network as of April 2019.

Another example of a utility token is the CVC token used by Civic, an identity management application that keeps track of encrypted identities on the Ethereum blockchain. Other popular utility tokens are Filecoin, Storj, Golem, etc.

3.4.2.2 Security Token

Security tokens, also referred to as Equity tokens, entitle the holder to a share or equity in the company/startup. In other words, they can be considered as a digitized, tokenized form of the traditional securities. Security tokens are issued during **STO** or **Security Token Offering**, which is a variation of ICO with more regulatory and due diligence controls fostering investor confidence.

They are seen as an investment opportunity. As the tokens represent percentage ownership of the company, including voting rights and decision-making power, they are subject to regulatory restrictions.

Countries like the USA, Canada, Switzerland, Malta, and Estonia have established regulations around security tokens, while many others are in the process of doing so. SEC classifies tokens as securities if they pass all the following criteria set by the 'Howey Test' to classify the security as an investment contract:

1) There is an investment of money or assets.
2) There is an expectation of profits from the investment.
3) The investment is in a common enterprise.
4) The expectation of profit is solely from the efforts of a promoter or third party.

Due to the regulations, launching security tokens is time-consuming. However, the same restrictions also protect investors from fraud and scams. BCap of Blockchain Capital, 22X Fund, and Spice VC are a few examples of security tokens.

Though tokens can be broadly classified as utility and security, there is the emergence of hybrid tokens that takes the best of both worlds. It can combine the advantages of both utility tokens and security tokens, as incentives are distributed across all participants. In contrast, the application of securities laws can weed out frauds and scams. An example of the hybrid token is BNB (Binance) that is used to pay transaction fees on the Binance Exchange. When using this token, the token holder also gets a discount on the transaction fees. Also, the Binance Exchange reduces/controls the token supply by 'burning' and redistributes the profits to the token holders.

Thus, it is not a straight-forward task for regulators to differentiate the various aspects of coins and tokens. Bitcoin and Ether are classified as not securities and are hence exempt from regulatory oversight. However, many others, like Ripple's XRP, are still under review.

3.4.3 Popular Coins and Tokens

Bitcoin still dominates the crypto market and is the key factor for the rise in crypto assets' market value in 2019. According to CoinMarket, Bitcoin outperformed all major financial asset classes in 2019, including equities, commodities, and bonds.

Aside from Bitcoin, the following are some of the popular coins/tokens in the cryptocurrency world today.

3.4.3.1 Ethereum

Ethereum is the second-largest coin by market capitalization as of January 2020. Like Bitcoin, Ethereum is a decentralized distributed public blockchain network. While Bitcoin functions only as a peer-to-peer electronic cash payment system, Ethereum is a smart contract platform that enables developers to build decentralized applications

(DApps) without the need of a centralized development platform. It was conceptualized by Vitalik Buterin in November 2013 and launched in July 2015 with the PoW consensus protocol. To address scalability issues, Ethereum is planning to transition into the PoS blockchain called Ethereum 2.0 in 2020. The native currency for the Ethereum platform is Ether (ETH).

In 2016, a venture capital fund smart contract called DAO (Decentralized Autonomous Organization) that ran on the Ethereum blockchain was hacked, resulting in the loss of about 3.6 million Ether. To refund the stolen Ether, a hard fork of Ethereum was created. The current running Ethereum of today is the hard fork. However, some believed forking Ethereum went against the immutability code of blockchain and continued to operate on the original Ethereum blockchain. To differentiate between the two, the original Ethereum is referred to as Ethereum Classic, and its crypto, though called Ether, is traded as ETC.

3.4.3.2 Ripple (XRP)

Released in 2012, the Ripple blockchain was developed by Ripple Labs Inc. XRP is a cross-border digital payment system that is powered by its altcoin XRP. Unlike Bitcoin that uses PoW consensus or NXT that uses PoS consensus protocol to mine coins, the XRP coins are already pre-mined. There are 100 billion XRP coins, of which around 43 billion are in circulation, and the rest are periodically released as set by an in-built smart contract. Validation on the Ripple network is done through a pBFT (Practical Byzantine Fault Tolerance) consensus algorithm where the nodes verify and agree on the authenticity of a transaction by conducting a poll. The transactions have to be approved by 66% of the validators to reach consensus. The transaction confirmation is done within seconds (approx. 3–5 sec), and as there is no mining, the energy consumption and related costs are very low. Users are charged a small XRP fee for their transactions.

The Ripple platform is mainly used as a global payment settlement, asset exchange, and remittance system. It can seamlessly transfer any form of currency, i.e., both cryptocurrency like BTC, LTC, etc. as well as fiat currency like USD, YEN, EUR, etc. Ripple is the third-largest cryptocurrency by market capitalization as of January 2020 and mainly used by banks and financial institutions. For example, RippleNet is a consortium of over 300+ financial institutions worldwide to facilitate low-cost cross-border payments.

3.4.3.3 Bitcoin Cash (BCH)

Bitcoin Cash is the first hard fork of the Bitcoin blockchain that appeared in August 2017. It uses the same PoW protocol as Bitcoin. The primary purpose of forking was to increase the block size limit from 1 MB to 8 MB to improve transaction times and scalability issues. A drawback to increasing the block size limit is the higher costs of running a full node on a blockchain-based network, potentially leading to centralization as only a few groups or corporations will have the resources to run a full node.

In November 2018, Bitcoin Cash was hard-forked and split into Bitcoin Cash ABC (BCHABC) and Bitcoin SV (BSV). Bitcoin ABC uses the BCH ticker as it is the dominant chain having a majority of nodes in the network. In many exchanges, Bitcoin Cash and Bitcoin Cash ABC listings are merged and represented only as Bitcoin Cash (BCH). It remains one of the top 10 cryptocurrencies by market capitalization.

3.4.3.4 Litecoin (LTC)

Litecoin is the earliest altcoin released in October 2011. It is a fork of the Bitcoin core client. It was created by Charlie Lee, an MIT graduate, to improve upon Bitcoin's block generation rate. Litecoin is similar to Bitcoin except in the hashing algorithm used for PoW. It uses memory intensive 'scrypt' proof-of-work mining algorithm. The average time taken per transaction of Litecoin is 2.5 minutes as compared to 10 minutes for Bitcoins. The number of Litecoins that can be mined is capped at 84 million, as compared to Bitcoins' 21 million. Litecoin is also one of the top 10 cryptocurrencies in the market.

3.4.3.5 Eos (EOS)

Launched in January 2018 by Block.One company, EOS is one of the top 10 cryptocurrencies by market cap. EOS is a smart contract blockchain development platform similar to Ethereum to develop decentralized applications (DApps) and development tools. EOS blockchain protocol is powered by the native cryptocurrency EOS. It was built to merge the positive aspects of both Bitcoin and Ethereum. It uses DPoS (delegated proof of stake) where the holder of the coins has a say (voting rights) in selecting the 'block producers,' i.e., those who can validate/verify the transactions. EOS does not charge any transaction fees.

3.4.3.6 Monero (XMR)

Monero is another popular cryptocurrency that was launched in April 2014. It is a completely private, secure, and untraceable cryptocurrency where no one is able to see anyone else's transaction or balances unless the person makes it public, unlike in Bitcoin, where transactions are public on the blockchain. Monero has no pre-set block size limit. Therefore to prevent malicious miners, a block reward penalty is built into the system where the block reward gets reduced. If a block mined is greater than the median of the last 100 blocks (say M100) mined, the percentage of increase reduces the block reward. For example, if the block size is 10% more than M100, then the block reward is reduced by 1% and if 50% more, then reduced by 25% and so on. Generally, blocks greater than two times M100 are not allowed.

Monero maintains its privacy through **ring signatures, RingCT**, and **stealth addresses**. Ring signatures are anonymous digital signatures where the signature of the sender is clouded with other public keys in the blockchain, making it computationally impossible to know the identity of the sender. Also, one-time random addresses called stealth addresses are generated for each transaction, thereby concealing the destination address and the identity of the recipient. Also, the Ring Confidential Transaction (RingCT) feature in the Monero protocol hides the transaction values. The combination of the three features in the protocol make Monero transactions completely private and untraceable.

The mass appeal of Monero's non-traceability and privacy features is also its drawback. It has attracted money launderers and other illegal activities leading to many crypto exchanges to drop the coin to mitigate compliance risk.

3.4.3.7 Stellar (XLM)

Stellar is a cross-border payment network system aimed at providing affordable financial services to people of all income levels. Conceptualized by Ripple co-founder Jed McCaleb with Joyce Kim in July 2014, Stellar can handle both fiat and cryptocurrency exchanges. The platform is open-source, decentralized, and non-profit. It connects financial institutions and drastically reduces the cost and time required for cross-border transactions. It utilizes the Federated Byzantine Agreement (FBA) algorithm called SCP or Stellar Consensus Protocol, which enables faster processing of transactions. A variant of the pBFT, in FBA, each node selects another node(s) in the Stellar network that they trust to form a quorum slice. A quorum slice is a subset of a quorum. Transactions are approved if all the nodes within the quorum agree. The native currency of Stellar is Lumens (XLM) and used as an intermediary currency on the platform for currency transfer/exchange.

3.5 Cryptocurrency Usage

3.5.1 Ecosystem Players

A blockchain is only as strong as its community. If the players in the community are honest and engaged, it makes for a sturdy and sustainable cryptocurrency ecosystem. The different players or actors that directly or indirectly constitute the blockchain ecosystem or community are:

1) Programmers/Developers

The success of the blockchain project is established by a strong team of blockchain programmers and developers. Blockchain programmers/developers develop, deploy, and maintain the blockchain protocols, applications, and interfaces. They design different levels of functionality of the blockchain such as blockchain protocols, the architecture, the consensus design, etc. DApps developers work with decentralized applications and smart contracts that can run on blockchains to create far-reaching applications for the industries. Developers continue to monitor and maintain the blockchain technology as per needs and requirements.

2) Miners

Miners validate new transactions and record them on the blockchain. Based on the consensus mechanism used, the miner who can mine the specific block can be predetermined, or miners compete for the privilege. They contribute their effort and computing power for solving the cryptographic problem and are rewarded with block rewards or transaction fees. In PoS, they are referred to as **forgers** as they stake a claim to add a block. The miners operate the infrastructure by mining the coins.

3) Users

A blockchain user is a consumer who uses the blockchain or cryptocurrency for the purpose it was designed for, e.g., doctors and patients using Veris Foundation Blockchain, traders or investors using Kraken exchange. A user can also be a miner or validator or investor. A large, engaged user base makes for a secure blockchain that is hack-proof. Users are represented as nodes that have a copy or partial copy of the ledger related to their needs for initiating, reading, or creating transactions.

4) Merchants

Merchants are retail or other business entities that accept cryptocurrency as a form of payment for their goods or services.

5) Traders

Cryptocurrency traders speculate and buy/sell cryptos based on the price movements via a decentralized exchange. A typical Exchange is designed to compare and find the cheapest rate of exchange between any two coins or tokens, making it more affordable to trade in cryptocurrency. Traders may sometimes also use **brokers** to help them in getting the best market rate and minimal slippage for long-term profitability.

Other players form part of the community as well, like the Board of Directors or owners or administrators, typically in private blockchain, who decide on the governance or protocol rules, lawyers, auditors and regulators, promoters, media, etc. Also, the bad actors with criminal intent who try to break the blockchain should not be ignored.

3.5.2 Cryptomining

Miners generate wealth through mining. But it is not an easy process. A miner needs to have some level of technical knowledge and expertise in setting up computing software and equipment. Blockchains vary in the mining systems that they use. However, they all have some form of a consensus algorithm and an incentive system. Bitcoin uses proof-of-work consensus mining algorithm, and the successful miner gets a block reward of 6.25 BTC, while for Ethereum, it is 2 ETH + additional fees per block. Miners are rewarded with the native cryptocurrency of the blockchain, e.g., BTC in Bitcoin, LTC in Litecoin, XMR in Monero. Blockchains like Hyper Ledger Fabric exist without a mining concept as it is not a cryptocurrency-based blockchain. In these cases, there are no rewards, and the consensus is reached through the logic in-built inside the Fabric Blockchain.

Miners use computing power to validate and produce a block. There are two types of miners; those that go solo called solo mining and those that collaborate with others, referred to as pool mining.

Solo mining: As the name suggests, here, the miner works independently to find the block. They typically have large mining equipment and do not rely on any third party. The miner connects to the network via their wallet. Once they can create a block successfully, i.e., the hash submitted is accepted by the network and added to the block and block is confirmed, they are rewarded in cryptocurrency. The advantage of solo mining is that the miner is credited with the full block reward. The disadvantage is that it may take a long time to find a block as compared to if you were in a mining pool that has more resources and hardware power.

Pool mining: Pool mining is when a group of miners join their collective computational resources over the network towards faster processing of the hash function. The miners submit the block hashes into the pool. If any of the hashes are successful in creating the block in the blockchain network, the mining pool distributes the reward between all the miners of the pool proportional to their contributed processing power or as per the terms agreed. Here, all miners who contributed to the pool benefit irrespective of whether they produced the winning hash. The success rate of finding a block in pool mining is much higher than when such mining is done individually, and it consumes fewer resources making it economically viable. The downside is that the rewards are divided and hence lower. Also, the system is not autonomous as the terms of the mining pool bind miners. Antpool, BTC.com, and F2Pool are some popular mining pools.

There are different types of mining based on the processors or equipment used by the miner:

CPU mining: CPU (Central Processing Unit) mining is the earliest form where a miner utilizes central processors to mine cryptocurrency. Here anyone with a laptop or desktop with the required software installed can mine. It was feasible earlier for mining with CPU processors at the time of Bitcoin induction. But since the 'difficulty' has increased, the process is exceedingly slow, consuming large amounts of electricity, cooling, and other resources, making it financially unviable.

GPU mining: GPU (Graphics Processing Unit) mining is where a miner utilizes a graphics processor to perform block validation. GPU mining is more efficient and cost-effective, offering higher processing power than CPU. GPU is upgradable, has high resale value, and the hardware can be used to mine several coins like Ethereum, Zcash, Lincoln, etc. The GPU equipment size is large and mostly preferred by cloud mining companies.

ASIC mining: A miner utilizes an Application Specific Integrated Circuit (ASIC) to perform block validation in ASIC mining. They are more efficient, faster, and can mine more coins in less time than any other kind of processor. Compared to CPU and GPU, they have the least power consumption and comparatively small physical size. Also, mining algorithms can be performed in parallel here, hence the potential for high-profit margin. AISC is custom-designed to mine specific currencies and hence not useful for miners who want the flexibility to mine various currencies. Also, they have low resale value as they are not upgradable and become obsolete when a new version comes out.

Cloud mining: Cloud mining is the mechanism by which a miner or a cloud mining company rents out cloud mining services to other miners for a pre-assigned period. It is more cost-effective for the miners as they do not have to pay for setting up the mining equipment. The miner will open an account with the company and agree to the contract period, including the hashing power, and can immediately start on the cloud mining services. The cloud mining hosting company has huge mining rigs and facilities, the cost of which is taken from the customer/miner as either an upfront subscription payment or a percentage of the monthly reward, including the cost for electricity and maintenance. IQMining, Genesis Mining, and FlyMining are some of the popular cloud mining farms.

3.5.3 Airdrop

A strong community is the backbone of a value-creating blockchain network. User engagement, along with cost and energy efficiencies, makes for a sturdy, secure blockchain network. **Airdrop** is a promotional activity aimed at spreading awareness among the blockchain community. It is a distribution event where a blockchain project distributes free coins or tokens to wallet addresses to create a market for the project and a buzz among investors.

Did you know?

In the cryptocurrency world, a **Snapshot** refers to the act of recording the state of a blockchain at a particular block height. The snapshot records the contents of the entire blockchain ledger, which includes all existing addresses and their associated data like transactions, fees, balance, metadata, and so on.

Snapshots are commonly used during airdrop events before each round takes place. Snapshots are also important during blockchain hard forks, as they mark the block height at which the main chain will be recorded before giving birth to the new chain. For instance, when the Bitcoin Cash hard fork took place on 01 August 2017, every blockchain address that had Bitcoins at block 478,558, had the balance copied on the Bitcoin Cash blockchain. This is because both blockchains share the same historical data before the fork. After the split, each blockchain network will operate independently of each other.

Reference: Binance Academy

3.5.3.1 Benefits of Airdrop

Airdrops are free money typically given out to early adopters, in the form of free coins or free tokens. Startup companies use airdrops as an effective marketing strategy to spread the word about their upcoming product, coin, or platform to encourage research of the project and adoption of the cryptocurrency. It is primarily aimed at investors for funds, increasing the user base, and for wider distribution of coins/tokens.

The various benefits of airdrops to the blockchain enterprise are:

Marketing and hype: Airdrop is an excellent way to get people interested in cryptocurrency, especially to create the much-needed hype around ICOs and projects. Dropping free coins in a wallet can lead to free advertising for the blockchain project or cryptocurrency through word of mouth, social media, websites, and forums, leading to a strong user community and network allowing for the cryptocurrency to appreciate in time.

Rewarding loyalty supporters and investors: Airdrops are used by some cryptocurrency companies as exchange and trading platforms to incentivize users to continue to buy and use tokens/coins on such platforms. If a user keeps substantial tokens of the company in their wallet, the company rewards them with free tokens. Also, tokens are given for providing loyalty services like promoting the cryptocurrency in social media, writing reviews, signing up on the platform, etc.

Wider and even distribution of tokens: Airdrops are an effective method of achieving wide and even distribution, especially for utility tokens projects. It helps these projects to gain new users and contributors, and at the same time, retain existing users, which ultimately leads to a high level of engagement within the community. Airdrops discourage whaling and promote an even distribution of tokens proportional to the users' participation in the ICO or project or holding of existing coins.

Lead database generation: Airdrops are a great mechanism adopted by blockchain-based companies for lead generation. Users are requested to fill online forms to claim their free coins/tokens. This valuable user information can be used by companies to develop targeted marketing campaigns and strategies.

3.5.3.2 Types of Airdrop

Cryptocurrency airdrops are done in two ways – a surprise airdrop or a planned airdrop. An established blockchain company does a surprise airdrop. The company takes a snapshot of the blockchain, i.e., the addresses and their related data like transactions, balances, metadata, etc., at a particular date and time and distributes free coins to the wallets based on the balance in each blockchain address. This method for airdrops is mostly used in blockchain hard forks, especially the Ethereum standard tokens.

Another method is the **planned airdrop** that is adapted by new enterprises. Here the blockchain start-up announces the date and time of the airdrop beforehand. In this instance, some prior action from the users is required to qualify for the airdrops like filling a KYC form, bringing in a new member, promoting on social media, etc.

The following are some of the eligibility requirements that blockchain companies base their airdrops on.

- The users have to register by providing their basic information like name, email id, and other related data to qualify for the airdrop.
- When a hard coin fork occurs, coin holders of the original protocol are airdropped the new coins, for example,, bitcoin cash airdrop on bitcoin holders.
- Users or participants who are loyal to a specific project or blockchain network and meet pre-set eligibility criteria may be provided with free coins unknown to the rest of the network.
- Users are expected to promote or advertise about the project in the specified media to be eligible for the airdrop.
- Users (Holders) who hold a particular token in their wallet will get free tokens. E.g., EOS tokens airdrop to EOS holders.

As airdrops are considered 'free money,' many investors may sign up for airdrops. This makes it a top candidate for fraudsters and scammers. Some top frauds are for information trolling, obtaining the private key, and pump and dump schemes. Information trolling involves convincing people to believe a false information as true. Airdroppng happens when people, when promised free currency, voluntarily spread false information about the coins, thus increasing its value automatically. 'Pump and dump' involves airdropping of relatively unknown cryptocurrency in mass, which in turn increases the value of the particular cryptocurrency coin. This enables (forces) other investors to buy. When these (second round of investors) buy, its value increases further. At its peak value, the first round of investors (who got the coins for free) will sell and come out, gaining huge profits. Whereas the second round of investors will still live with the worthless coin.

The onus is on the user to not fall for such schemes and to observe due diligence before accepting airdrops. Also, a request for your private key should immediately raise a red flag as the public key is sufficient for the company to send you tokens.

3.5.4 Token or Coin Burning

Token or Coin burning is a process of permanently removing coins out of circulation to reduce the total supply. If the primary aim of airdrop is to increase user awareness and engagement, the token burn is done to reduce the existing supply of coins/tokens and thereby to increase their value.

The coin is considered to be burned when it is sent to an unspendable public address, known as an **eater address**, which does not have an operable private key associated with it. The transaction is transparent on the blockchain for anyone to confirm that the coin(s) were sent to the address, but at the same time, the address does not belong to anyone and has no practical value. This is referred to as proof of burn wherein anyone can check and verify that the coins were legally destroyed or burned, i.e., in effect put out of circulation.

Coin burn is initiated when the user calls a burn() function with the amount of coins one wants to burn. A smart contract then verifies whether the user has the stated amount or not. If available, the coins are sent to the eater address, and the burn is completed.

The key purpose of burning coins is to increase the value of the remaining coins that are in circulation. It is done for the following reasons.

1) To Reward Investors

The main aim for sending the cryptocurrency to an unspendable address is to create a scarcity of supply that is ideally expected to increase the value of the coin/token, assuming the demand for the currency remains constant or increases. Thus the coin holder will benefit from the appreciated value of the coins. Companies adopt this method as an alternative to dividend payments to avoid being classified as security tokens that are subject to regulatory oversight. For example, Binance buys back their BNB tokens using a percentage of their quarterly profits and burns them so that everyone can benefit from the price appreciation.

2) To Destroy Unsold ICO Tokens

Tokens or coins generally appreciate after a token sale. ICOs that do not meet their hard cap, i.e., have excess tokens, burn their unsold tokens instead of keeping the coins/tokens for themselves. The act of distributing value back to the token holders instead of keeping the excess coins/tokens for themselves for personal benefit aims to demonstrate the commitment of the project team towards the long-term success of the project.

3) To Decentralize Mining Opportunity (Proof-of-Burn)

In proof-of-work (PoW), the miners need to expend a large amount of computational power to crack the cryptographic puzzle for the right to mine and earn block rewards. Proof-of-burn (PoB) is an alternative distributive consensus mechanism to PoW. In the PoB consensus protocol, miners are granted the right to mine and validate transactions based on the amount of coins that they burn. The more coins a user burns for the benefit of the network, the more virtual mining power he/she gains, thus increasing his/her chances of being the next block validator. Hence every user has the opportunity to mine blocks and is not constrained by a lack of powerful mining hardware and resources. Similar to PoW blockchains, miners of the PoB blockchain will receive block rewards that, in due time, is expected to be more than enough to cover the initial investment of the burned coins.

There are, however, some disadvantages associated with coin burn. In PoB, mining power and hence control go to users who have more resources to burn to create a 'rich get richer' scenario. This creates a mining centralization problem. Also, there is no guarantee that the user will recover the full value of the burned coins making proof-of-burn an inherent risk.

3.5.5 Investing and Trading

Investing and trading is perhaps the easiest way for the public to generate wealth through cryptocurrency. The common principle is 'buy low, sell high.' There are basic differences between investing and trading. Investors are in it for long-term profits while traders are focused on the short-term goals. Compared to the stock market, the cryptocurrency market is extremely volatile, with upward and downward market trends occurring in very short periods, in many cases within one year of release. So there are equal chances of making huge profits or incurring a high loss.

Investing in Cryptocurrency

Investors are generally risk-averse, and they take a stance that cryptocurrency is still in its infancy and that it will take some time for the technology/application/platform to be trusted and accepted in the mainstream. Investors study the inherent value of the company by evaluating the financial statements, research reports, and other economic factors. They bet on the long-term potential of the cryptocurrency and hold on to it for several years, hoping to sell it at a profit. Investors keep an eye out for potential ICOs, hard forks, and airdrops for investments. They also stand to profit from dividends or coin burns.

Trading in Cryptocurrency

The frequency of cryptocurrency trade can vary from a few seconds/minutes to a couple of weeks. Traders are speculators and have a higher risk appetite than investors. They aim to buy low and sell high within short periods. The trader is also willing to drop coins if the market trend looks bad or does not seem to pick up, even if it means a small loss. Traders are not typically interested in the intrinsic value of their assets but on the potential for percentage gains. They base their decisions by analyzing hourly price movements as well as historical price data. Since cryptocurrency is a very volatile market, traders tend to make huge profits if they do it right. They also capitalize on airdrops, hard forks, and ICOs and sell them on the exchange for immediate profits.

3.5.6 Cryptocurrency Safety

Currency is only as safe as the custodian. In fiat or traditional currency where the bank or financial institution is the custodian, in the event of an account getting hacked or compromised, credit/debit card stolen or bank account broken into, the related authorities take the responsibility to track down the culprits. In some instances, the customer is reimbursed, or transaction reversed if the customer can prove that a fraud has occurred. However, in the cryptocurrency world, you are the sole owner of the money. Transactions, once executed, cannot be reversed. Verification is done with private keys, and if the private keys are compromised or money is sent to the wrong address, then it is lost forever.

Hence safety measures must be followed while storing and transacting in cryptocurrencies.

1) Best Practice in Using Exchanges

- Choose regulated exchanges that have safety and security measures in place. Binance, Bittrex, and Coinbase are popular crypto exchanges.
- Do not store your cryptocurrencies on exchanges for more than a day or two. In 2019 alone, ten crypto exchanges were hacked, which resulted in a loss of nearly $170 million. Cryptocurrency Exchanges store private keys on behalf of the user, and the accounts are further encrypted with private keys. Unfortunately, exchanges are

easy targets for hackers, and one crack is enough for a hacker to swindle hundreds of millions of dollars.

- Conduct smaller trades instead of one large trade to avoid drawing the attention of malicious actors.
- Use a unique email when opening accounts on exchanges as a person can get your information if your regular email is hacked. Create a separate email for every exchange you use.

2) Storing Cryptocurrency

- Store your crypto in desktop or mobile wallets if short-term and in paper and hardware wallets if long-term.
- Use wallets from reputable sources. Avoid wallets where you do not have control of your private key.
- Carry only small amounts, equivalent to what you can afford to lose, in your mobile or online wallets.
- Keep your crypto assets spread over several devices/wallets to safeguard your investment and minimize the impact of any loss.
- Maintain backups of your wallets and back-up phrases to ensure coins are not lost in case the main device fails. Take multiple backups and keep them in different locations.
- Update your back-up after every transaction.
- Generate the seed phrase, write it on paper and keep it in safe custody.
  Note: Seed phrase or Seed recovery phrase is a wallet recovery phrase that consists of 12 words typically generated by any reputable hard wallet. It is a mechanism wallets use to back-up and store your coins. If, for any reason, your computer crashes or hard wallet is lost/damaged, the same wallet software can be downloaded and your coins retrieved using the seed phrase.
- Back up the data in your hardware wallet, including the passwords and seed phrases. Do not store them online.

3) Transaction Safety

- Study the transaction requirements of a cryptocurrency carefully as they may indicate the security precautions to be taken. For example, in Ripple, there could be two parts of the address: the wallet address and the destination tag. If the destination tag is not correctly entered, you could lose your coins.
- Be wary of phishing sites as they are made to look like the real one and steal your information. Infected links via emails are known to provide the attacker access to a victim's computer and wallets. It is recommended to manually type the URL of the cryptocurrency site you wish to check out. Ignore the email if the source is unknown to you or looks suspicious.
- Do your research to avoid falling prey to Ponzi schemes and other ICO scams.

- Do not share your password
- Do not openly publish on the media regarding your interest in crypto or the amount of crypto you have as it makes you a target. Keep a low profile.
- Be vigilant of the wallet address. If you enter the wrong address, you could lose your coins forever.

4) Enable Security Measures

- Protect wallets and backups with strong passwords. Passwords with at least 15 characters that include number, uppercase, and lowercase letters and symbols are recommended. Ideally, change your password every quarter.
- Use different passwords on different accounts so that if one account is compromised, the others are safe.
- Enable two-factor authentication (2FL) or multi-signature to secure your transactions.
- Avoid public Wi-Fi and use only trusted secure networks to avoid any malware from penetrating and reading your transactions.
- Allow only authorized devices to access your wallet. It is advisable to do online cryptocurrency transactions on a dedicated personal computer or device to ensure that no malware has entered during general browsing.
- Ensure that your software is up-to-date with the latest upgrades and security enhancements. This also includes the OS upgrades and software running on your computer/mobile devices, including virus protection software.

3.5.7 Regulations Around Cryptocurrency

The year 2019–2020 saw increased activity on policies and regulations around cryptocurrencies and crypto exchanges. With blockchain start-ups as well as established technology organizations penetrating various industry sectors with innovative solutions to real-world problems, the governments can no longer ignore this cutting-edge technology. Banks such as Fintech need to consider the growing demand for cryptocurrency and take necessary steps to stay involved, if not take an active role.

In June 2019, the G20 acknowledged the recommendations and the work done by various regulatory and standard-setting bodies like the FSB (Financial Stability Board), IOSCO (International Organization of Securities Commission), FATF (Financial Action Task Force) and requested for a global regulatory framework on cryptocurrencies and the virtual asset industry to better manage the benefits and challenges thereof.

Global regulations and standards are essential

a) around monetary policies with respect to cryptocurrencies to ensure economic stability
b) to curb illicit activity like money laundering, financing terrorism, drug and human trafficking using cryptocurrency transactions
c) for token standards, especially privacy tokens where sender/receiver information is kept completely private. This could be misused and poses a risk to national security and law enforcement.

The following table gives the current status of cryptocurrency regulations in some of the countries around the world.

Table 3.1 Cryptocurrency regulations in a few countries

	Countries
Cryptocurrency not considered as a legal tender	Canada, Estonia, Gibraltar, Lithuania, Luxembourg, Malta, Singapore (treated as goods and hence taxable), South Korea, United States, UK **Note**: Bitcoin is legal in the US
Cryptocurrency accepted as payment, exchanges and may or may not be regulated	Argentina, Belarus, Brazil, Canada, Chile, Germany, Holland, Japan, Malta, Mexico, New Zealand, South Korea, Switzerland, Venezuela
Legal crypto-exchanges regulated by financial regulators	Australia, Canada, Estonia, Gibraltar, Lithuania, Luxembourg Malta, Singapore, South Korea, Switzerland, United States, United Kingdom,
Cryptocurrency and crypto exchanges effectively illegal but global/federal regulations being considered	China, India (banned by banks), Kyrgyzstan, Russia, UAE
Cryptocurrency and crypto exchanges banned	Algeria, Bangladesh, Bolivia, Equador, Macedonia, Morocco, Nepal, Pakistan

Though cryptocurrencies or altcoins are illegal in some countries, countries like China, Equador, and India, among others, are looking to develop their national altcoin. Hence, it should be noted that the countries' stance on cryptocurrency is not permanent and ever-changing.

Summary

The need for a decentralized banking system picked up pace following the 2008 financial crisis. Many attribute it to the failure of the government and financial institutions to not have heeded the early warning signs and taken the necessary preventive measures. Some felt the solution was to create cryptographically encrypted currency where the public or users validate the accuracy and authenticity of the currency transactions without depending on an external agency. Thus, cryptocurrency was born. Cryptocurrency is digital money that can be used to purchase goods and services, and as an investment. Blockchain is the underlying technology that enables the creation and control of the cryptocurrency ecosystem.

Due to its decentralized nature and global access via the internet, the government or central authority like banks do not have control over cryptocurrency. There is full transparency as the transaction ledger (blockchain) is open to the public. In addition, there is the added advantage of user anonymity as transactions are linked to the person's cryptocurrency addresses and not to the person's identity. Cryptocurrency also assures secure transfer of money that is both cheaper as well as quicker as compared to traditional money transfers.

The total number of cryptocurrency that can be released is defined at the time of the blockchain project build; hence the value of the cryptocurrency is solely dependent on supply and demand. There is no government agency regulating it. However, the government can order banks, financial institutions, and businesses under their control to ban cryptocurrency. Countries like Saudi Arabia, Algeria, and Pakistan have completely banned cryptocurrency. But as cryptocurrencies are digital and live on the internet, people can work with an off-shore cryptocurrency exchange.

Bitcoin is the first cryptocurrency that was released, and thousands more have followed since then. These coins are referred to as altcoins (alternative coins). Some altcoins are derived from the Bitcoin blockchain like Litecoin and Dogecoin, while others are native to their blockchain, like ether (Ethereum) or XRP (Ripple). Altcoins operate on their unique blockchain. Tokens are another category of cryptocurrency that resides on top of an existing blockchain that facilitates the creation of decentralized applications (DApps). Hence here, the cryptocurrency need not be built from scratch, unlike altcoin. Tokens are created using standard templates from blockchain platforms such as Ethereum, Lisk, or Waves platform. They are typically used as a utility or asset that gives the holder specific rights to participate in the blockchain network. There are two kinds of tokens – utility tokens and security tokens. Table 3.2 lists the variances between altcoins and tokens.

Table 3.2 Comparison of altcoins with tokens

	Altcoins	**Utility tokens**	**Security tokens**
Origin	They are built from scratch, either using the open-source code of an existing cryptocurrency or using original code. They exist on their separate blockchain.	They are built from templates of an existing blockchain via smart contracts.	They are built from templates of an existing blockchain via smart contracts.
Purpose	Means of payment	Created for funding blockchain projects/companies. Distributed during an ICO.	Created for funding blockchain projects/companies. Distributed during an STO.

(*Continued*)

Table 3.2 (*Continued*)

	Altcoins	Utility tokens	Security tokens
Usage	Acts as currency or cash	Represents the right to use a product or service in the future	Acts as an investment product, it represents the shares of the company
Value	Value is dictated by market supply and demand	Used as a utility and not linked to the performance of the company. The user gets voting and other usage rights within the specific blockchain it operates.	Gives part ownership to the token holders. Its value is directly linked to the valuation of the company.
Security	Protection as strong as the strength of the blockchain network protocols and user diligence	Susceptible to ICO scams and frauds	The strict regulations aim to protect investors from scams and frauds.
Regulation	Typically not regulated or controlled by any bank or central authority	Typically unregulated	Strictly regulated by a financial market authority.

At present times, the value of cryptocurrency is quite volatile, making them excellent investment assets for smart investors or traders. People can make money from cryptocurrencies by mining, investing, or trading. A miner can choose to mine cryptocurrencies independently by setting up his mining equipment. This process of mining is called solo mining. As an alternative, the miner can pool his resources with like-minded miners through a process called pool mining. Working with a mining pool saves on initial set-up costs and ongoing maintenance costs, but end benefit is less compared to solo mining as the block rewards have to be shared with the whole pool. However, the frequency of getting a block reward in solo mining is unpredictable.

Cryptocurrency transactions cannot be done without a digital wallet. A cryptocurrency wallet is a software program that stores the private and public keys enabling the user to transact and store crypto assets. Cryptocurrency wallets are categorized as hot wallets (online or full-time connected to the internet) and cold wallets (stored offline). Hot wallets are used for day-to-day transactions, while cold wallets are used for long-term storage. Table 3.3 shows the two types of cryptocurrency wallets.

Table 3.3 Unique features of bitcoin wallets

	Hot wallets	Cold wallets
Connectivity	Online storage; connected to the internet	Offline storage; not connected to the internet
Purpose	Recommended for short-term storage. Convenient for regular day-to-day transactions	Recommended for long-term storage
Accessibility	Easily accessible and preferred by traders and new users.	Not easily accessible.
Cost	Free usage	Need to be purchased
Security	Susceptible to cyber-attacks and malware	No risk of hacks or malware from internet connections. However, it should be physically protected from loss or breakage.
Types	Software wallets, online wallets	Hardware wallets, paper wallets

Whether transacting with or trading in cryptocurrency, the onus is on the coin holder to keep it safe. Basic safety measures and best practices include not storing your assets in exchanges, using the right wallets for fund storage, ensuring that assets are encrypted and protected with strong passwords, diligently avoiding phishing sites, using dedicated laptop or devices and dedicated email. Use of secure online trading platforms may help to protect one from malicious attacks.

EXERCISES

Multiple Choice Questions

1. **The term 'cryptocurrency' evolved from the words 'crypto': Its meaning is:**
   A. Secret
   B. Currency
   C. Transaction
   D. Transparent
   Answer: A
   Explanation: The term 'cryptocurrency' evolved from the words 'crypto' meaning concealed or secret and 'currency' indicating a system of money. It refers to all encrypted decentralized digital currencies. Bitcoin was the first cryptocurrency. Cryptocurrencies are digital currencies designed to be faster, cheaper, and more reliable than money.

2. **This was the earliest form of trading where people exchanged goods and services for other goods and services:**
   A. POS Terminals
   B. Metal Coins
   C. Cowry shells
   D. Bartering
   Answer: D
   Explanation: Bartering was the earliest form of trading where people exchanged goods and services for other goods and services according to their needs. For instance, animal skins and tools were traded for barley, salt, spices, or other commodities/ services. Historians believe that the Mesopotamians used the earliest barter system at around 6000 BC.

3. **Around 1000 BC, China, India, and Africa used this as money.**
   A. POS Terminals
   B. Lydian Coin
   C. Cowry shells
   D. Bartering
   Answer: C
   Explanation: The difficulty in agreeing to a standard deal as to the fair value of the commodities or services being exchanged, as well as the issues in storage and transportation, made the bartering system impractical, long-term. Hence for practical purposes, around 1000 BC, China, India, and Africa used cowry shells as money. The Chinese also used miniature replicas of the commodity being traded that later evolved into circular coins made of bronze. Thus shell money, coin money, and other forms of commodity money were used as a medium of exchange for goods and services across all trade continents.

4. **The first known currency coin that was minted in 600 BC is called as:**
   A. POS terminals
   B. Lydian coin
   C. Cowry shells
   D. Bartering
   Answer: B
   Explanation: The first known currency coin was minted in 600 BC in Lydia, which is now part of Western Turkey. The Lydian coin was made of electrum, a naturally occurring alloy of gold and silver with the image of a lion head imprinted on it. Since then, coins became associated with money and commerce.

5. **The first credit card was called as**
   A. Visa Card
   B. Diners' Club
   C. Master card
   D. Charge-it

Answer: D

Explanation: The first credit card, as we know it now, came about in 1946. It was called 'Charge-it' and introduced by New York banker John Biggins. Other credit card types that could be used in a variety of establishments soon followed, such as the Diners' Club card in 1950 and the travel/entertainment card established by the American Express Company in 1958. Visa credit cards came out in 1958 and Mastercard in 1966.

6. **First fully decentralized cryptocurrency is called as:**
   A. Bitcoin
   B. Ethereum
   C. Altcoin
   D. Tokens

   Answer: A

   Explanation: The concept of a secure decentralized digital currency that can be operated by anyone who has an internet facility is truly revolutionary. This concept became a reality with the introduction of Bitcoin in 2008 as a fully decentralized cryptocurrency. Currently, there are over 1600 cryptocurrencies (as of 2018) and the number is growing. With the growing acceptance of cryptocurrencies and virtual money, Fintechs are focusing on disruptive technologies like blockchain, IoT, and AI. The combined effort of Fintechs, banks, and governments could bring about a new era of currency economics.

7. **Which of the following is not one of the key requirements that eventually brought about the birth of the Bitcoin:**
   A. A system where one can directly transact with another person without involving a third party.
   B. A system which is completely centralized ensuring 100% security for the money.
   C. A system that is not backed and controlled by a central authority and can assure the value of the money is maintained.
   D. A system where there is transparency in transactions, while still maintaining the users' privacy.

   Answer: B

   Explanation: The three key requirements that eventually brought about the birth of the Bitcoin was the need for a monetary system that could address the following: 1) A system where one can directly transact with another person without involving a third party, like a bank, to verify and validate the transaction and thus avoiding the cost of mediation. 2) A system that is not backed and controlled by a central authority and can assure the value of the money is maintained. 3) A system where there is transparency in transactions, while still maintaining the users' privacy

8. **The hashing algorithm used in the Bitcoin protocol is:**
   A. SHA256

B. SHA186
C. MD5
D. CRC32
Answer: A

Explanation: The Bitcoin system or protocol solved the problem of trust by allowing for willing parties to transact with one another using cryptography directly. The hashing algorithm used in the Bitcoin protocol is the hash function SHA256 (Secure Hash Algorithm 256-bit).

9. **Bitcoins are released at:**
A. Increasing rate
B. Decreasing rate
C. Constant rate
D. Random rate
Answer: B

Explanation: Bitcoins are released at a decreasing rate, i.e., as the number of bitcoins in circulation increases, fewer the number of bitcoins that can be created, thus appreciating its existing value. The value of the bitcoin is dependent only on the market supply and demand and free of intervention from any central authority.

10. **The value of the bitcoin is dependent only on:**
A. Government regulation
B. Market supply and demand
C. Miners
D. Inflation
Answer: B

Explanation: Same as for Question 10.

11. **The process of duplication from an existing blockchain is called as:**
A. Copy writing
B. Consensus mechanism
C. Forking
D. Mining
Answer: C

Explanation: Cryptocurrency and blockchain projects all have open-source code to foster trust and safety. Developers can copy the open-source software and modify it to either build compatible applications such as wallets on the existing blockchain, or build a completely new cryptocurrency network, e.g., Litecoin. This process of duplication from an existing blockchain is called forking.

12. **This kind of fork occurs when the protocol upgrade results in a split in the blockchain that is not backward compatible.**
A. Hard fork
B. Soft fork

C. Mandatory fork

D. Discretionary fork

Answer: A

Explanation: A hard fork occurs when the protocol upgrade results in a split in the blockchain that is not backward compatible, i.e., the software that validates according to the old protocol will see the new protocol as invalid and vice versa. Hence, all clients need to upgrade to the new version if they want to continue participating in the network.

13. This fork is backward compatible. Here, the old software will recognize the blocks made by the new protocol as valid.

A. Hard fork

B. Soft fork

C. Mandatory fork

D. Discretionary fork

Answer: B

Explanation: A soft fork is a protocol upgrade that is backward compatible. Here, the old software will recognize the blocks made by the new protocol as valid. Hence, it does not require all the nodes in the network to upgrade to maintain consensus as the soft-forked chain follows both the old as well as the new set of protocol rules.

14. Cryptocurrencies have distinct characteristics that set them apart from the traditional fiat currency. Which of the following is not a characteristic of cyrptocurrency?

A. Global access

B. Limited supply

C. Decentralized

D. Centralized

Answer: D

Explanation: Fiat currencies are issued by the government and regulated by the Central Bank. Thus, government-issued policies and control of supply can affect the value of the currency. Cryptocurrency, on the other hand, is not backed or controlled by any bank or central government. All the nodes/computers in the network work together in mining or processing a transaction. This decentralized feature means no central body can control or influence the value of the cryptocurrency.

15. One of the key features of cryptocurrency is:

A. Unlimited supply

B. Limited supply

C. Physical form

D. Central Bank controlled

Answer: B

Explanation: A key feature of cryptocurrency is that the supply is limited. The maximum supply of cryptos that can ever be generated or mined is defined when the genesis block is created.

16. **Cyrptocurrency transactions are reversible and this fosters trust.**
    A. True
    B. False
    Answer: B
    Explanation: Cryptocurrency transactions are irreversible, unlike traditional currency transactions. This feature has its pros and cons. Once a transaction is recorded and verified, it is irreversible and impossible to change. While this fosters trust, integrity, and auditability, it also means that if a cryptocurrency is sent to the wrong address, the coin is lost forever. Hence, users must be vigilant against cryptocurrency scams or frauds.

17. **Which of the following is not a type of hot wallet?**
    A. Mobile wallet
    B. Desktop wallet
    C. Hardware wallet
    D. Web wallet
    Answer: C
    Explanation: A cold wallet is a digital wallet that is not connected to the internet. Unlike hot wallets, they are not free of cost. Being offline, they are more secure and used for storing cryptocurrencies, long-term. Hardware wallets and paper wallets are the two types of cold wallets.

18. **Cold Wallets are more secure and used for storing cryptocurrencies for long term.**
    A. True
    B. False
    Answer: A
    Explanation: Same as for Question 17.

19. **Which of the following is not an example for online wallets?**
    A. MyEtherWallet
    B. GreenAddress
    C. MetaMask
    D. TrustWallet
    Answer: D
    Explanation: Bitcoin Core, Electrum, and Exodus are examples of desktop wallets. MyEtherWallet, GreenAddress, and MetaMask are examples of online wallets. Well-known examples of mobile wallets are Coinbase Wallet, Mycelium, and TrustWallet.

20. Which of the following is not an example for desktop wallets?

A. Bitcoin Core

B. Electrum

C. MetaMask

D. Exodus

Answer: C

Explanation: Same as for Question 19.

21. This type of wallet is offline and printed out in the form of QR Codes.

A. Paper wallet

B. Hardware wallet

C. Software wallet

D. Web wallet

Answer: A

Explanation: Paper wallets are offline and, as the name suggests, it is a piece of paper with the crypto address and its private key is physically printed out in the form of QR codes. These codes can then be scanned to execute cryptocurrency transactions.

22. This coin is not directly derived from Bitcoin:

A. Namecoin

B. Peercoin

C. Litecoin

D. Nxtcoin

Answer: D

Explanation: Many variants of the Bitcoin have risen using its open-source protocol, creating a new currency coin with a different set of features that operates on its blockchain. Namecoin, Peercoin, and Litecoin are examples of Bitcoin-derived blockchains. Some coins are not derived from Bitcoin but created as a native currency of an original blockchain and protocol, namely Monero, XRP (Ripple), and NxtCoin.

23. This facilitates creation of Decentralized Applications (DApps) and benefit from the technology and security of the native blockchain.

A. Tokens

B. Altcoins

C. Wallets

D. NxtCoin

Answer: A

Explanation: Tokens are a category of cryptocurrency that resides on top of an existing blockchain that facilitates the creation of decentralized applications (DApps), unlike the altcoin that operates on its blockchain. Tokens and their underlying application benefit from the technology and security of the native blockchain as they can leverage their existing mining strength and inherent resistance to malicious attacks.

24. Unlike coins that need to be built from scratch, these are created using standard templates:

A. Tokens

B. Altcoins

C. Wallets

D. NxtCoin

Answer: A

Explanation: Unlike coins that need to be built from scratch, tokens are created using standard templates on blockchain platforms like Ethereum, Lisk, or Waves platform. The functional variances are possible through the use of smart contracts.

25. Which of the following is a well-known ERC20 token?

A. ERC721

B. ERC223

C. OmiseGo(OMG)

D. ERC777

Answer: C

Explanation: Developers widely use smart contracts compliant with ERC20 standard due to their ease of implementation and wide adoption in the Ethereum platform. Status (SNT) and OmiseGO (OMG) are well-known ERC20 tokens. Aside from ERC20, a few other Ethereum token standards are ERC223, ERC 721, and ERC 777. Compliance to standard enables the storage of different types of tokens in a single wallet.

26. Tokens are created and distributed to the public through a crowd-funding method called:

A. Public funding

B. Initial coin offering

C. Initial public offering

D. Crowd sourcing

Answer: B

Explanation: Tokens are created and distributed to the public through a crowd-funding method called Initial Coin Offering, where tokens are issued in exchange for funding a potential blockchain project.

27. Tokens can be bought with coins; similarly, coins can be bought with tokens.

A. True

B. False

Answer: B

Explanation: Tokens can be bought with coins, but coins cannot be bought with tokens.

28. The most common type of utility token is:

A. ERC721
B. ERC223
C. ERC20
D. ERC777

Answer: C

Explanation: The most common type of utility token is the ERC20 token, which is the technical standard used for all smart contracts on the Ethereum blockchain for token implementation.

29. This type of token is used by Civic (an identity management application)

A. Filecoin
B. Storj
C. Golem
D. CVC Token

Answer: D

Explanation: The CVC token is a utility token used by Civic, an identity management application that keeps track of encrypted identities on the Ethereum blockchain. Other popular utility tokens are Filecoin, Storj, Golem, etc.

30. Security tokens are also referred to as:

A. Equity tokens
B. Utility tokens
C. Private tokens
D. Public tokens

Answer: A

Explanation: Security tokens, also referred to as Equity tokens, entitle the holder to a share or equity in the company/startup. In other words, they can be considered as a digitized, tokenized form of the traditional securities. Security tokens are issued during STO or Security Token Offering, which is a variation of ICO with more regulatory and due diligence controls to foster investor confidence.

31. Security tokens are issued during:

A. Initial Coin Offering
B. Security Token Offering
C. Initial Public Offering
D. Crowd Sourcing

Answer: B

Explanation: Same as for Question 30.

32. SEC classifies tokens as securities if they pass all the following criteria set by:

A. SEC Test
B. Currency Test

C. Howey Test
D. Security Test
Answer: C

Explanation: SEC classifies tokens as securities if they pass all the following criteria set by the 'Howey Test' to classify the security as an investment contract:

1) There is an investment of money or assets.
2) There is an expectation of profits from the investment.
3) The investment is in a common enterprise.
4) The expectation of profit is solely from the efforts of a promoter or third party.

33. Which of the following is not an example of Security Tokens?

A. Storj
B. BCap
C. 22x Fund
D. Spice VC
Answer: A

Explanation: Due to the regulations, launching security tokens is time-consuming. However, the same restrictions also protect investors from fraud and scams. BCap of Blockchain Capital, 22X Fund, and Spice VC are a few examples of security tokens.

34. Which of the following is an example of Hybrid Token?

A. BNB
B. BCap
C. 22X Fund
D. Spice VC
Answer: A

Explanation: Though tokens can be broadly classified as utility and security, there is the emergence of hybrid tokens that takes the best of both worlds. It can combine the advantages of both utility tokens and security tokens, as incentives are distributed across all participants. In contrast, the application of securities laws can weed out frauds and scams. An example of the hybrid token is BNB (Binance) that is used to pay transaction fees on the Binance Exchange.

35. 'Scrypt' proof-of-work mining algorithm is used in:

A. Bitcoin
B. Litecoin
C. BCH
D. XRP
Answer: B

Explanation: Litecoin (LTC) is the earliest altcoin released in October 2011. It is a fork of the Bitcoin core client. It was created by Charlie Lee, an MIT graduate, to improve upon Bitcoin's block generation rate. Litecoin is similar to Bitcoin except in the hashing algorithm used for PoW. It uses memory-intensive 'scrypt' proof-of-work mining algorithm. The average transaction time of Litecoin is 2.5 min/transaction compared to Bitcoin's 10 min/transaction. Litecoin has a supply limit of 84 million LTC compared to Bitcoin's 21 million. Litecoin is also one of the top 10 cryptocurrencies in the market.

36. The average transaction time of Litecoin is:

A. 2.5 min/transaction
B. 10 min/transaction
C. 10 sec/transaction
D. 2.5 sec/transaction

Answer: A

Explanation: Same as for Question 35.

37. This is used as a global payment settlement and asset exchange:

A. Litecoin
B. Ripple
C. Bitcoin
D. Ethereum

Answer: B

Explanation: The Ripple platform is mainly used as a global payment settlement, asset exchange, and remittance system. It can seamlessly transfer any form of currency, i.e., both cryptocurrency like BTC, LTC, etc. as well as fiat currency like USD, YEN, EUR, etc. Ripple is the third-largest cryptocurrency by market capitalization as of January 2020 and is mainly used by banks and financial institutions.

38. This is a cross-border payment network similar to Ripple.

A. Ethereum
B. Litecoin
C. Monero
D. Stellar

Answer: D

Explanation: Stellar is a cross-border payment network system aimed at providing affordable financial services to people of all income levels. Conceptualized by Ripple co-founder Jed McCaleb with Joyce Kim in July 2014, Stellar can handle both fiat and cryptocurrency exchanges. The platform is open-source, decentralized and non-profit, and connects financial institutions to drastically reduce the cost and time required for cross-border transactions.

39. This is a smart contract blockchain development platform similar to Ethereum to develop decentralized applications (DApps) and development tools.

A. XLM
B. XMR
C. EOS
D. LTC

Answer: C

Explanation: Launched in January 2018 by Block.One company, EOS is one of the top 10 cryptocurrencies by market cap. EOS is a smart contract blockchain development platform similar to Ethereum to develop decentralized applications (DApps) and development tools. EOS blockchain protocol is powered by the native cryptocurrency EOS. It was built to merge the positive aspects of both Bitcoin and Ethereum. It uses delegated proof-of-stake (DPoS) where the holder of the coins has a say (voting rights) in selecting the 'block producers', i.e., those who can validate/verify the transactions. EOS does not charge any transaction fees.

40. In ASIC mining, ASIC standards for:

A. Application Specific Integrated Circuit (ASIC)
B. Array String Integrated Consensus (ASIC)
C. Application Specific Integrated Cyrpto (ASIC)
D. Application Smart Integrated Currency (ASIC)

Answer: A

Explanation: A miner utilizes an Application Specific Integrated Circuit (ASIC) to perform block validation in ASIC mining. They are more efficient, faster, and can mine more coins in less time than any other kind of processor. Compared to CPU and GPU, they have the least power consumption and comparatively small physical size. Also, mining algorithms can be performed in parallel here, hence the potential for high-profit margin. AISC is custom-designed to mine specific currencies and hence not useful for miners who want the flexibility to mine various currencies. Also, they have low resale value as they are not upgradable and become obsolete when a new version comes out.

41. A bug in Ethereum DAO code was exploited in the year __________, causing the theft of $__________ in ether

A. 2016, $250 M
B. 2016, $50 M
C. 2006, $150 M
D. 2006, $50 M

Answer: B

Explanation: In 2016, a bug in the Ethereum blockchain was exploited to hack the DAO code resulting in a loss of $50 M in ether

Short-answer Questions

1. Explain the term 'cryptocurrency'.

The term 'cryptocurrency' evolved from the words 'crypto' meaning concealed or secret and 'currency' indicating a system of money. It refers to all encrypted decentralized digital currencies. Bitcoin was the first cryptocurrency. Cryptocurrencies are digital currencies designed to be faster, cheaper and more reliable than money.

2. What is the need for bitcoin currency?

The three key requirements that eventually brought about the birth of the Bitcoin was the need of a monetary system that could address the following:

1. A system where one can directly transact with another person without involving a third party, like a bank, to verify and validate the transaction and thus avoiding the cost of mediation.
2. A system that is not backed and controlled by a central authority and can assure that the value of the money is maintained.
3. A system where there is transparency in transactions, while still maintaining the users' privacy.

3. How Bitcoin solved the problem of Trust?

The Bitcoin system or protocol solved the problem of trust by allowing for willing parties to directly transact with one another using cryptography. The hashing algorithm used in the Bitcoin protocol is the hash function SHA256 (Secure Hash Algorithm 256-bit). The term 'Bitcoin' is used intermittently as both a currency and a technology, but basically the Bitcoin is powered by the blockchain technology.

4. What is hard fork?

A hard fork occurs when the protocol upgrade results in a split in the blockchain that is not backward compatible, i.e., the software validating according to the old protocol will see the new protocol as invalid and vice versa. This results in two versions (the original version and the new version) running independently of each other. All clients need to upgrade to the new version if they want to continue participating in the network and the miners/nodes need to decide on the version of the blockchain they want to run with. In some instances, the old chain is eventually abandoned while in other instances both chains remain active, e.g., Bitcoin–Bitcoin Cash fork, Ethereum Classic–Ethereum fork.

5. What is soft fork?

A soft fork is basically a protocol upgrade that is backward compatible. Here the old software will recognize the blocks made by the new protocol as valid. It does not require all the nodes in the network to upgrade to maintain consensus as the soft-forked chain follows both the old as well as the new set of protocol rules. However, a majority of the miners need to agree on the upgrade for any blocks made with the old set of rules to be rejected and the old network eventually abandoned. E.g., Bitcoin protocol upgrade to SegWit.

6. What is cryptocurrency wallet?

Bitcoin or any other cryptocurrency transactions can be done only using a digital wallet. A digital wallet or cryptocurrency wallet is a software program that stores your private and public keys enabling the user to transact crypto assets. It is a management system that interacts with various blockchains to enable users to send and receive digital currency and monitor their balance.

7. What is hot wallet?

A hot wallet is designed for online day-to-day transactions. It is connected to the internet at all times and hence an easy target for hackers. For this very reason, it is not advisable to use a hot wallet for long-term storage. Hot wallets are typically free, easy to set up and convenient to use.

8. What is desktop wallet?

Desktop wallets, as the name suggests, can be downloaded and installed on the user's desktop or laptop. It comes with a variety of features including built-in exchange platforms, multi-currency support, etc. As they are locally stored, the user has complete control of the wallet. However, the downside is that the wallet is dependent on the security features installed in the user's computer. If the computer gets hacked or virus infected, there is the risk of losing funds.

9. What is mobile wallet?

Mobile wallets are designed to operate on smartphone devices. Once installed, the smart phone applications operate just like their desktop counterparts, but they may not have all the savvy features. However, it circumvents the constraints of the desktop wallet with its ease of accessibility. It is portable and the user can make purchases from merchants who accept cryptocurrency payments, using QR codes.

10. What is online wallet?

Online wallets are also called web wallets. They run on the cloud and hence, the user does not need to download or install any application. They can be accessed from any computing device via a web browser. They are the most convenient of the hot wallets and preferred by fresh cryptocurrency users. Online wallets are hosted and controlled by third party and hence are vulnerable to threats of online fraud.

11. What is cold wallet?

Cold wallet is a digital wallet that is not connected to the internet. Unlike hot wallets, they are not available free of cost. Being offline, they are more secure and used for storing cryptocurrencies long-term. Hardware wallets and paper wallets are the two types of cold wallets.

12. Write notes on hardware wallet.

Hardware wallets are physical electronic devices that use Random Number Generator (RNG) to generate the public/private key that is stored in the device. It connects to the internet whenever the user needs to send or receive payments and disconnects once the transaction is executed. Transactions are confirmed through the private keys that are saved offline.

13. What is paper wallet?

Paper wallets are offline and, as the name suggests is a piece of paper with the crypto address and its private key physically printed out in the form of QR codes. These codes can then be scanned to execute cryptocurrency transactions. Paper wallets are completely secure as the question of hacking does not arise with paper wallets. However, a big disadvantage is that the paper wallet contains only a single public/private pair and hence they can be used only once for the whole amount in one transaction.

14. What are Altcoins?

Cryptocurrency alternatives to the Bitcoin are referred to as 'Alternative Cryptocurrency Coins', abbreviated as Altcoins or simply Coins. Many of the altcoins come from a fork of famous and durable cryptocurrencies like Bitcoin, Litecoin and Ethereum. While some altcoins are similar to their predecessor, most attempt to improve or set themselves apart by bringing in additional features or security like improved block times, different parameters, transaction management, scripting language, consensus mechanism, etc.

15. What is security token?

Security tokens, also referred to as Equity tokens, entitle the holder to a share or equity in the company/startup. In other words they can be considered as a digitized, tokenized form of the traditional securities. Security tokens are issued during STO or Security Token Offering and is a variation of ICO with more regulatory and due diligence controls to foster investor confidence.

16. What is Litecoin (LTC)?

Litecoin is the earliest altcoin released in October 2011. It is a fork of the Bitcoin core client. It was created by Charlie Lee, a MIT graduate, to improve upon Bitcoin's block generation rate. Litecoin is similar to Bitcoin except in the hashing algorithm used for PoW. It uses memory intensive 'scrypt' proof-of-work mining algorithm. The average transaction time of Litecoin is 2.5 min/transaction as compared to Bitcoin's 10 min/transaction. Litecoin has a total supply limit of 84 million LTC as compared to Bitcoin's 21 million. Litecoin is also one of the top 10 cryptocurrencies in the market.

17. What is EOS?

Launched in January 2018 by Block.One company, EOS is one of the top 10 cryptocurrencies by market cap. EOS is a smart contract blockchain development platform similar to Ethereum to develop decentralized applications (DApps) and development tools. EOS blockchain protocol is powered by the native cryptocurrency EOS. It was built to merge the positive aspects of both Bitcoin and Ethereum. It uses DPoS (delegated proof of stake) where the holder of the coins has a say (voting rights) in selecting the 'block producers', i.e., those who can validate/verify the transactions. EOS does not charge any transaction fees.

18. What is Monero (XMR)?

Monero is another popular cryptocurrency that was launched in April 2014. It is a completely private, secure, and untraceable cryptocurrency where one is not able to see anyone else's transaction or balances unless the person makes it public, unlike in Bitcoin where transactions are public on the blockchain. Monero has no pre-set block size limit. Therefore to prevent malicious miners, a block reward penalty is built into the system where the block reward gets reduced.

19. What is Stellar (XLM)?

Stellar is a cross-border payment network system aimed at providing affordable financial services to people of all income levels. Conceptualized by Ripple co-founder Jed McCaleb with Joyce Kim in July 2014, Stellar can handle both fiat and cryptocurrency exchanges. The platform is open-source, decentralized and non-profit, and connects financial institutions to drastically reduce the cost and time required for cross-border transactions.

20. What is solo mining?

As the name suggests, in solo mining the miner works independently to find the block. They typically have large mining equipment and do not rely on any third party. The miners connect to the network via their own wallet and once they are able to create a block successfully, i.e., the hash submitted is accepted by the network, added to the block and the block is confirmed, they are rewarded in cryptocurrency. The advantage of solo mining is that the miner is credited the full block reward. The disadvantage is that in solo mining it may take a long time to find a block, unlike in a mining pool that has more resources and hardware power.

21. What is surprise airdrop?

Cryptocurrency airdrops are done in two ways – a surprise airdrop or a planned airdrop. A surprise airdrop is done by an established blockchain company. The company takes a snapshot of the blockchain, i.e., the addresses and their related data like transactions, balances, metadata, etc. at a particular date and time and distributes free coins to the wallets based on the balance in each blockchain address. This method for airdrops is mostly used in blockchain hard forks, especially in the Ethereum standard tokens.

Essay-type Questions

1. Write an essay on the evolution of currency.
2. Write an essay on the birth of Bitcoin as a currency.
3. Explain the various characteristics of cryptocurrency.
4. Explain the various types of cryptocurrency wallets in detail.
5. Explain the various types of cryptocurrency.
6. What is Altcoin? What are the main features that are common to all altcoins?
7. How do tokens differ from Altcoins?
8. Write an essay on the different players who constitute the blockchain ecosystem.
9. What is crypto mining? What are the different types of mining based on the processors or equipment used by the miner?
10. What is Airdrop? What are the benefits of airdrops?
11. What are the types of Airdrops?
12. Write an essay on coin burning.
13. What are the safety measures followed while storing and transacting in cryptocurrencies?

CHAPTER 4

Public Blockchain System

LEARNING OBJECTIVES

This chapter explains public blockchain concepts in detail and dwells at length with how public blockchain enables people to tackle business challenges in a new way.

Bitcoin and Ethereum are magic names in the field of blockchain technology. Ethereum is a type of secured cryptocurrency that is being used in multiple countries across the globe. Both can be used very quickly. This chapter will explain about Bitcoin and Ethereum in detail.

4.1 Introduction

As per Gartner (2018), Blockchain is one of the top 10 strategic technology trends of 2019. According to IDC (premier intelligent marketing firm), global spending on blockchain solutions is expected to reach $9.7 billion by 2021. Public blockchain plays a crucial role in this. As described in the previous chapters, a blockchain is considered either public or private based on whether the network is open or accessible to anyone with an internet connection. In public blockchain, anyone can join the network. They can download a copy of the ledger and initiate, broadcast, or mine blocks. Users are anonymous. Whereas in a private blockchain, just anyone cannot join the blockchain. One has to meet certain pre-requisite conditions to be a member of the blockchain network.

4.2 Public Blockchain

The public permissionless blockchain is synonymous with a public blockchain. One does not need any permission to read/access a transaction, initiate a transaction, or participate in the consensus process (PoW) to create a block. Participants or nodes remain anonymous through high cryptographic protocols. Consider it as a Google excel/spreadsheet which is shared by every computer in the world and is connected to the internet. Whenever a transaction happens, it is recorded onto a row of this spreadsheet.

Table 4.1 Key characteristics of public blockchain

Characteristics	Public Blockchain
Organization Type	Public
Users	Anonymous; but web tracking and cookies pose a risk to privacy
Access	Open and transparent to all
Network Type	Decentralized; zero points of failure
Operation	Anyone can read or initiate or receive transactions
Verification	Anyone can be a node and take part in the consensus process to validate transactions and create a block
Immutability	Secured by hashing
Consensus Mechanism	PoW, PoS, etc.
Incentivization	Incentivizes miners to grow the network
Security	Security based on consensus protocols and hash functions. Higher the security, lower the performance
Trust	Trust-free system; trust is enforced via cryptographic proof
Transaction speed	Slow; takes about ten minutes for creating a block
Energy consumption	Very high
Scalability	Limited; as the network grows, the node requirements of bandwidth, storage, and computational power exponentially increases.

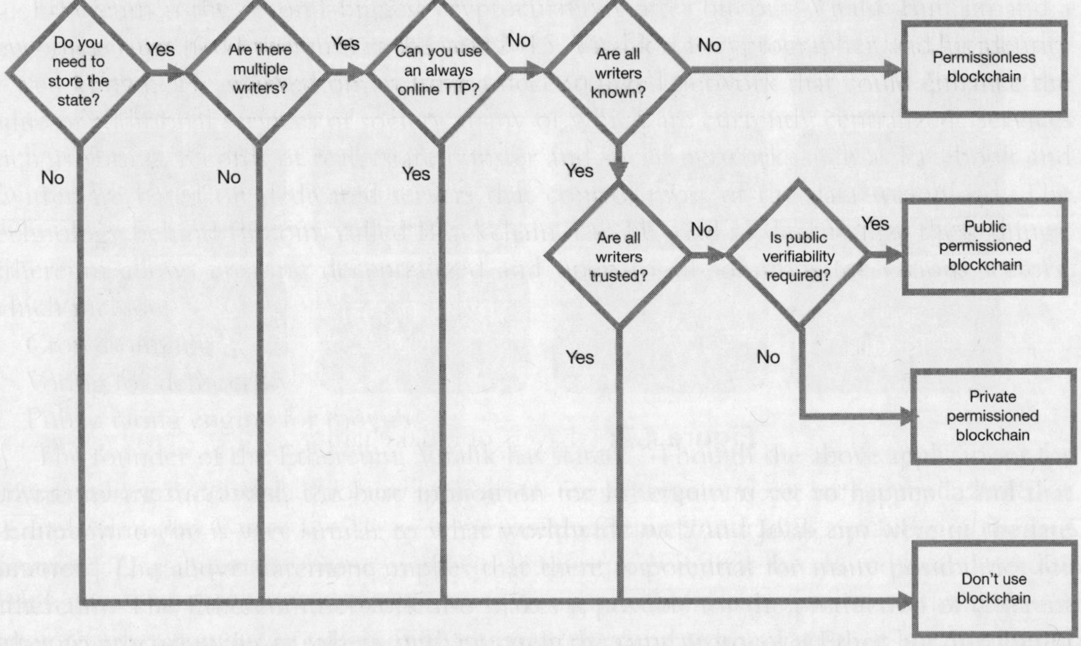

Figure 4.1: When to use permissionless blockchain

Table 4.1 summarizes the key characteristics of the public blockchain. In this type (public blockchain), the details of the users are unknown. The network is open and transparent to all. The network type is decentralized and has zero point of failure. Anyone can read, initiate, or receive transactions. It incentivizes miners to grow the network. The consensus mechanism used is PoW (most of the time) and PoS (occasionally). Security is based on consensus protocols and hash functions. As the network grows, the node requirements of bandwidth, storage, and computational power exponentially increase. Cryptocurrencies are the best example of a public blockchain.

Figure 4.1 explains the condition in which we need to use permissionless blockchain network. We need to use **Permissionless Blockchain** when the state (and date) needs to be stored in the database, if there are multiple writers for a transaction, when we cannot use always-online TTP and when not all the writers are known.

We need to use **Public Permissioned Blockchain** when the state (and date) needs to be stored in the database, if there are multiple writers for transaction, when we can't always use online TTP and when all the writers are known, when all the users cannot be trusted and when public verifiability is required.

4.2.1 Blockchain Layers

There are three different layers in a blockchain, which help to understand the relationship between blockchain, cryptocurrency, protocols, and tokens.

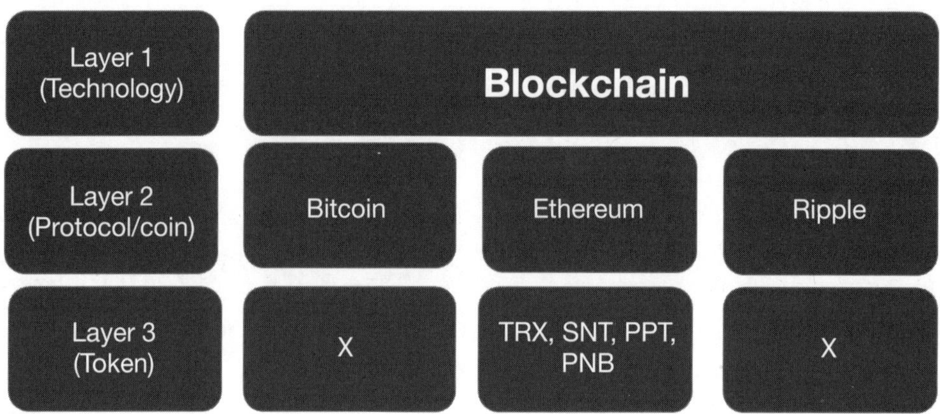

Figure 4.2: Layers of blockchain

Blockchain is the technology (Layer 1) behind various cryptocurrencies (Layer 2) including Bitcoin, Ethereum, and Ripple (refer Fig. 4.2). Layer 2 not only defines different coins but also describes the protocol of the respective currencies. A protocol is a standard set of rules that allow people to communicate or transact data with each other to reach consensus and validate transactions within the given network. Famous protocols that we already know, are 1. Http protocol 2. Https Protocol 3. TCP Protocol 4. Internet Protocol. Protocol in blockchain defines various consensus mechanisms, signature mechanisms, public key usages and cryptography related rules. Layer 3 consists of tokens. Tokens are the idea behind what people are building and serve as the symbol for an underlying contract. ICO refers more to token rather than coin. For example, Ethereum has 100s of tokens, which include TRX, SNT, PPT, and PNB. Bitcoin and Ripple have no tokens; hence they do not facilitate creating smart contracts.

4.3 Popular Public Blockchains

Bitcoin and Ethereum are the two famous public blockchain systems.

Bitcoin, the first blockchain created, is a public permissionless blockchain. The thought process behind the design was to create a 'decentralized digital currency,' that is free from government regulation, whereby two people can confidentially trade directly with one another, without the need for middlemen or intermediaries. The success of bitcoin can be measured by the number of transactions happening daily. As of March 2019, nearly three lakh daily transactions were reported worldwide. Bitcoin is a magic name in the blockchain field. People started knowing about blockchain only because of bitcoin, since blockchain technology is the concept behind the bitcoin. Bitcoin is a type of cryptocurrency and is currently used in multiple countries across the globe and in the coming days, it may determine the economic future of many nations. Even though there are more than a thousand cryptocurrencies available around the world, bitcoin is the primary form of cryptocurrency and is the first of its kind. The value of bitcoin is the highest across all other cryptocurrencies available.

Ethereum is the second-biggest cryptocurrency after bitcoin. Vitalik Buterin and a few others launched Ethereum in the year 2015. Vitalik is a cryptographer, and his identity is well known. He worked on creating a decentralized network that could enhance the value of additional services of society, many of which are currently centralized. Services such as voting, records of real-estate transfer and social networks such as Facebook and Twitter are based on dedicated servers that control most of the data we upload. The technology behind Bitcoin, called Blockchain, can be used to decentralize these things. Ethereum allows creating decentralized and open-to-all solutions for various sectors, which include:

- Crowdsourcing
- Voting for democracy
- Public rating engine for movies.

The founder of the Ethereum, Vitalik has stated: "Though the above applications for Ethereum are successful, the best application for Ethereum is yet to happen", and that "Ethereum today is very similar to what worldwide web and Javascript were in the late nineties." The above statement implies that there is potential for many possibilities for Ethereum. The Ethereum network also makes it possible for the production of different other cryptocurrencies or tokens, with precisely the same protocol as Ether, but distributed on distinct blockchains, which may be private or public. This means that they may be produced by associations to represent stocks, voting rights, or as a way of demonstrating authorization or identity certificate. Ethereum is popular among big corporates. Barclays is using Ethereum smart contracts in a way to trade derivatives.

4.4 The Bitcoin Blockchain

On 18 August 2008, a website domain in the name of Bitcoin.org was registered. In the same year, on 31 October, Satoshi Naga, a software developer in Japan, published a research paper on bitcoin. He introduced Bitcoin to the world, on 3 January 2009. He coded a special software to create bitcoin. This software is based on blockchain technology. Anybody across the globe can use this. Bitcoins can be created only using this software program. People who create bitcoin are called as Miners. Bitcoin touched its peak value in 2017, which is 19,498 USD. As one rupee is equivalent to 100 paise, so ten crore Satoshi is equivalent to one bitcoin (BTC). Those who are unable to buy bitcoin may buy groups of Satoshi to suit their needs.

Satoshi	USD($) Equivalent
1	0.00000001
10	0.0000001
100	0.000001
1,000	0.00001

Satoshi	USD($) Equivalent
10,000	0.0001
100,000	0.001
1,000,000	0.01
10,000,000	0.1
100,000,000	1

As of May 2019, its value was 7,952 USD. In January 2018, its value was 13,000 USD. The value of the bitcoin was reduced to this level due to various reasons. In fact, its value is volatile and keeps changing every second. When the bitcoin was created, the value of one bitcoin was lesser than one USD (its value was considered as 0.0003 USD). Below is the symbol of Bitcoin.

Bitcoin is a secured currency and can be used very quickly. Nobody controls this currency, and hacking the bitcoin network is very difficult. While there are regulatory bodies and governance structures to control the physical currency of countries, there are no such bodies or forms of governance to regulate the bitcoin currency.

4.4.1 Need for Bitcoin and Its Value

The dollar ($) is the currency of the USA. The pound is the currency of Britain. The Euro is the currency of Europe. Indian Rupees is the currency of India. To use foreign currency, it should be first converted to the local currency. A commission is to be given to the currency exchange for this conversion since banks spend much money to print the currencies physically. Moreover, a savings bank account is necessary to send currencies across banks/countries. In these bank-to-bank transactions, a commission is to be given to the banks. On the other hand, cryptocurrency is a direct peer-to-peer system without any intermediaries (banks). The bitcoin is the common currency that can be used in many countries across the world. There is no need to spend in millions to create bitcoins. Also, since only 21 million bitcoins can be built across the world, it will not lead to inflation.

The bitcoin was introduced in 2009. It was not famous until 2010. It started its momentum of monster growth from 2010. The growth of bitcoin is depicted in the below table.

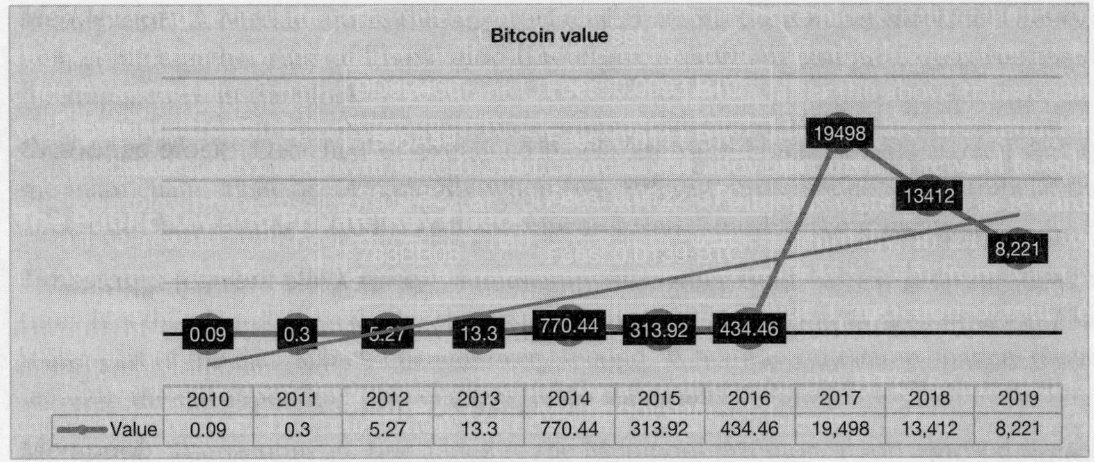

	2010	2011	2012	2013	2014	2015	2016	2017	2018	2019
Value	0.09	0.3	5.27	13.3	770.44	313.92	434.46	19,498	13,412	8,221

Figure 4.3: Bitcoin value over the years

The above graph (refer Fig. 4.3) indicates the equivalence of a single bitcoin to USD over a period, starting 2010. In 2010 (the starting period), the value of one bitcoin was equal to 0.09 USD, and it started growing steadily from there, touching a peak value of 19,948 USD in the year 2017. Its value has not been static ever since and in 2019, the value of one bitcoin was equivalent to 8,221 USD (approximately).

A few years back, a virus by name "Ransom" was spread maliciously over the internet. Because of this, computers from 150 countries across the globe got affected. Due to this, multi-million dollars were lost. The people who spread the virus demanded thousands of bitcoins to remove this virus. People who brought bitcoins in the early days (2009, 2010) are billionaires today.

4.4.2 Common Terminologies

Mining

Creation of bitcoin blocks is called mining. Mining is the mechanism whereby nodes called "miners" in the Bitcoin world validate the new transactions and add them to the blockchain ledger. Miners compete/fight with one another to solve the complex mathematical problem of coming up with a 64-digit hexadecimal number (called a hash) that is less than or equal to the target number (target hash) using massive computing power, energy, and time. The first miner to create the winning hash will receive rewards in the form of transaction fees or new bitcoins. This process of computing the hash is called proof-of-work or consensus mechanism and it provides integrity to the blockchain. All nodes in the network could participate in mining and earn mining rewards. This activity needs a lot of computing resource, and often, many nodes do not create a block: this is because validating transactions and finding nonce involve huge computing expense.

Block Frequency

Bitcoin transactions are being registered into blockchain once in ten minutes. Miners will first check the transaction. After reviewing the transaction, the software will give a complex target hash for the miners to solve. If miners can solve the hash, then the computer will give bitcoins as a reward to the miners.

Industrial Mining

When all the nodes are of uniform size and power, every node gets equal opportunity. It is a fair competition to get a reward. However there are some nodes, which do industrial-sized mining and connect to a massive set of computers, consume enormous power, and use complicated software. They themselves act as a network.

Mining Pools

To counteract the high time and energy consumption in transaction validation, some miners group together in mining pools to combine their mining resources for more efficiency and savings. Miners do not work for themselves; they work together in mine pooling. They share the mining power and processing power. Mining pools share the services of the individual miners so that the miners do not repeat the work done by others and waste time.

Halving Policy

Block frequency and Halving are the monetary policies of bitcoin. Nowadays, miners get 6.25 bitcoins as a reward to solve the hash. From the year 2009 to 2012, 50 bitcoins were given as rewards. The reward will be reduced to half for every four years. At last, in 2140, it will decrease to zero.

Block

Each block (of transactions) has three necessary informations, namely:

1. Block header
2. Hash of previous block header
3. Merkle root

Block header: In a bitcoin blockchain, the hash function of the previous block header is stored as a reference in the next subsequent block to ensure the blocks are correctly connected.

Hash: A hash function can be considered equivalent of fingerprint of a data, similar to a fingerprint of a person. Using a fingerprint, we can identify a person since it is unique. A hash function converts the data of arbitrary size to data of a fixed size. For example, the function keccak256 (parameter) converts the parameter given into a Hash Function. keccak256 (parameter) and SHA2-256(parameter) are now mainly used in Bitcoin and Ethereum projects. SHA stands for "Secured Hash Algorithm." For a small change in the parameter of SHA2-256 parameter, there are drastic changes in the Hash function output. Number 256 denotes the number of bits it takes to store into the memory. NSA (National Security Agency of America) developed SHA 256.

Merkle root: A Merkle root is the fingerprint of all transactions in the block. It is created by hashing together pairs of Transaction IDs to give a short and unique fingerprint for all the transactions in the block

Orphaned block: Detached or orphaned blocks are valid blocks, which are not part of the main chain. They occur naturally when two different miners successfully mine at the same time. Sometimes, a hacker may attempt to reverse/modify the transactions.

Timestamp (current UNIX time): Timestamp is another field, which indicates UNIX time. It is the seconds passed after the first of January 1970 and is a 10-digit number. This is also part of the data, which changes every second. When the timestamp changes (every second), the corresponding data changes, as do the results.

Mempool: Transactions are first added to the Mempool (Memory pool) attached to each node. They are unconfirmed transactions, which are to be included in the block once the combination becomes a valid transaction. Blocks are added every 10 minutes, but transactions happen all the time. Mempool is the staging area for the transactions. A mempool can have around 10,000 transactions at a time. When a miner successfully mines, a set of transactions are added from mempool to the block, and into the blockchain. Once the transactions are added to the blockchain, the same set of transactions is removed from the mempool.

Block propagation: When a miner announces a block, the block needs to propagate to all the nodes in the network, using gossip protocol (transactions are propagated using gossip protocols). The concept of gossip communication is similar to employees spreading rumours in a company. A node would perform the following functions:

a. Validate transaction
b. Make sure the nonce is valid (nonce is within an acceptable range)
c. Check if each transaction in the block is valid.

If all of the above three conditions, namely, a, b and c are valid (True), then the block is eligible for further propagation within the network; if the conditions are not met, the node will discard the block. Hence, the block would not be propagated further.

SEGWIT: The size of the signature and public key is so large within the transaction that it occupies more than 60 % of the overall size of the transaction. In the bitcoin protocol, this portion (signature and public key) is kept out of the block and is distributed separately. This segregation of the signature from the block is called as Segregated Witness (SegWit). Because of SegWit, more (almost double) transactions can be added into the block. SegWit will be sent separately in the network.

Nonce: Nonce is a number that can be used just once in the cryptographic communication. Adding a nonce to a transaction's identifier makes it additionally unique, thus reducing the chance of duplicate transactions. The nonce is key to creating a block within a blockchain database.

4.4.3 Bitcoin Mining

Creation of bitcoin blocks is called mining (refer Fig. 4.4). Mining is the mechanism whereby nodes called "miners" in the Bitcoin world validate the new transactions and add them to the blockchain ledger. In general, Mining refers to the extraction of various minerals or other geological materials from the earth. Similarly, unearthing the bitcoin from a computer is called mining. To mine bitcoin, we need computers with high processing power and computer graphics cards. While using the graphics cards, the computer consumes lots of electricity power; this, in turn, produces lots of heat in the overall system. Sophisticated, customized machines are used while mining the bitcoins. We also need the relevant software for mining.

Figure 4.4: Bitcoin mining components

4.4.3.1 Types of Bitcoin Mining

Miners compete to solve a complex mathematical problem based on a cryptographic hash algorithm referred to earlier. Mining comprises of hashing a block and then introducing a nonce to the hashing function and running the hash all over again.

In a hash, each digit can have 16 types of numbers ranging from zero to nine and A to F. Assuming there is a total of N different types of overall possible hash (let us call it as "population hash set").

Population hash set = 16^{64} which is roughly 1.16×10^{77}.

If the target hash consists of one leading zero, then it will reduce the overall target population by 1/16 (as each digit can have 16 possible values). Assume that the current target value of hash needs have "eight leading zeros" in the 64 digit hexadecimal hash.

$$\text{Target hash set} = 16^{(64-18)} \text{ which is roughly } 2 \times 10^{55}$$

$$\text{The probability that random hash value is within limit} = \frac{\text{Target Hash Set}}{\text{Population Hash Set}}$$

The probability that random hash value is within limit = 2×10^{-22}

= 0.00000000000000000000002%.

Therefore, the chances of identifying the next value of the hash by changing the nonce are negligible.

Note: The maximum value possible in the above calculation would be a hexadecimal (64 digits), all with the value F. In reality, there was no original (initial) target value. The maximum value of Target hash (initial value set in the initial day) was to identify a hash, which should be lesser than or below the value of the hash (maximum target): 0000000 0FFFF000000000000000000000000000000000000000. The above value was set to increase the complexity; otherwise, everybody will win the reward without mathematical calculation. As of 2019, it was needed to identify the hash value with 18 leading zeros (current target), which further increased the mining difficulty. Current

$$\text{Mining Difficulty} = \frac{\text{Current Target}}{\text{Maximum Target}} = 3 \text{ Trillion.}$$

This means it is 3 trillion times harder now to find the nonce than when it was started.

Bitcoin mining is of two types, namely:

1. Software mining
2. Hardware mining.

Big corporates of the world use software mining, whereas individuals use hardware mining.

There are two types of software mining, namely:

1. Pool mining
2. Cloud mining.

Pool mining is done by a network of people from different places through a computer network. They share the value of bitcoin at a ratio that is pre-specified and agreed upon. Cloud mining, in particular, is used by big organizations. This is similar to cloud projects, where the hardware setups and software are provided by third party agents (cloud providers). The miners need to just use that setup for mining and pay per usage. Cloud providers use remote data centre with high-power computers but the processing powers are shared across many miners. Cloud providers provide this service on some cost (pay per use) and because of this, the return on investment is lower compared to other types of mining.

We can register and pay at specified websites for software mining. The value of the mined bitcoin will be shared based on the investment ratio. However, individuals can earn bitcoin only through hardware mining.

There are four types of hardware mining (this was explained in Chapter 03 with respect to cryptocurrencies in general), namely:

1. CPU mining (Central Processing Unit)
2. GPU mining (Graphical Processing Unit)
3. FPGA mining (Field Programmable Gate Array)
4. ASIC mining (Application Specific Integrated Circuit).

To do hardware mining, the bitcoin mining, specific software has to be installed on the computer.

CPU mining is done on an ordinary computer. In early days, mining was done only through this method. Speed is the primary concern in this type of mining. Only a small number of bitcoins could be created using this method since it is capable of producing less than 10 million hashes per second (MH/s). With the advent of new and dedicated mining hardware, CPU mining became irrelevant.

In the next version of CPU mining, high-power graphics card were inserted into the computer. This is called as GPU bitcoin mining. GPU mining is a prominent form of mining, and some miners have even produced a mining farm using GPU mining. This is a hundred times faster than CPU mining. More bitcoins were created using this method as compared to CPU mining. However, since high-profile graphics cards were used for this method, it consumed a lot of electricity, which in turn produced a lot of heat. Hence, service cost was higher in GPU mining method. It is capable of producing less than 1 gigahash per second (GH/s).

In the next version, called FPGA, a dedicated circuit board (motherboard) was created for mining. This circuit was connected to the USB port for the mining function. More bitcoins were mined using this method. It also consumed lesser electricity with lower maintenance cost than the GPU method.

ASIC mining method was introduced after this. The hardware for ASIC mining looks similar to that of the UPS machine and can be connected with computers to do mining. The ASIC method is a hardware improvement where manufacturers design the hardware for a specific purpose (bitcoin in our case). This gives an advantage over CPU mining and in some cases, over GPU mining as well. Nowadays many people use the ASIC machine to mine bitcoins. It is capable of producing greater than 1,000 GH/s.

4.4.4 Proof of Work (PoW) and Hashcash in Bitcoin

In Chapter 02, we discussed about Proof-of-work algorithm (Section 2.4.3.1), which is the most well-known consensus mechanism and used by the first blockchain Bitcoin in 2009. Here, several nodes of the distributed ledger called miners compete to solve a complicated mathematical problem based on a cryptographic hash algorithm. The solution found is called Proof-of-Work or PoW. Without proof of work, adding blocks to the blockchain would be too easy making it vulnerable to hackers.

Let us discuss Proof of Work (PoW) and Hashcash in detail.

Bitcoin is a decentralized network and allows the transfer of coins between nodes in the network. The approach of two nodes sending money seems undoubted; it is like two friends sending money in today's world, outside of the computer network, and this could be in exchange for gold or another commodity. However, this transaction of payment between two nodes can be attacked by a hacker and may sometimes lead to double spending. This brings us to the essential aspect of transaction validation within a decentralized environment.

The first miner to create the winning hash will receive rewards in the form of transaction fees or new bitcoins. This process of computing the hash is called proof-of-work or consensus mechanism and it provides integrity to the blockchain.

Bitcoin comes with a default mining rule: hashcash-based proof of work. There are two components in mining here: proof of work and hashcash.

4.4.4.1 Proof of Work

Proof-of-Work (PoW) uses Hash and Nonce. As discussed earlier, miners compete to solve a complex mathematical problem based on the cryptographic hash. Mining comprises of hashing a block and then introducing a nonce to the hashing function and running the hash all over again. A nonce is a number that can be used just once in the cryptographic communication. Adding a nonce to a transaction's identifier makes it additionally unique, thus reducing the chance of duplicate transactions. The nonce is key to creating a block in a blockchain database.

Bitcoin Monetary Policy

Block frequency and Halving are the monetary policies of bitcoin.

Block frequency: Bitcoin transactions are being registered into the blockchain once in ten minutes. Miners will first check the transaction. After reviewing the transaction, the software will give a complex problem for the miners to solve. If the miners can solve the problem, then the computer will give bitcoins as a reward to the miners.

Halving policy: Today, miners get 6.25 bitcoins as a reward to solve the complex problem of coming up with a hash lower than or equal to the target hash (refer Fig. 4.5). From 2009 to 2012, 50 bitcoins were given as rewards. It was reduced to half in 2013 and miners received 25 bitcoins from 2013 to 2016. In 2017, the reward was further reduced to half, from 25 bitcoins to 12.5 bitcoins. The next halving took place in May 2020 and the current block reward for miners is 6.25 bitcoins. The block reward will continue to be at this value until 2024 and be reduced by half every four years until eventually, in 2140, it will come down to zero.

Year	Reward Value ($)
2009-2012	50
2013-2016	25
2017-2020	12.5
2021-2024	6.25
2025-2028	3.125
2029-2032	1.5625
2033-2036	0.78125
2037-2040	0.390625
2041-2044	0.1953125
2045-2048	0.09765625
2049-2052	0.048828125
2053-2056	0.024414063
2057-2060	0.012207031
2061-2064	0.006103516
2065-2068	0.003051758
2069-2072	0.001525879
2073-2076	0.000762939
2137-2140	0

Figure 4.5: Bitcoin block reward over the years

After the year 2140, the production/creation of bitcoin will be stopped (refer Fig. 4.5). At that point of time, the count of bitcoins will have touched 21 million, which is the maximum limit. Until May 2019, 17.7 millions of bitcoins (roughly) were created, which leaves about 3.27 million bitcoins to be mined in the coming years. Around 1,800 bitcoins are being mined every day.

Proof of Work: Creating a Block

In the proof-of-work algorithm, the miners need to compete with other miners and solve a puzzle, which is the proof of work. The first miner to solve the problem will be awarded the bitcoin block reward. To maliciously add a block (solve a puzzle), the miner's computer needs to be 51% more potent than other machines. To counteract the high time and energy consumption in transaction validation, some miners group together the mining pools to combine their mining resources for more efficiency and savings. Mining pools go against the basic principle of distributed ledgers as a person or group gaining control of over 50% of the computing power of the network, usually referred to as a 51% attack, can control the validation process.

Proof of work is like a race where nodes compete with each other. In this process, they pick up transactions within a block, validate them, find nonce for hashcash, and propose a block. The second part of hashcash involves finding a nonce, a cryptographic solution that is added into the block. Without proof of work, adding blocks to the blockchain would be too easy and could make it vulnerable to hackers. The mining node releases the proof-of-work to the other nodes for verification to reach the consensus. The solution to the problem is challenging to produce but easy for the network to verify. The process of mining is exceptionally computation-intensive. So the first miner who manages to produce the PoW will be rewarded a bitcoin.

Figure 4.6 represents the simplified bitcoin blockchain network. Each block (of transactions) has three necessary informations, namely,

- Block header
- Hash of previous block header
- Merkle root

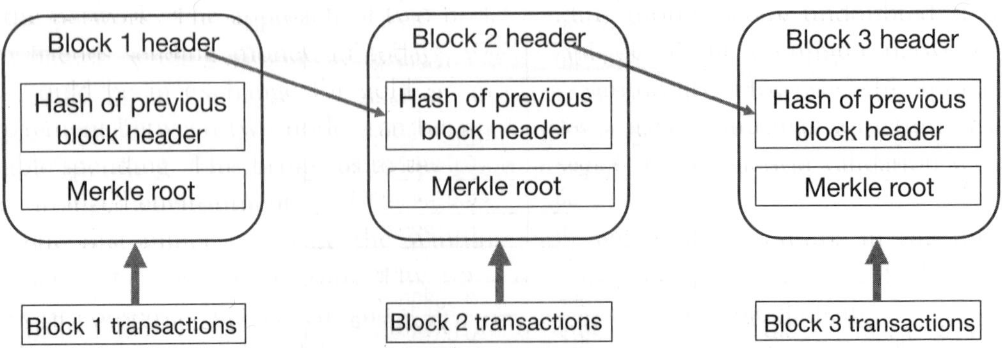

Figure 4.6: Simplified bitcoin blockchain

Block Header

A block header is a unique number, consisting of 80 bytes, which identifies the block.

Split up details	Block Header Contents (80 Byte)
4 Byte	Bitcoin Version number
32 Byte	Previous block hash
32 Byte	Merkle Root
4 Byte	Time Stamp
4 Byte	Difficulty Target
4 Byte	Nonce used by miners.
80 Byte	**Total**

Figure 4.7: Contents of a block header

It consists of the Bitcoin Version number (4 Bytes), Previous block hash (32 Bytes), Merkle root (32 Bytes), Timestamp (4 bytes), Difficulty target (4 Bytes), and Nonce used by miners (4 Bytes) as shown in Figure 4.7

Hash Function (to create a hash of previous block header)

A hash function can be considered equivalent of a fingerprint of a data (which is of any format), similar to a fingerprint of a person. Using a fingerprint, we can identify a person since it is unique. However, reverse engineering is not possible with fingerprint, which means that we cannot draw the picture of a person using his or her fingerprint.

Let us first understand the purpose of the hash function. A hash function converts the data of arbitrary size to data of a fixed size. For example, the function keccak256 (parameter) converts the parameter given into a hash function. keccak256 (parameter) and SHA2-256(parameter) are now mainly used in bitcoin and Ethereum projects.

As we can see from Fig.4.8, for a small change in the parameter (refer "mouli" and "Mouli" as a parameter) of the keccak256 parameter, there are drastic changes in the hash function output.

Parameter	keccak256 (parameter)
Chandramouli	e0eebe2736fec6510eac23b2ac2797212a984f1cb8a18449681912afe8f3e853
Chandramouli S	1245c1047bffbb181f8d7a2e1c4e23055b13cb26597ce23634caaf92e758ff7e
mouli	b39c1e4d76603545dda8a815dcba77f49cd852e187bc51b54948e53e1487c75
Mouli	6fecc02bcdcc44dca54f686a51608cca253daad7eedbf20637310129cc95b100

Figure 4.8: Examples for hash function using keccak256 (parameter)

Parameter	SHA3-256 (parameter)
Chandramouli	9907a52b76d3a33f08a218b09b5460f004a489bc726c2a3580eb214c8219cd43
Chandramouli S	1d15557c7295adcce7f1db2c3ff2c719cb2728f005e43d66acdd577aeb0ad3d0
mouli	3c99f645e4f058ec9fd106f4452ad4cf196f83c02541508ad7e72f1a71679d66
Mouli	fdc8d3460aff5f1da1cebf9362fa4d73e7f8d0a5947d1cf160f625da6bdbfab6

Figure 4.9: Examples for hash function using SHA 3-256 (parameter)

Similarly, as seen in Fig. 4.9, for a small change in the parameter (refer "mouli" and "Mouli" as a parameter) of SHA3-256 parameter, there are drastic changes in the hash function output. SHA stands for "Secured Hash Algorithm." Number 256 denotes the number of bits it takes to store into the memory. NSA (National Security Agency of America) developed SHA 256 algorithm. The algorithm is entirely open, available for all. It is the core building block of the blockchain. The output of SHA 256 is 64 characters long. It contains hexadecimal numbers, which consists of decimals (0 to 9) and Alphabets (A to F), representing the numbers 1 to 16 in that order. This hash function can take any data as input; the data format includes words, letters, text, PDFs, audio files, video files, executable files, etc.

Now, let us look into a few characteristics of the hash algorithm:

1. It has to be one way only, and reverse engineering is not possible. It can produce a unique hash from the data. However, it cannot provide the data from the hash.
2. The same document always produces the same hash. This shows the deterministic characteristics of hash
3. Fast computation is another characteristic of any hash function. It provides the results in a fraction of a second.
4. Avalanche Effect is a key characteristic of the hash function. Avalanche effect indicates that a small (tiny) change in the input produces drastic changes in the hash output. A change in output is not proportional to the change that is done in the input.
5. It must withstand collisions. It means that the same hash will not be produced for two different types of document.

In a bitcoin blockchain, the hash function of the previous block header is stored as a reference in the next subsequent block to ensure that the blocks are correctly connected. That means, the contents of the last block are used as the parameter of SHA3-256 (parameter) function and the resultant (hash output) is stored in the next block for reference.

The type of hash differs from one protocol to the other. Thus, the type of hash used in bitcoin is different from that used for Ethereum. Also, the complexity of forming the hash will vary across the different protocols.

4.4.4.2 Merkle Tree and Merkle Root

A Merkle root is synonymous to a fingerprint of all the transactions in the block, which is created by hashing together pairs of Transaction IDs, to a short and unique **fingerprint for all the transactions in a block**. This is also a field in a block header (refer Fig. 4.10).

Sometimes, the Transaction ID (TXID) gets hashed twice using SHA256 protocol. This is done for enhanced security.

A Merkle tree (binary hash tree) involves taking large amounts of transaction data and constructing it in a way that is more convenient to process.

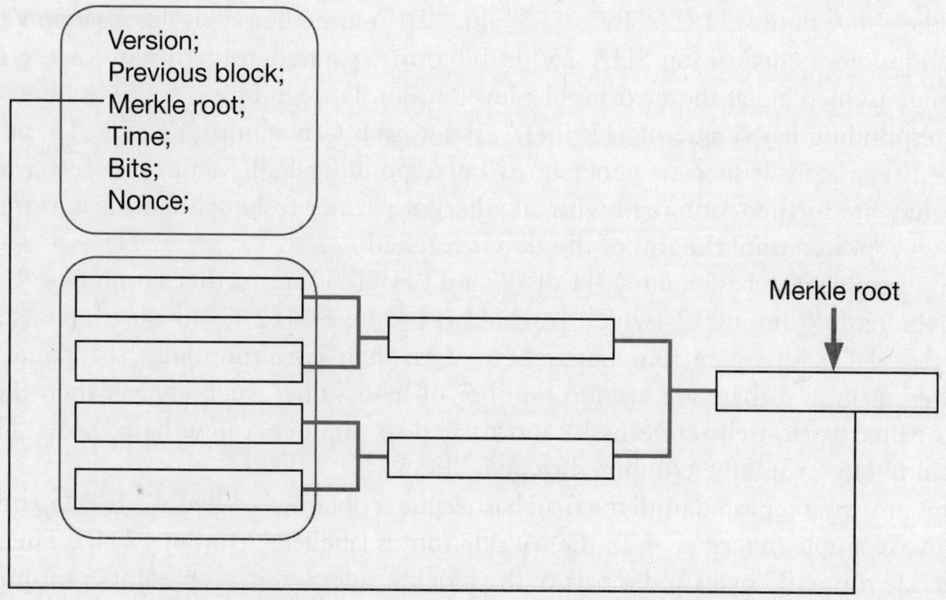

Figure 4.10: Merkle root creation: Example 1

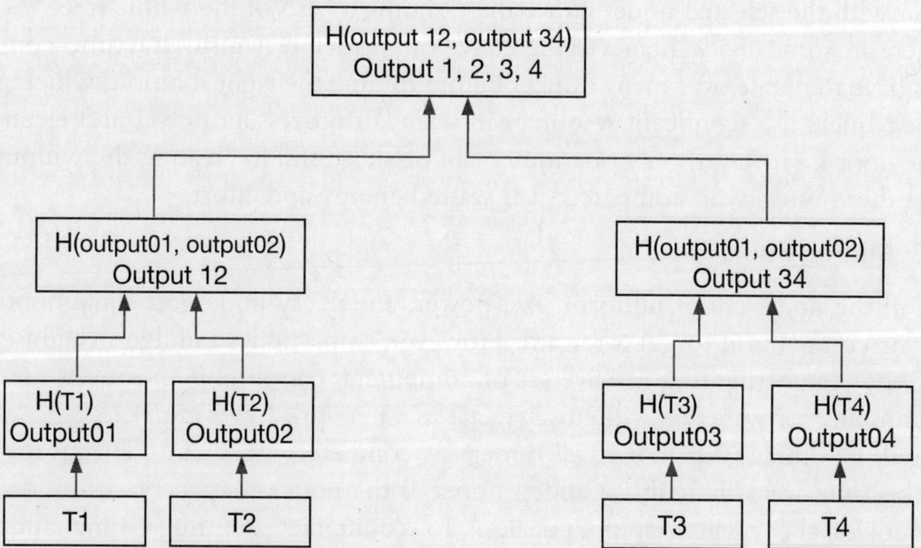

Figure 4.11: Merkle root creation Example 2

Merkle trees have nodes and leaves. Original transactions present in the block acts as the leaves of the Merkle tree (refer Fig. 4.11), and nodes in the tree act as a hash of leaves. Further, moving up the tree, new nodes are hashes of the lower nodes, and this process is repeated until the top of the tree is reached. The hash value at the top (peak) of the tree acts as a hash of the overall block (It is also called as Merkle Root).

In the above figure, "T1", "T2", "T3" and "T4" represent a typical transaction. These transactions are hashed (using SHA-256 in bitcoin) separately to get their corresponding hash value (which are at the next higher-level node). For example, "T1" is hashed to get its corresponding hash value of "H (T1) ". After each transaction (T1, T2, T3, and T4) has been separately hashed to generate its corresponding hash value, the new resultant hash values are further combined with an adjacent partner to be hashed once more. This process is repeated until the top of the tree is reached.

For example, the hash values "H (T1)" and "H (T2)" are further combined (hashed) to get the hash "Output 12 (which is the H (H (T1), H (T2))". In the above example (refer Fig. 4.11), there are four transactions with their corresponding hash value pairs. However, assume if there are an odd number of hash values, such as five, then the fifth hash is paired with itself (same hash) and hashed to produce a new hash value. That is, "H5" and "H5" would be combined to give "H55".

This process is repeated until the final hash value is obtained (called as **Merkle root).** In the above example (refer Fig. 4.7), the Merkle root is labelled "Output 1234". The size of the Merkle root is 32 bytes and is part of the block header, which represents a summary of all transaction data. By comparing, the top-level hash ensures that the integrity of the data has not been compromised. To validate a transaction in a tree, we can selectively compare the hash with the selected nodes rather than finding a hash of the whole tree.

There are some disadvantages to the PoW consensus mechanism. It is time-consuming: Miners have to iterate over many nonces before finding the right solution, which is a slow process. Miners use significant resources in terms of processing power and electricity to find the nonce. However, only one miner can be successful in creating the winning hash. For all other miners who competed, it is wasted energy and effort.

4.4.4.3 Mining Pools

When all the nodes are of uniform size, power, and every node gets equal opportunity, it is a fair competition to get a reward. However, some nodes indulge in industrial-size mining, by connecting to a massive set of computers, consuming enormous power, and using complex software. They themselves act as a network (refer Fig. 4.12).

It will be difficult for individual miners to compete with them. Getting the golden nonce becomes very difficult. . Golden nonce is the nonce that is created by the miner, which solves the cryptographic puzzles. To counteract the huge time and energy consumption involved in transaction validation, some miners group together in mining pools to combine their mining resources for more efficiency and savings. Miners do not work by themselves; they work together in mine pooling. They share the mining power and processing power. Mining pools provide the service; they share the services so that the individual miners do not repeat the work, thus avoiding double work and waste of time. Nonce values (cryptographic puzzles) are distributed among the individual miners within the pool.

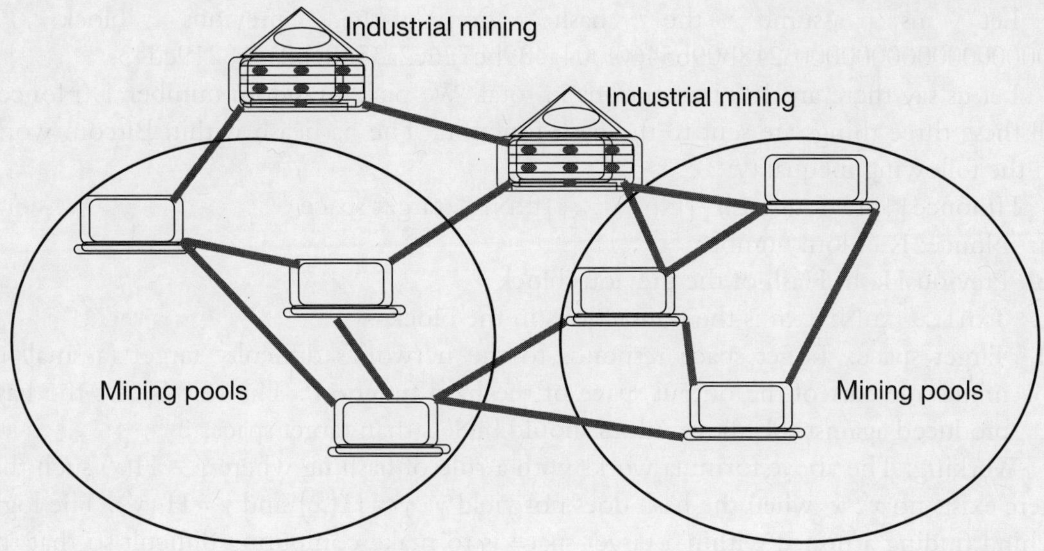

Figure 4.12: Mining pools

When one of them finds the golden nonce, the corresponding mining pool wins the reward for that block. They share/split the reward based on the computing power (hash rate) of the individual miners. A mining pool shares feed for the entire work. Any individual can join the mining pool.

However, mining pools can go against the basic principle of distributed ledgers as someone or a group gaining control of over 50% of the computing power of the network, usually referred to as a 51% attack, can control the validation process. 51 % attack is not designed to tamper with the blockchain. In order to attack a block within the blockchain, the hacker needs to change all the subsequence blocks of the blockchain (which is very difficult). They also need to do the same in many other computers (nodes), which is even more difficult in nature.

4.4.4.4 Hashcash

Hashcash is a cryptographic hash-based proof-of-work algorithm, which was initially used to identify email spammers. For example, the number of seconds/minute of CPU time taken to write an email is hashed and is written in the header of the email. For example, let us say on average, to write an email, it takes a minimum time of 20 seconds. When an email is received, the hash value is checked, and if the value is lesser than 20 seconds, then that email can be classified as Email Spam.

The same logic is used nowadays in Bitcoin. Let us say that we have the following three values to start computing:

1. Hash of the previous block
2. Transaction
3. Nonce (random number)

Let us assume the hash from the previous block as 000000000000000000a218b09b5460f004a489bc726c2a3580eb214c8219cd43.

Let us say there are 100 transactions in total. We pick a random number 1 (Nonce). All these three things are sent to the hash function. The hashcash within Bitcoin works on the following inequality:

H(nonce||previous hash||txn||....||txN) < target space

a. Nonce: Random number
b. PreviousHash: Hash of the previous block
c. Txn1... txnN: Txn is the transactions in the block.
d. Target space: Target space responds to the network's difficulty target (a small or minuscule part of the output space of the hash function). The rule is that the hash produced against the above values should be less than target space.

Working: The above formula works with a rule of hashing where $y = \mathrm{H}(x)$ such that there exists no x': x' when the hash does not yield y. $y = \mathrm{H}(x)$ and $y \neq \mathrm{H}(x')$. The logic behind finding a nonce within a target space is to make computing difficult so that the advantage due to hardware advancement is negated. To find a nonce, a miner needs to attack the inequality with brute force, i.e., keep trying until he succeeds. Every fortnight, the bitcoin raises the target space requirement, thus making it harder for the miner to find a nonce. Target space is a tiny space.

As we can see from Fig. 4.13, the SHA-256 hash function is not yielding the right string for Nonce 1, as the resulting string starts with only four zeros whereas a string starting with 18 zeros is required. Therefore, we choose another random number and start another calculation. This time, let us choose number 2 as nonce.

This time, we get a string (refer Fig. 4.14) that starts with exactly 18 zeros. We verify this block and if we are the first ones to get this result, we will get a reward for that block.

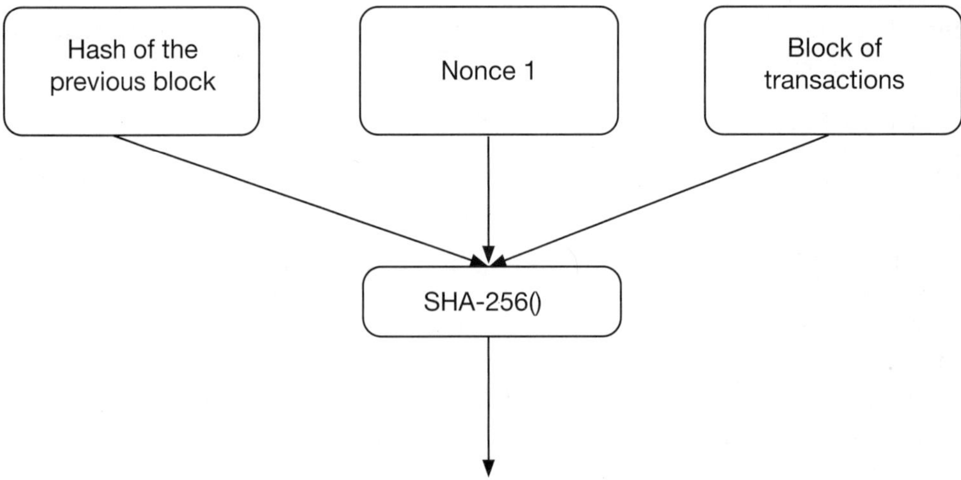

Figure 4.13: Hashcash mapping Example 1

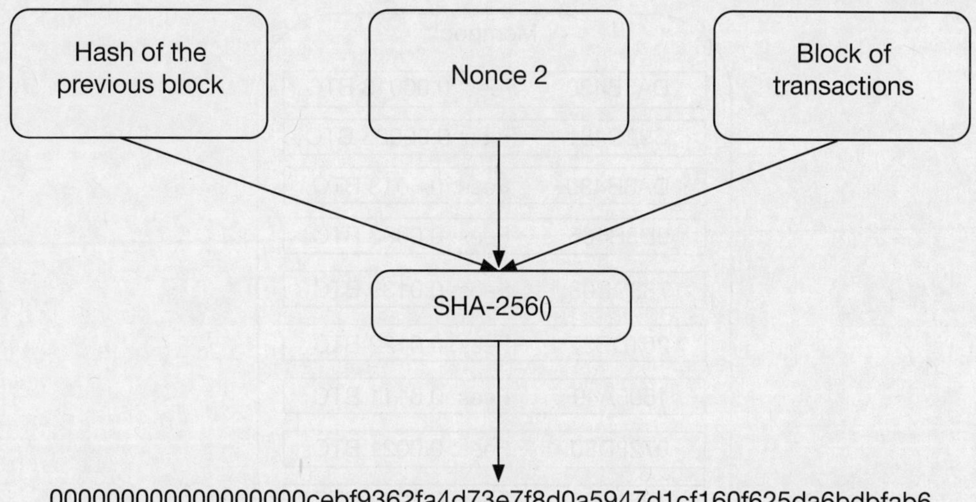

Figure 4.14: Hashcash Mapping Example 2

A nonce is a finite number with a 32-bit unsigned integer.
The maximum nonce = $2^{32} = 4 \times 10^9$
It has a value between zero and 4 billion (approximately).
Probability that one of them becomes valid = (4 billion) × (2×10^{-22})
$= 8 \times (10^{-33}) = (10^{-22}) = 0.0000000001\%$

A modest miner can create 100 million hashes per second. To traverse 4 billion hashes, it takes just 40 seconds.

Timestamp (Current UNIX time): Timestamp is another field, which indicates UNIX time. It is the seconds passed after the first of January 1970 and is a 10-digit number. This is also part of the data, which changes every second. When the timestamp changes (every second), the corresponding data changes; when the data changes, the resultant also gets altered. The miner's computer should keep incrementing the nonce until it finds the right one as per the algorithm (< target space). That means, the computer needs to generate millions of hashes per second to create the one that will have the same number of starting zeros as defined. It is time-consuming to execute a PoW for a single block, but at the same time, it is easy for someone to verify if some block is correct.

Block of transactions: Transactions are added to the block in the form of data, which is used for hash mapping.

Transactions are first added to the Mempool (memory pool) attached to each node (refer Fig. 4.15). These are unconfirmed transactions, which are to be included in the block once the combination becomes a validated. Blocks are added every 10 minutes, but transactions happen all the time. Mempool is the staging area for the transactions.

Mempool	
DA3B430	Fees: 0.00013 BTC
C92C481	Fees: 0.00023 BTC
DA3B430	Fees: 0.0013 BTC
9D8B466	Fees: 0.0093 BTC
783BB08	Fees: 0.0139 BTC
261DB27	Fees: 0.0122 BTC
1505A49	Fees: 0.0111 BTC
9728D80	Fees: 0.0021 BTC
1051C02	Fees: 0.1233 BTC

Figure 4.15: Mempool example

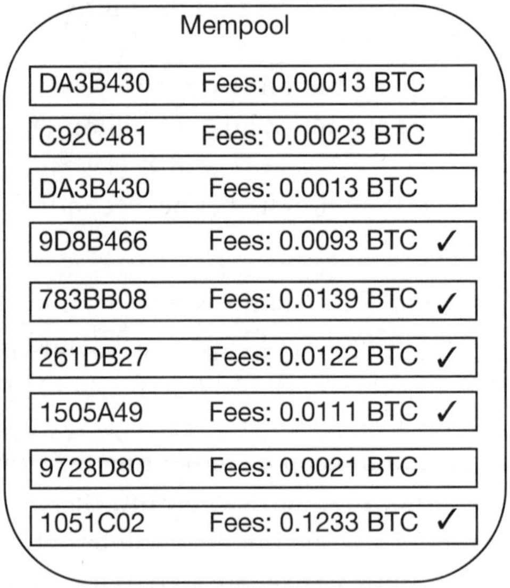

Mempool	
DA3B430	Fees: 0.00013 BTC
C92C481	Fees: 0.00023 BTC
DA3B430	Fees: 0.0013 BTC
9D8B466	Fees: 0.0093 BTC ✓
783BB08	Fees: 0.0139 BTC ✓
261DB27	Fees: 0.0122 BTC ✓
1505A49	Fees: 0.0111 BTC ✓
9728D80	Fees: 0.0021 BTC
1051C02	Fees: 0.1233 BTC ✓

Figure 4.16: Mempool selected transactions

Each transaction has a Transaction ID. Each transaction has fees for the miner. The miner will get these fees to add these transactions into the block. Based on the limit (the number of transactions that can be inserted into the block at a time), the miner will pick the highest fees transactions. A miner can add 1 MB of the transaction at a time into the block, which is equivalent to 2,000 transactions. Usually, 1000 to 3500 transactions are added per second. Sometimes, it reaches 7000 transactions per second. Let us assume that a miner can add up to 5 transactions into the block from the mempool. Five transactions will get selected based on the fees (highest first) (refer Fig. 4.16).

Check data set 1	
9D8B466	Fees: 0.0093 BTC
1505A49	Fees: 0.0111 BTC
261DB27	Fees: 0.0122 BTC
783BB08	Fees: 0.0139 BTC
1051C02	Fees: 0.1233 BTC

Figure 4.17: Mempool data set 1

Check data set 1	
9728D80	Fees: 0.0021 BTC ✓
1505A49	Fees: 0.0111 BTC
261DB27	Fees: 0.0122 BTC
783BB08	Fees: 0.0139 BTC
1051C02	Fees: 0.1233 BTC

Figure 4.18: Mempool data set 2

As discussed, nonce value ranges from zero to 4 billion. There is a timestamp attached to the block (UNIX time) which changes every second. A miner may not go through the entire nonce in one second (as the timestamp changes every second). The miner changes the combinations of transactions (to the data) and then calculates the test hash.

For example, the check data set 1 from the mempool, and target hash can be calculated as shown in Fig. 4.17.

Assume that the test hash does not match with the target hash, then the top transaction (9D8B466) is replaced with the next highest fees transaction (refer Fig. 4.18).

Assume once again that this test hash also does not match with the target hash. Then the top transaction (9728D80) is replaced with the next highest fees transaction as shown in Fig. 4.19. If this combination identifies the resultant hash (test hash), then it is added to the block, and removed from the mempool. When it is removed from the mempool, the copy of the same is broadcast to all other nodes.

This happens algorithmically. In a miner pool, the pool will allocate the combination of transactions to each miner to check and verify.

Note: If the fees for a transaction is given as zero, the transaction may be stuck in the memory pool itself and chances of it to be selected for addition into the block will be minimal. Usually, transactions with the highest transaction fees will be selected for the transaction.

Check data set 3

DA3B430	Fees: 0.0013 BTC	✓
1505A49	Fees: 0.0111 BTC	
261DB27	Fees: 0.0122 BTC	
783BB08	Fees: 0.0139 BTC	
1051C02	Fees: 0.1233 BTC	

Figure 4.19: Mempool data set 3

How mempools work: A mempool is attached to each node in the distributed network. As discussed, mempool is not part of the blockchain; it is just the staging area for transactions (refer Fig. 4.20).

When one person (one node) sends a transaction to another person, it is not only added in the receiver's mempool; it is also duplicated in all other mempools after the validity of the transaction is checked. In a similar way, more and more transactions are added to the mempools. A block can have approximately 2000 transactions. A mempool can have around 10,000 transactions at a time. When a miner successfully mines, a set of transactions are added from mempool to the block, and into the blockchain. Once the transactions are added to the blockchain, the same sets of transactions are removed from the mempool.

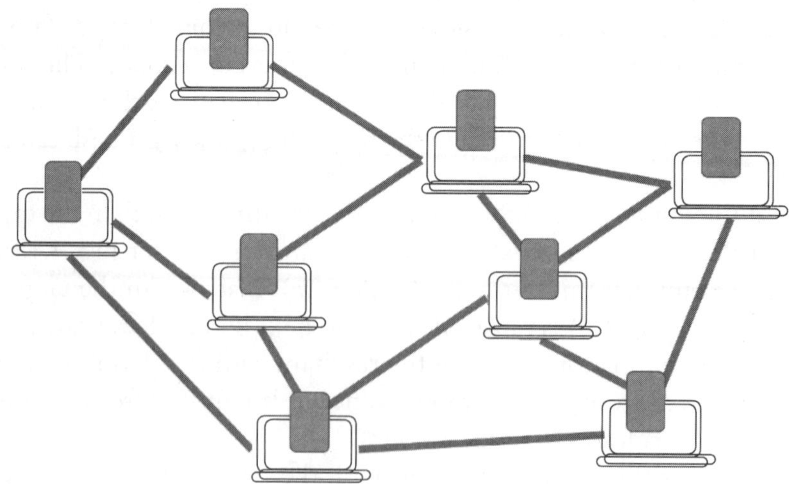

Figure 4.20: Mempool distribution

Essential aspects of the hash puzzle

There are three crucial aspects of this hashcash puzzle:

a. Difficult to compute
b. Steady block insertion rate
c. Easy to verify.

Challenging to compute: As discussed in the previous paragraph, the target space is a small or minuscule part of the output space of the hash function. If the target space were 1 percent of the overall output space, a miner would have to perform 100 iterations before arriving at a nonce. However, the difficulty level goes down by only 1 percent. Thus a large number of iterations are needed before a nonce is arrived at.

Configurable difficulty: Every fortnight, i.e., after 2016 blocks, the bitcoin server is reconfigured such that the average time taken to create the next block is 10 minutes.

Easy to verify: As seen earlier, it took 10^{20} transactions in 2014 to find a nonce. The process is time-consuming and resource intensive. However, verifying the nonce is not complicated; a node needs to find a hash and validate against the inequality.

4.4.5 Block Propagation and Relay

When a miner announces a block, the block needs to propagate to all the nodes in the network, using gossip protocol (transactions are propagated using gossip protocols).

The concept of gossip communication is similar to employees spreading rumours in a company. Let us say, each hour the employees assemble in the Coffee Corner. Each employee joins another employee chosen at random (casually) and shares the latest gossip of the office. At the start of the day, Anan starts a new rumour: he comments to Babi that he believes that Monika dyes her hair. At the next meeting, Babi tells Dani, while Anan repeats the idea to Elan. After each water/coffee meeting, the number of individual employees who have heard the said rumour roughly doubles (though this doesn't mean gossiping twice to the same person – perhaps Anan tries to tell the story to Fathima, only to find that Fathima already heard it from Dani). Gossip protocol also works in the same way from one node to the other.

4.4.5.1 Nodes

Nodes in the network do the following:

a. Submit transaction to other nodes
b. Validate transaction from other nodes
c. Maintain the full block of the bitcoin network
d. Come up with a nonce for the block of transaction
e. Publish your block to the network
f. Hope other miners accept your block.

Of these steps, Steps b, c and d are collectively called mining for the network. All nodes in the network could participate in mining and earn mining rewards. This activity needs a lot of computing resource, and often, many nodes do not create a block. This is because validating a transaction and finding a nonce involves massive computing expense. Some people have invested in server farms to compute so that they could solve the problem faster and claim for the block and prize money before others.

There are three types of nodes:

a. Full node
b. Mining node
c. SPV node

Full node: These are the nodes that do all the activities (a to f) that are listed above, diligently (they accept transactions, validate transactions and relay them to further full nodes). This needs lots of computing and there is some chance of creating a block and earning a mining reward. A node that operates in the bitcoin network would have full node rights; all nodes in the network are equal, as stated by the Satoshi Nakamoto. Many people and organizations volunteer themselves to run full nodes using spare computing and bandwidth resources in their computer—but more volunteers are needed to allow Bitcoin to continue to grow.

Mining nodes: Mining nodes are the set of nodes with mining capability, but they do not hold the entire blockchain.

SPV node: This stands for "Simplified Payment Verification." The vast majority of nodes today are SPV nodes. SPV nodes do not involve in mining; however, they are bound to validate the block using block scripts; this feature can be easily achieved even via handheld devices.

Bitcoin network, as of May 2019:

Total no of nodes in the network: 82,640

No. of full nodes in the network: 9,384

No. of SPV nodes in the network: 2,312

The number of nodes keeps varying; also, the above figure of the total number of nodes is an approximation. The number of full nodes cited varies from source to source; it hovers between 9000 and 10000 on an average.

Miners would create a block and propose a block. A proposed block needs to travel to all the nodes in a network. This is often a bottleneck considering that data in the block has to be validated by all nodes in the network. During this journey, once a block reaches a node, the node would validate the block; this is again a resource-intensive task.

In general, a node would perform the following functions:

a. Validate the overall block
b. Make sure the nonce is valid (nonce is within an acceptable range)
c. Validate each transaction in the block.

If the outcome of all of the above three functions, namely: a, b, c are valid (True), then the block is eligible for further propagation within the network. If the conditions are not met, the node will discard the block and not propagate it further. Studies have shown that the time taken for a solid block to travel to all nodes is close to 30 seconds for large blocks. Being a decentralized network, bitcoin does not devise a shortest path mechanism nor keep a record for the shortest path. Nodes that validate the block (entirely logical blocks) would need to require high storage (the entire chain starting from 2009 needs to be downloaded and used) and a high-speed connection. Besides, the new transactions coming to the network also need to be recorded.

Let us look into a simple transaction: Let us assume, Panos decides to send one bitcoin to Murray as commission for a work done. Murray is notified about the transaction by Panos. Should Murray send the material the moment he is notified about the transaction? Well, as we saw in previous sections, Murray could be subjected to double spending attack. Murray should be smart enough to wait for the transaction to be validated (the transaction has to be listed in the block, and all other nodes need to add their block to the block where the transaction is listed). As the transaction is submitted, all other nodes would start to validate the transaction. The higher the number of valid outcomes, the higher the probability of validation.

Zero confirmation transaction: A scenario where a user submits a transaction and the other party accepts the same — there is no additional confirmation from any other node.

One confirmation transaction: A scenario where receivers wait until at least one node has confirmed the transaction as valid.

Six confirmation transaction: A scenario where receivers wait until at least six nodes have confirmed the transaction as valid.

The general rule of thumb (heuristics) is to wait for at least six confirmations on the transaction. This assures with a high degree of certainty that the transaction would be part of the block. In the bitcoin network, the blocks are created after a delay of 10 minutes, as mandated by Bitcoin. Let us compare the transaction time between bitcoin and popular schemes for money transfer. Refer to the following table to compare bitcoin with Visa and MasterCard. Bitcoin can transact only seven transactions per second; in essence, bitcoin is a time-consuming process.

	Bitcoin	**Visa**	**MasterCard**
Transactions per second	7	65,000	40,000
Total no of transaction in year 2018	81 Million	124 Billion	74 Billion

4.4.5.2 Orphaned Block

Detached or orphaned blocks are valid blocks, which are not part of the main chain. They occur naturally when two different miners successfully mine at the same time. Sometimes, a hacker may attempt to reverse/modify transactions (refer Fig. 4.21).

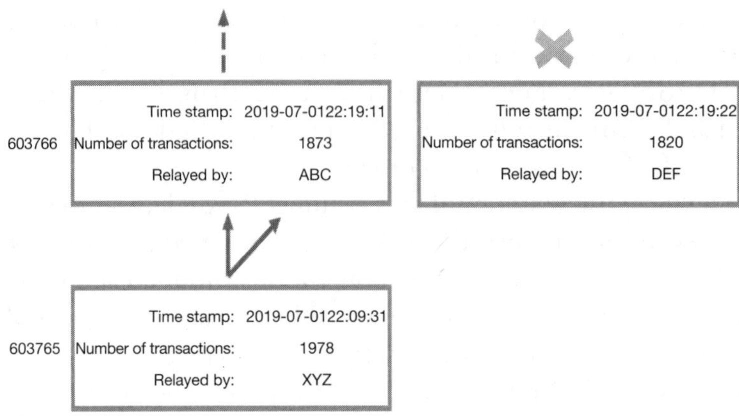

Figure 4.21: Orphaned block

In the above example, XYZ added a block on 1 July with the time stamp of "22:09:31". ABC added the next block on the same day, with the time stamp of "22:19:11". Eleven seconds after that, DEF also added a block with the time stamp of "22:19:22". Two miners (ABC, DEF) claim the same block number now. Very clearly, they are not identical blocks as the number of transactions within those blocks are different (1873 and 1820 respectively).

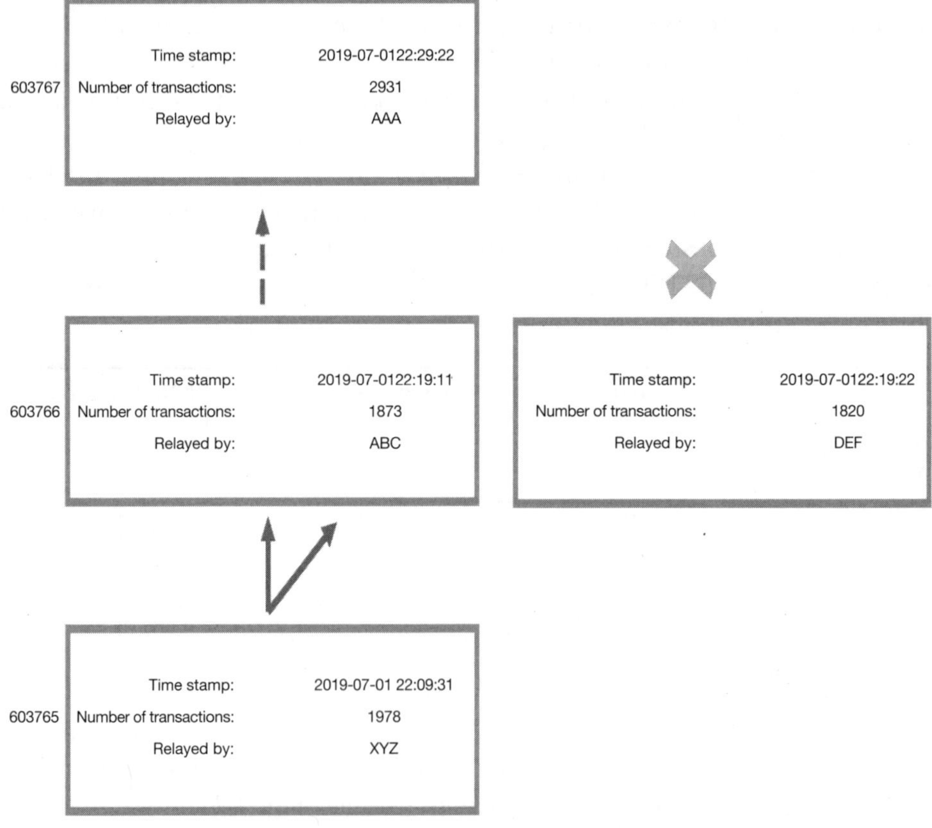

Figure 4.22: Orphaned block continuation

The block with ABC provider will continue to grow and the block with DEF will also continue to grow. Nodes close to ABC will grow/relate to ABC block. Similarly, nodes close to DEF will grow/relate to that block. Therefore, we will have two competing blocks at the same time. The golden rule is: the longest chain wins. Therefore, whoever builds the next block first will take over the chain (refer Fig. 4.22).

Assuming the next block is added first on top of ABC, then ABC will grow and will be part of the blockchain. The rule of thumb (heuristics) is to wait for six confirmations before releasing a block back to the mempool. In the above example, the block added by DEF is called an orphaned block. All the transactions of block DEF (1820 transactions, that are not part of 1873 transactions) are released back to the mempool.

4.4.6 Bitcoin Script

A programming language called "Forth" is used for writing bitcoin scripts. It is very easy to write and supports cryptographic operations. This language supports stack-based programming. The bitcoin scripting language has a limited set of instructions: there are only 256 instructions, and each one is represented by one byte. Out of those 256 instructions, fifteen are currently not in use on the bitcoin script and 75 others are reserved for a particular purpose, which means that they have not yet been assigned any significance. The scripting language directly executes instructions. This language supports conditional statements; however, no looping is allowed. The famous conditional statements: `if` statements and `if else` statements are supported. Another necessary instruction is MULTISIG; this instruction is accessible on platforms where validation by multiple nodes are mandated.

Common Characteristics of Bitcoin Script

a. Simple to execute
b. Lightweight (does not rely on other software packages).
c. Less computing energy.
d. Built for bitcoin
e. Uses cryptography (hashes, signature verification)
f. Stack-based
g. No loops
Conditional statements supported (`If` statements)
h. Validation of multiple nodes supported (MULTISIG instruction)

Majority of the instruction set in Bitcoin is the same as any of those in common software design languages such as C, C++, and JAVA. For example, basic arithmetic operations such as addition, subtraction, multiplication, division, and logic blocks such as if-else, throwing of errors, and returning function value. There are two types of instructions:

1. **Data instructions:** If data instructions are encountered in a bitcoin script, the data corresponding to the particular instruction is simply hard-pressed on top of the stack data structure. Data instructions are given with angle brackets on either side. Data instructions such as <Sig>, <PubKey> and <PubKeyHash> are examples.

Opcode instruction	Purpose
OP_DUP	Duplicates the top item of the stack
OP_HASH160	Hashes twice, first using SHA-256 and then RIPEMD-160
OP_EQUALVERIFY	Returns true if the inputs are equal. Returns false and marks the transaction as invalid if they are unequal
OP_CHECKSIG	Checks if the input signature is a valid signature using the input public key for the hash of the current transaction
OP_CHECKMULTISIG	Checks if the k signatures on the transaction are valid signatures from k of the specified public keys

Figure 4.23: Opcode instructions (Sample)

<Sig> indicates Signature
<PubKey> indicates the public key used to verify the signature.
<PubKeyHash> computes the cryptographic hash

2. **Opcode instructions:** Whenever an opcode is used in a bitcoin script, it executes a particular function, which takes the input data that usually resides on the top of the stack. Opcode instructions start with "OP." For example, OP_DUP, OP_HASH160, OP_EQUALVERIFY, and OP_CHECKSIG (refer Fig. 4.23).

Sample Bitcoin Script: Example 1

Now, let us try to write a simple bitcoin script using the above-known functions.

<Sig>
<PubKey>
OP_DUP
OP_HASH160
<PubKeyHash>
OP_EQUALVERIFY
OP_CHECKSIG

Let us evaluate the above script systematically:

Step 1

In this example, <sig> implies instruction to start (scriptSig is used to unlock transactions). The first two instructions of the script, namely, <Sig> and <PubKey> represent the signature and the corresponding public key used to verify the signature (refer Fig. 4.24)

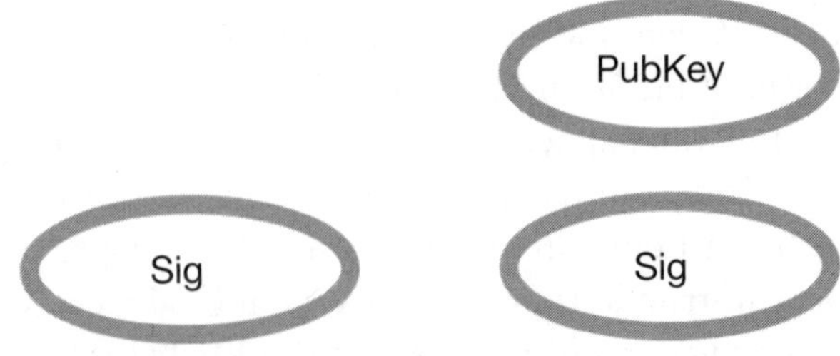

Figure 4.24: Step 1 of Bitcoin script (Example 1)

Step 2

The second step is the execution of the operation code titled "OP_DUP" on top of the Step 1 output. We already know that this command duplicates the item on top of the stack (refer Figs 4.24 and 4.25).

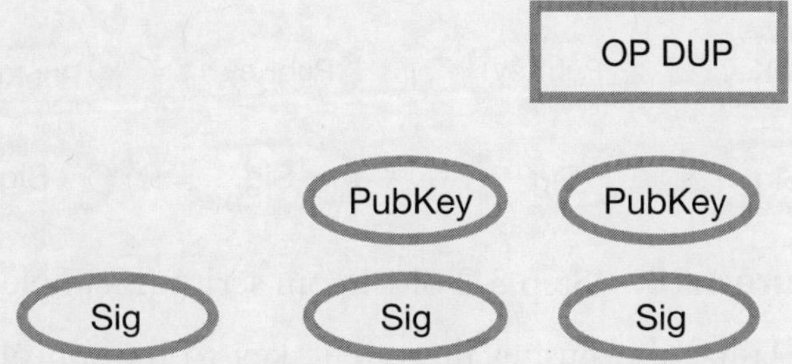

Figure 4.25: Step 2.1 of Bitcoin script (Example 1)

We use OP_DUP to duplicate the instruction, which is used to push a copy of the public key onto the overall stack data structure (refer Fig. 4.26).

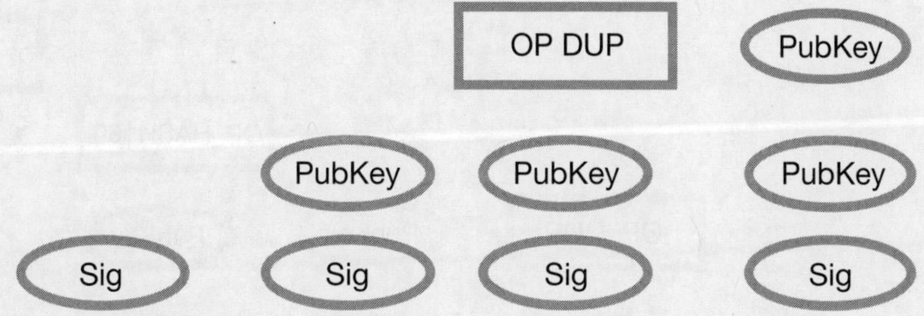

Figure 4.26: Step 2.2 of Bitcoin script (Example 1)

Step 3

OP_DUP instruction is followed by another opcode instruction, namely: OP_ HASH160 (refer Fig. 4.27).

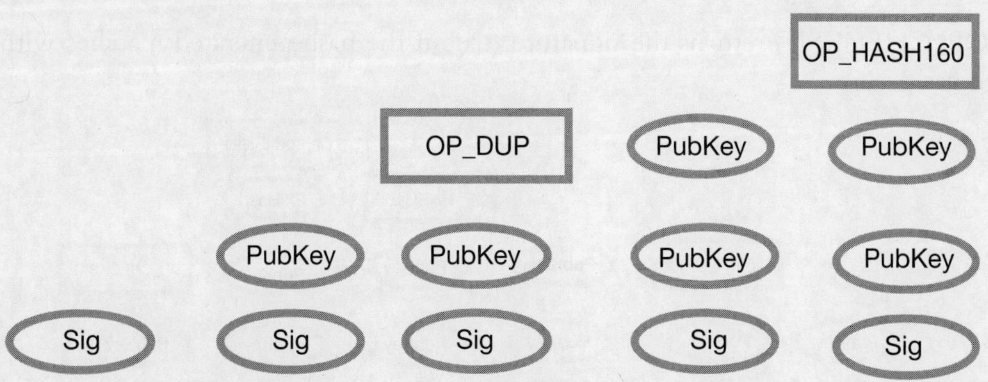

Figure 4.27: Step 3.1 of Bitcoin script (Example 1)

OP_ HASH160, which pops (takes out) the top value from the stack, computes the cryptographic hash (Hash160) and then pushes the result onto the stack (refer Fig. 4.28).

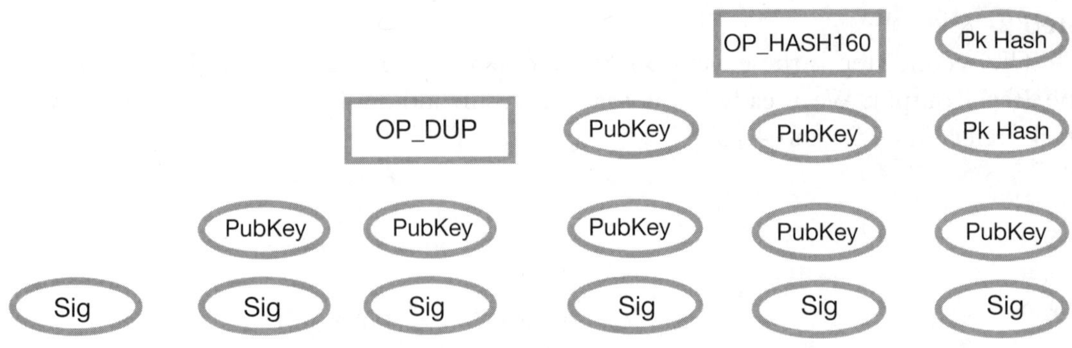

Figure 4.28: Step 3.2 of Bitcoin script (Example 1)

Note: <PubKeyHash> also pushes the public key to the top of the stack, ensuring the duplicate as above.

Step 4

OP_EQULVERIFY matches the hash generated with the given hash (refer Figs 4.28 and 4.29).

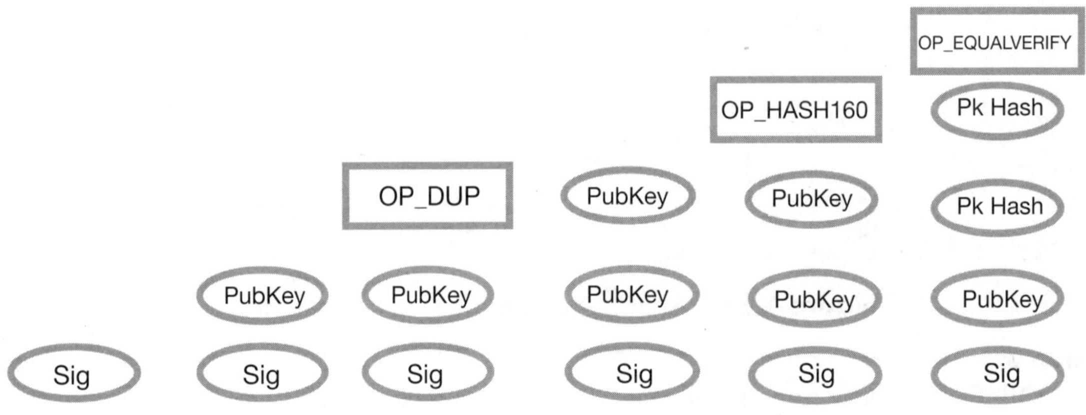

Figure 4.29: Step 4.1 of Bitcoin script (Example 1)

OP_CHECKSIG returns the signature (true if the hash generated matches with the given hash).

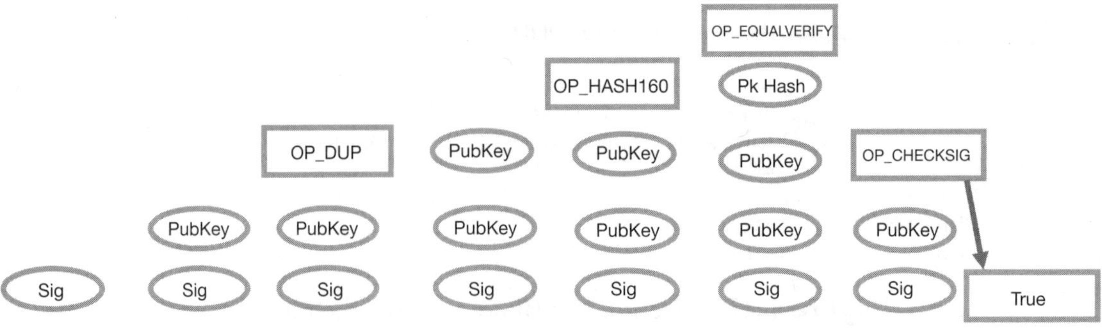

Figure 4.30: Step 4.2 of Bitcoin script (Example 1)

OP_CHECKSIG then pushes the value *true* onto the top of the stack when the signature (Sig) matches the "public key."

Summary of the bitcoin script execution

This script is called as pay-to-pubkey-hash. In this example, <sig> implies instruction to start (scriptSig is used to unlock Transactions). The first two instructions of the script, namely, <Sig> and <PubKey> represent the signature and the corresponding public key used to verify the signature. Then we use OP_DUP to duplicate the instruction, which is used to push a copy of the public key onto the overall stack data structure. This instruction is followed by another Opcode instruction, namely OP_ HASH160, which pops (takes out) the top value from the stack, computes the cryptographic hash (Hash160) and then pushes the result onto the stack. OP_EQULVERIFY matches the hash generated with the given hash. OP_CHECKSIG returns the signature (True if the hash generated matches with the given hash). This transaction is called as "The P2PKH" which means "The pay to the public key hash."

Sample Bitcoin Script: Example 2

OP_3
OP_2
OP_ADD
OP_5
OP_EQUAL

Step 1

Number 3 is added to the top of the Stack.

Figure 4.31: Step 1 of Bitcoin script (Example 2)

Step 2

Number 2 is added to the top of the Stack.

Figure 4.32: Step 2 of Bitcoin script (Example 2)

Step 3

OP_ADD is executed

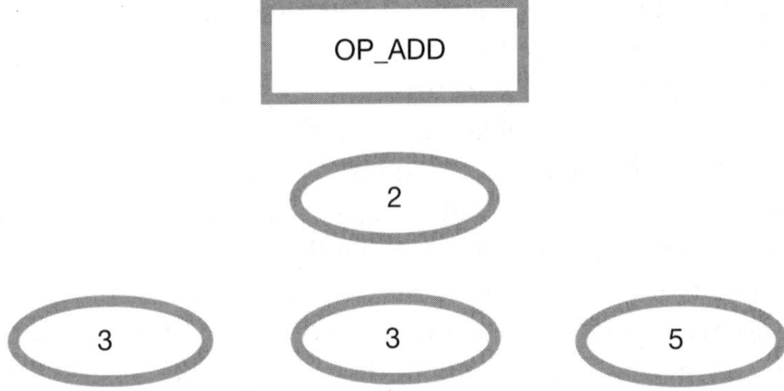

Figure 4.33: Step 3 of Bitcoin script (Example 2)

Step 4

Number 5 is added on top of the Stack

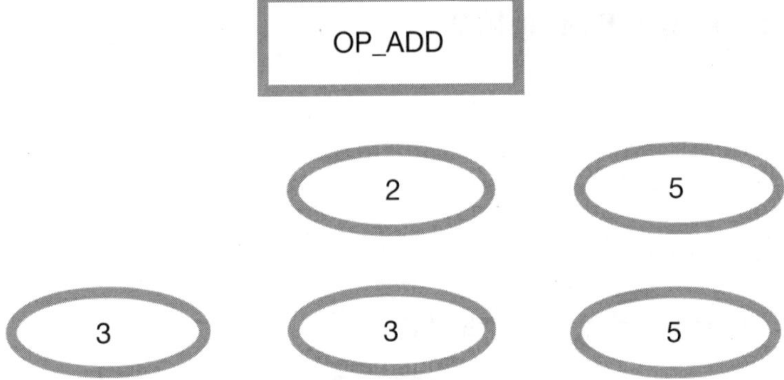

Figure 4.34: Step 4 of Bitcoin script (Example 2)

Step 5

OP_EQUAL is executed

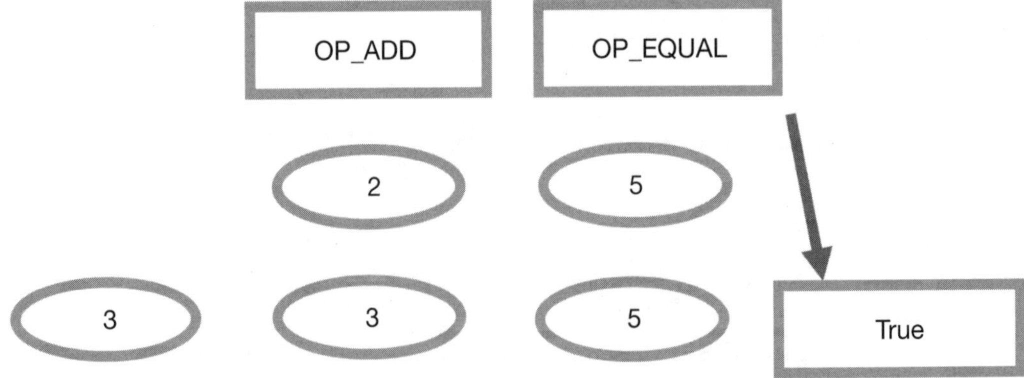

Figure 4.35: Step 5 of Bitcoin script (Example 2)

Therefore, finally TRUE is returned (present at the top of the Stack) (refer Figs 4.31 to 4.35 for the steps).

Overall, 256 instructions are grouped under the following headings:

- Flow control, Constants, Stack,
- Crypto, Arithmetic
- Bitwise logic, Pseudowords
- Lock time, Splice, and Reserved words.

Flow control related opcode instructions

WORD	OP CODE	EXPLANATION
OP_NOP	97	Does nothing.
OP_IF	99	If the top stack value is not False, the statements are executed. The top stack value is removed.
OP_NOTIF	100	If the top stack value is False, the statements are executed. The top stack value is removed.
OP_ELSE	103	If the preceding OP_IF or OP_NOTIF or OP_ELSE was not executed then these statements are and if the preceding OP_IF or OP_NOTIF or OP_ELSE was executed then these statements are not.
OP_ENDIF	104	Ends an if/else block. All blocks must end, or the transaction is invalid. An OP_ENDIF without OP_IF earlier is also invalid.
OP_VERIFY	105	Marks transaction as invalid if top stack value is not true. The top stack value is removed.
OP_RETURN	106	Marks transaction as invalid. Since bitcoin 0.9, a standard way of attaching extra data to transactions is to add a zero-value output with a scriptPubKey consisting of OP_RETURN followed by data. Such outputs are provably unspendable and specially discarded from storage in the UTXO set, reducing their cost to the network. Since 0.12 , standard relay rules allow a single output with OP_RETURN, that contains any sequence of push statements (or OP_RESERVED [1]) after the OP_RETURN provided the total scriptPubKey length is at most 83 bytes.

Constants

WORD	OP CODE	EXPLANATION
OP_0, OP_FALSE	0	An empty array of bytes is pushed onto the stack. (This is not a no-op: an item is added to the stack.)
OP_PUSHDATA1	76	The next byte contains the number of bytes to be pushed onto the stack.
OP_PUSHDATA2	77	The next two bytes contain the number of bytes to be pushed onto the stack in little endian order.
OP_PUSHDATA4	78	The next four bytes contain the number of bytes to be pushed onto the stack in little endian order.
OP_1NEGATE	79	The number -1 is pushed onto the stack.
OP_1, OP_TRUE	81	The number 1 is pushed onto the stack.
OP_2-OP_16	82-96	The number in the word name (2-16) is pushed onto the stack.

Stack related instructions

WORD	OP CODE	EXPLANATION
OP_TOALTSTACK	107	Puts the input onto the top of the alt stack. Removes it from the main stack.
OP_FROMALTSTACK	108	Puts the input onto the top of the main stack. Removes it from the alt stack.
OP_IFDUP	115	If the top stack value is not 0, duplicate it.
OP_DEPTH	116	Puts the number of stack items onto the stack.
OP_DROP	117	Removes the top stack item.
OP_DUP	118	Duplicates the top stack item.
OP_NIP	119	Removes the second-to-top stack item.
OP_OVER	120	Copies the second-to-top stack item to the top.
OP_PICK	121	The item n back in the stack is copied to the top.
OP_ROLL	122	The item n back in the stack is moved to the top.
OP_ROT	123	The top three items on the stack are rotated to the left.
OP_SWAP	124	The top two items on the stack are swapped.
OP_TUCK	125	The item at the top of the stack is copied and inserted before the second-to-top item.
OP_2DROP	109	Removes the top two stack items.
OP_2DUP	110	Duplicates the top two stack items.
OP_3DUP	111	Duplicates the top three stack items.
OP_2OVER	112	Copies the pair of items two spaces back in the stack to the front.
OP_2ROT	113	The fifth and sixth items back are moved to the top of the stack.
OP_2SWAP	114	Swaps the top two pairs of items.

Crypto related opcode instructions

WORD	OP CODE	EXPLANATION
OP_RIPEMD160	166	The input is hashed using RIPEMD-160.
OP_SHA1	167	The input is hashed using SHA-1.
OP_SHA256	168	The input is hashed using SHA-256.
OP_HASH160	169	The input is hashed twice: first with SHA-256 and then with RIPEMD-160.
OP_HASH256	170	The input is hashed two times with SHA-256.
OP_CODESEPARATOR	171	All of the signature checking words will only match signatures to the data after the most recently-executed OP_CODESEPARATOR.
OP_CHECKSIG	172	The entire transaction's outputs, inputs, and script (from the most recently-executed OP_CODESEPARATOR to the end) are hashed. The signature used by OP_CHECKSIG must be a valid signature for this hash and public key. If it is, 1 is returned, 0 otherwise.
OP_CHECKSIGVERIFY	173	Same as OP_CHECKSIG, but OP_VERIFY is executed afterward.
OP_CHECKMULTISIG	174	Compares the first signature against each public key until it finds an ECDSA match. The process is repeated until all signatures have been checked or not enough public keys remain to produce a successful result. All signatures need to match a public key. Because public keys are not checked again if they fail any signature comparison, signatures must be placed in the scriptSig using the same order as their corresponding public keys were placed in the scriptPubKey or redeemScript. If all signatures are valid, 1 is returned, 0 otherwise. Due to a bug, one extra unused value is removed from the stack.
OP_CHECKMULTISIGVERIFY	175	Same as OP_CHECKMULTISIG, but OP_VERIFY is executed afterward.

Arithmetic related opcode instructions

WORD	OP CODE	EXPLANATION
OP_1 ADD	139	1 is added to the input
OP_1 SUB	140	1 is subtracted from the input
OP_ADD	147	a is added to b.
OP_SUB	148	b is subtracted from a
OP_BOOLAND	154	If both a and b are not "" (null string), the output is 1. Otherwise 0.
OP_BOOLOR	155	If a or b is not "" (null string), the output is 1. Otherwise 0.
OP_NUMEQUAL	156	Returns 1 if the numbers are equal, 0 otherwise.
OP_NUMEQUALVERIFY	157	Same as OP_NUMEQUAL, but runs OP_VERIFY afterward.
OP_NUMNOTEQUAL	158	Returns 1 if the numbers are not equal, 0 otherwise.
OP_LESSTHAN	159	Returns 1 if a is less than b, 0 otherwise.
OP_GREATERTHAN	160	Returns 1 if a is greater than b, 0 otherwise.
OP_LESSTHANOREQUAL	161	Returns 1 if a is less than or equal to b, 0 otherwise.
OP_GREATERTHANOREQUAL	162	Returns 1 if a is greater than or equal to b, 0 otherwise.
OP_MIN	163	Returns the smaller of a and b.
OP_MAX	164	Returns the larger of a and b.
OP_WITHIN	165	Returns 1 if x is within the specified range (left-inclusive), 0

Bitwise logic

WORD	OP CODE	EXPLANATION
OP_EQUAL	135	Returns 1 if the inputs are exactly equal, 0 otherwise.
OP_EQUALVERIFY	136	Same as OP_EQUAL, but runs OP_VERIFY afterward.

Pseudo words

WORD	OP CODE	EXPLANATION
P_PUBKEYHASH	253	Represents a public key hashed with OP_HASH160.
OP_PUBKEY	254	Represents a public key compatible with OP_CHECKSIG.
OP_INVALIDOPCODE	255	Matches any opcode that is not yet assigned.

Reserved words

WORD	OP CODE	EXPLANATION
P_RESERVED	80	Transaction is invalid unless occuring in an unexecuted OP_IF branch
OP_VER	98	Transaction is invalid unless occuring in an unexecuted OP_IF branch
OP_VERIF	101	Transaction is invalid even when occuring in an unexecuted OP_IF branch
OP_VERNOTIF	102	Transaction is invalid even when occuring in an unexecuted OP_IF branch
OP_RESERVED1	137	Transaction is invalid unless occuring in an unexecuted OP_IF branch
OP_RESERVED2	138	Transaction is invalid unless occuring in an unexecuted OP_IF branch
OP_NOP1, OP_NOP4-OP_NOP10	176, 179-185	The word is ignored. Does not mark transaction as invalid.

LOCK TIME: This attribute is one of the most used in new bitcoin platforms where settlement is involved. LOCK TIME is added in the metadata. LOCK TIME implies a payment is locked until a particular time: no miner will validate the transaction.

MULTISIG: This is commonly used in escrow transactions done using bitcoin, as financial transactions are getting refined. A transaction with MULTISIG requests for approval from two or three third parties.

4.4.7 Transaction in the Bitcoin Network

Bitcoin network is an open network; any node within the network can submit a transaction to the network. Like all transactions, a bitcoin transaction needs to be recorded in a ledger. Since it is a public network, a ledger needs to be shared with all nodes. Hence, there will be a large number of nodes (computers) having a copy of the ledger. This feature distinguishes the transaction block from other computer record-keeping systems as here one cannot fudge a record; any attempt to fudge would necessitate making a change in all the nodes where the transaction was written.

A transaction consists of the Version number, Inputs, Outputs and Lock time as shown in Fig. 4.36

Outputs of the previous transaction will be the inputs for the next transaction as shown below. Transaction 0's outputs are the inputs for the Transaction 1. Outputs indicate Unspent Transaction Output (UTXO) from previous transactions (refer Fig. 4.37).

Figure 4.36: A simple transaction block

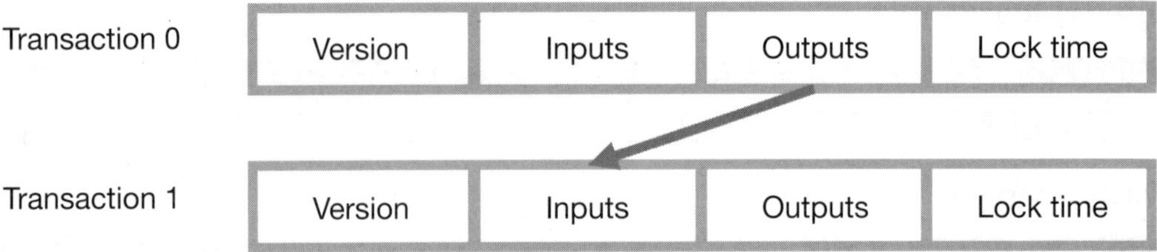

Figure 4.37: A simple transaction block: Outputs as Inputs

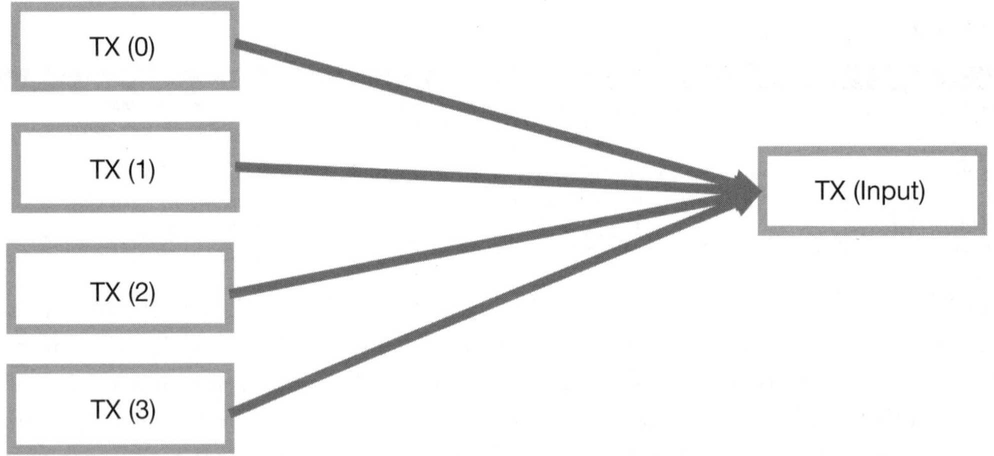

Figure 4.38: TX(INPUT) Creation from previous transactions.

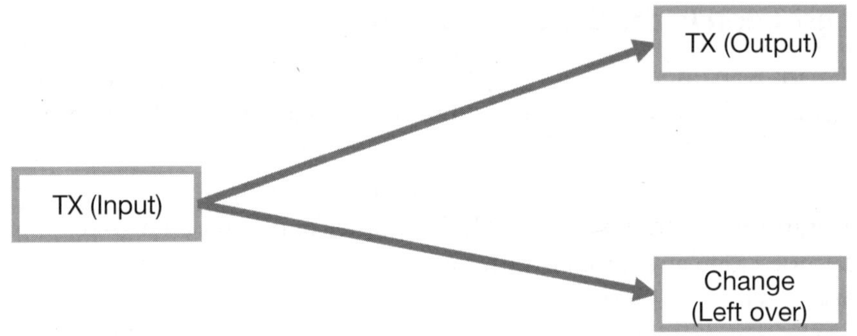

Figure 4.39: A simple transaction block: INPUTS, OUTPUTS, and CHANGE

So, suppose Anan (sender) needs to pull out bitcoins from the following four transactions, namely: TX (0), TX (1), TX (2) and TX (3). These four transactions will be added/combined, yielding the input transaction represented as TX (Input): refer Fig. 4.38.

In our example: TX (INPUT) = TX (0) + TX (1) + TX (2) + TX (3)

Now, let us look into the output, which is basically the number of bitcoins that Babi will possess, post-transaction (after the transaction) and any remaining change is then sent back to Anan This change then becomes his (Anan's) input value for all his future transactions (refer Fig. 4.39).

This is how a simple transaction, with single output (TX (OUTPUT)), looks like. A transaction may also have multiple outputs.

Note: Each transaction (TX (INPUT) and TX (OUTPUT)) will consist of a Version number, Inputs, Outputs, and Lock time.

For a valid transaction, TX (INPUT) > TX (OUTPUT)

That means Anan can send only lesser money to Babi that what he already has.

For example, if Anan in total has 500 $, he can only send money, which is lesser than 500$ in total.

Note: A transaction includes Transaction cost (Transaction fees) and hence the greater than condition. We are not using ≥ symbol in the equation.

Transaction fees (bitcoin) = TX (INPUT) – (TX (OUTPUT) + Change (left over)

In our example: TX (INPUT) = TX (0) + TX (1) + TX (2) + TX (3)

If TX (INPUT) < TX (OUTPUT) it would reject the transaction.

Transactions and UTXOs

UTXOs stand for "Unspent Transaction Output" (refer Fig. 4.40). Let us take an example. Let us assume that the below four transactions happened from Murray, Anan, Babi and Karin to Monika. They are called as "unspent Transaction output" of Monika.

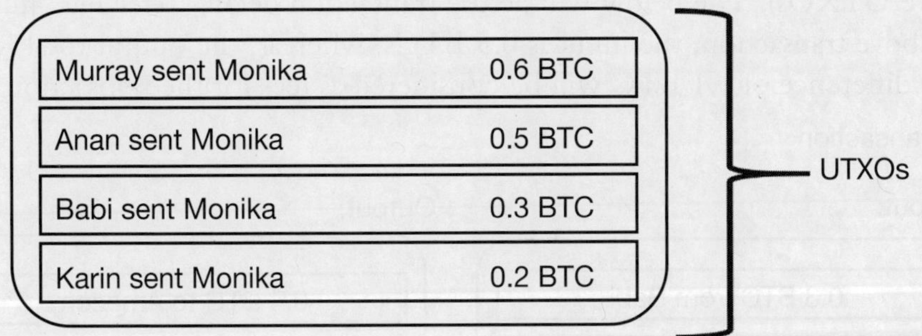

Figure 4.40: UTXOs

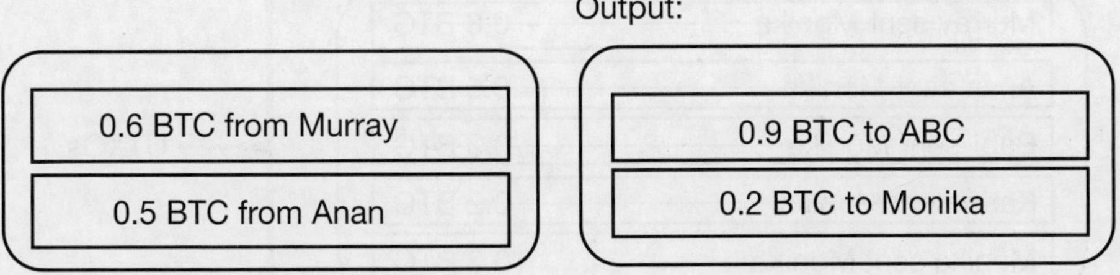

Figure 4.41: UTXOs and Transaction

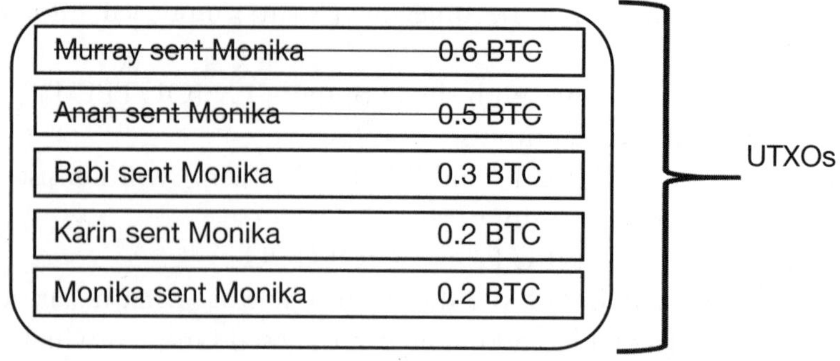

Figure 4.42: UTXOs and Transaction 1

Let us assume now, Monika decided to buy a small car for 0.9 BTC from a vendor called ABC Motors. The transaction will be as given above (refer Fig. 4.41).

In any transaction, the input BTCs will be equal to output BTCs. Any difference between the output and the input will be taken as "Fees" for the transaction. In the above case, 1.1 BTCs are there in both the input and output. And so, Fees is considered as 0 BTC. The resultant UTXOs of Monika will appear as in Fig 4.42).

How the Fees Are Calculated for a Transaction

Let us assume that Monika decided to buy some goods on Amazon for 0.4 BTC (based on the above UTXOs). The below will be the transaction details (refer Fig. 4.43).

In the above transaction, the input is 0.5 BTCs, whereas, the output total is only 0.4 BTCs. The difference of 0.1 BTC will be considered as fees for the transaction.

Transaction

Input:

0.3 BTC from Babi

0.2 BTC from Karin

Output:

0.4 BTC to Amazon

Figure 4.43: UTXOs Transaction 2

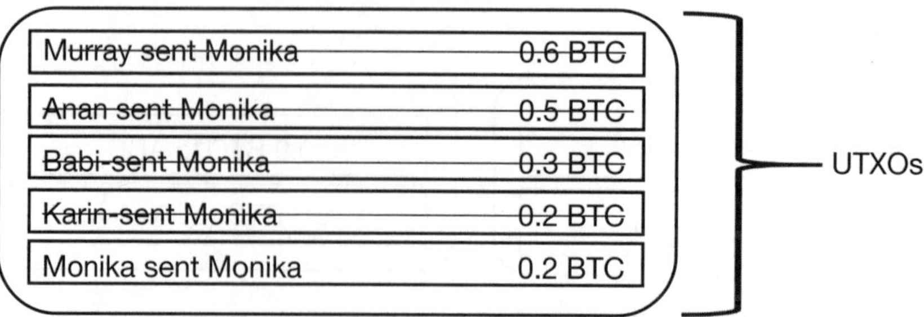

Figure 4.44 UTXOs and Transaction 3

As we had seen earlier, the selection of transactions into the blockchain depends on the fees. In other words, miners will select the transaction with the highest fees to be added into the blockchain. The resultant UTXOs of Monika, after the above transaction, will appear as shown (refer Fig. 4.44).

The wallet of Monika will have 0.2 BTC as per the above picture.

The Public Key and Private Key Check

Let us assume that the following four transactions happened (refer Fig. 4.45) from Murray, Anan, Babi, and Karin to Monika. They are called as "unspent Transaction output" of Monika.

Let us assume now, Monika decided to buy a small car for 0.9 BTC from a vendor called ABC Motor. The transaction can be shown as below (refer Fig. 4.46).

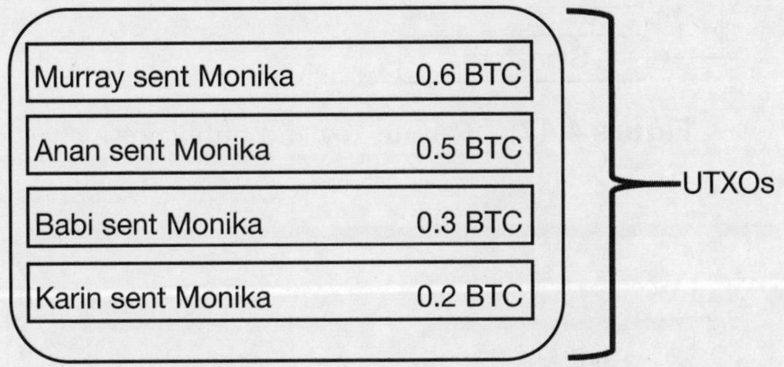

Figure 4.45: UTXOs and Transaction 4

Transaction

Input:

0.6 BTC from Murray

0.5 BTC from Anan

Output:

0.9 BTC to ABC

0.2 BTC to Monika

Figure 4.46: UTXOs and Transaction 5

If the transactions are maintained in the above format, then it can be easily hacked by a third party. It clearly reveals the identity of the sender and receiver apart from details of the amount involved in the transaction. There is no privacy. A signature, consisting of Private key and Public key, comes handy in the above situation. A private key is a unique identifier, which is assigned to each node. This will be kept private, as the name suggests, and is like a password to one's bank account. If one needs to send money to another person on the other hand, the public key (refer Fig. 4.47) of the concerned person is shared (the bank account number, in this case). A message will always go with the signature of the sender (which is produced using the sender's private key).

After adding the signature, the public key (of the receiver) determines to whom the money is sent (refer Fig. 4.48).

How can the receiver check whether the sender sends a particular transaction? In cryptocurrency, there is a verification function which takes the signature and the public key as input message to check whether or not a particular transaction has been sent by a person. The output of the verification function is a YES/NO Value (refer Fig. 4.49).

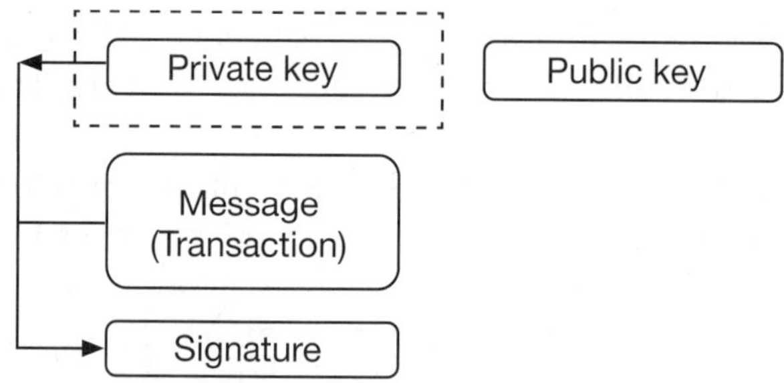

Figure 4.47: Private key and public key

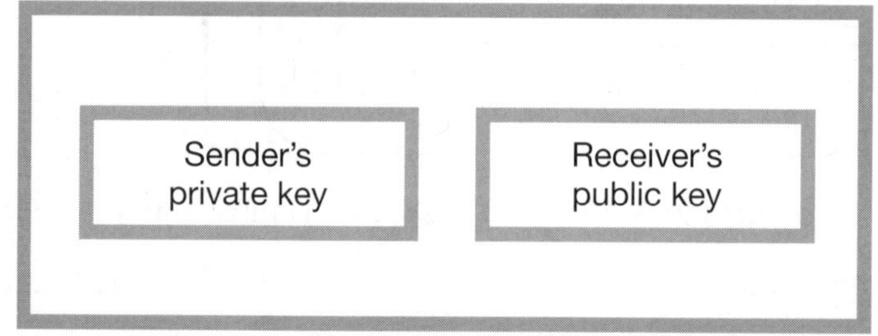

Figure 4.48: Sender's private key and receiver's public key

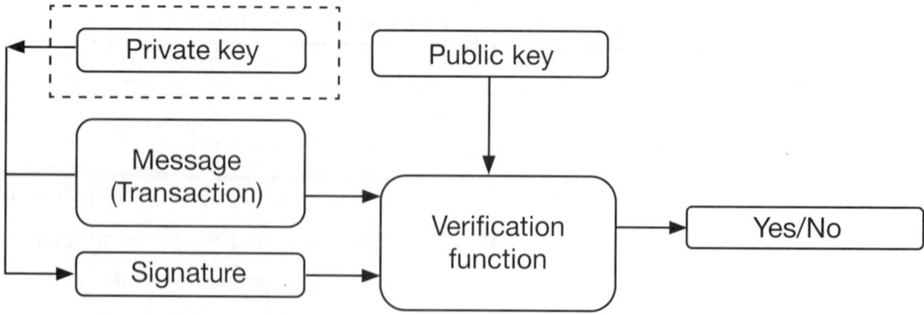

Figure 4.49: Verification function

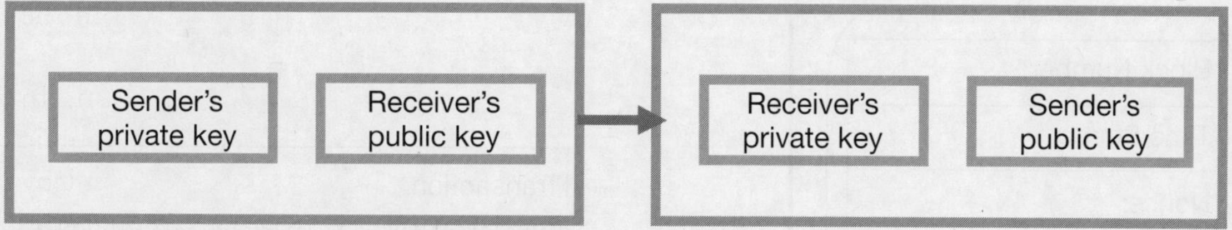

Figure 4.50: Public and private key

The input and the output data are hashed together using the SHA 256 algorithm, and the resultant is called as "output hash".

In another case, let us say Anan is trying to send 100 $ to Babi for a specific work. Anan needs to verify if he has the necessary permission to send the bitcoin. The way he (Anan) does that is by signing off his transaction with his digital signature (i.e., his private key). Anybody can decode this by using his (Anan's) public key to check whether Anan (refer Fig. 4. 49) sends the transaction. The public key is known to everyone. However, the private key is confidential and known only to the concerned person (refer Fig. 4.50). This is called signature data. Anan will lock the transaction using Babi's public key.

Note: Only Babi, using his private key, can unlock this transaction. P2PKH (The pay to a public key hash) can be used to check the transaction.

An output has an inferred index number (signifying the location in the transaction). It also has a dollar equivalent in USD (bitcoin value) along with the public key script. Inputs will have Signature Script (which is used for verification). Transactions mined are kept in a block. Among other things, a block would have a nonce (explained later in this chapter) and the miner hopes the other nodes will accept his block as part of block propagation (explained in the next section of this chapter). Data in bitcoin can be written only once to a block; it cannot be changed or deleted. A node in the network submits a transaction: e.g., a payment towards another node in the bitcoin network. The node propagates the transaction to all nodes that it is paired with and the transaction moves via the gossip protocol. Each node would run a check to verify if the transaction needs to be accepted or not, and the acceptance will be propagated to other nodes.

Segregated Witness (SEGWIT)

The size of the block in any cryptocurrency is limited. In the case of bitcoin, it is limited to 1 MB size, which enables the bitcoin to do seven to ten transactions per second (on average). This is the optimal size of a single block based on the original design. If the block size is kept smaller, then people will be waiting at the transaction stage for a longer period. If the block is size is kept higher, the network bandwidth may become an issue making the data vulnerable to attackers.

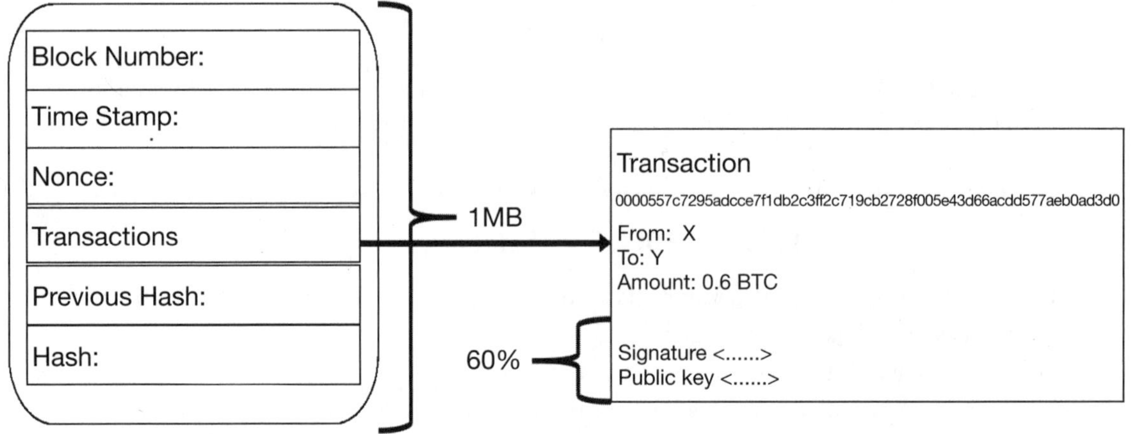

Figure 4.51: Transaction within a block

A block consists of block number, time stamp, Nonce, transactions, previous hash, and the current hash. In addition, each transaction consists of the message (actual transaction) along with the signature and the public key. The size of the signature and public key is so large within the transaction that it occupies more than 60 % of its overall size (refer Fig. 4.51). In the bitcoin protocol, this portion (signature and public key) is kept out of the block and is distributed separately. This segregation of the signature from the block is called as Segregated Witness (SegWit). Because of the removal of SegWit, more (almost double) transactions can be added into the block. SegWit will be sent separately in the network.

Public Key vs. Bitcoin Address

Bitcoin address provides one more security layer on top of the public key using the SHA256 algorithm. It is not that the public key is kept a secret like a private key. However, instead of the public key, the bitcoin address is exposed to all to send the money (bitcoin). As we already know, it is very difficult to get the public key from a bitcoin address even though the address corresponds to the public key (refer Fig. 4.52).

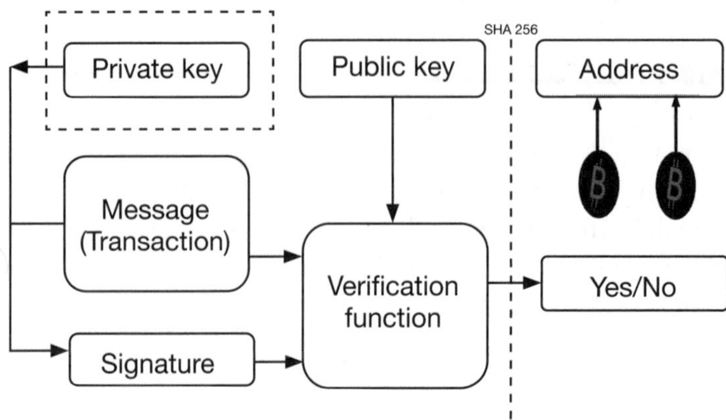

Figure 4.52: Address of a bitcoin

Hierarchically Deterministic Wallet (HD Wallet)

If more privacy is desired, people may create more private keys for the transaction (refer Fig. 4.53).

In such cases, it is tedious to keep track of the private keys securely. Hence, Bitcoin introduced Bitcoin Improvement Plan, BIP 32, which proposed HD Wallet. In this, a master key is generated, which can be used to create more private keys (refer Fig. 4.54).

This helps to track the wallet hierarchically and hence it is called as HD Wallet. This HD wallet can create private key dynamically, thereby making it very difficult for someone to look into the transaction to identify the owner. This is also very helpful for an organization. It can use the master private key to generate private keys for individual departments and use it to track the overall transactions.

Transaction details of bitcoin can be checked in:

1. Bitcoin Explorer: https://explorer.bitcoin.com/btc
2. Bitcoin block chair: https://blockchair.com/bitcoin/

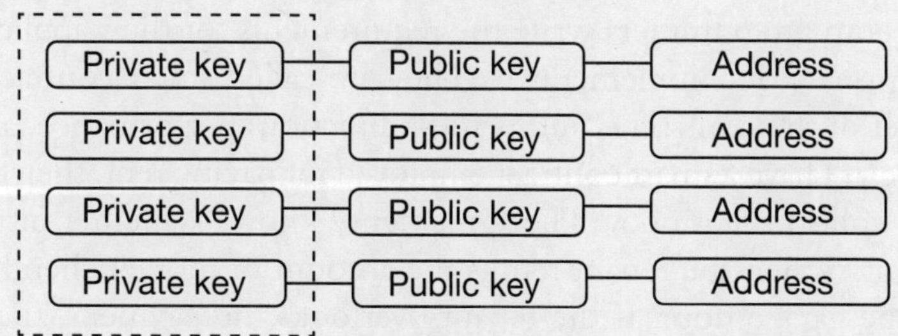

Figure 4.53: Group of private keys

Figure 4.54: Master private key

4.5 Ethereum Blockchain

Ethereum could be a do-it-yourself platform for most decentralized programs. It is also referred to as DApps - de-centralized programs. If you would like to produce a brand new application that no person controls (perhaps not you personally though you may have written it), then all you need to learn the Ethereum programming language named Solidity.

The Ethereum platform includes 1000s of computers, meaning it is fully decentralized. Once a program is set up in the Ethereum system, the computers, also called nodes, are likely to make sure it implements as composed. Ethereum could be your infrastructure for executing DApps global. It may not be just a currency; it is a platform.

People can purchase cryptocurrencies from others without needing a market, which may be subject to unethical hacking. Ethereum's programming terminology, Solidity, can be utilized to publish "Smart Contracts," which can be used as the logic to execute DApps.

4.5.1 Introduction to Smart Contracts

In real life, a contract is a set of "Ifs" and "Then." In other words, it is a collection of conditions and actions. For example, if I pay my property owner $1,500 on the first of the month, then he lets me use his apartment. That is how smart contracts work with Ethereum. Ethereum programmers write the requirements for their application or DApp, and the Ethereum network implements it. They are called smart contracts because they handle every facet of the contract – authorities, direction, performance, and payment.

As an instance, if I have a smart contract employed for paying rent, then the homeowner does not have to collect the money. The agreement "knows" whether or not the amount of money was sent. When the tenant remits the amount of money, then he will have the ability to open the garage door. If the tenant overlooks the payment, then he would be locked out of his door. Thus "smart" contract can be used to lock a non-paying tenant out of the flat as opposed to physically evicting him. It could also be made to consider other factors like extenuating circumstances, the place where the arrangement was written and create exceptions when justified. To put it differently, it might function as a smart judge.

A "smart contract" from the circumstance of Ethereum is not intelligent. Its programming is uncompromisingly strict. It follows the rules to the letter and cannot make secondary considerations to uphold the "spirit" of the law, similar to what occurs typically in real-life contracts. Once a smart contract has been set up on the Ethereum system, it cannot be corrected or edited, even by its first author. It is immutable. The only means to improve this arrangement is to convince the full Ethereum system; a change needs to be produced, and that is almost impossible. This brings up a severe problem as, unlike Bitcoin, Ethereum was assembled with all the capability to come up with complex contracts that are incredibly tricky to secure.

The harder a contract, the tougher it would be to enforce it as there would be more room for interpretations, or longer exemptions may have to be written to manage contingencies. With smart contracts, the security methods are designed to deal with these with perfect accuracy; every feasible term in the contract can be implemented in just the way the author (coder) had planned it.

4.5.2 Code Is Law and Ethereum Classic

Ethereum started with the notion that "code is a law." In other words, a contract on Ethereum could be your ultimate authority, and nobody can overrule the contract. "Dow" or DAO means "Decentralized Autonomous Organization," enables users to deposit funds and receive returns dependent on the DAO investments.

A DAO is an electronic digital organization that works with no hierarchical direction; it functions in a decentralized and democratic way. Hence, the decisions of the DAO are not in the control of a centralized authority but instead at the command of certain designated authorities or some set of chosen people who are part of a body. It is based on a blockchain system, where it is regulated by the protocols inserted inside a smart contract, and consequently, DAOs count on smart contracts for decision-making. The conclusions themselves are crowd-sourced as well as decentralized. The creation of the DAO was an unexpected success since it managed to raise 12.7 M ether (worth around $150 M at that time). Ether was trading around $20 and the total ether from the DAO was over $ 250 M. However, the code was not secured nicely and someone succeeded in finding a loophole to drain the DAO. Today one might state that the man or woman who drained the DAO was a "hacker." But some might assert that he was someone who was taking good advantage of those loopholes of the DAO's Smart Contract.

Ethereum split into ETH and ETC after the DAO Attack. As discussed earlier, the problem was with DAO code and not in the Ethereum network. In July 2016, Ethereum introduced a new rule to reverse the logic at block number 1919999, which previously allowed the hacker to steal Ether from DAO. They reversed the funds and returned them to the people who had invested. The Ethereum chain that continued to function after this was called hard fork. This is the rule for hard forks: They loosen up rules. A majority of the miners switched to ETH.

At the same time, some miners who were unhappy with the changes wanted to follow the old rule. They were not ready to change the software and retained the old software. They said it is very sad that the money was stolen, but at the same time wanted to stick to the old rules that were formulated. These miners continued mining the old way, they did not upgrade. This old chain of series is called as Soft Fork (or) Ethereum Classic. This is the rule for soft forks: They tighten up rules.

4.5.3 Ethereum Components

Ethereum blockchain is a design consisting of multiple components, and these components interact and function with each other. Some of the critical Ethereum components are

- o Miner and mining node
- o Ethereum virtual machine
- o Ether
- o Gas
- o Transactions
- o Accounts
- o Swarm and whisper
- o Ethash

4.5.3.1 Miner and Mining Node in Ethereum Network

A miner is responsible for writing a transaction to the Ethereum series. A miner's job is very much like that of an accountant. For example, an accountant is in charge of both writing and taking care of the ledger; likewise, a miner is solely accountable for writing transactions to the Ethereum ledger. A miner is enthusiastic about adding paper transactions to the ledger due to the reward associated with it. Miners get two kinds of rewards: benefit for writing a block into the series and accumulative gas prices (this is a unique concept in Ethereum blockchain and is not available in the Bitcoin blockchain) from all transaction in the block (cube).

Every miner needs to have a backup of these blockchain transactions, and that includes features such as, code (this is a unique concept in Ethereum blockchain and is not available in Bitcoin blockchain) and other things that are within those transactions. Data is not stored directly in the blockchain because it is too big. Let us say we are storing movies in a blockchain; these are big files and occupy excessively huge memory. No one will want to download all movies to watch a single movie. To fix this issue, the data in stored in some distributed hash table, like IPFs (Interplanetary File System) (refer Figs 4.55 and 4.56).

The below diagram explains how typical IPFs work. Let us look into how a PDF is stored in an IPF. IPF gives back the address where the PDF is getting stored, in the form of a hash number.

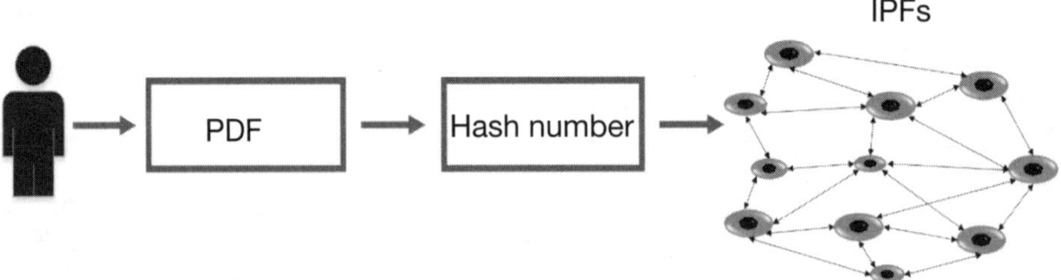

Figure 4.55: PDF to IPFs

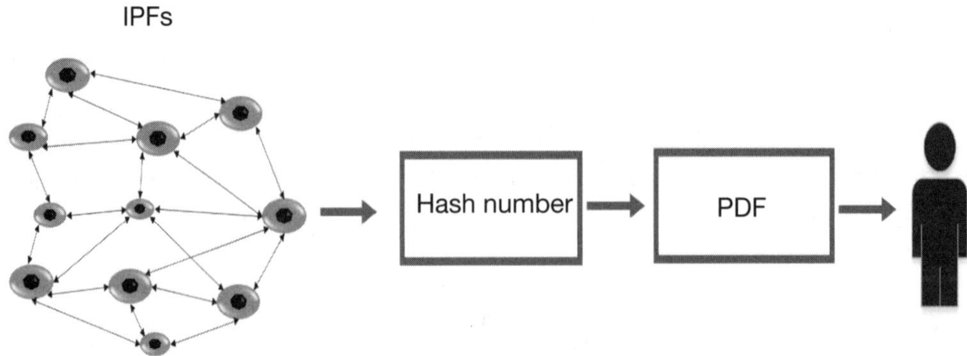

Figure 4.56: IPFs to PDF

IPFs give back a hash or content address for all of the content and help to retrieve the file easily from the data storage system. Thus, hash points of the data are stored in the blockchain.

There are many miners available in a blockchain network, each competing and trying to write transactions. Nevertheless, just a single miner can write the block into the ledger (after solving the problem given); the remaining miners may be unable to write in the existing block, and the conclusion of a miner who will eventually compose the block is a challenge. The problem is awarded to each node and every miner attempts to resolve the same problem with the computing power available to him/her. Finally, the miner who solves the problem is allowed to write the block comprising transactions to the ledger and receives five ethers as a reward.

There are two types of nodes (computers) in the Ethereum network.

1. Mining nodes and
2. Ethereum virtual machines.

These types were designed to clarify the concepts of Ethereum. Generally, in the majority of cases, you can find no devoted EVM nodes. Instead, all nodes will function as a miner in addition to EVM node. Mining nodes refer to the nodes that belong to miners. Figure 4.57 depicts the difference between the mining node and EVM. EVM has a unique code attached to it, which is called as smart contract.

Mining nodes are part of the same network where EVM is hosted. Eventually, the miners develop a brand new block and amass all of the transactions out of the transaction pool and add them to the recently constructed block. This new block is then inserted into the chain series. Each mining node maintains its version of Ethereum ledger, and the ledgers could be identical. It is the job of the miners to ensure that their ledger is always upgraded with the latest blocks.

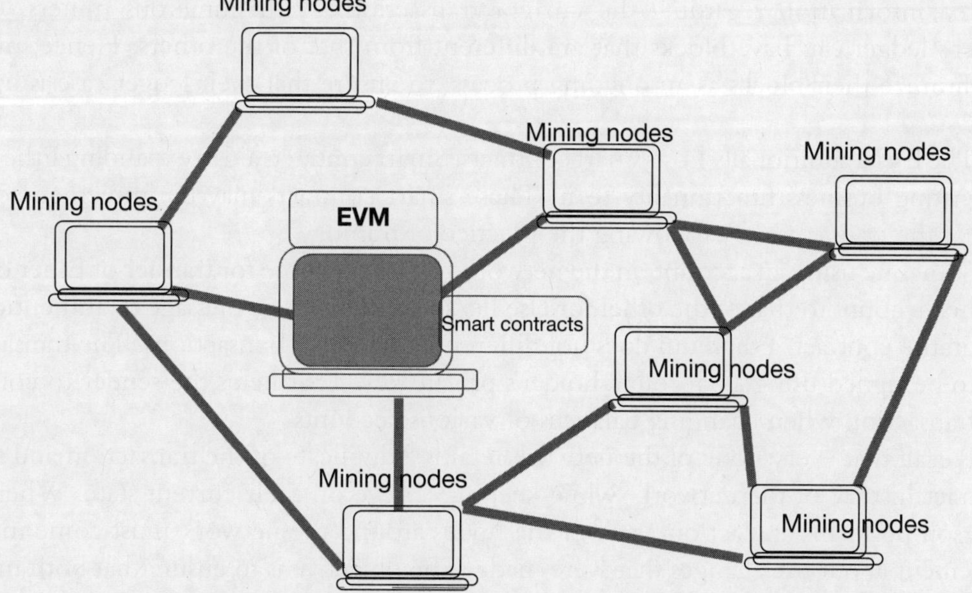

Figure 4.57: Mining nodes and EVM network

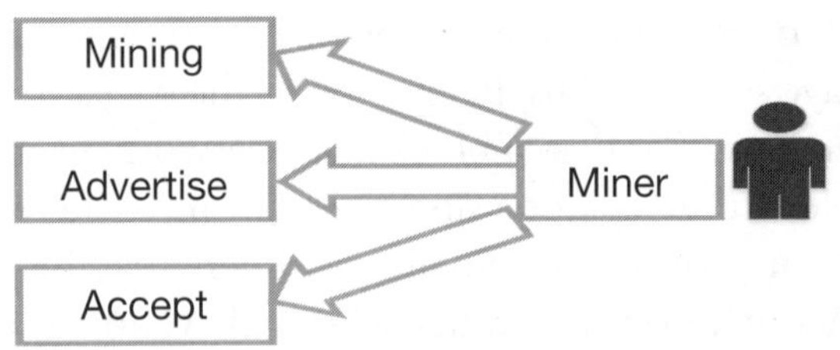

Figure 4.58: Functions performed by a miner.

There are three fundamental functions (refer Fig. 4.58) performed by miners on mining nodes.

1. Mine or make a fresh new block together with the transaction and then compose the same into Ethereum Ledger
2. Promote and send a recently mined block to other miners
3. Accept new blocks created by other miners and maintain them in the ledger.

4.5.3.2 Ethereum Virtual Machine (EVM)

A blockchain system is made of numerous nodes belonging to both miners and also a few who do not want to mine but function as aids for the implementation of smart contracts. These nodes are called Ethereum Virtual Machines (EVM). Each node is connected with different nodes on the network system and work on a peer-to-peer protocol to communicate with one another. Also, by default, the nodes utilize the 30303-port number to communicate among themselves.

EVM can be considered as the implementation run-time of the Ethereum community. EVMs are mostly in charge of supplying a run time that could perform code compiled from smart contracts. It does not need access to the ledger; however, it contains only limited information regarding the current transaction. With numerous miners, every miner's ledger can have blocks that are different from that of the others. Hence, miners synchronize their blocks as an ongoing process, to ensure that their ledger case is up-to-date.

The EVM additionally hosts smart contracts. Smart contracts aid in expanding Ethereum by writing business functionality to it. These smart contracts may be implemented as a part of the transaction by following the practice of mining.

Someone using an account on the network may send a note for transfer of Ether out of his/her account to that of the other, or else he/she could place a message to authenticate a role in the contract. Ethereum does not differentiate them as transactions. The transaction has to be signed using an account-holder's private key. That helps the sender to confirm the transaction when changing balances of various accounts.

Recall that every node of the network retains a duplicate of the transaction and smart contract history of the network, while keeping an eye on their current state. Whenever a person performs any action, each of the nodes around the network must come into an agreement about the changes that were made. The objective is to ensure that both miners and nodes take responsibility and are directly accountable for changing the transaction states instead of relying on some third party such as pay pal or a financial institution. The EVM implements a contract together with all these programming rules that are initially written by the programmer.

Genuine computation about the EVM is accomplished via a stack-based byte-code terminology (ones and zeroes a machine may read). However, programmers may write smart contracts using high-tech languages like Solidity. They may also use Serpent, which is less complicated for individual programmers to write and read.

4.5.3.3 Ether

Ether is a type of cryptocurrency used in Ethereum. Figure 4.59 shows the symbol of Ether.

Ethereum is a massive group of computers that work with each other just like a super-computer used in coding, and they power DApps. However, this costs money: Money to get the machines, to power them, save them, and cool them whenever essential. That is precisely why Ether was invented. The price of Ethereum is the Ether. It incentivizes people to follow the Ethereum protocol in their PC. The incentive is similar to the manner in which Bitcoin miners are paid for keeping up the Bitcoin blockchain. To set up a smart contract using the Ethereum platform, the contract author needs to pay in the form of ether. That is done so people might write optimized codes that do not waste the Ethereum network's computing ability on unnecessary tasks.

Ether was distributed in Ethereum's authentic Initial Coin Offering (ICO) in 2014. It charged approximately 40 cents to buy one Ether. Now, one Ether equals hundreds of dollars, considering that usage of the Ethereum network has increased tremendously.

The reference code of Ether is ETH, and as of June 2019, one ETH is comparable to USD 269.55. Every task onto Ethereum that modifies its state costs Ether. Miners that are successful in writing and generating a new block at the chain series are rewarded Ether. Ether can readily be transformed into dollars or alternative conventional currencies through crypto-exchanges. Ethereum comes with a metric system of denominations. The smallest denomination (also known as foundation component of ether) is popularly named Wei. Table 4.2 gives a list of these termed denominations, along with their value in Wei.

Figure 4.59: Ethereum (Ether) symbol

Table 4.2 Units of Wei

Wei	Unit	Ether
Wei	1 Wei	10^{-18} ETH
Babbage	1,000 Wei	10^{-15} ETH
Lovelace	1,000,000 Wei	10^{-12} ETH
Shannon	109 Wei	10^{-9} ETH
Szabo	10^{12} Wei	10^{-6} ETH
Finny	10^{15} Wei	10^{-3} ETH
Ether	10^{18} Wei	1 ETH

Growth of Ether

Ethereum is the second-biggest cryptocurrency behind Bitcoin. Vitalik Buterin and a few others launched Ethereum in the year 2015. It started its momentum of monster growth from 2017. The growth of Ether is depicted in the below table.

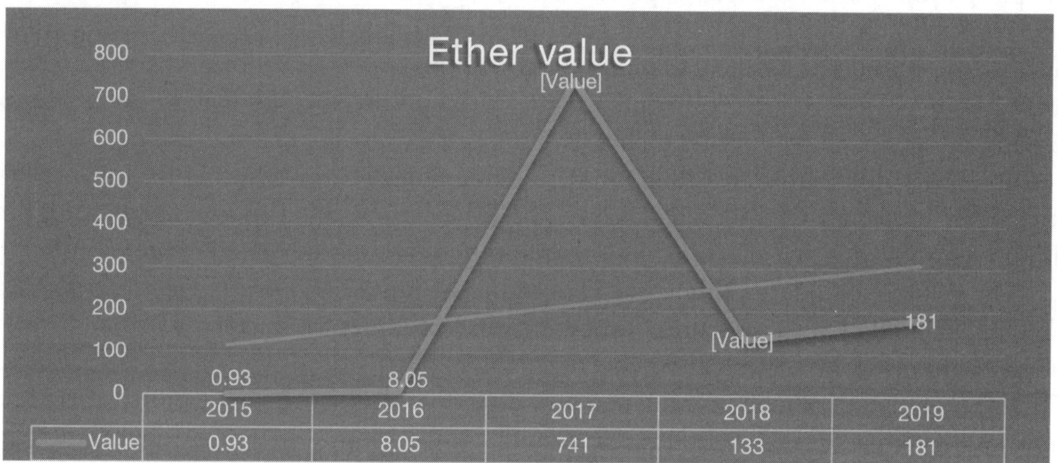

Figure 4.60: Ether value over the years

Figure 4.60 shows the equivalence of a single ether to USD over a period starting 2015. In 2015 (in the starting period), the value of one ether was equal to 0.93 USD, and it grows steadily from there. It touched the peak value of 741 USD in the year 2017. In 2019, the value of one ether was equivalent to 181USD (approximately in June 2019).

4.5.3.4 Gas

Ether is paid as commission for any execution that affects the state in Ethereum. Ether (ETH) is traded on people exchanges and its particular selling price value changes each day. Its value could be high on certain days, when it is used for payment and lower on other days. Individuals may wait for the price of Ether to fall to perform their transactions. This condition is not ideal for the Ethereum platform.

The gas helps in alleviating this problem. Gas is the internal currency of Ethereum. The execution and resource utilization costs are predetermined in Ethereum in terms of Gas units. This is also known as Gas Cost.

Additionally, a gas price may be corrected to lower the costs whenever the amount tag on Ether increases and increase the price as soon as the amount tag on Ether decreases. For instance, to set up a function at a transaction that simplifies a series will probably cost predetermined gas, and consumers need to pay in the denomination of gas to guarantee implementation of the transaction.

4.5.3.5 Transactions in Ethereum

We know that Blocks are related and linked to each other. Let us now examine how transactions are connected to Blocks. Ethereum stores transactions within Blocks. A transaction (Contract) is simply a set of agreements between parties; there would be an exchange of assets, products, or services in place of currency, cryptocurrency, or some other asset either in the present or in the future. Ethereum helps in executing a transaction.

4.5.3.6 Ethereum Accounts

Accounts are the primary building block for the Ethereum ecosystem. Ethereum supports two kinds of accounts. Every account comes with a balance that yields the present value saved in it. Externally owned accounts are possessed by men and women around Ethereum.

Accounts are perhaps not known by names within Ethereum. Once somebody creates an externally owned account on Ethereum, a public-private key is produced. The private key is maintained safe with the individual, whereas the public key gets to be the identity of the externally owned account. The public key consists of 256 characters. However, Ethereum uses the initial 160 characters to represent the identity of an externally owned account. Private Key is a unique identifier, which is assigned to each node. This will be kept private as the name suggests and is like a password to one's bank account. The public key of the beneficiary is required for sending money to him/her (it is like an account number). When a transaction is made using the public key, a message always goes with the signature of the sender (which is produced using the sender's private key).

In case John creates an account on the Ethereum network, he will possess a private key for himself. The first 160 characters of this public key are now his identity. Other accounts on the network may subsequently send ether or alternative cryptocurrencies to this account. An externally owned account may keep Ether in its balance and not have any code related to it. Such an account can transact with other externally owned accounts by invoking functions (refer Fig. 4.61) within smart contracts. Normal contract accounts are similar to externally owned accounts. They are identified by the public address and do not hold any private keys. They can carry ether very similar to externally owned accounts. However, they have a programming code for smart contracts comprising of state variables and functions.

The account consists of four components (refer Fig. 4.62) which are present regardless of the type of account:

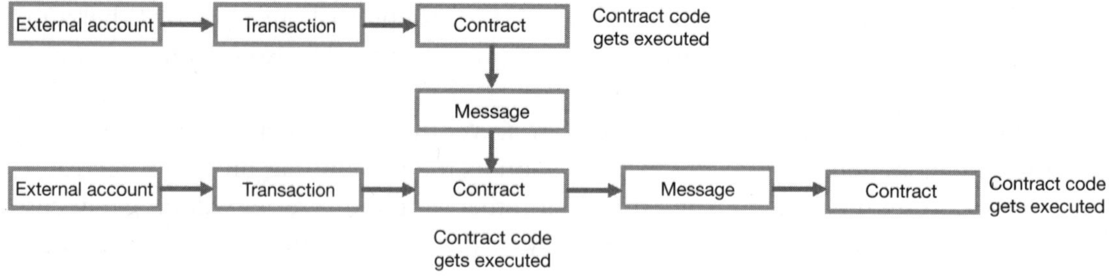

Figure 4.61: External account vs. Contract account

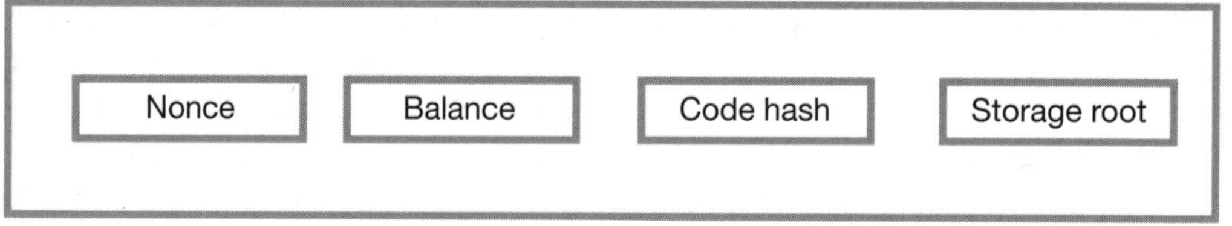

Figure 4.62: Components of accounts

Nonce: When the account is externally owned, this signifies the number of transactions delivered (sent) out of that account. In the case of the contract accounts, this means the number of contracts generated via this account.

Balance: The total balance of the accounts in Wei.

Storage root: A storage root has the main root node of the Merkle Patricia Tree. This tree encodes the hash of all the storage contents of the account, and it is empty by default.

Code hash: It is the hash code of the EVM code of the particular account. For contract accounts, this code is hashed and stored as the code hash. For externally owned accounts, the code hash field is not applicable. Code hash will have the empty string.

4.5.3.7 Swarm and Whisper

Personal computers, in general, calculate, store, and communicate data. For Ethereum to realize its vision of a censorship-resistant, self-sustaining, decentralized world, it must have these three things (Calculate, Store, and Communicate) efficiently. Swarm is a peer–to-peer document sharing platform, similar to Bit Torrent, incentivized with micro-payments of ETH. File records are divided into small chunks, dispersed, and stored with volunteers. The nodes which save and also function (calculate, store, and communicate) on those pieces are all paid with ETH from people using the information. This will be a storage point without even counting upon a central server. Whisper is an encrypted messaging protocol that enables nodes to send out messages directly to each other, in a secure way that additionally hides the sender and recipient identity from third-party snoopers.

4.5.3.8 Ethash

Ethash is a proof-of-work algorithm that is a modified variant of a precursor algorithm called Dagger-Hashimoto. Together with Ethash, the outcome in the hashing process has to result in a hash value that is below a particular threshold. This notion is called 'difficulty'. It involves the Ethereum system increasing or diminishing the threshold, as a way to control the speed and time at which blocks are mined on the Ethereum network. In case the rate at which blocks are found rises, then the system will automatically boost the difficulty, i.e., it will diminish the system threshold, thereby reducing the number of valid hashes with the capacity to be detected. Conversely, when the amount of exposed blocks declines, the system threshold increases to create a higher number of hash values, which are available. Using the Ethereum algorithm, the difficulty adjustment, typically one block, is produced by the system every 12 seconds.

A miner who discovers a block that can be inserted to the blockchain receives a block reward comprising of three ethers. All the gas spent on the block, to put it differently, all the gas that has been absorbed while implementing the transactions, are included within the block. This gas price is credited to the miner's accounts included as proof-of-work mining procedure — an additional reward for adding transactions as a portion of this block.

Ethash is predicated over a sizable, randomly created data collection termed Directed Acyclic Graph (DAG). The DAG is upgraded for every 30,000 blocks, and the current DAG dimension of Ethereum at the point of writing, is 2.84 GB. The DAG will be the last to rise while the blockchain grows.

Mining on Ethash involves drawing arbitrary data from the data set (DAG), calculating some randomly chosen transactions from some other block, and then returning the hash of this outcome. This usually means that a miner will soon be asked to save the full DAG as a way to fetch arbitrary data and calculate randomly chosen transactions. Because of this, the proof-of-work mining process around Ethereum is the one that will be distinguished to be memory-hard or even memory-bound. A memory-hard process identifies a scenario in the time required to finish a specified computational problem and is primarily determined by the total amount of memory necessary to carry data.

It follows that most of the miners' attempts will probably be focused on scanning the DAG rather than calculating data pulled out of this. This is designed to produce Ethash ASIC (application-specific built-in circuit) resistance, because of the huge memory necessity for mining. Ethereum usually means that largescale miners achieve benefit from packaging terabytes of memory in their gadgets because bigger miners may get precisely the same result investing in memory-carrying apparatus such as ASIC processors and other similar physical devices capable of storing memory.

4.5.3.9 End-to-End Transaction in Ethereum

Transaction in Ethereum is completely different from the transaction in Bitcoin. There are four different types of transactions that can be executed on the Ethereum system to transfer Ether across accounts. The accounts can be (1) Externally owned accounts or (2) Smart contract account (refer Fig. 4.63).

Following are the possible cases:

a. An externally owned account can send ether to another externally owned account in a transaction
b. An externally owned account can send ether to a contract account in a transaction
c. Contract-to-contract ether transaction
d. A contract to externally owned account transaction.

Externally owned accounts may invoke a contract through a transaction in EVM (Ethereum Virtual Machine) or invoking a function in a smart contract. If executing a function does not affect state, it does not need a transaction. A transaction contains some mandatory characteristics.

Equipped with all the comprehension of fundamental theories of blockchain and Ethereum, let us now observe end-to-end transaction and the way that it flows through numerous components and gets stored finally in the ledger. In this example, Steve wants to send 2 ETH to Mark. Steve generates a transaction message containing From and To value fields and sends it across to the Ethereum network.

"From Account" refers to the account that is originating the transaction and symbolizes an account that is ready and about to send gas or ether. "From account" could be either an externally owned or a contract account.

"To Account" refers to an account that is obtaining ethers. In the event of transaction regarding the deployment of the contract, then the "To field" remains an empty string. This may again be an externally owned or a contract account. Value denotes the quantity of ether that is moved from one account to another (refer Fig. 4.64).

Input identifies contract byte-code and is traditionally used throughout contract deployment in EVM. It is utilized for storing data associated with smart contract functions combined with parameters.

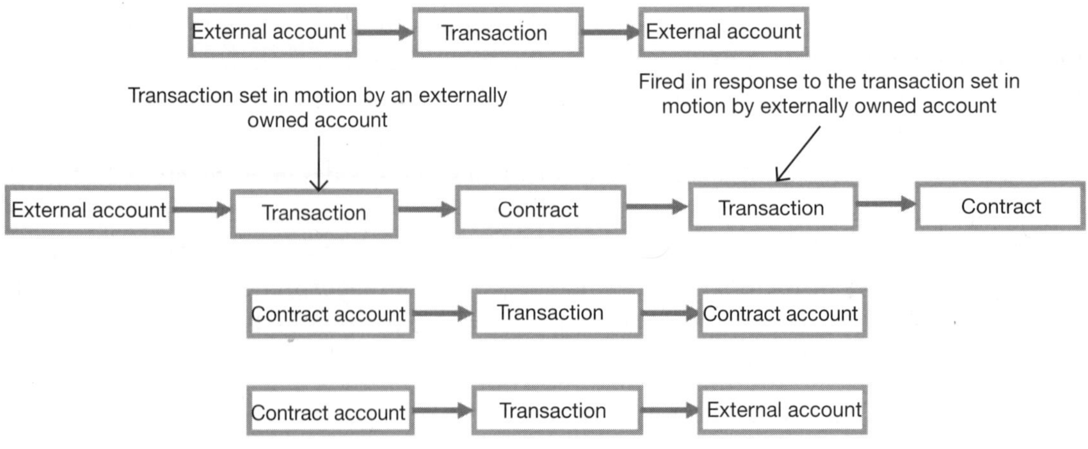

Figure 4.63: Four types of transactions

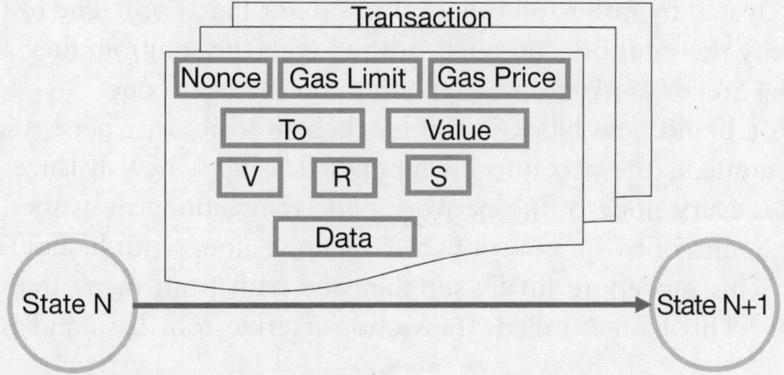

Figure 4.64: Components of transactions

The transaction is not written into this ledger instantly; it is set in a transaction pool. The mining node results in a brand new block, carries all transactions out of the pool based on the gas limitation standards, and adds them to the block. Every block includes an upper gas limit, and just about every transaction desires a specific quantity of gas to become a member of its implementation. The accumulative gas from many transactions still not written in the ledger cannot exceed the block gas limit. This guarantees that all of the transactions do not become stored in just one block. Once the gas limitation is reached, the transaction is taken out of the block, and mining commences.

Suppose your vehicle has a mileage of 20 kilometres per litre, and the price of petrol is $1.5 per litre. Then driving a car for 60 kilometres would cost you three litres of petrol, which is worth $4.5. Similarly, to execute code on Ethereum, you need to obtain a certain amount of gas (petrol), and the per-unit price is called as gas price. If the user provides lesser than the amount of gas to run a particular code, then the entire process will fail, and the user will be given the message "out of gas." Gwei is the lowest denomination of ether used for measuring a unit of a gas price.

Let us look into a transaction. Let us say the gas limit is mentioned as 11,500. The gas used by the transaction is 11,500, and the gas price is 20 Gwei, which is the lowest denomination of ether. So 20 Gwei × 11,500 gives the actual transaction fees.

$$\text{Transaction Fees} = \text{Gas Limit} \times \text{Gas Price}$$

In this example, Steve wants to send 2 ETH to Mark. Steve generates a transaction message containing From and To value fields and sends it across to the Ethereum network.

The transaction fee is given to the miner, who has successfully validated the transaction. All miners on the network do this activity. The miners contend, attempting to fix the problem thrown to them. The winner would be a miner who can fix the problem first. After a brief timespan (approximately 10 seconds in Ethereum), one among the miners will advertise that he has solved the problem, and he could be the winner. In addition, he composes the block into a series. The winner sends the problem solution along with the new block created by him to all other miners (refer Fig. 4.65). The rest of the miners validate and verify the solution, and once satisfied with the solution, they accept the new block containing Steve's transaction to append in their ledger copy.

This creates a brand new block in the blockchain series that persists over space and time. At that moment, the accounts are upgraded using a new balance. The block is duplicated across every node of the network. The transactions are hashed and stored in the block. The hashes of two transactions are obtained along with hashes farther to build a second hash. This procedure finally supplies one hash from many transactions stored inside the block. This hash is called Transaction Merkle root hash and is stored in the block's header.

A small alteration in any transaction causes a big change in its hash and also changes the root transaction hash. Additionally, it has a cumulative influence since the hash of this block will probably shift; further, the child block must alter its hash for the reason that it stores the parent hash. This makes transactions immutable. Block hash refers to the hash of the block to which this transaction belongs. Block Number is the block to which this transaction belongs. Gas refers to the amount of gas supplied by the sender which is executing this transaction. Gas Price refers to the price per gas the sender was willing to pay in Wei/Gwei. Total Gas is calculated as, Gas units × Gas price. Nonce refers to the transactions made by the sender before the current transaction. Adding a nonce to a transaction's identifier makes it additionally unique, thus reducing the chance of duplicate transactions. The nonce is key to creating a block to a blockchain database.

Transaction Index refers to this consecutive serial number of the current transaction. In block, the Value identifies the quantity of ether moved in Wei. The space denoted as V, R, and S (refer Fig. 4.64) relates to the digital signature of the transaction. A standard transaction in Ethereum is when an externally owned account sends several ethers into another externally owned account.

Notice that the input field is not used here. Since two ethers were sent in the transaction, the value field is showing the value accordingly in Wei.

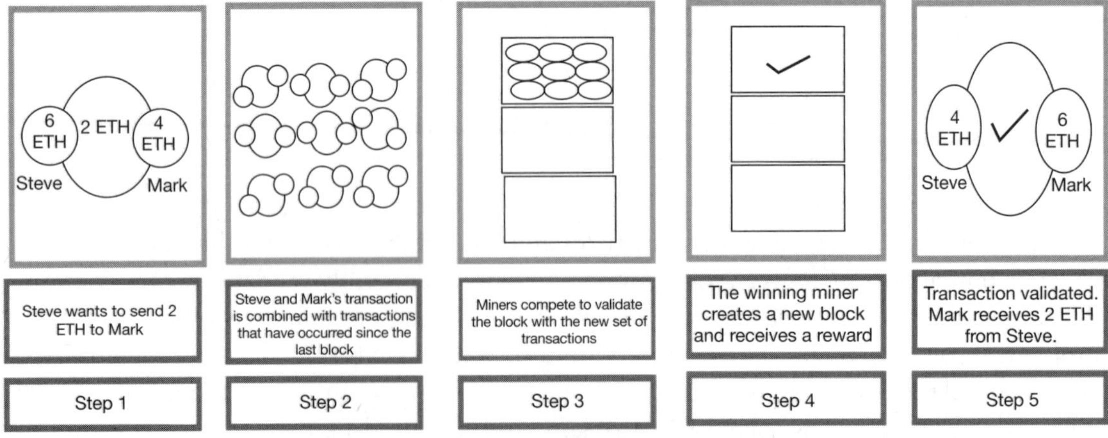

Figure 4.65: Complete transaction steps

Table 4.3 Intrinsic gas fees

Detail	Intrinsic gas fees
Pre-define Cost	21,000 gas for executing the transaction
Data or code that equals zero	4 Gas/Byte
Every non-zero byte of data or code	Additional 6 Gas/byte
Contract-creating transaction	Additional 32,000 Gas

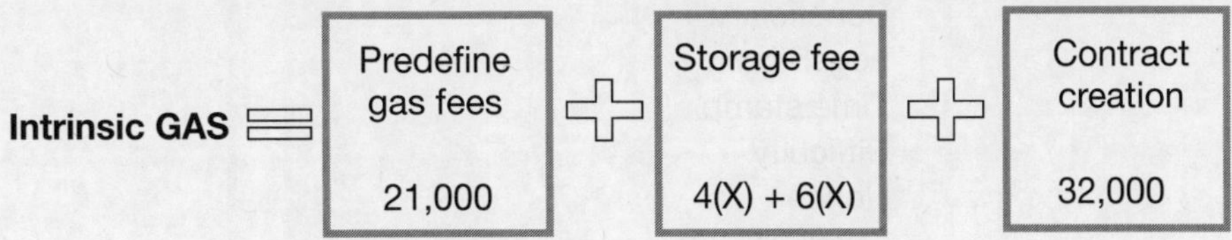

Figure 4.66: Intrinsic gas fees formula

All transactions must meet the below set of requirements to be executed. These include:

- The transaction must be formatted correctly in the form of RLP. "RLP" means "Recursive Length Prefix." It is a data format used to encode nested arrays of binary data and is used by Ethereum to serialize objects.
- The transaction should have a legal transaction signature.
- The transaction should have a valid transaction nonce. A transaction nonce should be the sender's account nonce.

The sum of the transaction's gas limit must be equal to or greater than the overall intrinsic gas (refer Table 4.3 and Fig. 4.66) used by the transaction.

4.5.4 How Mining Works in Ethereum

The mining principle of Ethereum is different from that of bitcoin. Ethereum miners always look forward to mine new blocks and listen actively to receive new blocks from other miners. All miners gather the transactions from the transaction pool. The miner constructs a brand new block and adds all transactions to it. Before the transactions are added, they are assessed to check if any transaction is not yet written within a block that is possibly obtained from different miners. Any such transactions, if detected, are dropped. The miner will then add his coin-based transaction to get the reward of mining the block. The endeavour of miners will always be to build on the block header (refer Fig. 4.67) and perform the subsequent task.

Parent Hash (Previous block Hash): The miner also identifies the hash of the previous block. The last block will become the parent to the current block, and its hash will be added to the block header.

LogsBloom: It is a Bloom filter (data structure) that consists of log information.

Block header

Previous block hash
Transaction root hash
Receipt root hash
State root hash
Ommers hash
MixHash
Beneficiary
LogsBloom
Time stamp
Difficulty
Nonce
Gas limit
Gas used
Extra data
Number

Figure 4.67: Contents of a block header

Transaction, state, and receipt hash: The miner hashes all transactions that are inside the block; the hashes are further combined in pairs to create a brand new hash. This approach is continued until there is only one hash for all transactions in the block. The hash is also known as Root transaction hash or Merkle Root transaction hash. The miner, in the same manner, computes the State transaction hashes and Receipt transaction hashes and adds to the block header. Ethereum blockchain simplifies the Merkle tree, to save not a single Merkle tree but about three trees, to get three different forms of items, namely, Transactions, Receipts, and State.

Nonce: A nonce and timestamp are also added to the block header. A nonce is a random number that can be used just once in the cryptographic communication. Adding a nonce to a transaction's identifier makes it additionally unique, thus reducing the chance of duplicate transactions. The nonce is key to creating a block in a blockchain database.

The mining procedure starts when the miner shifts the nonce value and attempts to get a hash, which could solve the problem. It must be kept in mind that every single miner from the ethereum system implements everything mentioned here. Eventually, one of the miners could manage to fix the problem and advertise the same to the other miners of the system. The other miners check the proposed solution and if it is found accurate, would verify and confirm every transaction while accepting the block, and then add the block to the blockchain.

Timestamp (Current UNIX time): The timestamp is another field, which indicates the UNIX time. It is the seconds passed after the first of January 1970 and is a 10-digit number. This is also part of the data, which changes every second. When the timestamp changes (every second), the corresponding data changes; when data changes, the resultant also has a changed value. The miner's computer should keep incrementing the nonce until it finds the right one as per the algorithm (< target space). That means the computer needs to generate millions of hashes per second to generate the one that will have the same number of starting zeros as defined. It is time-consuming to execute a PoW for a single block, but at the same time, it is easy for someone to verify if any given block is correct.

Ommer block: An ommer block is a block, whose parent corresponds to the current block's parent's parent. The block time of Ethereum is about 15 seconds. This is lower than that of many different blockchains, such as Bitcoin (~10 minutes) and empowers faster transaction processing. But indeed, one of those drawbacks of competing block times will be that even more competing blocks solutions are all located by miners. These blocks are also referred to as "orphaned blocks" (i.e., mined blocks that do not make it into the main chain). The aim of ommers is to help miners to add these orphaned blocks significantly. Even the ommers need to be "legitimate" or "valid," which means it should be within the sixth generation or smaller of the present block. After six children, orphaned blocks can no longer be retrieved (since older transactions would complicate matters). Ommer blocks have a smaller benefit compared to the usual complete block. There is still some incentive for miners to incorporate these orphaned blocks and experience an advantage.

MixHash: MixHash can be a hash which, when together with all nonce, demonstrates that a block has enough computation. Miners generate a mixhash until the outcome is below the desired target hash. After the output meets the condition, this nonce is deemed legitimate, and the block may be added into the blockchain series.

Difficulty: It decides the intricacy (complexity) of the puzzle/challenge awarded to miners of a particular block. The gas limit of the block decides the most gas allowed for the block. This aids in ascertaining the number of transactions that might be a part of the block.

Number: It refers to the number of helps received in solving the challenge. Miner property is the account identifier of the miner. It is also known as coin base or Ether base. The number is the sequential number of a block on the chain. Parent Hash refers to parents' blocks hash. Transactions refer to transactions that are part of the block (refer Fig. 4.68). Total Difficulty refers to the total difficulty of the chain until the given block.

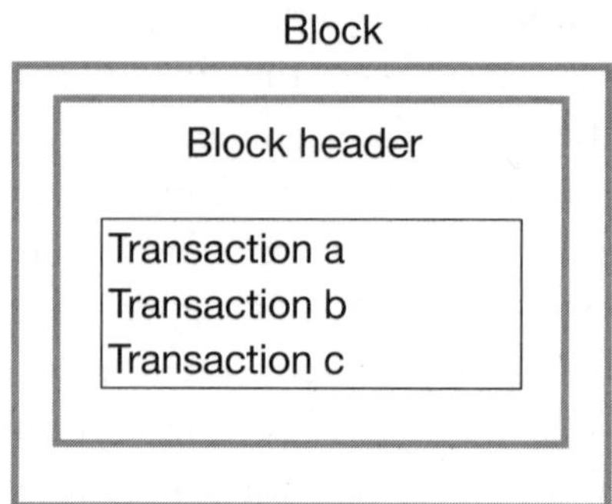

Figure 4.68: Simple structure of a block

This whole procedure is additionally called proof-of-work, through which a miner offers evidence that he has worked on computing the answer to the problem. There are other similar algorithms like Proof-of-Stake (POS) and Proof-of-Authority (POA).

4.5.5 Merkle Patricia Tree

Merkle Patricia tree is different from the normal Merkle Tree. Inside of each Ethereum block, there is a tree called a Merkle tree. It is a glorified, singly linked list. It has a pointer to the previous hash of the last block. It has a nonce, and then it has a Merkle root.

Why does it have to store data like a tree? When we store data as a huge list, it could create massive scalability problems. Thus, to avoid this, the Merkle tree data structure can be employed. This is a way to hash a large number of chunks of data together, which relies on splitting the pieces into buckets, and each bucket only contains a few chunks. Therefore, it hashes down the chain, like a kind of file directory.

Figure 4.69: Patricia tree

You have the root directory. You also have got child directories. Also, it keeps moving. The rationale why we make use of a Merkle tree is the fact that it enables us to get Merkle proof, which comprises the hash root of this tree, and also the branch comprising of all hashes moving up across the path, from piece to the root. Anyone who reads this proof may verify if the hashing moves continuously all the way up the tree. So the Bitcoin blockchain employs an uncomplicated Merkle tree. The limitation of the Merkle tree is that it cannot prove anything about the current state of the contract or the transaction since all the hashes keep moving up across the path. It is difficult to identify the details of who is currently holding the particular digital asset (currency) and the execution status of the financial contract (who is selling to whom?).

In the case of Ethereum blockchain, it modifies the Merkle tree to store three trees for three kinds of objects (refer Fig. 4.69), transactions, receipts and state. A decentralized computer can store a state.

4.5.6 Architecture of Ethereum

Let us discuss the architecture of Ethereum (refer Fig. 4.70). It has the consensus layer, the economic layer, the blockchain services layer, the interop layer, browsers, and the DApps Layer.

If we had a server, there is no need for consensus, since there is no democracy; it is a dictatorship and the server controls everything. However, in a decentralized application, we need consensus, and the blockchain was that missing ingredient to reach consensus in a distributed decentralized way. So, for consensus (Layer 0), we need some way to agree upon all of these application-level constructs.

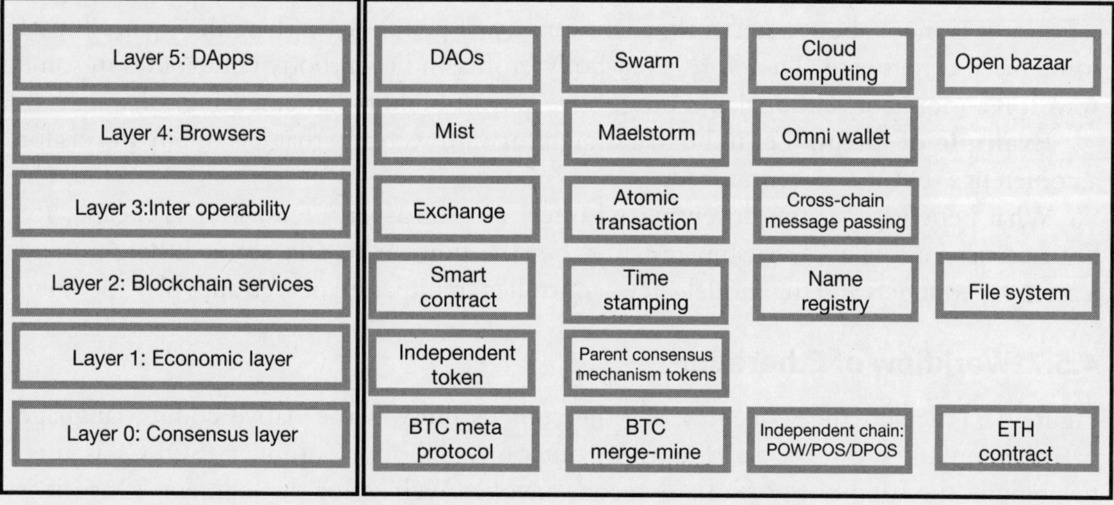

Figure 4.70: Architecture of Ethereum

In addition to the consensus mechanics, we want an economic token (Layer 1) to incentivize each one of these nodes, either to do the computation, or perform the storage operation (Layer 2), or even to do anything that is needed. This is where the crypto tokens get involved. We generally utilize AWS or Google Cloud. There are certainly a lot of other providers for keeping data from the cloud which are controlled by an individual (single entity); however, in this scenario, we desire a brand new form of decentralized storage, which is peer-to-peer reviewed, along with IPFs (interplanetary file system). In addition, we also have smart contracts. Smart contracts are code snippets that will live on the blockchain. They are used for decentralized computation. While IPFS decentralizes storage and content delivery, the Ethereum blockchain's smart contracts decentralize computation.

On top of that, we have interoperability (Layer 3). In this decentralized world, where we have all of these different apps, each with their tokens, how is one supposed to exchange value between these tokens? The answer is to write a universal wrapper around all cryptocurrencies. This exchange protocol (of Layer 3) would transfer the amount or transmit value between all of these different tokens as one uses the various services. You can think of this as kind of a stack of decentralized API's, that all use their tokens.

You can pay the top API with any token, and in turn, it pays all the other APIs and the tokens.

Next, we have the browser (Layer 4) to access the decentralized applications. The mainstream browsers such as Chrome, Firefox or Opera would accept these decentralized protocols natively, so we would not have to create another browser. Some browsers are made for decentralized applications, like Mist, Omni wallet, and Maelstrom. After this layer, we build DApps (Layer 5), where we start adding application-level constructs. This does not depend on any specific existing party, and it is not about any one party selling its services; it is more about a network, a community of people who share the ownership of some piece of software. Therefore, everybody profits and everybody contributes in some way. It is a more conventional and progressive way of building an emerging software.

Example of DApp: Let us build a straightforward web app that lets us buy tickets for a conference, but it is decentralized.

What benefits does the decentralized offer? No third-party fees, so it is cheaper for both parties, and it is censorship-resistant, but the point is to talk about Ethereum and learn how contracts and the model-view-controller architecture of web apps play together.

4.5.7 Workflow of Ethereum

Figure 4.71 depicts the workflow of Ethereum. Solidity is the native coding language of Ethereum. It creates a .sol file as an outcome. Solidity Compiler (SOLC) is used to compile the solidity file (.sol). Bytecode invokes Web3js for deployment. Once it is deployed, it returns the Contract and the Application Binary Interface (ABI). When this contract is initiated, it can invoke contract methods and signs, and pass them to Ether to perform the Operation.

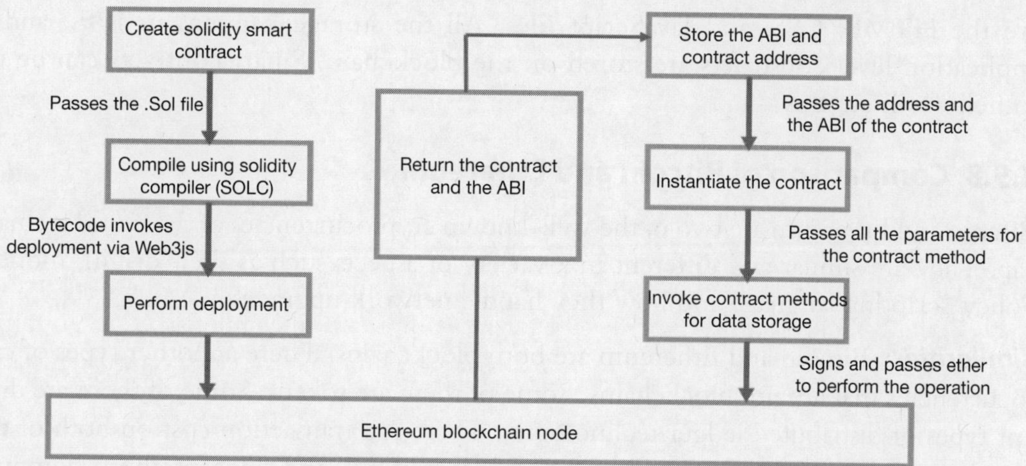

Figure 4.71: Workflow of Ethereum

If we look at the Ethereum GitHub, there are so many repositories there. We have the Ethereum virtual machine (EVM), which calculates elements of the contract logic. Ethereum virtual machine allows for a centralized computation like IPF for storage.

Then, we have a swarm, which is the storage layer, which we can use as IPFs. Then, a whisper is for all the nodes being able to message each other. Again, we can also use IPFs. So really all we need are Ethereum and IPFs.

To run Ethereum, we can download the client and can connect to the ethereal network. We can run smart contracts and mine new blocks. Therefore, the client is our gateway into the Ethereum network. A smart contract is written using solidity, which is very similar to JavaScript. You have the client, eath, that is in C++, and you have pi, which is written in Python. However, the best one is the language called Google "Go", which is upgraded from C++. It was designed with many of the elements of distributed computing and even decentralized computing in mind. Currently, the Ethereum solidity is the most popular, and it is very similar to JavaScript.

Workflow for Deploying Smart Contracts

- The first step is to download the Ethereum node.
- Then, write the solidity code.
- Compile it using a framework like a truffle.
- Deploy the smart contract code to the network.
- Call the deployed contract using Web3js, which is the front-end client that speaks to the Ethereum blockchain.

In web apps, we have the Model View controller architecture. So, in all these software stacks of decentralized applications, the controller, instead of speaking to a server, talks to blockchains and to distributed hash tables in terms of a model that is remarkably similar. However, there is also another type of model like a smart contract. You can think of smart contracts like models that the controller will speak to. Views are the HTML, CSS, and JavaScript files. All the storage happens in IPFs, and the application level constructs are stored on the blockchain. Smart contracts can be used to help facilitate this.

4.5.8 Comparison of Bitcoin and Ethereum

Bitcoin and Ethereum are two of the well-known cryptocurrencies of the world by market cap. They are similar and different in a variety of aspects such as their origin, monetary policy, scripting language and how they handle network upgrades.

Similarities: Bitcoin and Ethereum are both blockchains. There are other types of cryptocurrencies that are not blockchains. Some of them are tokens. Some of them are different types of distributed ledger technologies. There is a transaction cost on both of these networks. Ethereum is also a "distributed ledger (DL)" and "decentralized computing platform" with "smart contract" capabilities. Both Bitcoin and Ethereum provide anonymous transactions, and neither of these two is controlled or regulated by way of a centralized body. Both use proof-of-work (PoW) consensus algorithms at present. Energy and hardware costs have to be invested to secure the blockchain. The mining and hash power (BTC and ETH) can both be used for p2p transactions. Many software wallets support both of these popular coins. Both the projects are also open-source with large developer communities. They also have an extensive global network of independent nodes and are generally deemed as decentralized. They are both cryptocurrencies. They are both digital currencies. They both have value and allow sending currencies from one account to another. They can be purchased on websites such as coinbase.com or other exchanges where cryptocurrencies can be bought with traditional or fiat currencies.

Differences: Bitcoins were introduced in 2009 when Satoshi Nakamoto published the white paper. No fundraising was done. Also, miners had to mine the Bitcoin network to receive the very first bitcoins. Therefore, Satoshi himself has around 1 million Bitcoins, and this is supposed to be the most significant amount for any single owner. However, he never spent any and disappeared from the internet, without anyone knowing his identity. Ethereum, on the other hand, originated in 2013 when the founder Vitalik Buterin published the Ethereum white paper. An ICO was carried out in 2014, in which a pre-mined quantity of Ethereum was distributed to investors. Other early leaders include the Oreo Hoskinson, Alesi Lubin and Gavin Wood. Bitcoin and Ethereum have two different objectives. Bitcoins aim to be a digital currency or digital payment alternative or a digital store of value. It is designed for you to buy Bitcoin, hold it, and store the value there. You can also use it as just a payment alternative by sending Bitcoin from one person to another. You can do the same thing on Ethereum. You can buy an Ethereum and hold it like an asset. You can send Ethereum or ether back and forth to pay for things. However, Ethereum has an extra level that reveals its real purpose, which is supposed to run programs right. Ethereum is designed for programs with pieces of code called smart contracts that run on the blockchain. Smart contracts are the building blocks for blockchain-based applications. Block time is essentially how long it takes new transactions to be mined and put in blocks. Transactions differ in their block time. Ethereum may move away from proof-of-work to proof-of-stake.

Let us look at a bitcoins use case: It is designed to be peer-to-peer and optimized to transfer a value and store a value (and nothing else). Bitcoin is also the native currency on the network that facilitates this function. The fee in Bitcoin is for making transactions, and that is based on the data size of each transaction. Ethereum' s native coin is called as ether or ETH. It is not designed to be a digital dollar like Bitcoin, but rather fuel for a wide range of transactions and functions. The function of ether is more like payment for a utility, a fee to run the code on the network. Ethereum aims to be decentralized, it has distributed nodes all across the network, and this is called the Ethereum virtual machine. It can host smart contracts on which decentralized apps can use the function. It enables tokens such as ERC20, with many different functionality in their apps. Current notable usages of the apps include decentralized finance and gaming. People have also been using ether as a digital currency.

In terms of scripting language, bitcoin keeps it simple; it uses the script language that keeps the code robust and also less vulnerable to bugs. It can only support simple, smart contracts like multi-signature and escrow. More advanced smart contracts on Bitcoin are doable, but they need a second layer solution or a side chain solution. The main focus of bitcoin is on centralization, censorship, resistance, and security. These are of utmost importance to the Bitcoin community. Ethereum scripting languages were designed with the primary purpose of building smart contracts and decentralized apps platform. This means Ethereum scripting language needs to support a more complex logic. This also leads to a broad range of bugs in the code, which are hard to be detected in review/audit, Specific bugs have led to the loss of millions of dollars in ether. This is the inherent risk factor. However, the additional benefits of this dominant logic outweigh the risks and methods are being developed by the Ethereum space to minimize these risks.

In the monetary policy, the inflation rate of the Bitcoin supply is hard-coded into the protocol itself (maximum of 21 million bitcoins). Currently, over 17 million Bitcoin has already been mined. With roughly ten minutes per block, this means 75 bitcoins are produced per hour. Roughly speaking, for each year, the reward is cut in half. This is called the halving or happening. In May 2020, it was reduced to 6.25 bitcoins per block of bitcoins. The monetary policy is considered to be fixed and non-inflationary. For Ethereum, there is no upper limit set. The emission rate is occasionally reduced during significant network upgrades.

Ethereum produces the block roughly every 15 seconds. This equates to 480 ethers per hour. There is no fixed maximum supply for ethers. In terms of block x, why would one choose something less like a few seconds (15 sec), as compared to Bitcoin, which takes about 10 minutes?

ASIC (Application Specific Integrated Circuit) is meant to do one specific type of computational task, and that is why it dominates general-purpose chips like CPUs or GPUs. In the case of Bitcoin, ASICs are built for mining their algorithm better than anything else. ASICs provide the Bitcoin network with a ton of hash power. Because of this, the blockchain is extremely secure. It would take many resources for someone to make a 51% attack.

Most forks in the Bitcoin world are preferably softworks; this maintains backward compatibility. That is, the nodes do not have to upgrade their software to continue to work properly, and the network stays intact. Hard forks often update Ethereum, and this could increase the risk of chain splits because these are not backward compatible.

Ethereum has a more centralized leadership and developer community coordination. Therefore, there is more agreement in terms of decisions made for upgrading the network. Hence, hard forks do not usually lead to chain splits in the Ethereum world.

In terms of transaction models, Bitcoin has an unspent transaction output model or UTXOS. Bitcoin only tracks transactions from address to address. Therefore, the wallet has a software, which aggregates the addresses controlled by the same private key. One private key can control a bunch of different addresses and can show the total balance that an account holder has. A wallet usually collects multiple addresses that contain an amount of unspent bitcoin.

Therefore, when a bitcoin transaction is made, you spend the entire amount in those addresses. The remainder is sent to a chained address. You can think of this as paying ten bucks in cash to buy an $8 item and receiving $2 back. Ethereum, on the other hand, uses an account model. This is a much less complicated model. You get one Ethereum address, which works similarly to any account. A user would use one address for everything. You can have two addresses or make a third one as well. However, you can use them and continually use them for perpetuity. Wallets do not create new addresses. This simplified model, however, has less privacy and anonymity because the history of the entire user accounts is viewable.

In Ethereum, the pricing value is named gas, and the cost of transaction fluctuates based on storage requirements, bandwidth usage, and sophistication. In the case of Bitcoin, the total cost of the transaction is dependent upon block size. Ethereum, on the other hand, has higher-level computing features embedded into it; this sophistication makes it more susceptible to cyberattacks as compared to Bitcoin (refer Table 4.4).

The Bitcoin blockchain has a block limitation of just one MB, which means that the number of transactions that squeeze to one block cannot exceed 1 MB. The time that is required to make or mine a brand new block onto Bitcoin blockchain is about 10 minutes. This effectively implies that the Bitcoin system is designed for 34 trades per minute. In Ethereum, the number of transactions which can be placed to a block is decided by the miners. Each new block is mined in 12 to 14 seconds, and the range of transactions per second is approximately 15.

Table 4.4 Bitcoin Vs. Ethereum

	Bitcoin	**Ethereum**
Founder	Satoshi Nakamoto(unknown)	Vitalik Buterin and team
Purpose	Cryptocurrency	Network software
Release date	January 2009	July 2015
Scripting language	Turing incomplete	Turing complete
Coin release method	Early mining	Through ICO
Average block time	Approx. 10 minutes	Approx. 15 seconds
Transaction model	UTXO	Account
Coin symbol	BTC	ETH
Tokens	Not available	Available
Monetary policy	Hardcoded	Not hardcoded
Emission rate	Halving policy followed	Occasional
Backward compatibility	Available	Not available
Block limitation	1 MB per block	No limit

Summary

As per Gartner (2018), Blockchain is one of the Top 10 Strategic Technology Trends of 2019. The public permissionless blockchain is synonymous with a public blockchain. We need to use "permissionless blockchain" when the state (and date) needs to be stored in the database, if there are multiple writers for a transaction, when we cannot use always-online TTP and when not all the writers are known. We need to use "public permissioned blockchain" when the state (and date) needs to be stored in the database, if there are multiple writers for transaction, when we cannot use always-online TTP, when all the writers are known, when all the users cannot be trusted and when public verifiability is required.

Bitcoin was the first blockchain created, and it is a public permissionless blockchain. The idea behind the design was to create a decentralized digital currency that is free from government regulation whereby two people can trade directly with one another without the need for middlemen or intermediaries. The success of the bitcoin can be measured based on the number of transactions happening daily. The creation of bitcoin blocks is called mining. Bitcoin mining is of two types, namely, software mining and hardware mining. Further, there are two types of software mining, namely, pool mining and cloud mining. Miners compete to solve a complex mathematical problem based on a cryptographic hash algorithm. Mining comprises of hashing a block and then introducing a nonce to the hashing function and running the hash all over again. A Merkle root is synonymous to a fingerprint of all the transactions in the block, which is created by hashing together pairs of Transaction IDs.

Ethereum is a type of secured cryptocurrency and is being used in multiple countries across the globe. Both Ethereum and Bitcoin can be used very quickly. Ethereum could be a do-it-yourself platform for most decentralized programs, also referred to as DApps or de-centralized programs. Ethereum programmers write the requirements for their application or DApp, and the Ethereum network implements it. They are called smart contracts because they handle every facet of the contract – authorities, direction, performance, and payment. Ethereum features a Genesis Block, also called the very first block. This block is made automatically whenever a series of blockchain is first initiated.

There are two types of nodes (computers) in the Ethereum network – Mining nodes and Ethereum virtual machines. Gas is the internal currency of Ethereum. The execution and resource utilization costs are predetermined in Ethereum in terms of Gas units. This is also known as Gas Cost. The architecture of Ethereum has the consensus layer, the economic layer, the blockchain services layer, interop layer, browsers, and DApps layer. In 2015 (the starting period), the value of one ether was equal to 0.93 USD, and it has grown steadily from there. It touched the peak value of 741 USD in the year 2017. In 2019, the value of one ether touched 181 USD (approximately in June 2019).

EXERCISES

Multiple Choice Questions

1. **Bitcoin transactions are being registered into blockchain once in:**

   A. 10 minutes
   B. 2.5 minutes
   C. 5 minutes
   D. 30 seconds

   Answer: A

   Explanation: Block frequency: Bitcoin transactions are registered into blockchain once in ten minutes. Miners will first check the transaction. After reviewing the transaction, the software will give a complex problem for the miners to solve. If miners can solve the problem, then the computer will give bitcoins as a reward to the miners.

2. **In early 2020, a miner received ____________ bitcoins as a reward to solve complex problems.**

   A. 100
   B. 50

C. 25
D. 12.5
Answer: D
Explanation: Halving policy: Block frequency and Halving are the monetary policies of bitcoin. In early 2020, miners received 12.5 bitcoins as a reward to solve complex problems. From the year 2009 to 2012, 50 bitcoins were given as rewards. The reward will be reduced by half every four years. Eventually, in 2140, it will decrease to zero.

3. **In bitcoins, the reward per mining will be reduced by half every:**
A. 10 years
B. 5 years
C. 4 years
D. 2 years
Answer: C
Explanation: Same as for Question 2

4. **From the year 2009 to 2012, a miner was getting ___________ bitcoins as a reward to solve complex problems.**
A. 100
B. 50
C. 25
D. 0
Answer: B
Explanation: Same as for Question 2

5. **In the year 2140, a miner will be getting ___________ bitcoins as a reward to solve complex problems.**
A. 100
B. 50
C. 25
D. 0
Answer: D
Explanation: Same as for Question 2.

6. **Detached block, which is a valid block and not part of the main chain is called as:**
A. New block
B. Orphaned block
C. Main block
D. Sub block
Answer: B
Explanation: Orphaned Block: Detached or Orphaned blocks are valid blocks, which are not part of the main chain. They occur naturally when two different

miners successfully mine at the same time. Sometimes, a hacker may attempt to reverse/modify transactions.

7. **Timestamp in bitcoin is a ___________ digit number**

   A. 256-digit
   B. 12-digit
   C. 10-digit
   D. 18-digit
   Answer: C

   Explanation: Timestamp (Current Unix time): Timestamp is a field which indicates the UNIX time. It is the seconds passed after the first of January 1970 and is a 10-digit number. This is also part of the data, which changes every second. When this changes (every second), the corresponding data changes; when there is a change in data, the resultant also has a changed value.

8. **Timestamp is a field in bitcoin block, which indicates the UNIX time. It is the number of seconds passed after:**

   A. 1st of January 2010
   B. 1st of January 2009
   C. 10th of january 2010
   D: 1st of January 1970
   Answer: D

   Explanation: Same as for Question 7.

9. **In bitcoin, unconfirmed transactions, which are to be included in the block once the combination becomes a valid transaction, are kept at:**

   A. Merkle Hash
   B. Mempool
   C. Block
   D. Merkle Root
   Answer: B

   Explanation: Mempool: Transactions are first added to the Mempool (Memory pool) attached to each node. They are unconfirmed transactions, which are to be included in the block once the combination becomes a valid transaction. Blocks are added every 10 minutes, but transactions happen all the time. Mempool is the staging area for the transactions.

10. **In bitcoin, when a miner announces a block, the block needs to propagate to all the nodes in the network, using this protocol:**

    A. TCP Protocol
    B. Http Protocol
    C. Gossip Protocol
    D. Internet Protocol
    Answer: C

Explanation: Block Propagation: When a miner announces a block, the block needs to propagate to all the nodes in the network, using gossip protocol (transactions are propagated using gossip protocols). The concept of gossip communication is similar to employees spreading rumors in a company. A node would perform the following functions: a. Validate transaction; b. Make sure the nonce is valid (nonce is within an acceptable range); c. Check if each transaction in the block is valid.

11. **In bitcoin, the size of the signature and public key is so large within the transaction and is kept out of the block and distributed separately. This segregation of the signature from the block is called as:**
    A. SEGWIT
    B. MemPool
    C. Gossip
    D. Mining
    Answer: A
    Explanation: SEGWIT: The size of the signature and public key is so large within the transaction that it occupies more than 60 % of the overall size of the transaction. In the bitcoin protocol, this portion (signature and public key) is kept out of the block and is distributed separately. This segregation of the signature from the block is called as Segregated Witness (SegWit). Because of SegWit, more (almost double) transactions can be added into the block. SegWit will be sent separately in the network.

12. **In a bitcoin, the block header consists of ___________ bytes:**
    A. 10
    B. 20
    C. 40
    D. 80
    Answer: D
    Explanation: A bitcoin block header is a unique number, consisting of 80 bytes. The block header is an 80-byte long string, which consists of Bitcoin Version number (4 bytes), Previous block Hash (32 bytes), Merkle root (32 bytes), Timestamp (4 bytes), Difficult target (4 bytes), and Nonce used by miners (4 Byte) as shown in Fig. 4.7.

13. **In a bitcoin, in the block header, Difficult Target consists of ___________ bytes:**
    A. 4
    B. 8
    C. 10
    D. 32
    Answer: A
    Explanation: Same as for Question 12.

14. **In a bitcoin block header, Nonce used by miners consists of ___________ bytes:**

A. 4
B. 8
C. 10
D. 32
Answer: A
Explanation: Same as for Question 12.

15. In a bitcoin block header, previous block hash consists of ___________ bytes:
A. 4
B. 8
C. 10
D. 32
Answer: D
Explanation: Same as for Question 12

16. Which of the following is not a characteristic of the Hash Algorithm?
A. It has to be one-way
B. It has to be two-way
C. Fast Computation
D. Avalanche Effect
Answer: A
Explanation: The following are the characteristics of the Hash Algorithm:
1. It has to be one-way only, and Reverse Engineering is not possible. It can produce a unique hash from the data. However, it cannot provide the data from the Hash.
2. The same document always produces the same hash. This shows the deterministic characteristics of Hash.
3. Fast computation is another characteristic of any hash function. It provides the results in a fraction of a second.
4. Avalanche Effect is another key characteristic of the hash function. Avalanche effect indicates that a small (tiny) change in the input produces drastic changes in the hash output. A change in output is not proportional to the corresponding change in the input.
5. It must withstand Collisions. This means that the same hash will not be produced for two different types of documents.

17. The type of hash used in bitcoin is different from Ethereum.
A. True
B. False
Answer: A
Explanation: The type of hash differs from one protocol to the other. That is, the type of hash used in bitcoin is different from Ethereum, and so, the complexity of forming the hash will vary across protocols.

18. Hashcash is a cryptographic hash-based proof-of-work algorithm, which was initially used to identify:

A. Blockchain hackers
B. Email spammers
C. Industrial miners
D. Mining pools

Answer: B

Explanation: Hashcash is a cryptographic hash-based proof-of-work algorithm, which was initially used to identify email spammers. For example, the number of seconds/minutes of CPU time taken to write an email is hashed and is written in the header of the email. For example: Let us say on average, to write an email, it takes a minimum time of 20 seconds. When an email is received, the hash value is checked, and if the value is lesser than 20 seconds, then that email can be classified as Email Spam.

19. A modest miner of bitcoin can create ___________ hashes per second:

A. 100 million
B. 1 million
C. 10 million
D. 0.5 million

Answer: A

Explanation: A modest miner can create 100 million hashes per second. To traverse 4 billion hashes, it takes just 40 seconds.

20. For your transaction, if the fees are given as Zero, it may be stuck in the memory pool itself, and chances of it to be selected and added into the block will be minimal.

A. True
B. False

Answer: A

Explanation: In a miner pool, the pool will allocate the combination of transactions to each miner to check and verify. For your transaction, if the fees are given as Zero, it may be stuck in the memory pool itself, and chances of it to be selected and added into the block will be minimal. Usually, transactions with the highest transaction fees will be selected for the transaction.

21. In bitcoin, there are ___________ types of nodes.

A. Two
B. Five
C. Three
D. Four

Answer: C

Explanation: In bitcoin, there are three types of nodes:

a. Full node.
b. Mining nodes
c. SPV node

22. In bitcoin, SPV node stands for

A. Strange Payment Validation
B. Strong Path Validation
C. Simplified Path Validation
D. Simplified Payment Verification

Answer: D

Explanation: SPV node: This stands for "Simplified Payment Verification." The vast majority of nodes today are SPV nodes. SPV nodes do not involve in mining; however, they are bound to the validate block using block scripts; this feature can be easily achieved even via handheld devices.

23. In bitcoin, the time taken for a solid block to travel to all nodes is close to ___________ seconds for large blocks.

A. 30 Seconds
B. 10 Minutes
C. 5 Minutes
D. 5 Seconds

Answer: A

Explanation: Studies have shown that the time taken for a solid block to travel to all nodes is close to 30 seconds for large blocks.

24. In bitcoin, the general rule-of-thumb is to wait for at least ___________ confirmations on a transaction:

A. 3
B. 6
C. 7
D. 8

Answer: B

Explanation: The general rule-of-thumb (heuristics) is to wait for at least six confirmations on a transaction; this assures a high degree of certainty that the transaction would be part of the block. In the bitcoin network, the blocks are created after a delay of 10 minutes, as mandated by Bitcoin.

25. Bitcoin can transact only ___________ transactions per second

A. 3
B. 6
C. 7
D. 8

Answer: C

Explanation: The general rule of thumb (heuristics) is to wait for at least six confirmations on the transaction, this assures with a high degree of certainty that the transaction would be part of the block. In the bitcoin network, the blocks are created after a delay of 10 minutes, as mandated by Bitcoin. Bitcoin can transact only seven transactions per second; in essence, bitcoin is a time-consuming process.

26. Which of the below programming language is used for writing bitcoin scripts?

A. C++

B. Forth

C. Python

D. Golang

Answer: B

Explanation: A programming language called "Forth" is used for writing bitcoin scripts. It is very easy to write and supports cryptographic operations. This language supports stack-based programming. The bitcoin scripting language has a limited set of instructions: only 256 instructions and each one is represented by one byte. Out of these 256 instructions, fifteen are currently not in use on the bitcoin script and 75 others are reserved, which means that they have not been assigned any significance.

27. This is commonly used in escrow transaction of the bitcoin script:

A. MULTISIG

B. SEGWIT

C. LOCK TIME

D. Pseudowords

Answer: A

Explanation: MULTISIG: This is commonly used in escrow transactions done using bitcoin, as financial transactions are getting refined. A transaction with MULTISIG requests for approval from two or three third parties.

28. Ethereum started with the notion that __________ is a law

A. Ether

B. Code

C. DAO

D. Cryptocurrency

Answer: B

Explanation: Ethereum started with the notion that "code is a law." In other words, a contract on Ethereum could be your ultimate authority, and nobody can overrule the contract. "Dow" or DAO means "Decentralized Autonomous Organization," which enabled users to deposit funds and receive returns dependent on the investments which the DAO made.

29. A DAO counts on __________ for decision-making

A. Smart Contracts

B. Protocol

C. Ethereum

D. Coder

Answer: A

Explanation: A DAO can be an electronic digital organization that works with no hierarchical direction; it functions in a decentralized and democratic way. Hence, DAO can be an organization where the decision is not in the control of a centralized authority but instead at the command of certain designated authorities or some set of chosen people who are part of a body. It is based on a blockchain system, where it is regulated by the protocols inserted inside a smart contract, and consequently, DAOs count on smart contracts for decision-making.

30. Ethereum split into ETH and ETC after DAO Attack:

A. True
B. False

Answer: A

Explanation: Ethereum split into ETH and ETC after DAO Attack

31. There are ____________ types of nodes in the Ethereum network

A. 3
B. 5
C. 2
D. 4

Answer: C

Explanation: There are two types of nodes (computers) in the Ethereum network. 1. Mining nodes and 2. Ethereum virtual machines. These node types were designed to clarify the concepts of Ethereum. Generally, in the majority of cases, you can find no devoted EVM nodes. As an alternative, all nodes will function as a miner in addition to being an EVM node. Mining nodes refer to the nodes that belong to miners. EVM has a unique code attached to it, which is called as Smart Contract.

32. ____________ is a type of cryptocurrency used in Ethereum.

A. Ripple
B. Ether
C. Bitcoin
D. Litecoin

Answer: B

Explanation: Ether is a type of cryptocurrency, digital currency, virtual currency, future currency, electronic currency, or an online currency. Ether is the cryptocurrency used in Ethereum.

33. The reference code of Ether is:

A. ETH
B. ETR
C. EHR
D. THER

Answer: A

Explanation: Ether was distributed in Ethereum's authentic Initial Coin Offering (ICO) in 2014. It charged approximately 40 cents for one Ether. Now, one Ether is valued at hundreds of dollars, considering that usage of the Ethereum network has increased tremendously.

The reference code of Ether is ETH, and as of June 2019, one ETH is comparable to USD 269.55.

34. This is the internal Currency of Ethereum:

A. Wei
B. Gas
C. Bitcoin
D. USD
Answer: B

Explanation: Gas is the internal currency of Ethereum. The execution and resource utilization costs are predetermined in Ethereum in terms of Gas units. This is also known as Gas Cost.

35. Which of the following is not a component of Account in Ethereum

A. Nonce
B. Balance
C. Code Hash
D. Transaction
Answer: D

Explanation: The account consists of four components which are present regardless of the type of account:

Nonce: When the account is externally owned, this signifies the number of transactions delivered (sent) out of that account. In the case of contract accounts, this means the number of contracts generated via this account.

Balance: The total balance of the accounts in Wei.

Storage Root: A hash of the main root node of the Merkle Patricia Tree. This tree encodes the hash of all the storage contents of the account, and it is empty by default.

Code Hash: The hash code of the EVM code of the particular account. For contract accounts, this code is hashed and stored as the code hash. For externally owned accounts, the code hash field is not applicable. Code hash will have the empty string.

36. In the monetary policy, the inflation rate of Bitcoin supply is hard-coded into the protocol itself.

A. True
B. False
Answer: A

Explanation: In the monetary policy, the inflation rate of Bitcoin supply is hard-coded into the protocol itself (maximum of 21 million bitcoins). Currently, over 17 million bitcoins have already been mined. With roughly ten minutes per block, this means 75 Bitcoins are produced per hour. Roughly speaking, every four years, the rate is cut in half. This is called the halving or happening.

37. Ethereum blockchain uses modified Merkle tree called as:

A. Modified Merkle Tree

B. Ethereum Tree

C. State Tree

D. Patricia Tree

Answer: D

Explanation: In the Ethereum blockchain, it modifies the Merkle tree to store three trees for three kinds of objects: transactions, receipts and state. A decentralized computer can store a state. Also, it is called a Patricia tree, not a Merkle tree because it is modified to store state.

38. Which of the following is not a type of transaction in Ethereum?

A. Externally owned account to another externally owned account

B. Externally owned account to contract account

C. Contract-to-contract transaction

D. State-to-state transaction

Answer: D

Explanation: There are four different types of transactions that can be executed on the Ethereum system to transfer Ether across accounts. Following are the possible cases:

a. An externally owned account can send ether to another externally owned account in a transaction.
b. An externally owned account can send ether to a contract account in a transaction.
c. Contract to contract ether transaction.
d. A contract to externally owned account transaction.

39. In the Architecture of Ethereum, Layer 0 represents:

A. Consensus layer

B. Economic layer

C. Blockchain services

D. Inter-operability

Answer: A

Explanation: In the architecture of Ethereum, Layer 0 represents Consensus layer, Layer 1 represents Economic layer, Layer 2 represents Blockchain services, Layer 3 represents Inter-operability, Layer 4 represents Browsers and Layer 5 represents DAPPs layer.

40. In the Architecture of Ethereum, Layer 1 represents:

A. Consensus layer

B. Economic layer

C. Blockchain services

D. Inter-operability

Answer: B

Explanation: Same as for Question 39.

41. In the Architecture of Ethereum, Time Stamping, Name Registry and File System are at:

A. Layer 0
B. Layer 1
C. Layer 2
D. Layer 3

Answer: C

Explanation: Same as for Question 39.

42. In Ethereum, this decides the intricacy of the challenge:

A. Difficulty
B. MixHash
C. Nonce
D. State

Answer: A

Explanation: Difficulty: It decides the intricacy (complexity) of the puzzle/challenge awarded to miners of a particular block. The gas limit of the block decides the most gas allowed for the block. This aids in ascertaining the number of transactions that might be a part of this block.

43. What does UTXO stand for?

A. Used Transaction Output
B. Unnecessary Transaction Output
C. Unspent Transaction Output
D. Unspent Trigger Output

Answer: C

Explanation: UTXOs stand for "Unspent Transaction Output". In terms of transaction models, Bitcoin has an unspent transaction output model or UTXO. Bitcoin only tracks transactions from address to address. Therefore the wallet has a software, which aggregates the addresses controlled by the same private key. One private key can control a bunch of different addresses and can show the total balance the user has.

44. What does IPFs stand for?

A. Interplanetary file system
B. Interprocess file system
C. Interprocess fork system
D. Interplanetary fork system

Answer: A

Explanation: We can store data in some distributed hash table, like IPFs (Interplanetary file system). IPFs give back a hash a content address for all of the content and helps to retrieve the file easily from the data storage system. This is what we store in the blockchain. The hash points to the data storage.

45. The name of the research paper related to the Bitcoin origin?

A. Orange paper
B. Pink paper
C. White paper
D. Black paper
Answer: C

Explanation: Bitcoins origin came in 2009 when Satoshi Nakamoto published the white paper. No fundraising was done. Also, miners had to mine on the Bitcoin network to receive the first bitcoins. Therefore, Satoshi himself has around 1 million Bitcoins, and this is supposed to be the most significant amount for any single owner. However, he never spent any and disappeared from the internet, without anyone knowing his identity.

46. While Bitcoin uses UTXO, Ethereum uses:

A. UTXO
B. Account Balance
C. Transaction Pools
D. Memory Pools
Answer: B

Explanation: Accounts are the primary building blocks for the Ethereum ecosystem. Ethereum supports two kinds of accounts. The account consists of four components, which are present regardless of the type of account: Nonce, Balance, Code Hash and Storage Root.

47. What is the maximum limit of supply of bitcoin?

A. 2 million
B. 2.1 million
C. 28 million
D. 21 million
Answer: D

Explanation: After the year 2140, the production/creation of bitcoin will be stopped. At that point of time, the count of bitcoins will touch 21 million, which is the maximum limit. Until May 2019, 17.7 millions of bitcoins (roughly) have been created. Only 3.27 million bitcoins are left behind, which are to be maintained. Around 1,800 bitcoins are being mined every day.

48. A block in blockchain, can have more than one Parent Block.

A. True
B. False

Answer: B

Explanation: A block in blockchain stores the hash of the previous block and hence every block can have only one parent block. Genesis block does not have a parent block as it is the first block in the blockchain.

49. What is a public key?

A. A key that opens secret account

B. A key not to be given to the public

C. A key given to the public

D. A key on the wallet.

Answer: C

Explanation: Public key is distributed and private key is not distributed. Private key is a unique identifier, which is assigned to each node. This will be kept private as the name suggests and is like a password to one's bank account.

50. Ethereum uses __________ characters to represent the identity of an external account

A. 36

B. 160

C. 256

D. 80

Answer: B

Explanation: The public key consists of 256 characters. However, Ethereum uses the initial 160 characters to represent the identity of an externally owned account. Private key is a unique identifier, which is assigned to each node. This will be kept private as the name suggests and is like a password to one's bank account.

Short-answer Questions

1. Write short notes on public permissionless blockchain.

The public permissionless blockchain is synonymous with a public blockchain. In a public blockchain, anyone in the world can access the blockchain, download a copy of the code, and run a node. It is a fully decentralized distributed network. One does not need any permission to read/access a transaction, initiate a transaction, or participate in the consensus process (PoW) to create a block. Participants or nodes remain anonymous through high cryptographic protocols. Anonymity, transparency, and immutability are valued over efficiency.

2. When is permissionless blockchain used?

We use "permissionless blockchain" when the state (and date) needs to be stored in the database, if there are multiple writers for a transaction, when we cannot use always-online TTP and when not all the writers are known.

3. What is bitcoin? How can we measure the success of the bitcoin?

Bitcoin, the first blockchain created, is a public permissionless blockchain. The idea behind the design was to create a decentralized digital currency that is free from government regulation whereby two people can confidentially trade directly with one another without the need for middlemen or intermediaries. The success of bitcoin can be measured based on the number of transactions happening daily. As of March 2020, nearly three lakh transactions were made per day.

4. What is the relationship between Satoshi and bitcoin?

One Rupee is equivalent to 100 Paise. Similarly, ten crore Satoshi is called as one bitcoin (BTC). 100,000,000 Satoshi = one BTC. Those who are unable to buy a single bitcoin can buy groups of Satoshi as desired. Bitcoin, which was created in 2009, gained momentum and started its monster growth from 2010.

5. What are the advantages of bitcoin?

Bitcoin is a secured currency and can be used very quickly. Nobody controls this currency, and hacking the bitcoin network is very difficult. While there are regulatory bodies and governance structure to control the physical currency of countries, there is no regulatory body or governance to control the bitcoin currency.

6. Why is bitcoin called as world currency?

Bitcoin is the common currency of the globe. Bitcoin is a type of cryptocurrency, digital currency, virtual currency, future currency, electronic currency, or an online currency.

Bitcoin can be used in all countries across the globe. That is why bitcoin can be called as World Currency. We cannot see the bitcoin, touch or feel it. There is no central governing body (controller) to control the currency. Bitcoin is connected through networks across the globe, and hence, it is called a distributed network.

7. What is bitcoin mining?

Creation of bitcoin blocks is called mining. Mining is the mechanism whereby nodes called "miners" in the Bitcoin world validate the new transactions and add them to the blockchain ledger. Unearthing the bitcoin from a computer is called mining. To mine bitcoin, we need high-processing computers. Computer graphics cards are used to mine bitcoins.

8. Name four types of hardware mining.

There are four types of hardware mining, namely:

1. CPU mining (Central Processing Unit)
2. GPU mining (Graphical Processing Unit)
3. FPGA mining (Field Programmable Gate Array)
4. ASIC mining (Application Specific Integrated Circuit).

9. What is PoW in bitcoin?

Proof-of-work is like a race where nodes compete with each other. In this process, miners pick up transactions within a block, validate them, find nonce for hashcash, and propose a block. The second part of hashcash involves finding a nonce, a cryptographic solution that is added into the block. Without proof of work, adding blocks to the blockchain would be too easy and could make it vulnerable to hackers.

10. Write main three components of a block.

Each block (of transactions) has three necessary informations, namely:

1. Block header
2. Hash of previous block header
3. Merkle root.

11. What is Merkle root?

A Merkle root is synonymous to a fingerprint of all the transactions in the block. It is created by hashing together pairs of Transaction IDs, giving a short and unique fingerprint for all the transactions in a block. This is also a field in a block header.

12. What is HashCash? How it can be used to identify email spam?

Hashcash is a cryptographic hash-based proof-of-work algorithm, which was used to identify emails spammers. For example, the number of seconds/minutes of CPU time taken to write an email is hashed and is written in the header of the email. Let us say, on average, it takes a minimum time of 20 seconds to write an email. When an email is received, the hash value is checked, and if the value is lesser than 20 seconds, then that email can be classified as Email Spam.

13. What are the essential aspects of a hashcash puzzle?

There are three essential aspects of a hashcash puzzle:

a. It is difficult to compute
b. It has configurable difficulty
c. It is easy to verify.

14. What are the types of nodes in a bitcoin?

There are three types of nodes:

a. Full node
b. Mining node
c. SPV node.

15. What is a full node in a bitcoin network?

Full Node: These are the nodes that act as a proper full node, doing all the activities of a node (they accept transactions, validate transactions and relay them to further full nodes). This needs lots of computing and some chance of creating a block and earning mining reward. A node that operates in the bitcoin network would have full node rights; all nodes in the network are equal, as stated by the Satoshi Nakamoto. Many people and organizations volunteer themselves to run full nodes using spare computing and bandwidth resources in their computer—but more volunteers are needed to allow Bitcoin to continue to grow.

16. What is Zero, One and Six confirmation Transaction?

Zero confirmation transaction: A scenario where a user submits a transaction, and the other party accepts it — no additional confirmation from any other node.

One confirmation transaction: A scenario where receivers wait until at least one node has confirmed the transaction as valid.

Six confirmation transaction: A scenario where receivers wait until at least six nodes have confirmed the transaction as valid.

17. List down the common characteristics of a bitcoin script.

Common characteristics of a bitcoin script:

a. Simple to execute
b. Lightweight (does not rely on other software packages).
c. Less computing energy
d. Built for bitcoin
e. Uses cryptography (Hashes, Signature verification)
f. Stack-based
g. No loops
h. Conditional statements supported (`If` statements)
i. Validation of multiple nodes supported (MULTISIG instruction)

18. What is data instruction in a bitcoin script? Give examples.

Data instructions: If data instructions are found in a bitcoin script, the data corresponding to the particular instruction is simply hard-pressed on top of the stack data structure. Data instructions are given with surrounding angle brackets. For example: <Sig>, <PubKey> and <PubKeyHash> are examples for Data Instructions.

19. What is opcode instruction in a bitcoin script? Give examples.

Opcode instructions: An opcode, in a bitcoin script, executes a particular function taking the input data, which usually resides on the top of the stack. Opcode instructions start with "OP." For example, OP_DUP, OP_HASH160, OP_EQUALVERIFY, and OP_CHECKSIG are examples of Opcode instructions.

20. What is lock time and multisig of a bitcoin script?

LOCK TIME: This attribute is one of the most used in new bitcoin platforms. Where settlement is involved, LOCK TIME is added in the metadata. LOCK TIME implies that a payment is locked until a particular time and no miner will validate the transaction for this period.

MULTISIG: This commonly used in the escrow transactions done using bitcoin. A transaction with MULTISIG requests for approval from two or three third parties.

21. List any three applications of Ethereum.

Ethereum allows creating decentralized and open-to-all solutions for various sectors, which include:

- Crowdsourcing

- Voting for democracy
- Public rating engine for movies

22. Write notes on Do-it-yourself platform of Ethereum.

Ethereum could be a do-it-yourself platform for most decentralized programs, also referred to as DApps – de-centralized programs. If you would like to produce a brand new application that no person controls, perhaps not you personally, though you wrote it, then all you need to learn the Ethereum programming language named Solidity.

23. Write notes on "Code is Law" principle of Ethereum.

Ethereum started with the notion that "code is a law." In other words, a contract on Ethereum could be your ultimate authority, and nobody can overrule the contract. "Dow" or DAO means "Decentralized Autonomous Organization," which enabled users to deposit funds and receive returns dependent on the investments which the DAO made.

24. Write notes on DAO of Ethereum.

A DAO can be an electronic digital organization that works with no hierarchical direction; it functions in a decentralized and democratic way. Hence, DAO can be an organization where the decision is not in the control of a centralized authority but instead at the command of certain designated authorities or some set of chosen people who are part of a body. It is based on a blockchain system, where it is regulated by the protocols inserted inside a smart contract, and consequently, DAOs count on smart contracts for decision-making.

25. What is block in Ethereum?

Block is an essential concept in Ethereum. Blocks are containers for the transaction. A block contains multiple numbers of transactions. Also, these blocks are all linked together. These transactions include state and other programmable parameters. It is stored on every miner's computer. It (Ethereum) currently uses the proof-of-work algorithm.

26. How is a miner's role similar to that of an accountant?

A miner is responsible for writing a transaction to the Ethereum series. A miner's job is similar to that of an accountant. For example, an accountant is in charge of both writing in and taking care of the ledger alike; a miner, on the other hand, is solely accountable for writing transactions to Ethereum ledger. A miner is enthusiastic about transferring paper transactions to the ledger due to the prospect of gaining a reward. Miners get two kinds of reward: benefit for writing a block into the series and accumulating gas prices from all transaction in the block (cube).

27. What are the types of nodes in Ethereum?

There are two types of nodes (computers) in the Ethereum network.

1. Mining nodes and

2. Ethereum virtual machines.

These types of nodes were designed to clarify the concepts of Ethereum. Generally, in the majority of cases, you can find no devoted EVM nodes. As an alternative, all nodes will function as a miner in addition to EVM node. Mining nodes refer to the nodes that belong to miners.

28. What are the fundamental functional elements performed by miners on the mining nodes?

There are three fundamental functional elements performed by miners on mining nodes.

1. Mine or make a fresh new block together with the transaction and then compose it into Ethereum Ledger
2. Promote and send a recently mined block to other miners
3. Accept new blocks created by other miners and maintain them in the ledger.

29. What is ether?

Ether is a type of cryptocurrency, digital currency, virtual currency, future currency, electronic currency, or an online currency. Ether is the cryptocurrency used in Ethereum. Ether was distributed in Ethereum's authentic Initial Coin Offering (ICO) in 2014. It charged approximately 40 cents for one Ether. Now, one Ether is valued at hundreds of dollars, considering that usage of the Ethereum network has increased tremendously.

30. Write notes on Gas of Ethereum.

Ether is paid as commission for any execution that affects the state in Ethereum. Ether (ETH) is traded on people exchanges and its particular selling price value changes each day. As an example, if Ether is used for paying fees, the cost could be high on certain days and lower on other days. Individuals may wait for the price of Ether to fall to perform their transactions. This is not ideal for the Ethereum platform. The gas helps in alleviating this problem. Gas is the internal currency of Ethereum. The execution and resource utilization costs are predetermined in Ethereum in terms of Gas units. This is also known as Gas Cost.

31. What is an Account in Ethereum?

Accounts are the primary building blocks of the Ethereum ecosystem. Ethereum supports two kinds of accounts. Every account comes with a balance that yields the present value saved in it. Externally owned accounts are possessed by men and women around Ethereum.

32. Write notes on Swarm.

Swarm is currently a peer-to-peer document sharing network, similar to Bit Torrent; however, it is incentivized with micro-payments of ETH. Whisper is an encrypted messaging protocol that enables nodes to send out messages directly to each other in a secure way by hiding the sender and recipient identity from third-party snoopers.

33. What is Ethash?

Ethash is a proof-of-work algorithm that is a modified variant of a precursor algorithm called Dagger-Hashimoto. Together with Ethash, the outcome in the hashing process has to result in a hash value that is below a particular threshold.

34. What are the four types of transactions in Ethereum?

There are four different types of transactions that can be executed on the Ethereum system to transfer Ether across accounts. Following are the possible cases:

a. An externally owned account can send ether to another externally owned account in a transaction.
b. An externally owned account can send ether to a contract account in a transaction.
c. Contract-to-contract ether transaction.
d. A contract-to-externally owned account transaction.

35. What are the conditions of a transaction in Ethereum?

All transactions must meet the below set of requirements to be executed. These include:

- The transaction must be formatted correctly in the form of RLP. "RLP" means "Recursive Length Prefix." It is a data format used to encode nested arrays of binary data and is used by Ethereum to serialize objects.
- The transaction should have a legal transaction signature.
- The transaction should have a valid transaction nonce. A transaction nonce should be the sender's account nonce.

Essay-type Questions

1. Explain the various characteristics of public blockchain in detail.
2. When is use permissionless blockchain used? Explain using a flowchart.
3. What is bitcoin? What is the thought process behind it? What is satoshi? What are the main futures of bitcoin?
4. Describe the characteristics of the bitcoin network.
5. Describe growth of the value of bitcoins over the years.
6. What is bitcoin mining? What are all the types of bitcoin mining?
7. Write an essay on Merkle Tree and Merkle Root. How is Merkle tree created?
8. Explain the components of a block and block header of bitcoin in detail.
9. What is hashcash? What are the essential aspects of a hashcash puzzle?
10. Explain the three types of nodes in a bitcoin network in detail.
11. Explain two types of instructions of a bitcoin script in detail.
12. Explain the steps of the P2PKH script of bitcoin.
13. What are the flow control-related Opcode instructions in a bitcoin script?
14. Write an essay on transactions block of a bitcoin.
15. Write an essay on the public key and private key check of bitcoin transaction.

16. Describe the characteristics of the Ethereum network.
17. Describe growth of the value of Ether over the years.
18. Write an essay on Merkle Tree and Merkle Root. How does it differ in Ethereum network?
19. Explain the components of a block and block header of Ethereum in detail.
20. Explain the two types of nodes in an Ethereum network in detail.
21. Explain the do-it-yourself characteristics of Ethereum.
22. Explain "Code is Law" principle of Ethereum.
23. What is DAO? How is it related to the formation of Classic Ethereum?
24. Explain the structure of block in Ethereum in detail.
25. Explain the function of IPIFs in detail.
26. Explain the types of nodes in ethereum. What are the three fundamental functional elements performed by miners on mining nodes?
27. Explain Ethereum accounts and components of accounts in detail.
28. Compare and contrast Bitcoin with Ethereum.
29. Explain the architecture of Ethereum.
30. Explain the workflow of Ethereum.

CHAPTER 5

Smart Contracts

LEARNING OBJECTIVES

Smart contracts are like contracts in the real world. However, these are entirely digital and stored as a small program inside a blockchain. With smart contracts, it is possible to build a system where we need not have to trust a third party to do the transactions. This chapter covers the basics of the Smart Contracts, its features, and how it works. It also discusses the types of Smart Contracts and use cases in different industries.

5.1 Introduction

According to Nick Szabo (1996), a Smart Contract is a set of promises, specified in digital form, including protocol related procedures within which all the parties perform on these promises. Multiple technological supports (languages) evolved over time, and this definition still holds good.

5.2 Smart Contract

A Smart Contract is a computer program working on top of a blockchain, which has a set of rules (conditions) based on which, the parties to that Smart Contract, agree to interact with each other (refer Fig. 5.1). The agreement will be automatically executed if and when the pre-defined rules are met. The Smart Contract Program will facilitate, verify, and ensure the agreement or transaction. This can be treated as the purest form of automation. Real-life agreements to record the purchase and sale of assets can be automated using Smart Contracts.

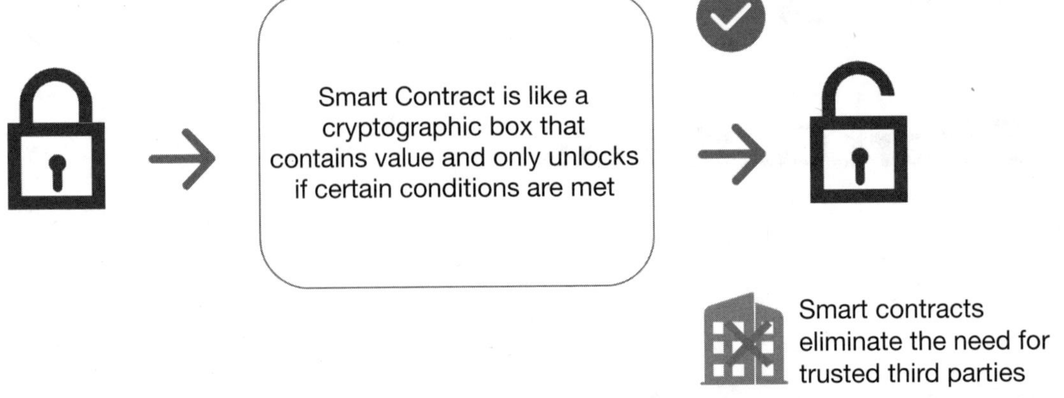

Figure 5.1: Smart Contract concept

When the smart contracts are programmed, compiled, and migrated on to the blockchain, they are provided an address depicting the location of the contract on the blockchain, and the application is recorded in a block and distributed across all the nodes. Hence, the applications are called decentralized applications. Decentralized applications interact with external users through a web browser.

A Smart Contract reduces transaction costs drastically. Smart Contract code, which is typically an auto enforceable code at the protocol level or application level, will be able to standardize transaction rules, thereby reducing the transaction costs of:

1. Reaching an agreement – Agreement of a smart contract typically defines the conditions, rights and obligations to which the parties agree.
2. Formalizing an agreement – Smart contracts can formalize the relationships between people, institutions, and the assets they own by executing an agreement. The rule is formalized in machine-readable code.
3. Enforcing an agreement – This can be automatically executed by a computer or a network of computers once the rules or conditions are met.

Smart contracts have allowed a large-scale innovation to be triggered where all centralized organizations conducting businesses offering a variety of services to end consumers could be modeled without a centralized control structure and aggregation. End users can now conduct businesses among themselves in a peer-to-peer manner without any human intervention. There are several critical areas where smart contracts are superior to their traditional counterparts.

The benefits of smart contracts are:

- Simple, faster to execute, and availability of updates in real-time
- No requirement of mediators and centralized entities
- Lesser cost because there is no need to pay fees to middlemen
- There is no delay in delivering outcomes.

5.2.1 Smart Contract Example

If X and Y do not know each other, and hence do not trust/believe each other, they need a trusted third party as an intermediary (third party) to verify transactions and enforce them, which is audited by another trusted third party. With Smart Contracts and Blockchains, there is no need for such trusted intermediaries for clearing or settlement of transactions.

Let us take the example of buying and selling a piece of land.

If Parker (refer Fig. 5.2) wants to purchase land from Michael, a series of trusted third parties are required to verify and authenticate the land documents and sale agreement. The process differs from state to state or among countries but it always involves at least one, but usually more, trusted third parties such as land revenue office, government registrar's office, and local authorities in combination with lawyers' organizations. It is a highly complicated and time-consuming process, and considerable (high amount) fees for these intermediaries apply.

Once all the concerned authorities and government agencies are on a blockchain, a smart contract could be used to define all the rules of a valid sale (refer Fig. 5.3). If Parker wanted to buy the land from Michael for a consideration of, say 42000 USD, using a smart contract program on the blockchain, the transaction information would be verified/validated by each node in the overall Blockchain Network to see if Michael is the real owner of the land, and whether the respective government tax has been paid, encumbrance is valid, and if Parker has the credentials to buy the land.

If the blockchain network agrees that all the conditions are right, Parker automatically gets the access code to the land documents, and hence the land is transferred to his name. Now, the blockchain registers Parker as the owner of the land, and Michael has 42,000 USD more in his bank account. No intermediaries are required.

Figure 5.2: Traditional sale agreement

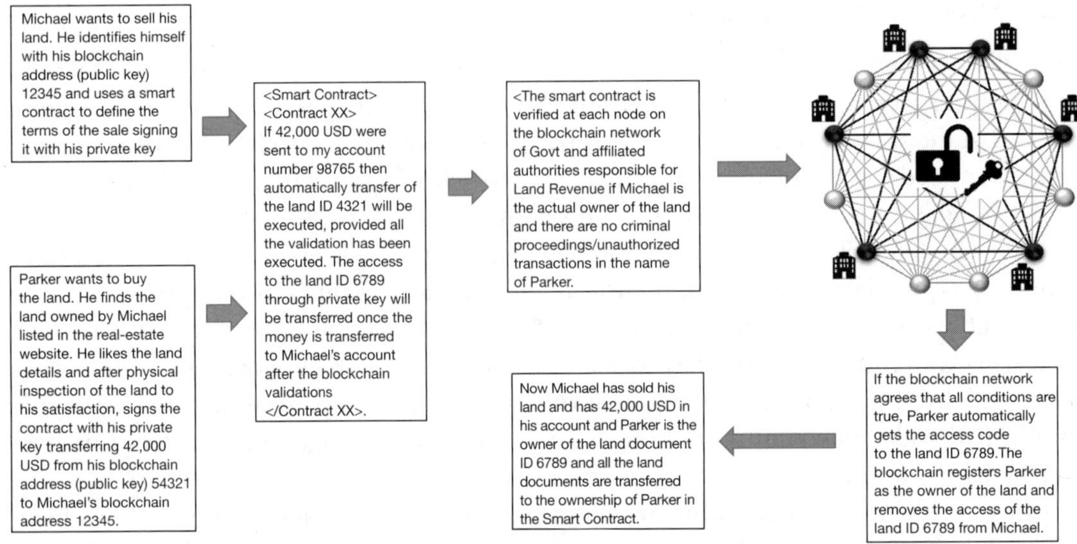

Figure 5.3: Sale agreement using smart contract

Every computer (node) running the blockchain program could check whether a specific person is the rightful (legal) owner of the land or not. The blockchain also ensures that no benami or illegal transactions are happening by checking the Seller's and Buyer's identity and security records, which are also stored in the blockchain.

With the smart contract, no one can illegally build houses on the land owned by Parker, since every transaction about the land, owner, seller, buyer and complete traceability of the land ownership and financial transactions are stored in the public blockchain. Anyone can inspect these records with the right access. The smart contract will hold the money paid by Parker until all the agreement conditions are met and will transfer the money to Michael only when all conditions are met. These terms of the agreement can be defined in the Smart Contract, including terms like installment payments based on the milestones agreed within a limited duration.

5.2.2 How Smart Contracts Work

A smart contract works through automated conditional performance. When a contractual obligation is met, the corresponding obligation is triggered.

For example, an obligation could be triggered by:

- a specific event ("if X happens, then action Y")
- a specific date or at the expiration of time ("at X date, action Y")

Different platforms can host smart contracts, but generally, smart contracts are implemented on a blockchain where the validation logic sits inside a "block". All the messages needed for a smart contract are bundled within the block. These messages can be either inputs or outputs or may interface to other outside computer codes.

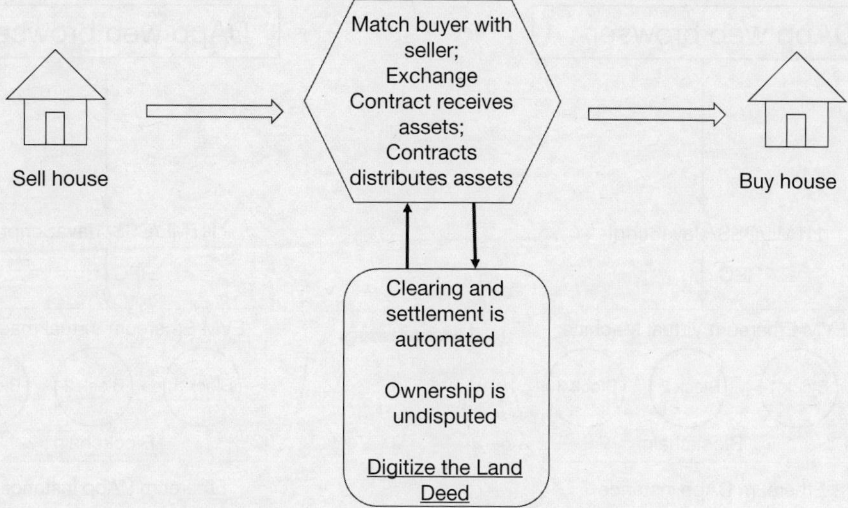

Figure 5.4: How smart contracts work

Ethereum revolutionized the concept of Blockchain. High-level programming languages like Solidity are used to create applications to automate business processes and record them on the blockchain in the form of smart contracts. Using Ethereum, we can automate real-life agreements to record the purchase and sale of assets. The assets are digitized, and the ownership transfer and traceability are captured in the blockchain records. These are done through complex logic, mirroring the real-life execution of the agreements. Smart contacts receive the address of the location of the contracts on the blockchain, just like a stack register in an ALU, and the application is recorded in a block and distributed across all the nodes which constitute decentralized applications.

Blockchain technologies use public-key encryption infrastructure (PKI). The initiator who wishes to participate in a smart contract (refer Fig. 5.4) hosted on a permissionless blockchain can download the software from open sources and publish the public key on the system (software) publicly.

When publishing the public key, the blockchain will also generate a corresponding private key for the initiator's address. This is a software key kept securely. If the initiator wants to trigger a smart contract transaction on the relevant ledger, the software will use its address to send an initiating message, encrypted with its private key to the other active participants. That message is picked up by the nodes (computer of the participants); however, it can be signed only by nodes having the private key. Participants with access to the public key received from the software can use it to verify the smart contract transaction and also to authenticate the message contents. Whenever a sufficient number of other participants or nodes is reached in a permissionless blockchain (typically more than 50%), the consensus protocol determines that message related to the smart contract should be added to the blockchain. In the case of a permissioned blockchain, such a determination may be arrived at by the administrator. Most of the times, smart contracts have to access external applications or external users through a web browser, as shown in Fig. 5.5.

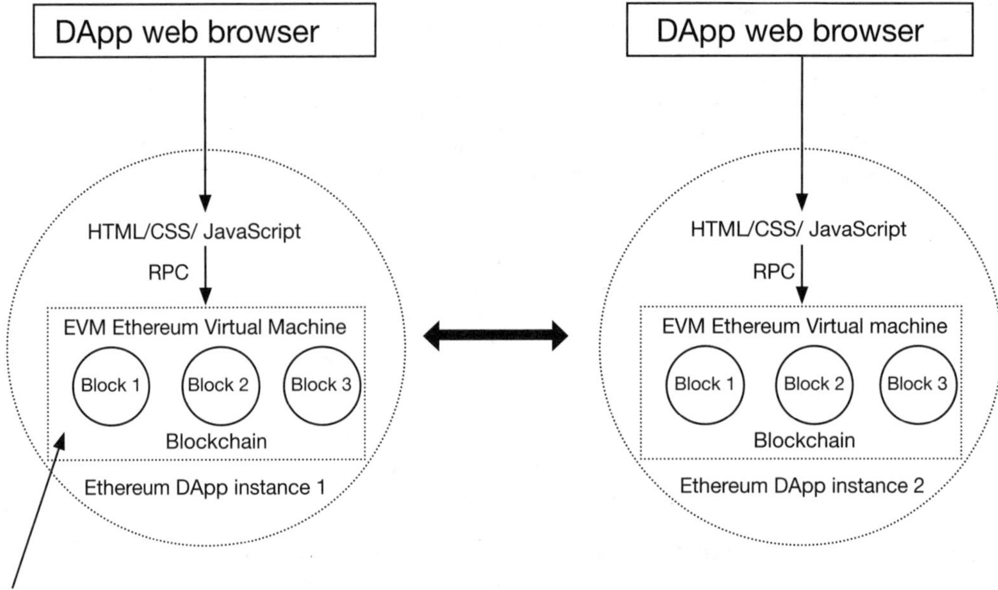

Figure 5.5: Decentralized applications interacting with blockchain through browser

The transactions in Ethereum blockchain are implemented through Smart Contract code, which is triggered by data and inputs from external accounts automatically. The list of all transactions and their Merkle roots are stored in all the blocks. How the smart contracts change the status through transactions is depicted in Fig. 5.6.

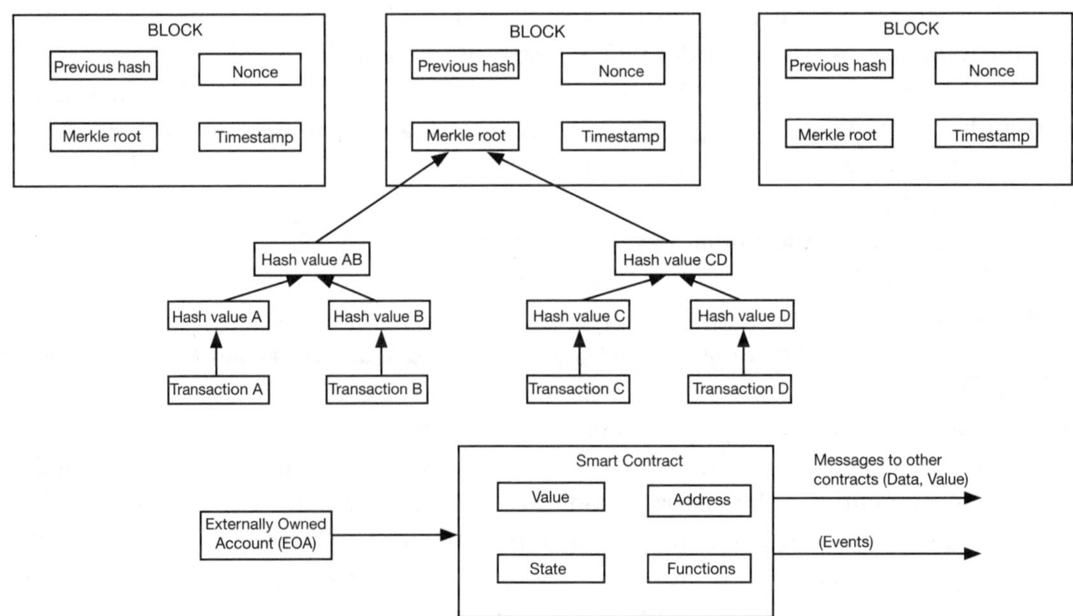

Figure 5.6: Smart contracts changing state through transactions

When smart contracts are triggered through an external account, the business rule inside the contract gets executed (refer Fig. 5.6), and the status of ledger balances of all the respective accounts gets updated. It is over-written on all the nodes of the blockchain. The data is written into all the nodes while writing; but while reading, the data is read-only from the nearest node. Therefore the effort and resources needed for writing and updating the information are more in an Ethereum blockchain. To incentivize this, Ethereum blockchain uses a concept called Gas that is used by the users to pay for the effort undertaken to execute any transaction conducted on the blockchain. Gas is expressed as a conversion unit of the native currency of Ethereum Blockchain, namely "ether", and keeps varying in time to adjust for the variation of the price of ether, thus keeping the transaction costs stable.

5.3 Characteristics of a Smart Contract

Smart contracts can track performance and transactions in real-time and can bring tremendous cost savings. To get any external information, a smart contract needs "information oracles", which feed the smart contract with the required external information. This external information can be made available by interacting with a web browser (refer Fig. 5.7).

Smart contracts can

- Verify itself
- Execute itself
- Be tamper-resistant
- Automate legal transactions in terms of rules
- Provide a high degree of security
- Reduce reliance on trusted intermediaries.
- Lower transaction costs

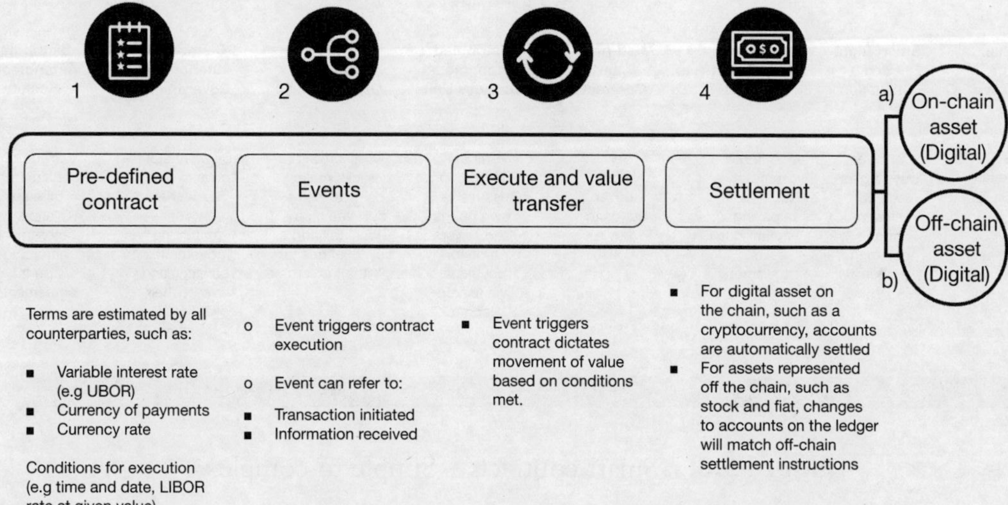

Figure 5.7: Smart contracts value chain

5.4 Types of Smart Contracts

There are different types of smart contracts as below:

Smart contracts distributed in a spectrum			
Contract entirely in code	Contract embedded in code using separate natural language version	"Split" natural language contract with encoded performance	Contract with encoded payment mechanism Encoding in natural language
Encoding natural language			**Automation**

Smart contracts have the potential to influence many industries. Smart contracts are currently used in fields such as banking, insurance, energy, e-government, telecommunication, media, manufacturing, education, and many more, ranging from simple to complex use cases (refer Fig. 5.8). Decentralized autonomous organizations are of the more complex type.

Smart Contracts can be categorized into four types based on the applications (refer Fig. 5.9).

- Smart legal contracts
- Distributed Applications (DApps)
- Decentralized Autonomous Organizations(DAO)
- Smart contracting devices

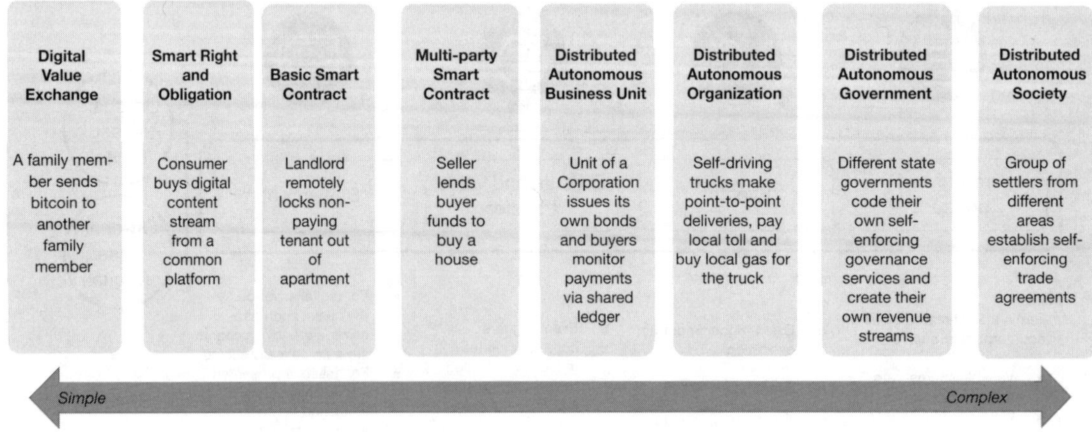

Figure 5.8: Smart contracts – Simple to complex

5.4.1 Smart Legal Contracts

They are smart contracts with various legal contract templates. They just execute the contracts as per the templates used.

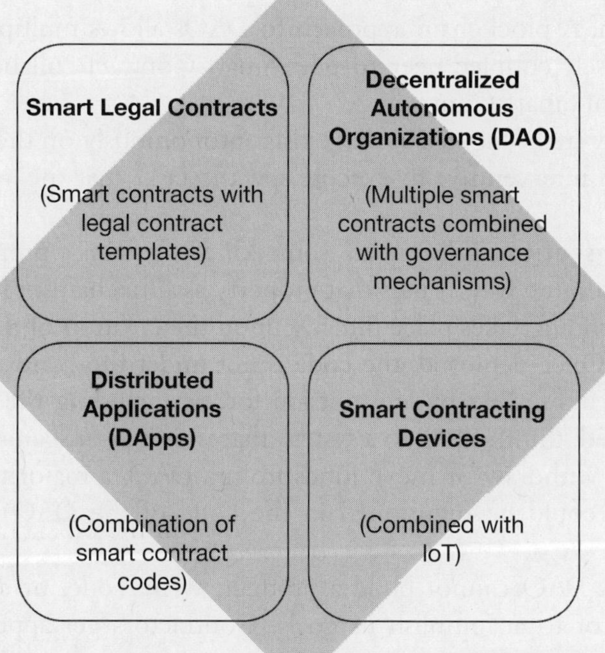

Figure 5.9: Types of smart contracts

5.4.2 DApps (Decentralized Applications)

DApps are applications which run point-to-point (P2P) network of computers instead of a single computer. DApps need not necessarily run on top of the blockchain network. Unlike simple Smart Contracts, which send bitcoin typically from X to Y, DApps have an unlimited number of participants from the market. DApps are a 'blockchain enabled' website, whereas the Smart Contract is what allows it to connect to the blockchain. DApps covers from end to end – both front-end and back-end. If we have to create a decentralized application on a smart contract system, several smart contracts must be combined and third-party systems have to be relied upon for the front-end.

5.4.3 DAO (Distributed Autonomous Organization)

A Decentralized Autonomous Organization, called as DAO, is an organization represented by logic encoded as a transparent program controlled by shareholders and not influenced by a central authority. A DAO's financial transaction records and program logics are maintained on a blockchain.

Tokens allow traceability to allow algorithmic measures of contribution utility. DAOs use blockchain technology to maintain a secure advanced record in the form of Digital Ledger (DL) to trace the financial transaction over the internet. It is hardened against fraud by trusted timestamping, and uses a distributed database. This approach disposes of the need for a trusted third party in the financial transaction, thus simplifying the overall financial transaction. A blockchain approach to DAOs allows multiple cloud computing users to enter a loosely coupled peer-to-peer Smart Contract collaboration. DAOs are a sophisticated form of Smart Contracts.

DAO is an entity (refer Fig. 5.10) that exists autonomously on the internet. However, it also needs human intervention to execute specific tasks that the automation, by itself, cannot do.

- **Tokens of Transaction**: DAO needs some kind of intrinsic property such as Tokens that is valuable in some way. It uses that property as a mechanism for rewarding certain activities. The funding takes place directly upon the creation of the organization.
- **Autonomous:** Once deployed, the code is not under the control of its creators, and so, it cannot be influenced by any outside forces, including the creators. DAOs are open-source based, thus leading to a system that is more transparent and incorruptible.
- **Consensus:** To withdraw or move funds from a DAO, a majority of its stakeholders (this percentage could be mentioned in the code of the DAO) must agree on the decision.
- **Contractors:** A DAO cannot build a product, write code, or develop hardware. It needs a contractor to accomplish its goals. Contractors get appointed via voting of token holders.
- **Proposals:** Proposals are the primary way of making decisions in a DAO. To avoid overloading the network with proposals, a DAO may require a monetary deposit that deters people from spamming the network.
- **Voting:** After submitting a proposal, voting takes place. DAOs allow people to exchange economic value with anyone in the world, through activities such as investing, money raising, lending and borrowing, without the need for an intermediary, just by trusting the code.

All cryptocurrencies which use public blockchains are DAOs (Bitcoin, Ethereum, Dash, Digix, etc.).

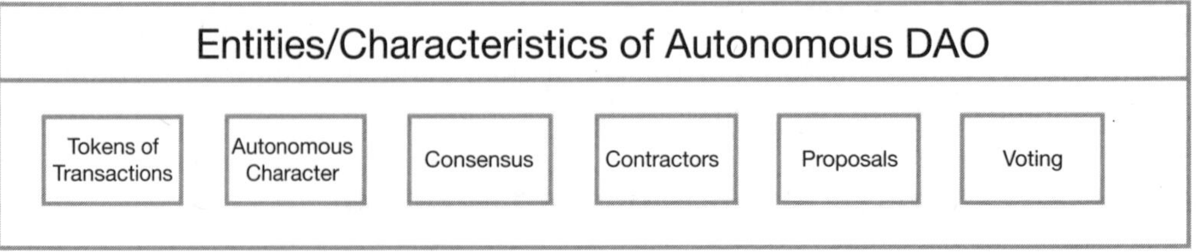

Figure 5.10: Entities of autonomous DAO

5.4.4 Smart Contracting Devices (Combined with IoT)

Currently, any failure in an IoT (Internet of Things) ecosystem on a web internet will reveal critical personal data such as our shopping data, census data, etc., that are being used by multiple devices. A security breach occurs predominantly in authentication, connection, and transaction. By using Blockchain to manage and access data from IoT devices, the hacker has to bypass an additional layer of security coded with some of the most robust encryption standards available. Also, there is no worry of a single-point failure, due to decentralized authority.

5.5 Types of Oracles

An Oracle, in the context of blockchains and smart contracts, is an agent that verifies real-world occurrences (off-chain) and submits the details to a blockchain (on-chain) to be used by smart contracts.

For example, the trustees can agree to release funds only when a particular set of conditions is met. To release any funds, an oracle has to sign the smart contract as well. There are different types of oracles based on usage. We differentiate between software oracles, hardware oracles, consensus oracles, and inbound and outbound oracles.

5.5.1 Software Oracles

Software oracles (refer Fig. 5.11) handle information that is generally available over the internet.

An example could be the temperature in a city, prices of commodities and goods, the details of flight or train delays, etc. The data originates from online sources, like railway reservation sites, e-commerce sites, etc. The software oracle culls out relevant information and sends it to the smart contract.

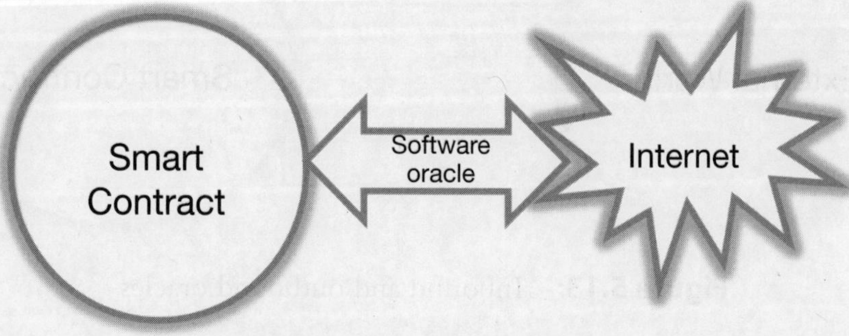

Figure 5.11: Software oracles

5.5.2 Hardware Oracles

Some smart contracts get input directly from the physical world (refer Fig. 5.12).

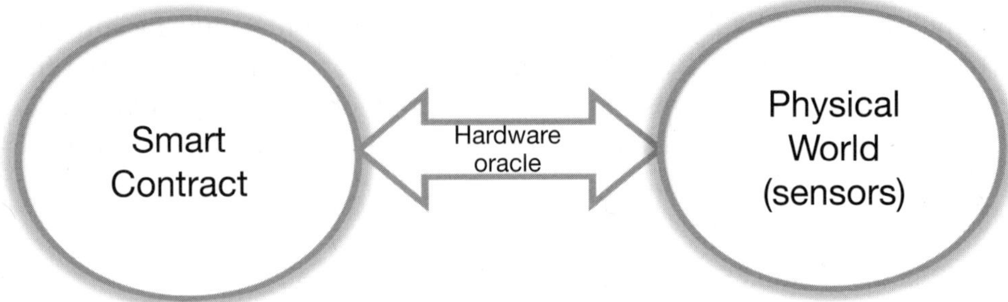

Figure 5.12: Hardware oracles

For example, it can be used to track the details of a car crossing a specific cross-junction (date, time, speed, direction, or location), where moving sensors help. Another use case is in logistics, where RFID sensors are used to track the movement of the trucks carrying goods.

5.5.3 Inbound and Outbound Oracles

Inbound oracles (refer Fig. 5.13) provide a smart contract with data from the external world. One such use case is an automatic purchase order for the spares if the spares stock value hits a certain inventory number.

Outbound oracles (refer Fig. 5.13) are like an outbound integration that allows smart contracts the capability to send data to the outside world.

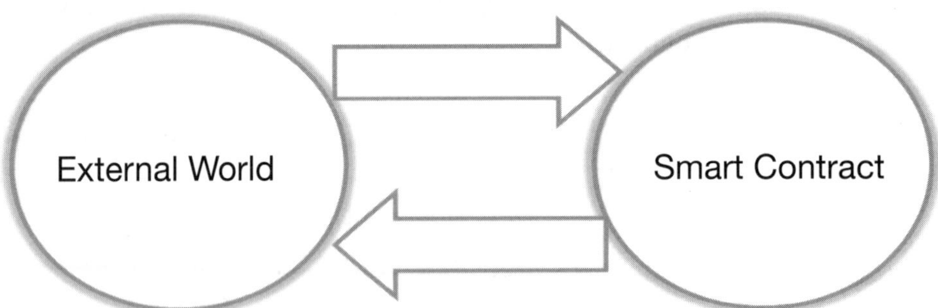

Figure 5.13: Inbound and outbound oracles

5.5.4 Consensus-based Oracles

To confirm future outcomes, the prediction markets rely heavily on oracle. It would be risky to depend on only one source of information. To avoid market manipulation, prediction markets implement a rating system for oracles. A combination of different oracles is used (refer Fig. 5.14) for enhanced security; for example, 6 out of 10 oracles (majority) could determine the outcome of an event.

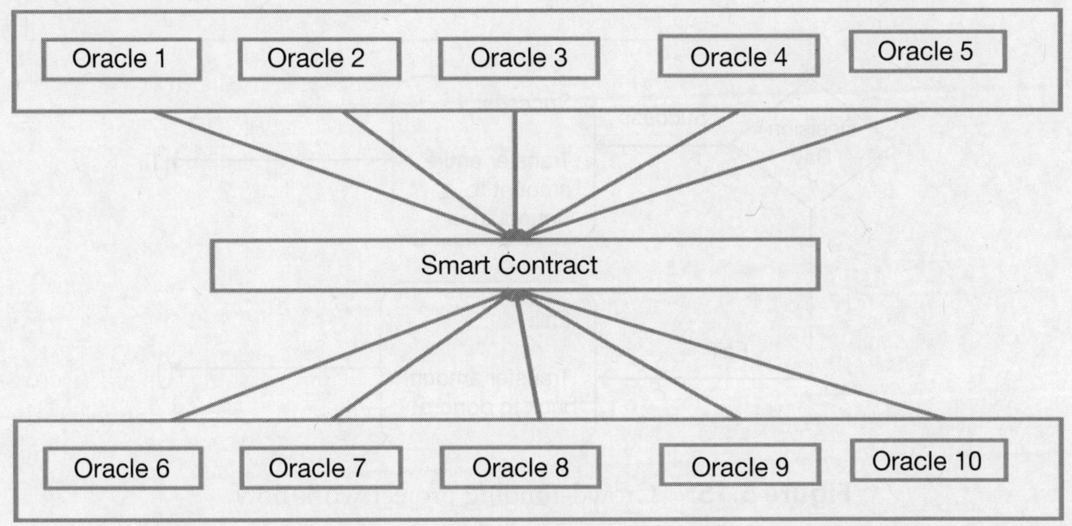

Figure 5.14: Consensus oracles

Connecting a smart contract to external data and existing applications, data feeds, APIs, bank payments, etc. is an essential step for creating useful smart contracts.

5.6 Smart Contracts in Ethereum

A contract is a legal record that binds two parties who intend to do a transaction immediately or later on. Since exemptions are valid, they are enforced and empowered by law. Examples of this type of contract include a person getting to a contract with an insurance agency to secure his wellness, an individual investing in a slice of property belonging to somebody else or even a business attempting to sell its shares to some other corporation. A smart contract will be a contract executed, deployed, and implemented in an Ethereum environment. Smart contracts result in the digitization of these legal contracts. They may store data. The data (information) stored may be utilized to capture data, facts, associations, accounts, or some additional information required to execute the logic to get real-world contracts. A smart contract may call another smart contract in an object-oriented environment to make and utilize objects of a different class. Think of a smart contract as a small program comprising of Functions. It is possible to create an instance of this contract and invoke contract functions to view and update contract data alongside the implementation of logic.

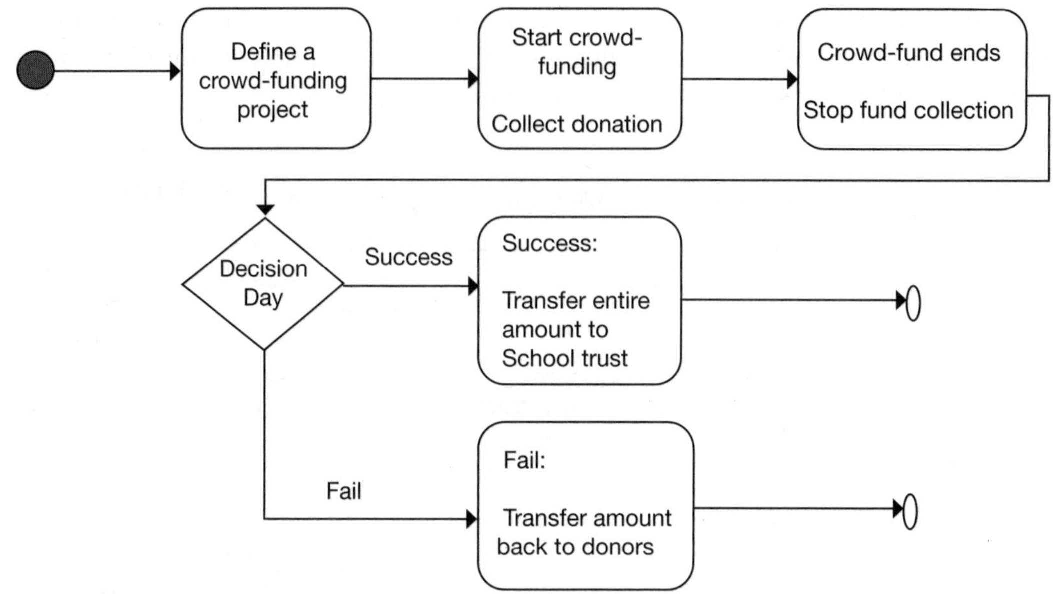

Figure 5.15: Crowd-funding project work flow

Smart contracts enable the blockchain to be useful in a variety of industries such as finance and trade, supply chain, derivatives and securities, and banking. Let us assume we are crowd-funding to build a school. (Crowd-funding is a new mechanism for raising funds from individuals or group of individuals, for a specific purpose). Let us raise the project, with the following features (refer Fig. 5.15).

Project: Crowd-funding for school

Total amount: 63,000 USD

Total duration: 30 days.

Decision: If the total amount is raised within the duration, it would be sent to "School Trust"; however, if the funding does not reach the total amount, funds would go back to the respective donor, i.e., each donor would get back the amount he/she has donated.

Step 1: Project for crowd-funding launched via Ethereum.

Step 2: This is public funding, it is open to all participants, and people can join the network, mine, or buy currency on the Ethereum network. People donate the Ethereum (Ether) for the project.

Step 3: Stop further funding for the project.

Step 4: Decision day: Did the crowd-funding meet the intended amount? If Yes, transfer the entire amount to school trust. If No, move the amount back to each of the donors, ensuring that the right amount goes to the right donor.

The same scenario of crowd-funding is also possible through a bank. However, the process is heavily centralized and subject to scrutiny by various agencies, which increases transaction costs and time. The above scenario of parking funds temporarily is available via escrow accounts. However, returning the amount, if needed, would need multiple approvals and time.

5.7 Smart Contracts in Industry

Smart Contracts in Blockchain is an emerging technology; it seems evident that industries such as banking, healthcare, supply chain and logistics, retail, government, etc., are ready to be influenced by Smart Contracts. Some of the use cases are briefly described below.

5.7.1 Healthcare Industry

In the Healthcare industry, there is a massive opportunity for Blockchain revolution to lead the digital transformation. From medical records to pharmaceutical supply chain to smart contracts for payment distribution, there are plenty of ways to leverage this technology. Here are some ways in which Blockchain and smart contracts can change the way healthcare services are delivered.

Health records are the backbone of every medicare system. However, medical records tend to get more extensive and involved with each visit to the doctor or medical consultant. Since every hospital, clinic, or doctor's office has a different way of storing them, it is not easy for healthcare providers to gain access to the records. There are individual companies like Medical Chain and Medicos, who have come up with solutions to this problem. The goal is to give patients the authority over the entire medical history and to provide one-stop access to patients and physicians, inherently bringing security to the health record access. Smart Contract functionality allows authorized parties to access, receive, and make changes to their records. These documentation changes are verified by the blockchain and authenticated cryptographically, which is more secure than traditional medical records. The prescriptions can be shared and managed with a proper timestamp, which cannot be tampered with. This offers better security of health records. There is no need to carry physical prescription, and the same can be accessed by physicians, patients, hospital or clinic staff with proper access.

Along with Big Data and Artificial Intelligence, this solution can predict and manage epidemics, lead to more informed research, and characterize our attempts toward improved health.

5.7.2 Manufacturing and Supply Chain

The supply chain industry has the highest standards of product safety, security and stability, right from inception. Pharma manufacturing and distribution partners are spread across multiple countries and regions; hence it is difficult to track and trace the information of each drug or medical device. Added to that, there are complex drug rules in different countries. Due to the current unavailability of real-time tracking, records can be easily altered. Additionally, non-legitimate vendors contribute to the black market for drugs with manipulation of the records.

A supply chain with smart contracts in the blockchain can be monitored securely and transparently. This can review time delays and human mistakes, which are very costly for the drug industry. It can be used to monitor costs, labor, even waste at every point of the supply chain. It can be used to verify the originality of the product, tracking it from origin to combat the counterfeit drug market, which causes losses to the market to the extent of about 200 billion USD annually. Companies like Blockpharma and Modum are working towards more efficient logistics solutions. Modum, in particular, works in compliance with EU laws that require proof that the medicine has not been exposed to certain conditions such as unusually high or low temperatures, which may change their quality. These solutions will have sensors that monitor the environmental conditions of the products while in transit and physically record them on the blockchain with smart contracts written on it.

5.7.3 Banking and Financial Services Industry

a) Syndicated Loans

Banking and financial services deploying smart contracts for managing their standard loans are no longer uncommon. In addition to this, syndicated loans, which involve multiple lenders providing loans to multiple borrowers on the same loan terms, have much to benefit from using smart contracts. All the steps, including syndication, diligence, underwriting, and servicing of syndicated loans, can be processed faster using smart contracts. With several entities involved in syndicated loans, establishing relationships, identities, and maintaining security becomes far more simplified with the on-chain/off-chain information that smart contracts can offer. Smart contracts thus act to drive effectiveness and efficiency in this scenario.

b) Securities

Smart contracts can enable the simplification of capitalization table management for private companies. In the current state, the issuer of the security has to transfer the title to many custodians who will have sub-custodians until it reaches the investor.

Various other challenges faced by the securities segment are listed below:

- Securities are paper-based
- Manual company registration procedures
- Increased cost
- Increased counterparty risk
- Increased latency.

Using smart contracts logic, the benefits are:

1. The security reaches the investor directly from the issuer.
2. Digitized complete workflows due to securities existing on a distributed ledger.
3. In private securities markets, benefits can be realized better than in public securities markets.
4. The cryptographic signature on the ledger entry takes the place of the State's seal on paper stock certificates.

5. Issuers will need visibility into who owns their securities, but some buy-side firms carefully protect this information.

c) Trade and Finance

Smart contracts help in the smooth transfer of goods across countries through smart credit and trade payment initiation, while also enabling higher liquidity of finance-related assets. The payment method provided by smart contracts is risk mitigated. It also improves process efficiencies of all parties involved including buyers, suppliers, and financial institutions.

The current challenges we face are:

- The letter of credit issuance is a costly and time-consuming process.
- There is multiple coordination across multiple parties due to paperwork and physical document management.
- These manual de-linked processes may lead to fraud and duplicate financing.

Using smart contracts logic, the benefits are:

- Quick payment and approval initiation through automated compliance and tracking of Letter of Credit conditions
- Enhanced efficacy in creating, changing and supporting trade, title, along with transport-related contract arrangements
- Increased liquidity of financial assets: Thanks to the simple transport process and decrease in fraud.

Some of the considerations while implementing smart contracts in Trade and Finance are:

- Standard templates for smart contracts should be implemented for each industry, and procedures must be implemented for wider acceptability and adoption.
- In particular, for defaults and dispute resolution, legal implications for potential smart contract execution fall-out must be decided and taken care of.
- To achieve full benefits, integration with settlement systems, off-chain ecosystems, and technology prerequisites (e.g., Internet of Things) must be implemented.

d) Derivatives

Post-trading has many redundant and time-consuming processes due to asset servicing being managed independently by each counter-party, for most over-the-counter (OTC) derivatives. Most of them are paper-based transaction agreements that contain terms, trade agreements, and/or post-trade confirmations.

Post-trade processes can be strengthened through smart contracts, eliminating the duplicate processes performed by each counter-party for recording and verifying trades, and executing appropriate trade levels and other lifecycle events. Implementing standard rules and regulations in smart contracts optimizes post-trade processing of OTC derivatives.

Some of the considerations while building smart contracts for derivatives are:

- Proper governance needs to be established to manage large-scale protocol changes to existing contracts due to regulatory reform, change in the contract or other unforeseen events.
- There should be proper agreement upon lifecycle events for OTC derivatives (e.g., an external source of data).
- There should be integration and governance of oracles required to feed smart contracts with information to/from the blockchain network.

5.7.4 Smart Contracts in Other Industries

There are many other use cases of smart contracts in other industries, as given below:

Legal: Smart Contracts stored in blockchain track contract parties, terms, transfer of ownership, and delivery of goods or services in the absence of the disadvantageous need for a legal intervention.

Government: Smart Contracts on the blockchain offers promise as a technology to store personal identity information, criminal backgrounds and "e-citizenship" authorized by bio-metrics.

Food: In the food supply chain industry, the blockchain concept can be used to trace the product origin, batch, processing, expiration, storage conditions, and shipping.

Insurance: In the insurance industry, when autonomous vehicles like driverless cars and smart devices communicate status updates with insurance providers via the blockchain, overall costs decrease as there is no need for auditing and authenticating data.

Education: Universities and educational institutions can utilize the blockchain to store grade or credentials data around assessments, degrees, and transcripts and for providing such data to the government or companies who want to employ the students.

Travel and Hospitality: Hotel customers and airline passengers can store their authenticated "single travel ID" on the blockchain. This will help them with the travel document identification cards, loyalty program, personal preferences, and payment data.

Summary

A smart contract is a computer program that directly controls the transfer of digital information, currencies, or assets between parties under specific rules or conditions. These contracts are stored on the blockchain in a decentralized manner.

Smart contracts are important because they exist in the blockchain network. So any entry made in the contract is immutable, and no changes can be made. This helps to resolve the trust issues between parties and business partners. Since smart contracts are self-executing, they do not need an external party to validate them, thereby reducing the cost and time of audits.

Smart contracts are legally binding contracts that obligate all parties to do what was agreed upon through the contracts. In case of a breach of contract, when one party does not act as per the agreement, the smart contract has the provision to enforce it legally.

A smart contract will reduce transaction costs drastically. Smart contract code, which is typically an auto enforceable code at the protocol level or application level, will be able to standardize transaction rules, thereby reducing the transaction costs of:

- Reaching an agreement – Agreement of a smart contract typically define the conditions, rights and obligations to which the parties agree.
- Formalizing an agreement – Smart contracts can formalize the relationships between people, institutions, and assets they own by executing an agreement. The rule is formalized in machine-readable code.
- Enforcing an agreement – This can be automatically executed by a computer or a network of computers once the rules or conditions are met.

The benefits of smart contracts are:

- Simple, faster to execute, and availability of updates in real-time
- No requirement of mediators and centralized entities
- Lesser cost because there is no need to pay the fees to middlemen
- There is no delay in delivering outcomes.

Different platforms can host smart contracts, but generally, smart contracts are implemented on a blockchain, where the validation logic sits inside a "block." All the messages needed for a smart contract are bundled within the block. These messages can be either inputs or outputs or may interface to other outside computer codes.

Smart contracts can

- Verify themselves
- Execute themselves
- Be tamper-resistant
- Automate legal transactions in terms of rules
- Provide a high degree of security
- Reduce reliance on trusted intermediaries
- Lower transaction costs

Smart contracts can be categorized into four types based on the applications:

- Smart legal contracts
- Distributed applications (DApps)
- Decentralized autonomous organizations (DAO)
- Smart contracting devices

The transactions in Ethereum Blockchain are also implemented through Smart Contract code, which is triggered by data and inputs from external accounts automatically. The list of all transactions and their Merkle root is stored in all the blocks.

Smart contracts need more research to make them near-perfect. The advent of smart contracts does not mean that lawyers, realtors, government staff will lose their jobs. They will still have many other jobs and responsibilities, which only human beings can handle.

It is the dynamics of the process that will change for the better. Smart contracts have the potential to revolutionize the world.

EXERCISES

Multiple Choice Questions

1. **This is a computer program working on top of a blockchain which has a set of rules (conditions) based on which, the parties agree to interact with each other.**

   A. Smart contract

   B. Oracle

   C. Protocols

   D. Consensus algorithms

   Answer A

   Explanation: A smart contract is a computer program that works on top of a blockchain based on a set of rules (conditions) according to which the parties to the smart contract agree to interact with each other. The agreement will be automatically executed if and when the pre-defined rules are met. The smart contract program will facilitate, verify, and ensure the agreement or transaction. This can be treated as the purest form of automation. Real-life agreements to record buy and sell assets can be automated using smart contracts.

2. **When the smart contracts are programmed, compiled, and migrated on to the blockchain, they are provided with this, depicting the location of the contract on the blockchain:**

   A. Account number

   B. Transaction

   C. Hash

   D. Address

   Answer: D

   Explanation: When the smart contracts are programmed, compiled, and migrated on to the blockchain, they are provided an address depicting the location of the contract on the blockchain, and the application is recorded in a block and distributed across all the nodes.

3. **Which of the following is not a benefit of smart contract?**

   A. Simple, faster to execute

   B. Interpreting a contract based on personal need

   C. Availability of updates in real-time.

   D. Lesser cost because there is no need to pay the fees to middlemen.

   Answer: B

Explanation: The benefits of smart contracts are:

- Simple, faster to execute, and availability of updates in real-time
- No requirement of mediators and centralized entities
- Lesser cost because there is no need to pay the fees to middlemen
- There is no delay in delivering outcomes.

4. **Smart Contracts can be categorized into ___________ types based on the applications.**
A. 2
B. 3
C. 4
D. 8
Answer: C
Explanation: Smart Contracts can be categorized into 4 types based on the applications.
- Smart legal contracts
- Distributed applications (DApps)
- Decentralized autonomous organizations (DAO)
- Smart contracting devices

5. **In a smart contract, the data originates from online sources, like railway reservation sites, e-commerce sites, etc. This is an example of:**
A. Software oracle
B. Hardware oracle
C. Railway oracle
D. Online oracle
Answer: A
Explanation: A smart contract utilizes the temperature in a city, prices of commodities and goods, the details of flight or train delays, etc. The data originates from online sources, like railway reservation sites, e-commerce sites, etc. This is an example of Software Oracle

6. **A smart contract getting the details of a car crossing a specific junction (date, time, speed, direction, or location) using a sensor, is an example of:**
A. Software oracle
B. Hardware oracle
C. Inbound oracle
D. Online oracle
Answer: B
Explanation: Some smart contracts get inputs directly from the physical world. For example, the details of a car crossing a specific cross-junction (date, time, speed, direction, or location) can be captured, where moving sensors help. One other use case is in logistics, where RFID sensors are used to track the movement of the trucks carrying goods.

7. **An automatic purchase order for spares if their stock value hits a certain inventory number using smart contract is an example of:**
   A. Software oracle
   B. Hardware oracle
   C. Inbound oracle
   D. Online oracle
   Answer: C
   Explanation: Inbound oracles provide a smart contract with data from the external world. One such use case is an automatic purchase order for the spares if their stock value hits a certain inventory number.

8. **In a smart contract, more than one oracle from multiple sources is used to take a decision. This is an example of:**
   A. Software oracle
   B. Hardware oracle
   C. Inbound oracle
   D. Consensus oracle
   Answer: D
   Explanation: To confirm future outcomes, the prediction markets rely heavily on oracle. It would be risky to depend on only one source of information. To avoid market manipulation, prediction markets implement a rating system for oracles. A combination of different oracles is used for enhanced security; for example, 6 out of 10 oracles (majority) could determine the outcome of an event.

Short-answer Questions

1. **Define Smart Contract.**

   According to Nick Szabo (1996), a Smart Contract is a set of promises, specified in digital form, including protocol related procedures within which all the parties perform on these promises. Multiple technological supports (languages) evolved over time, and this definition still holds good.

2. **List the benefits of Smart Contracts.**

   The benefits of smart contracts are:
   - Simple, faster to execute, and availability of updates in real-time
   - No requirement of mediators and centralized entities
   - Lesser cost because there is no need to pay the fees to middlemen
   - There is no delay in delivering outcomes.

3. **What are DApp Browsers?**

   Most of the times, smart contracts have to access external applications or external users through web browsers known as DApp Browsers.

4. List the characteristics of smart contracts.

Smart Contracts can

- Verify themselves
- Execute themselves
- Be tamper-resistant
- Automate legal transactions in terms of rules
- Provide a high degree of security
- Reduce reliance on trusted intermediaries
- Lower transaction costs.

5. What are the types of smart contracts based on the applications?

Smart contracts can be categorized into four types based on the applications:

- Smart legal contracts
- Distributed applications (DApps)
- Decentralized autonomous organizations (DAO)
- Smart contracting devices.

6. What are DApps?

DApps are applications which run on point-to-point (P2P) network of computers instead of a single computer. DApps need not necessarily run on top of the blockchain network. Unlike simple smart contracts, which send bitcoin typically from X to Y, DApps have an unlimited number of participants from the market.DApps are a 'blockchain enabled' website, where the Smart Contract allows it to connect to the blockchain. DApps cover end to end – both front-end and back-end. If a decentralized application has to be created based on a smart contract system, we have to combine several smart contracts and rely on third-party systems for the front-end.

7. What is DAOs?

A Decentralized Autonomous Organization, called as DAO is an organization represented by logic encoded as a transparent program, controlled by shareholders and not influenced by a central authority. A DAO's financial transaction records and program logics are maintained on a blockchain. Tokens allow traceability to allow algorithmic measures of contribution utility. DAOs use blockchain technology to maintain a secure advanced record in the form of Digital Ledger (DL) to trace the financial transaction over the internet, hardened against fraud by trusted timestamping, and using a distributed database.

8. Write notes on applications of Smart contracts in legal, government and food industries.

Legal: Smart contracts stored in blockchain track contract parties, terms, transfer of ownership, and delivery of goods or services in the absence of the disadvantageous need for a legal intervention.

Government: Smart contracts on the blockchain offers promise, as a technology, to store personal identity information, criminal backgrounds and "e-citizenship" authorized by bio-metrics.

Food: In the food supply chain Industry, the blockchain concept can be used to trace the product origin, batch, processing, expiration, storage conditions, and shipping.

9. Write notes on applications of smart contracts in the insurance industry.

In the insurance industry, when autonomous vehicles like driverless cars and smart devices communicate status updates with insurance providers via the blockchain, overall costs decrease as there is no need for auditing and authenticating data.

10. Write notes on applications of smart contracts in the education industry.

Universities and educational institutions can utilize the blockchain to store grade or credentials data around assessments, degrees, and transcripts and for providing such data to government or companies who want to employ the students.

11. Write notes on applications of smart contracts in travel and hospitality industry.

Hotel customers and airline passengers can store their authenticated "single travel ID" on the blockchain. This will help them with the travel document identification cards, loyalty program, personal preferences, and payment data.

12. What are the current challenges in trading and finance?

The current challenges we face in trading and finance are:

- The letter of credit issuance is a costly and time-consuming process.
- There are multiple coordinations across multiple parties due to paperwork and physical document management.
- These manual de-linked processes may lead to fraud and duplicate financing.

13. What are the applications of smart contracts in trading and finance?

Using smart contracts logic, the benefits are:

- Quick payment and approval initiation through automated compliance and tracking of letter-of-credit conditions
- Enhanced efficacy in creating, changing and supporting trade, title, along with transport-related contract arrangements
- Increased liquidity of financial assets, thanks to the simple transport process and decrease in fraud.

14. What are the challenges faced by securities?

- Securities are paper-based
- Manual company registration procedures
- Increased cost
- Increased counterparty risk
- Increased latency.

15. What are consensus-based oracles?

To confirm future outcomes, the prediction markets rely heavily on oracle. It would be risky to depend on only one source of information. To avoid market manipulation, prediction markets implement a rating system for oracles. A combination of different oracles is used for enhanced security; for example, 6 out of 10 oracles (majority) could determine the outcome of an event. Connecting a smart contract to external data and existing applications, data feeds, APIs, bank payments, etc. is an essential step for creating useful smart contracts.

Essay-type Questions

1. How do smart contracts reduce the transaction costs? List the benefits of smart contract.
2. Explain traditional contract and smart contract while transferring an asset.
3. How do smart contracts work?
4. Explain how smart contracts change the state through transactions.
5. Write about the types of smart contracts distributed in a spectrum.
6. Explain in detail the types of smart contracts based on applications.
7. Explain the various properties of DAO.
8. What is Oracle? What are the types of oracles?
9. Explain the applications of smart contracts in various industries.

CHAPTER 6

Private Blockchain System

LEARNING OBJECTIVES

Private blockchains are private, as all permissions for the blockchain are kept centralized. Centralizing permissions essentially bring out the critics. Now, these are the sort of blockchains banks are testing with. Another name of a private blockchain is Permissioned blockchain. They are closed ecosystems where all participants are well-defined. Only pre-approved entities can run nodes.

Let us look into private blockchain in detail in this chapter.

6.1 Introduction

In the previous chapters, we discussed the details of public blockchain using bitcoin and Ethereum network. In a public blockchain, anyone with reasonable computational power and compatible software like a CC miner can get on the network. Anybody in the world can download the blockchains and primarily read the data. The power of cryptography secures the public blockchain. It is part of the fully decentralized nature of blockchain technology.

However, Private Blockchains are private, as all permissions for the blockchain are kept centralized. Now, these are the sort of blockchains banks are testing with. Another name of a private blockchain is Permissioned or Centralized blockchain. They are closed ecosystems where all participants are well-defined. Only pre-approved entities can run nodes.

6.2 Key Characteristics of Private Blockchain

Private blockchains are a great place to start in a Business-to-Business (b2b) model. In the private blockchain, the degrees of decentralization (or centralization) and transparency are entirely up to the company or companies running and configuring the blockchain. There is no anonymity. It does not require mining to validate transactions or execute smart contracts. It does not require a crypto-economic incentive or tokens for those who are running the nodes (miners).

The private blockchain is a closed environment consisting of a set of nodes, which knows each other, but they may not trust each other. Before joining the blockchain network, the nodes have to go through some authentication or pre-authorization mechanism through which they can validate themselves.

Table 6.1 Characteristics of a private blockchain network

	Private blockchain
Organization type	Single entity or organization
Users	Known and trusted participants
Access	Access fully restricted
Network type	Centralized; single point of failure
Operation	Pre-approved participants can read and/or initiate transactions
Verification	Single validator node or central authority to create a block
Immutability	Secured by distributed consensus
Consensus mechanism	Voting or variations of PoW/PoS consensus algorithms
Security	Security is dependent on the blockchain architecture adopted
Trust	Entrusted; central control
Transaction speed	High; secs to create a block
Energy consumption	Low
Scalability	Better scalability as high storage and computational power is not required

Table 6.1 summarizes the key characteristics of private blockchain. In this type (private blockchain), the users are trusted participants and their details are known. The network is not open, and it is not transparent to all. The network type is centralized and is a single point of failure.

Operation: Only pre-approved participants can read and/or initiate transactions in a private blockchain. There is no miner, and so there is no incentivisation. It is secured by a distributed consensus mechanism such as PAXOS, RAFT, BFT, and pBFT. It has better scalability as high storage and computational power is not required. Also, the transaction speed is high as it just takes seconds to create a block.

6.2.1 Private Blockchain and Permissioned Blockchain

Private blockchain is often compared and deemed similar to the permissioned blockchain; it is a fact that almost all private blockchain implementations are permissioned; however, there is a small difference between a private blockchain and permissioned blockchain.

A private permissioned blockchain strictly controls who can do what within the IT system, and there is a well-defined mechanism to revoke and grant access to the nodes. Hyperledger Fabric is an example of such private permissioned blockchain. A permissioned blockchain can be different from a private blockchain. It is possible to create a blockchain network, which is open to the public, i.e., it may allow the public to authenticate itself, grant privilege to nodes and create a mechanism for proper audit trail; such a blockchain is public permissioned blockchain (Ripple blockchain).

Whereas, a blockchain set up by a private company or a bank is a network, with full privileges for all nodes (limited to a few nodes within the bank), is an example of the private non-permissioned blockchain.

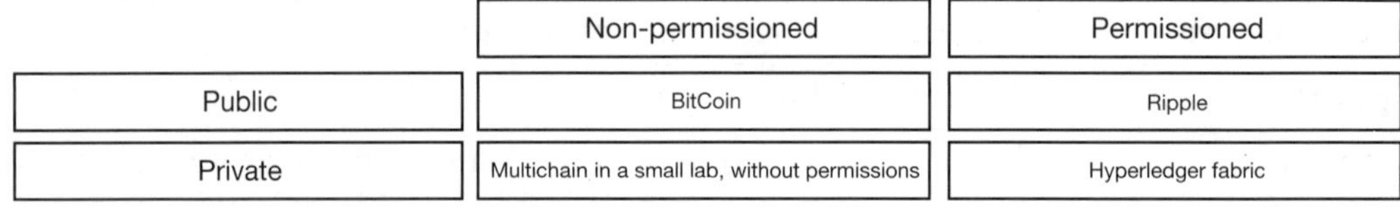

	Non-permissioned	Permissioned
Public	BitCoin	Ripple
Private	Multichain in a small lab, without permissions	Hyperledger fabric

Figure 6.1 a: Permissioned vs non-permissioned blockchain

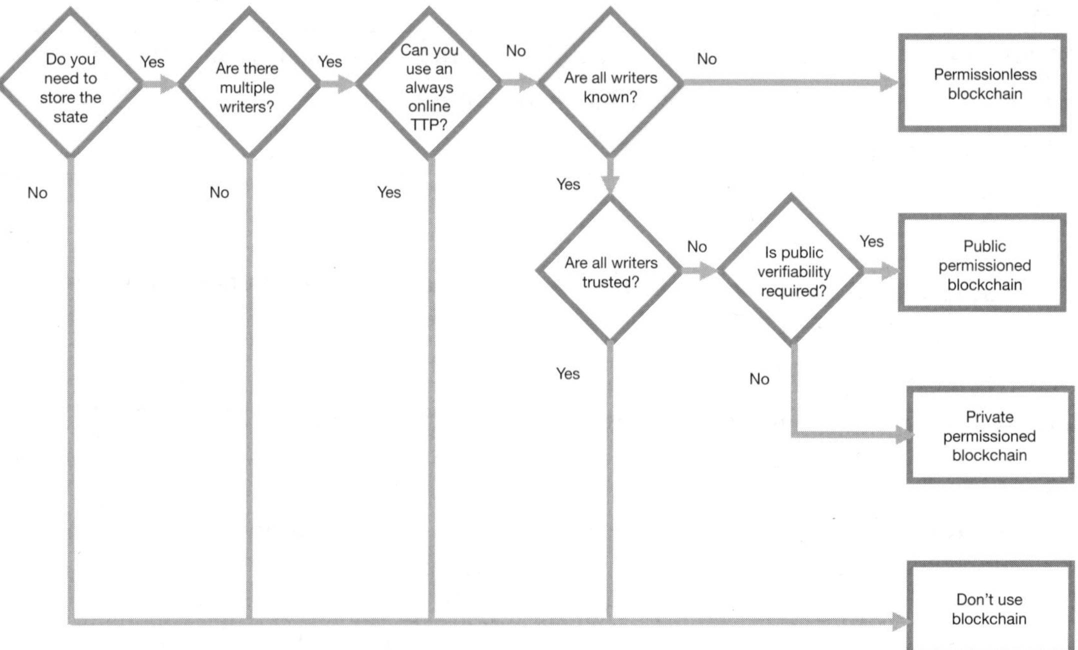

Figure 6.1 b: Permissioned vs. non-permissioned blockchain

The above diagram (refer Figs 6.1 a and b) explains the condition surrounding the use of permissioned blockchain network. We need to use "public permissioned blockchain" when the state (and date) needs to be stored in the database, if there are multiple writers for transactions, when we cannot always use online TTP (Trusted Third Party) and when all the writers are known, when all the users cannot be trusted and when public verifiability is required. We need to use "private permissioned blockchain" when the state (and date) needs to be stored in the database, if there are multiple writers for transaction, when we cannot always use online TTP (Trusted Third Party), when all the writers are known, when all the users cannot be trusted and when public verifiability is not required.

6.3 Why We Need Private Blockchain

Private blockchains are used by individual hobbyists, private enterprises or organizations for a specific purpose (e.g., an NGO may like to keep a record of money spent in various schools). Organizations may prefer to use a private blockchain, if they would like to control:

- Who can use the system
- Who can write to the system
- Who can read the system

Besides, organizations need a solution with a mechanism to ensure users are added via process, and user rights are created, changed, or deleted by an authorized user. These needs arose and gave birth to a need for private blockchain. In addition, the solution needs faster transactions, proper audit trail, and interactions with the organization's existing IT systems.

6.4 Private Blockchain Examples

Private blockchains are used as a company's internal blockchain. If we consider public blockchain as the internet, then private blockchain can be considered as the intranet. Private blockchains need read/write permissions; it is centralized to one organization. Permissioned is faster because there are fewer people on it. Permissioned blockchain uses the multi-signature method. Private blockchain has low energy consumption; it also has fewer people transacting on the network. The transaction approval frequency is short. There are many private blockchains like IBM working with the hyperledger. In the concept of "blockchain as a service," private blockchains can come in handy. Many companies do not want to create their blockchain but they use the blockchain of other companies offered as a service to suit their needs.

Private blockchain uses RAFT and PAXOS as its consensus mechanism. The following are the main features of permissioned blockchain:

1. Permission blockchain allows only actors who have the right to view the transactions.

2. Scalability: A permission blockchain can build a simplified proof-of-state model with established consensus.
3. Fine-grained access control.

Examples of Permissioned Blockchain

One compelling use case of permissioned blockchain environment is for business application, such as to execute specific contracts among a closed set of participants. Another interesting example of the use of permissioned blockchain is for provenance tracking of assets. Here, a specific asset is moved from one particular supplier to the distributors, to the vendors, to the market. At every stage, logs are maintained, where people make an entry, indicating that this particular asset has now (at that time) moved from location A to location B (or) Person X to Person Y. Now, the interesting question is: Why not to use a centralized server to maintain the log? (Refer Section 6.7.1).

6.5 Private Blockchain and Open Source

Open-source movement has picked up steam over the past three decades, with the explosive use of Java language. The internet in the late nineties gave massive importance to Java and other open-source software packages such as Python, Golang, etc. Use of such open-source packages has gained momentum, bringing the best minds of the world together for a common purpose. The output of the open-source is available for everyone on the planet.

When open-source software is purchased, it has (is expected to have) the following qualities:

- Purchaser has the ownership of design and source-code
- Purchaser has the right to modify the design and source-code
- Purchaser has the right to distribute source code as desired.

Public blockchains are less protected since, typically, they do not offer privacy services and are not generally acceptable for industries like finance, medicine, and law because of their open and relatively less protected nature. Private blockchains hold the promise to tackle these constraints and bring end-users one step nearer to reaping the advantages of blockchains, while fulfilling all privacy and security requirements.

Private blockchains permit the participants or a subset of participants to take complete charge of the system. Thus, private blockchains are useful in finance and other sectors, where control and privacy are needed. Enterprises have used ledgers since time immemorial, and with the advent of electronics, they have moved to electronic ledgers, which can now be shared on the computer network.

6.6 E-commerce Site Example

Let us take the example of an e-commerce site, which delivers and installs an electronic gadget, at the customer site (refer Fig. 6.2).

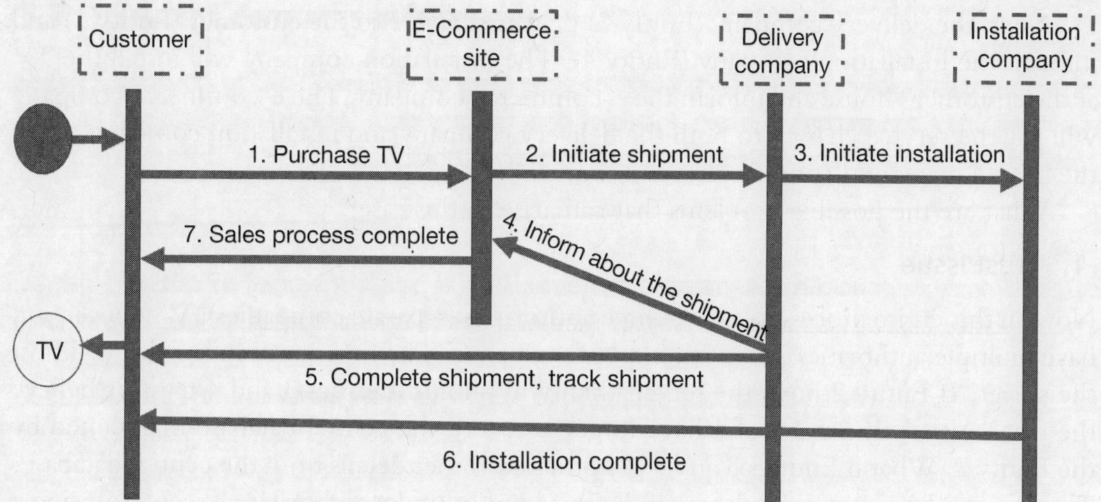

Figure 6.2: Transaction on an e-commerce site

Assume that the E-commerce company (E-commerce site) itself owns the delivery company and installation company. That means there is only one entity, namely, the e-commerce company. When there is only one entity, company, organization, or domain, then we can use a centralized server, instead of a permissioned blockchain.

The centralized server can have the details of all the seven tasks shown in Fig. 6.2, namely, 1. Purchase TV (payment) 2. Initiate shipment 3. Initiate installation 4. Share information regarding the shipment details 5. Complete shipment 6. Complete installation 7. Complete sales process (obtain signature from the customer).

Every single party of the entity makes an entry to the central database and everyone can access the log to find the details of where the particular asset (TV in this case) is moving. Using the bill number, the user can also track where the parcel (TV) is currently placed at, and when it is expected to reach the customer. It is good to use a centralized server in this case, because a single company (e-commerce website company) itself maintains the centralized server.

6.6.1 Problems with the Centralized Server

Think of a typical scenario where we have four separate entities (instead of a single entity scenario explained above):

1. Customer (Entity 1)
2. E-commerce company (Entity 2)
3. Delivery company, who also sells and installs TV for the Customer (Entity 3)
4. Installation company, who also sells TV to the Customer (Entity 4)

The customer (Entity 1) is ordering TV from the e-commerce company (Entity 2) through its website. After that, the e-commerce company gives orders to the delivery company (Entity 3) and installation company (Entity 4) to deliver the TV and install it for the customer.

Next, the delivery company (Entity 3) delivers the TV to the customer (Entity 1) and informs the installation company (Entity 4). The installation company will install the TV at the customer's house and inform the e-commerce company. The e-commerce company will pay the service charges to both the delivery company and installation company as per the agreed terms and conditions.

What are the possible problems that can arise in this case?

1. Trust Issue

Now, in this entire process, we have three authorities (separate companies). Whenever you have multiple authorities, there is a kind of trust issue. The first question is, who will host the server? If Entity 2 hosts the server, then why should Entities 3 and 4 trust or believe the data? Also, Entities 3 and 4 have to get access to the central database maintained by the Entity 2. What if Entities 3 and 4 get other customer details from the central database? Therefore, it becomes difficult to handle this kind of multi-authoritative scenario using a central database. Even if Entity 3 or 4 host the server, the same problem persists.

2. Dollar Issue

What if a third party cloud does the hosting? All parties need to pay a significant amount of money for data on the cloud. Also, there is no guarantee that the data, which is entered by one entity, is not tampered with or modified when it is entered by another entity.

Due to problems in the trust relationship and dollar issue involving multiple entities, people do not want to go for a centralized server. So, what is the solution?

6.6.2 Private Blockchain Concepts in an E-commerce Scenario

The beauty of the permissioned blockchain environment is that it does not require any centralized server, and the data would be in the hands of all the private entities, where everyone will be able to validate the other's data. There will be a closed set of participants who enter the blockchain environment. This maintains a certain kind of security to ensure that the data is not being tampered with while it is transferred from one authoritative domain to another authoritative domain.

A private blockchain can be established involving the above four entities. The customer related information could be kept in an e-commerce site and the three business entities (the e-commerce site, delivery company and installation company) can join and start a private blockchain system.

Maintaining a separate ledger is necessary for all the below stages.

- Purchase of TV
- Track shipment at various milestones.
- Notifying delivery
- Completing the installation
- Completion of the sale process.

Information relevant to each entity is shared and recorded in the ledgers. Let us look into the various steps in a blockchain, where a customer purchases a TV set from an e-commerce site, and has it delivered and installed at his premises, and compare each blockchain feature:

Step 1: Purchase of TV Set on E-commerce Site

When a user is browses for a TV set, he is makes a purchase of the TV set as well as its corresponding features (which includes additional purchases like Chromecast, etc.). The user provides the billing address and shipping address to the e-commerce site, often the billing address and shipping address remains the same. Once the end-user is convinced about the product (TV set), he proceeds to check out, where the amount is shown to the customer along with delivery timelines. When the user is happy with both the price and delivery schedule, he purchases the TV set. At this point, there two data points generated between an e-commerce site and user: Price for TV and delivery schedule. Users could choose the existing payment options – net banking, debit card, installment, or the new-age blockchain-based payment system. The second data point generated is about the delivery schedule: This would include the address to which the product is to be delivered and installed and the agreed date and time for delivery.

Step 2: Delivery of TV

The e-commerce site may have its physical offices in major cities and can also have preferred delivery partners for delivering customer purchases. An e-commerce site operating within the same city may have more than one delivery agent. Also, the selection of delivery partners depends on the many factors: some partners may prefer to deliver expensive purchases using a professional company, while opting to use small delivery companies for low-cost products.

The delivery company that handles the TV would have the following facilities: collection centre at many places, warehouses in significant cities of the country, logistics for inter-warehouse movement of goods, and a suitable site for delivery near the customer's address.

Assuming that the e-commerce site has handed over the TV to the delivery partner, the TV moves to the destination via the delivery channel. The TV moves from the collection centre to a warehouse; at the warehouse, the TV is clubbed along with other articles in transit and sent to another warehouse. From the blockchain point of view, the presence of TV is tracked at all places, and the blockchain is updated. Please note that there are blockchain solutions to track the movement of products even every minute, such as for perishable goods like milk and fish. For perishable goods, the container's temperature and pressure are tracked every minute. However in our solution, tracking a TV product at each point is enough. Once the TV is ready for delivery, the customer is notified and the schedule for delivery is shared with the customer. Finally, the product is delivered. All the above steps are added to the blockchain to track the supply chain process.

Step 3: Installation By a Trained Technician

Once the product is delivered, the status is updated to both the e-commerce site and the installation company. The installation company would, based on the schedule, fix an appointment with the customer, install the product and verify if the product and its associated services are working. The scheduled installation would be added to a blockchain. An associate from the company goes to the site and performs the necessary activities related to the installation of the product. This information is also added to the blockchain.

In the above example, successful completion of service involves co-ordination between the involved parties as follows:

- E-commerce company intimates sale of TV to both delivery and installation company.
- The delivery company notifies installation company when TV reaches the nearest warehouse and confirms the lead-time.
- The workforce manager plans a schedule after the TV is delivered.

The various characteristics of blockchain constructed for the e-commerce Web site are given in the Table 6.2.

Table 6.2 Characteristics of private blockchain in e-commerce example

	The E-commerce Example
Organization Type	In our example, multiple organizations coming together to achieve a particular goal. In our example, the goal is purchase, delivery, and installation of a TV set. We saw at least three organizations coming together; however, in a real-world scenario, there could be further subcontracts. For example, other companies can handle movement between warehouses.
Users	An essential aspect of the private blockchain is users who are known and trusted. In our example, we have demonstrated that people who work on the delivery of the TV set are trusted because they are associates of the company responsible for shipping the article. The person who purchased the TV would never know about him; the person is allowed to log in to the blockchain network and update the status of the TV set and its movement. An associate from the installation company is sent to the customer's residence based on the appointment. The scheduling manager would schedule weekly/daily schedule for associates: this would be updated in the blockchain. Similarly, before the associate comes to the site, the customer is informed about the associate's name and ID. The associate comes with a prior appointment and proves his identity to the customer. This info is captured in the blockchain. Once the associate completes his task, the task-related details are also captured on the blockchain.

(*Continued*)

Table 6.2 (*Continued*)

	The E-commerce Example
Access	Users have access to the organization's blockchain, and depending on the organization, the role a person plays in the organization. Access is given to a set of (group of) users. For example, Operator 1 and Operator 2 with the delivery company would be grouped as Supervisors while Operator 3 and Operator 4 would be grouped as Shipping agents. The blockchain maintains a list of approved activities a role can do. If a person submits a task which is not allowed in his role, the blockchain will mark this task as invalid. In all blockchains: there would be a mechanism to add a user to a role and to remove a user from a role. Activities a role could perform are based on the agreement between the organizations. The administrator of the blockchain handles any changes in the role and access.
Network Type	Centralized; single point of failure In our e-commerce example, we have a centralized network, where the members of the network work towards achieving the given tasks. This network is different from that of the public chain, where there is no central authority.
Operation	Pre-approved participants can read and/or initiate transactions. Operations allowed are clearly defined in the blockchain; each role is allowed to make a list of operations. Blockchains need to decide on whether a transaction is valid or not: this mechanism is popularly known as Consensus. Popular consensus mechanisms for private blockchain include PAXOS, RAFT, and the pBFT. In our example, an associate from a delivery company or an associate from the installation company submits transactions to the blockchain network. The blockchain would run the chain code or smart contract to validate the transaction using the consensus mechanism to decide whether this transaction is valid or not.
Verification	Single validator node or central authority to create a block. Only transactions that pass the consensus check are considered as valid transactions. In our example, the chain code, based on the input from associate, validates the transaction delivery done at customer address. This information is further validated using date timestamp provided by previous locations.

(*Continued*)

Table 6.2 (*Continued*)

	The E-commerce Example
Immutability	Secured by a distributed consensus mechanism. Immutability is the beauty behind the blockchain, transactions are encrypted and kept in the blockchain, and information that is available to the delivery company is limited and based on the streams that are set for the delivery company. Even if a person manages to get data from other streams, he may not be able to do much with data, as it is encrypted. Blockchain is a chain of blocks formed, one after another. Every block apart from the first block, would have the hash-code of the previous block. And since every peer in the block has a copy of the block, a person attempting to tamper data needs to tamper the hash-code for the block, in all the peers of the network, which may not be possible.

6.7 Various Commands (Instructions) in E-commerce Blockchain

Now, let us look back at the various activities involved in the above private blockchain. An e-commerce site can be a central authority that connects customers, the delivery company and the installation company. Assume that a private blockchain is created as soon as the customer orders a good (TV).

The core aims of blockchain are threefold:

(a) To ensure that the blockchain's activity is only visible to chosen participants
(b) To introduce controls over which transactions are permitted, and
(c) To enable mining to take place securely without proof-of-work and its associated costs.

Assume one block is created for each order of goods (TV in this case). Now a new block is created in the blockchain. The e-commerce site, delivery company and installation company share this block. Transaction in the block is completed only when the TV is installed at the customer's site and the delivery company and installation company get the service charges. The following are some of the permissions to be given by the e-commerce website for this private blockchain (refer Fig. 6.3).

- Admin – to change all permissions for other users, including issue, mine, activate, and admin. The e-commerce site usually keeps the Admin user.
- Activate – to change connect, send, and receive permissions for other users, i.e., sign transactions which change those permissions.
- Connect – to connect to other nodes and see the blockchain's contents. This is possible for all three parties in the above case.

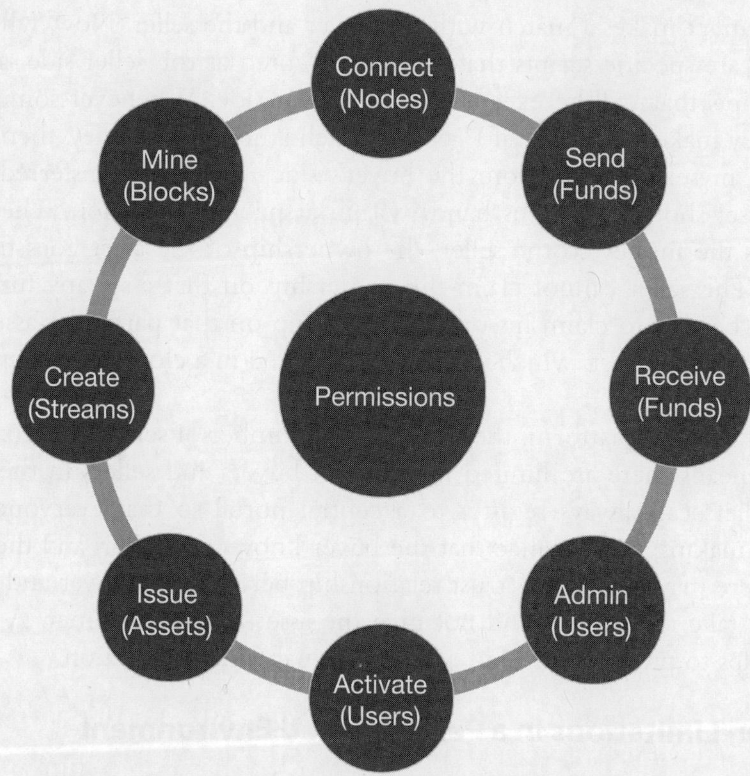

Figure 6.3: Sample commands in a private blockchain

- Send – to send funds, i.e., sign inputs of transactions. The customer sends money to the e-commerce site. The e-commerce site sends service charges to the delivery company and the installation company.
- Receive – to receive funds, i.e., appear in the outputs of transactions. The e-commerce site receives money from the customer. The delivery company and the installation company receive service charges.
- Issue – to issue assets, i.e., sign inputs of transactions, which create new native assets.
- Create – to create streams, i.e., sign inputs of transactions, which create new streams. In this case, the e-commerce site creates streams of transactions.
- Mine – to create blocks, i.e., to sign the metadata of coin-based transactions.

6.8 Smart Contract in Private Environment

A smart contract is a self-executing contract (auto contract) in which the agreement between the buyer and the seller is directly written into the lines of code. Therefore, the question arises as to how this kind of contract may be executed or how a blockchain can help to execute this type of contract.

One typical example is that you may execute a selling transaction when a specific condition is satisfied. Let us say you have a seller, you have a buyer, and you have individual assets. The contract makes a match with the buyer and the seller. Now, when a purchase is made, there are specific scripts that will be executed at the seller side, and there will be specific scripts that will be executed at the buyer side. Whenever someone is selling something, they make a contract: if I get this much amount of money, then I can transfer the asset from myself to you. From the buyer's side, if he has transferred the required money; the asset should be in his hands within a specified duration. The moment the buyer transfers the money to the seller, the ownership of the asset goes from the seller to the buyer. The seller cannot claim the ownership on that asset any further, whereas the buyer will be able to claim his or her ownership on that particular asset. This is the concept of a smart contract, which we can be executed in a closed environment with the help of a blockchain.

In an ideal business platform, there are a fixed number of sellers or a fixed number of buyers. That means there are limited numbers of buyers and sellers in the market. The buyers and sellers can always register to a central portal so that everyone knows each other. We are making an assurance that the buyer knows the seller, and the seller knows the buyer. There may not be any trust relationship between the buyer and the seller and the seller may take the money and not give the asset to the customer. A permissioned blockchain helps to prevent this kind of fraud or malicious transaction.

6.8.1 Design Limitations in a Permissioned Environment

There are certain design limitations in a permissioned environment, which are part of any permissioned blockchain. We need to first understand these limitations before identifying possible solutions for them.

6.8.1.1 Denial-of-Service Attack

In a typical environment, we execute the transaction sequentially based on consensus. If a specific transaction is verified or gets committed, it will be executed first. The transaction that is committed later on will be executed next. So, the request to the applications (smart contracts) is carried out by order of the consensus.

This kind of sequential order gives a bound on effective throughput since absolute consensus has to be ensured. We apply the proof-of-work based techniques in the case of the permissionless model. Where the network or the system throws a challenge (problem) to the individual users, every user tries to solve that particular challenge (problem) individually, and the nature of the challenge is such that it is difficult to find out a solution, but once the solution is found everyone can verify it very quickly. Thus, using the challenge-response based method, the nodes try to come to a consensus. As the challenge is very hard, it takes some time to arrive at a solution. If the order of transaction execution is serialized, it gives a bound on the effective throughput. Throughput is inversely proportional to the commitment latency. If the difficulty of the challenge is increased, the effective throughput will drop — throughput here is in terms of transactions that can be committed per second or per unit time.

This can be a source of possible attack on the smart contract platform. The attacker can introduce a contract, which will take a long time to execute. If a specific contract takes considerable time to execute, other contracts will not be able to execute any further since these can be taken up only when consensus for the previous contract has been reached. Therefore, the serialized nature of the order of transactions prevents execution of the next contract until the previous contract is executed. If a malicious contract is introduced in the system, it will take a considerable amount for execution, thus launching a denial-of-service attack.

6.8.1.2 Implementation Language of Smart Contract

The second problem is the identification of a suitable language for implementing a smart contract. To implement a smart contract, we need to use a language that gives more power. Bitcoin script is not a "Turing complete" language. For example, it does not support loops, and it has certain limitations in execution. Therefore, developers use different types of suitable languages; one example is `Golang`, which is widely used in the contract execution platform. In `Golang`, you have a construct called a map.

The map data structure: The map is a data structure, where there are specific keys and absolute values. Every value is associated with one key. Whenever there is a `for` loop over a map that just prints the key-value pair and this code is run multiple times, there will be multiple key-value pairs in the map. You will see for each run, the ordering of the key-value pairs is different. In `Golang`, every key is ordered in a certain way in the map data structure, so it uses its internal contracts matching a corresponding value, which is associated with that key. Different runs may have different ordering of the keys. Otherwise, the system may lead to an inconsistent state, or result in having many forks. So one particular user may execute the contract to get one result, or one ordering of result, while in another contract the user may get another ordering of results. If two different ordering of results are obtained, then it may be difficult for the user to ensure that his blockchain is the longest.

Even if it is ensured that the blockchain is the longest, there may still be multiple unnecessary forks in the system, which we want to prevent. The solution here is to use a domain-specific language, which will implement the smart contract platform. For example, Ethereum is a platform where you can implement the smart contract with the construct that they have supported.

6.8.1.3 State Synchronization in Smart Contract

Now, the third design limitation is the execution of smart contracts. Generally, we execute the smart contract at all nodes and propagate the state to others to reach a consensus. To ensure consensus, the problem that comes to the user has a sufficient number of cross-state nodes to validate the execution of smart contracts. If the number of trusted nodes is less than the number of malicious nodes, then the malicious nodes may take control over the entire environment. However, this can be prevented by using a proof-of-work based consensus mechanism.

In a proof-of-work (PoW) based consensus mechanism, the problem is that the user will be stuck to a particular block leading to a kind of starvation scenario where a contract takes a long time to execute, resulting in a backlog of all the other contracts. One interesting question that comes to mind regarding the consensus of permission blocks and model is whether it is always needed to execute the contacts at each node? Indeed, it is not necessary.

We need state synchronization across all the nodes. The contract in one node is executed and after executing it, the state of the contract is broadcast (refer Fig. 6.4) to the neighboring node, and that state is further propagated. In this way, every node in the system gets the same states of the contract, and they can be on the same page of the smart contract.

What if the node that executes the contract becomes faulty? Nodes become faulty when the system goes down and is unable to make any progress further. In a particular scenario, you can use the concept of state machine replication. That is, you execute the contract at a subset of a node, rather than a single node (refer Fig. 6.4).

You have multiple nodes, which can be selected to run the particular contract. In this diagram (refer Fig. 6.4), three nodes are executing the contract, and the result of the three nodes is broadcast to all other nodes in the network. This ensures that every node which is a part of the smart contract is on the same page.

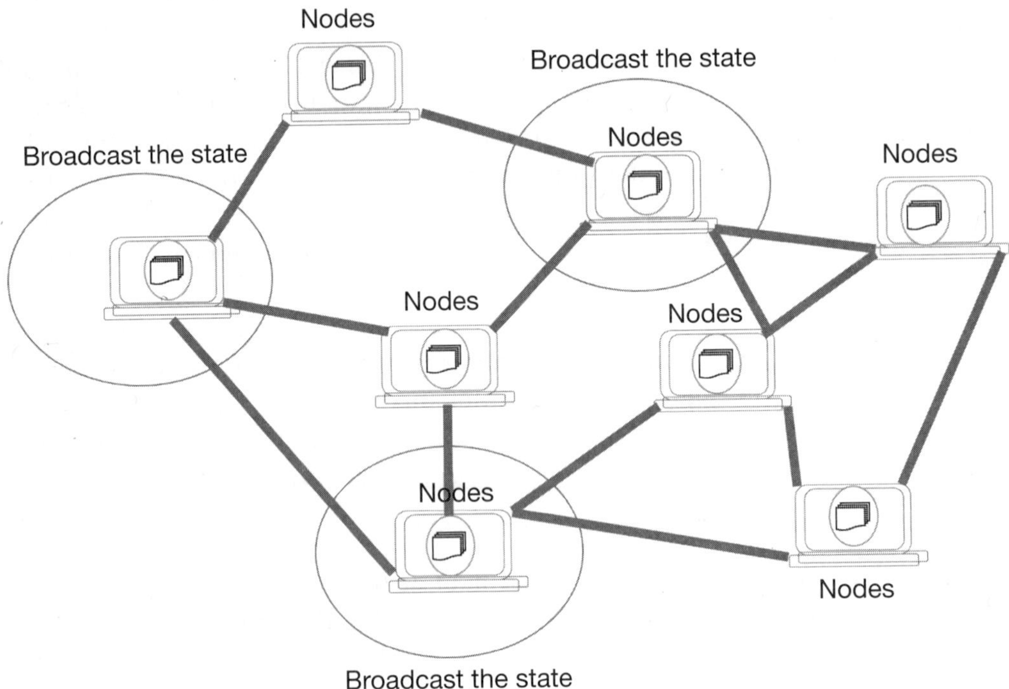

Figure 6.4: Contracts in a subset of the node and broadcasted state

6.8.2 The CAP Theorem

For distributed systems, as per Eric Brewer, it is impossible to simultaneously provide more than two out of the following three guarantees:

- Consistency
- Availability
- Partition Tolerance

Please note that consistency in distributed environment and consistency in RDBMS are different. From the distributed perspective, the above three properties can be visualized as shown in Fig. 6.5.

Consistency: Every node in the cluster will be provided with the latest and correct data set. All nodes see the exact data at precisely the same moment. In other words, performing a read operation will yield the result of their very recent write operation inducing all nodes to get precisely the exact data.

Availability: All requests to a node in the cluster will be provided with a response, although it does not mean this is the latest data written to the data set. This state ensures that every petition receives a response on success/failure. For accessibility in a distributed network, the machine has to remain operational 100 percent of the time. The end-user sees only Consistency and Availability. This view is formed when connectivity between nodes fails, which results in multiple transactions being lost; you would have to choose either of consistency or availability.

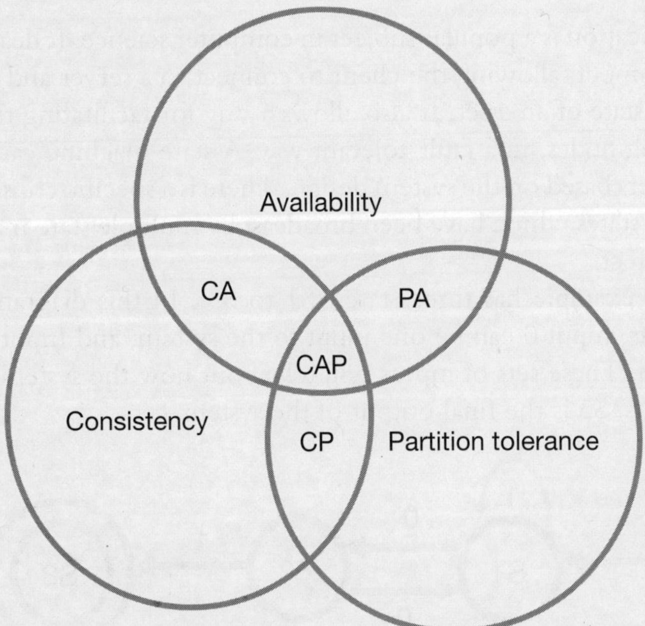

Figure 6.5: CAP theorem

Partition tolerance: The network will continue to operate even if messages were lost as node(s) or communication between nodes failed. A system that is partition-tolerant can withstand any amount of network failure that does not fail the entire network. Data records are sufficiently replicated across combinations of nodes and networks to keep the system up to date.

6.8.3 The BASE Theory

BASE - Basically Available Soft State, Eventual Consistency

- Basically available—Information is available with all nodes, so that availability at all nodes is ensured.
- Soft state—Data is retained in the ledger; there is no delete for data.
- Eventual consistency—Assuming a node goes off, data would be consistent once the node is up and running; the ledger would be built to ensure consistency is maintained with other nodes.

Blockchain is a distributed system; it is often evaluated based on CAP Theory, although Hyperledger fabric strives to bring in the ultimate consistency feature of BASE. Hyperledger Fabric is a permissioned blockchain infrastructure and is famous for its modular, scalable, secure architecture. The first production-ready version of hyperledger fabric was available in June 2017.

6.9 State Machine

State machine replication is a popular subject in computer science. It deals transactions in a distributed environment, allowing the client to connect to a server and make transactions which change the state of an asset. It also allows a way for facilitating the transfer of data among participating nodes, in a fault-tolerant way. A state machine can be characterized by a set of parameters based on the system design. There is a specific consensus mechanism, which ensures that states which have been broadcast by multiple state machines are on the same page and correct.

This particular example has three states (S1 to S3). In this diagram (refer Fig. 6.6), there are two inputs. Input 0 can be one input to the system, and Input 1 can be another input to the system. These sets of inputs will tell about how the system will behave. You have a set of outputs. S3 is the final output of the system.

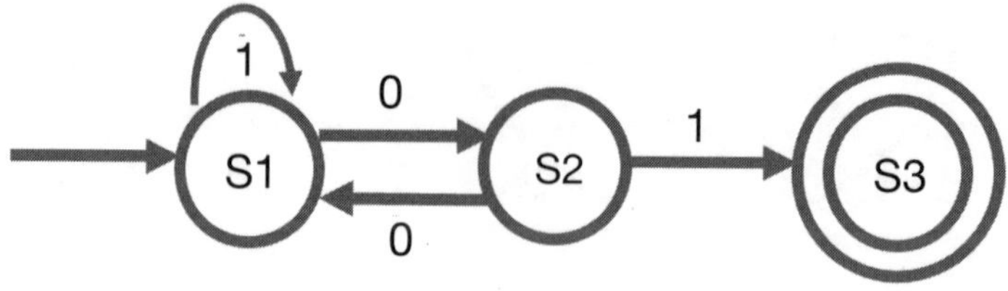

Figure 6.6: Example of a state machine

The transition function will take a state and input as a parameter and will produce a state as the output. In this diagram, State S1 takes 0 as an input and produces State S2 as the output; so this is a transition function. (S1, 0) → S2. The output function produces the output of the system. In this particular example, we do not have any output. In this diagram, S1 denotes the starting state of the system.

6.9.1 Example of State Machine Using a Real-time Example

State machines, in simple terms, can be explained with the help of an oven: The oven provides a facility to insert an ovenproof utensil, which would be subjected to high temperatures. The oven provides a facility to set a range of temperature and start the oven (refer Fig. 6.7).

- The oven provides a mechanism for a client to interact with oven, via switches (buttons)
- When a user clicks (touches) the start button, the state of the oven is changed to heating. The transition from non-heating mode to heating mode is an event.
- Similarly, when an oven is switched off, another event is fired, signifying the oven's return to the non-heating state.
- Another important aspect of the oven is the light that glows when it is being heated.
- The glowing light does not have anything to do with the heating process; however, a glowing lamp is a way to communicate to users: that the heating is in progress.

Coming to state machine replication (refer Table 6.3), in computer science parlance, it is a way for designing and operating a fault-tolerant service, replicating servers and ensuring smooth client interactions, with multiple servers acting as replica in a distributed environment. Distributed computing should be able to withstand faults. The most common solution for attacking this problem is the pBFT – Practical Byzantine Fault Tolerance.

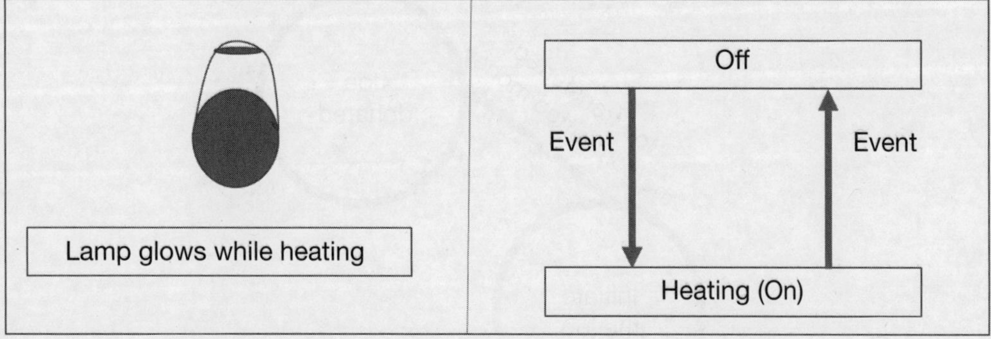

Figure 6.7: Example of a state machine (Oven)

Table 6.3 Oven vs. Blockchain

	An oven as a state machine	**Blockchain as a state machine**
Client library	Switches for customers to operate	Allows client libraries for customers to operate.
Event	Event – Heating starts and stops as requested by the client	Based on the customer requests an asset in the network will change its state.
Event notification	Light glows while heating	Events are notified to all members of the network.

6.9.2 Smart Contract in a State Machine

Let us look at an example where we can represent a smart contract as a state machine. In a crowd-funding platform, you have a set of people who have certain funds available, for donation to specific jobs that are being done.

The proposers propose certain jobs. The system has an initiation state, where some users have submitted a specific proposal. Here, let us assume that Joe has transferred the committed money to the system. You are listening to a state that Joe has donated the money (refer Fig. 6.8).

After that, if Ann has transferred the committed money, you are reaching a state that Ann has donated the money (refer Fig. 6.9).

It may happen that Ann donated the money first (before Joe), and after that, Joe has committed the money (refer Fig. 6.10).

When both of them have donated the money, in the order of either Joe–Ann or Ann–Joe, you have the money available in the system as below (refer Fig. 6.11).

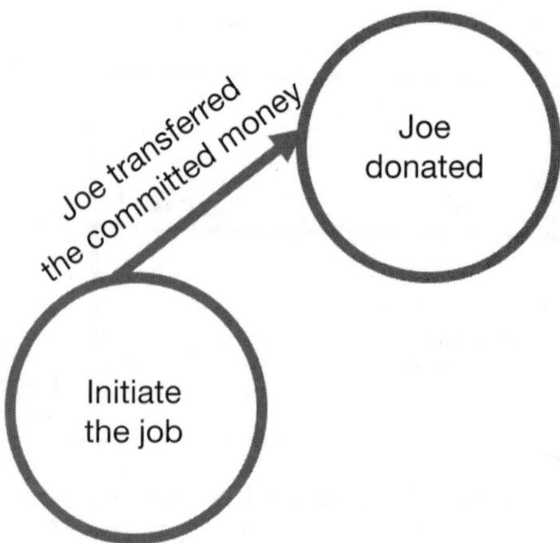

Figure 6.8: Smart contract in state machine (Step 1)

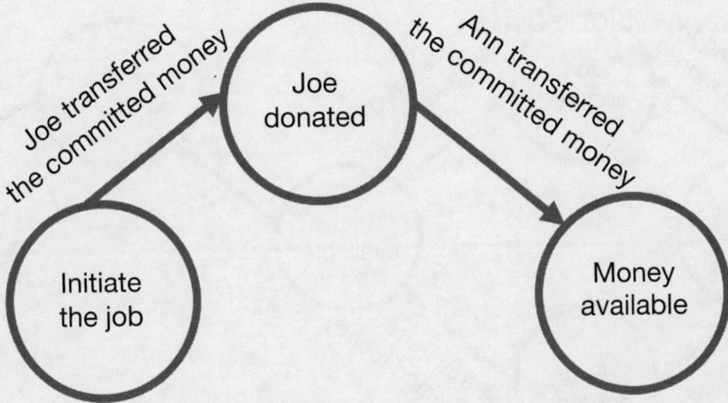

Figure 6.9: Smart contract in state machine (Step 2)

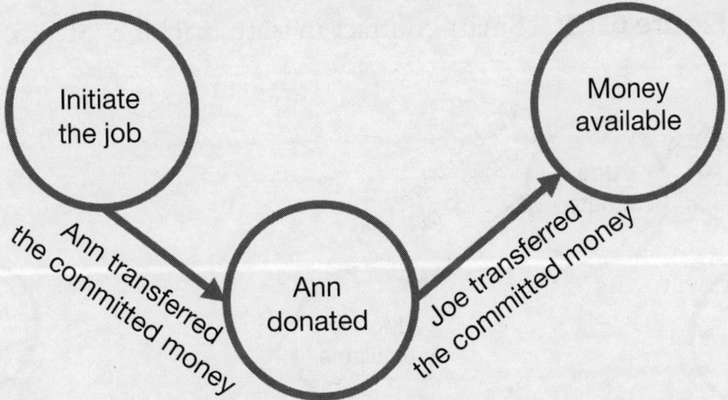

Figure 6.10: Smart contract in state machine (Step 3)

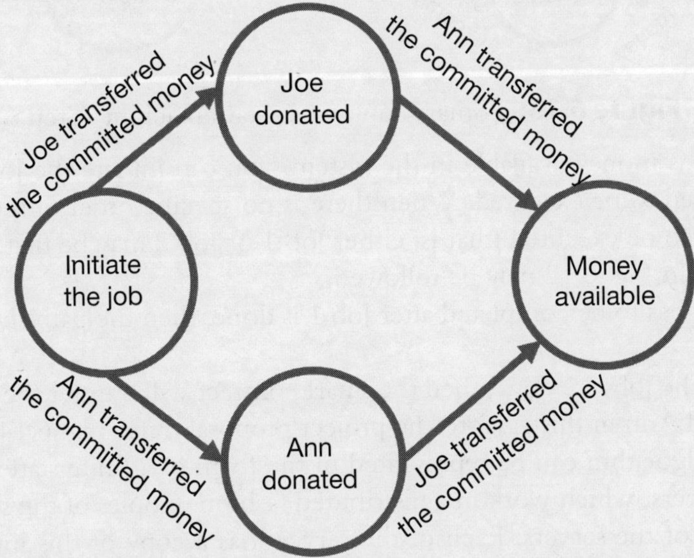

Figure 6.11: Smart contract in state machine (Step 4)

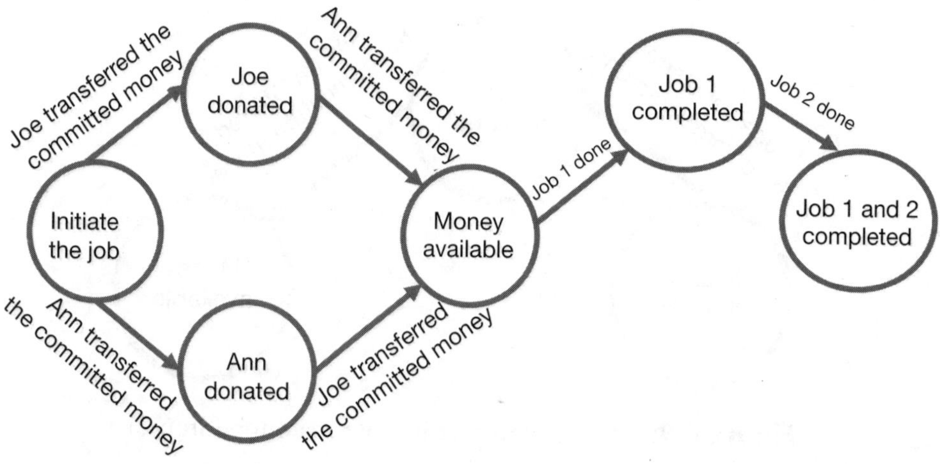

Figure 6.12: Smart contract in state machine (Step 5)

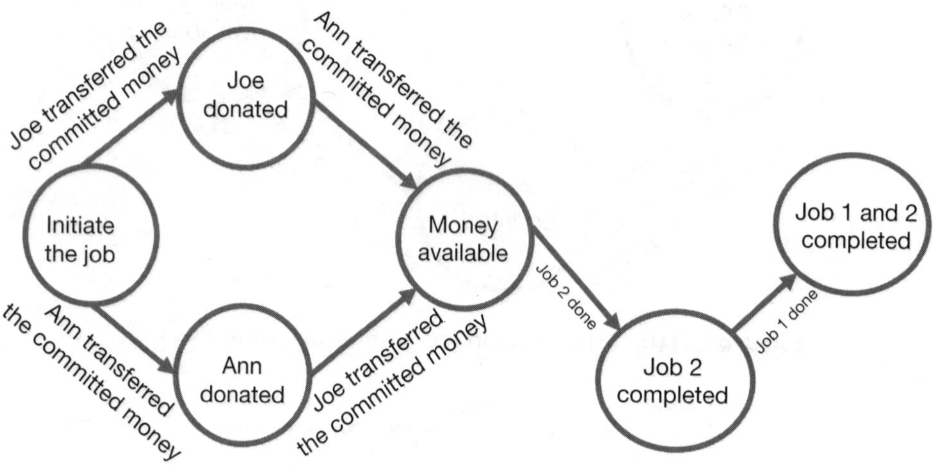

Figure 6.13: Smart contract in state machine (Step 6)

Once there is money available in the system, you can initiate the job. Two jobs (Job 1 and Job 2) need to be executed. When there is no specific order of sequence in which these jobs need to be executed (that is, either Job 1 or Job 2 may be first completed, then the path shown in Fig. 6.12 may be followed.

If Job 2 first has to be completed after Job 1 is done, then the path shown in Fig. 6.13 is followed.

Once both the jobs are done, then the smart contract will transfer the money from the funders (Joe and Ann in this case) to the project proposal (refer Fig. 6.14).

Any finite algorithm can be represented in the form of a finite state machine. There are multiple servers, which work in a distributed fashion. Copies of the state machine are placed on each of the servers. Each of the servers has a copy of this interstate machine (refer Fig. 6.15).

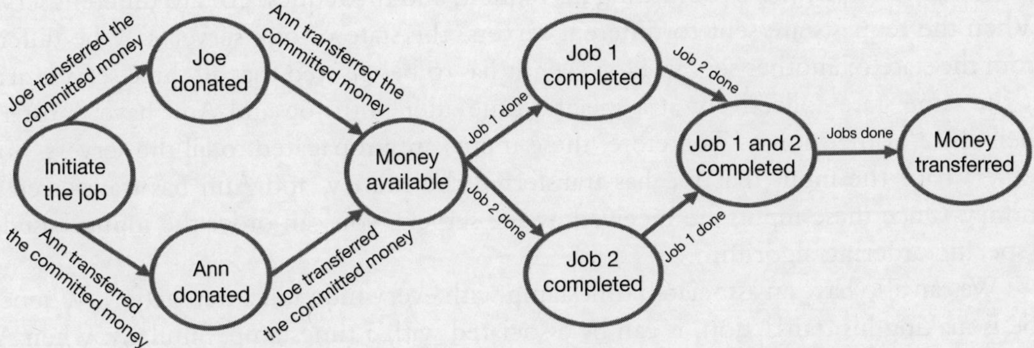

Figure 6.14: Smart contract in state machine (Step 7)

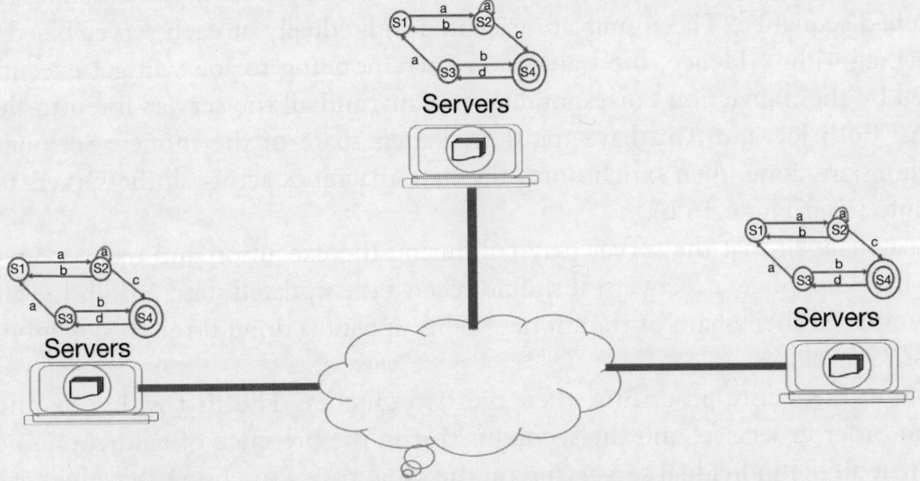

Figure 6.15: Servers with a copy of the state machine

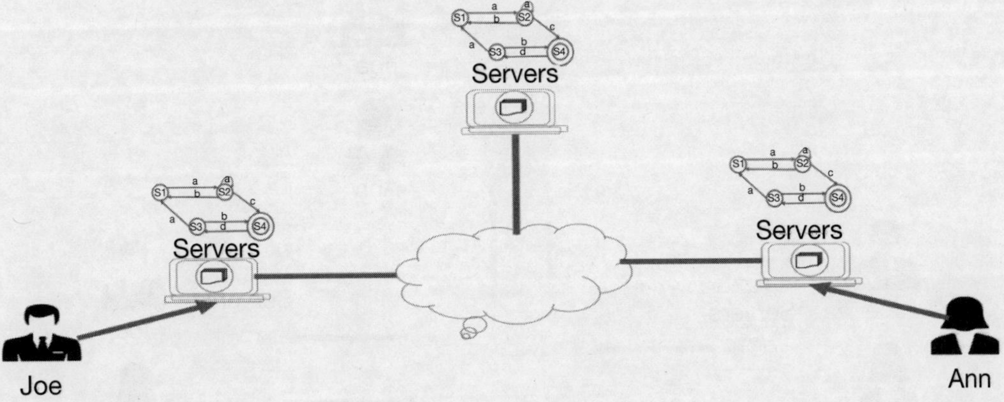

Figure 6.16a: Servers with a copy of the state machine: Client requests added

Then servers get the client requests. Refer Fig. 6.16 a.

In this example, Joe is submitting his request, and the request goes to different servers. When the requests are sent to different servers, the state of one server may be different from the state of another server. However, it has to be ensured that all three servers arrive at the same state collectively at a specific time after both Joe and Ann have transferred their share of the money. Therefore, these inputs are propagated to all the servers. So all servers have the input that Joe has transferred the money, and Ann has transferred the money. Once these inputs are received by the servers, you can order the inputs based on a specific ordering algorithm.

We can also have an associated timestamp with every individual transaction. Whenever Joe is making his transaction, it can be associated with a timestamp. Similarly, when Ann transfers her share of the money, that particular transaction may be associated with another timestamp. Based on the timestamp, we can arrange the transactions in any given order. Once the ordering is done, every individual system executes the inputs based on this pre-determined sequence. The inputs are executed individually at each server based on the ordering algorithm. Hence, the transaction corresponding to Joe will get executed first followed by the transaction corresponding to Ann, and all the servers listen to the same state that both Joe and Ann have transferred their share of the money. So, once these transactions are done, then synchronize the state machines across all the servers to avoid any failure (refer Fig. 6.16 b).

It may happen that the server may fail during the execution and recover after some time. Therefore, once it recovers, it should receive the updated state: "Both Joe and Ann have transferred their share of the money." This update is done through synchronization of the state machine.

Now in this entire procedure, there are two glitches. The first is that one needs to maintain order in service, and the second is that in the presence of failure, one needs to ensure that all the individual servers are on the same page. To ensure this, there is a need for a consensus algorithm in the system.

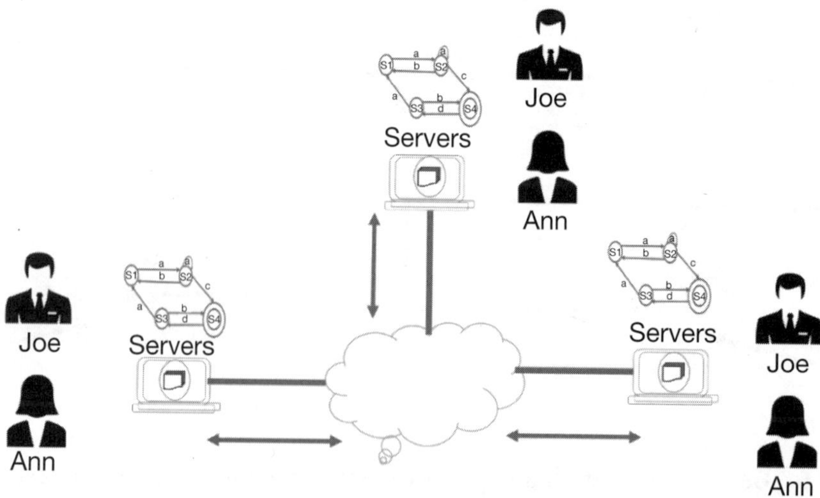

Figure 6.16 b: Servers with a copy of the state machine: State synchronization

6.9.3 State Machine Replication Consensus

Why do we use a state machine replication-based consensus algorithm instead of the challenge-response based consensus model, which we used in the permissionless setting? The first reason is that the network is closed, and when the network is closed, every individual node knows each other. Thus state replication is possible among the known nodes. If you know your peers, you can always replicate your state machine with the current state as that of your peers. The second reason is that the state machine replication-based scenario avoids the overhead of mining. Mining has significant overhead in terms of the system power used, the time consumed and the amount of Bitcoin that you may apply for this kind of mechanism. However, in case of a permission model, because the participants know each other, this kind of overhead is absent in case of a state machine replication.

The consensus is still required on top of the state machine replication because the machines can be faulty, or they can behave maliciously. We do not need a consensus in a single-node process. In the case of two nodes, if the network is faulty or even if there is a kind of crash fault, consensus cannot be arrived at. To reach a consensus, more than two nodes are required. Whenever there are multiple decision-makers, and they want to take specific decisions in a standard way, then there is a need for consensus. In the case of state machine replication, you replicate the current state so that all the processes have the same view of the state.

6.9.4 Applications of State Machine Replication

One typical application of the state machine replication is in the flight control system. This technique is applied when there are multiple flights that want to coordinate their positions among themselves.

A state machine replication-based algorithm for consensus is also applied for fund-transferring systems in a distributed environment. In an open environment, we have a challenge-response based consensus algorithm to achieve a consensus.

For other distributed applications like a distributed leader election, where all the nodes collectively need to elect one leader in the system, consensus is ensured when the entire node selects the same leader, or the same leader is elected at the end of the route. Distributed consensus algorithm is used for this kind of agreement protocols where the nodes collectively need to come to a specific agreement.

6.9.5 Three Common Types of Fault in a Distributed Environment

In a distributed environment, you can have primarily three types of fault:

1. Crash Fault

In this type of fault, a node crashes and is not able to send any message; it may recover after some time and start sending messages again. So this recall is a fail-stop behaviour (refer Fig. 6.17). In a fail-stop behaviour, all the requested messages arrive, and all are considered to be valid, and thus, problems are solved easily. This model does not consider the case where the server itself may crash. Thus, this model is not very commonly used.

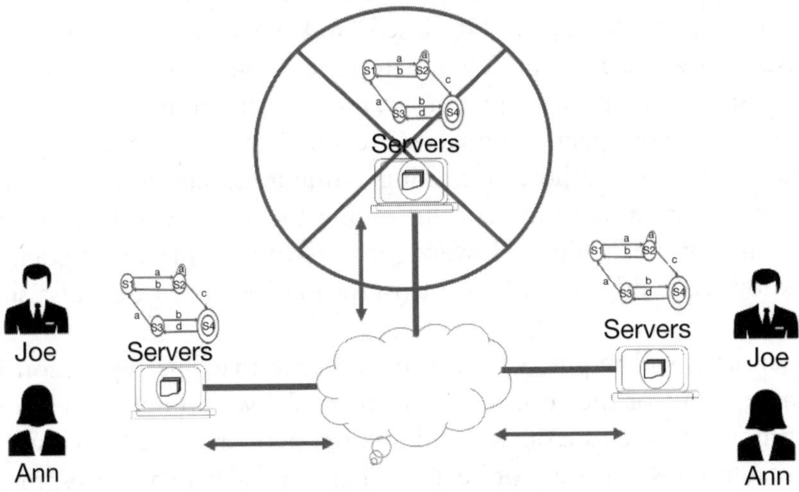

Figure 6.17: Crash fault in a distributed environment

2. Network or Partition Fault

Here, because of a network fault, the network is partitioned, and the message from one partition cannot be propagated to another partition (refer Fig. 6.18).

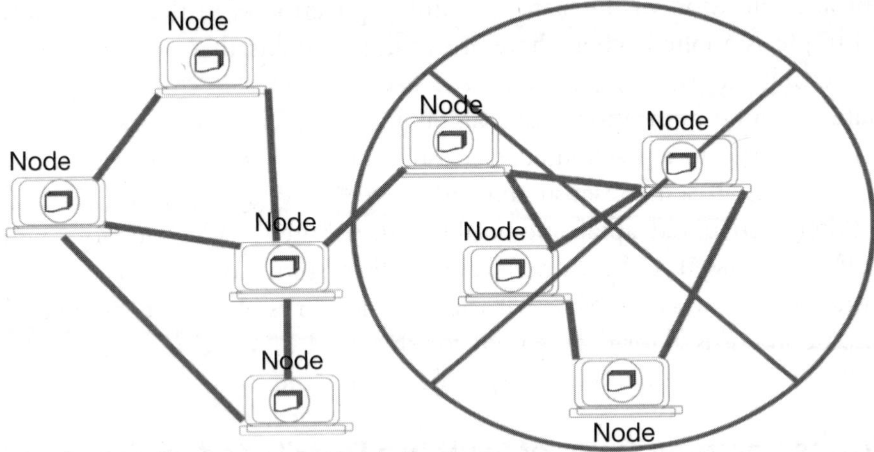

Figure 6.18: Network partition fault in a distributed environment

3. Byzantine Fault

The third type of fault, which is difficult to handle in a distributed environment, is a byzantine fault. It is a kind of malicious behaviour among the nodes, which may be due to hardware faults or software faults (refer Fig. 6.19).

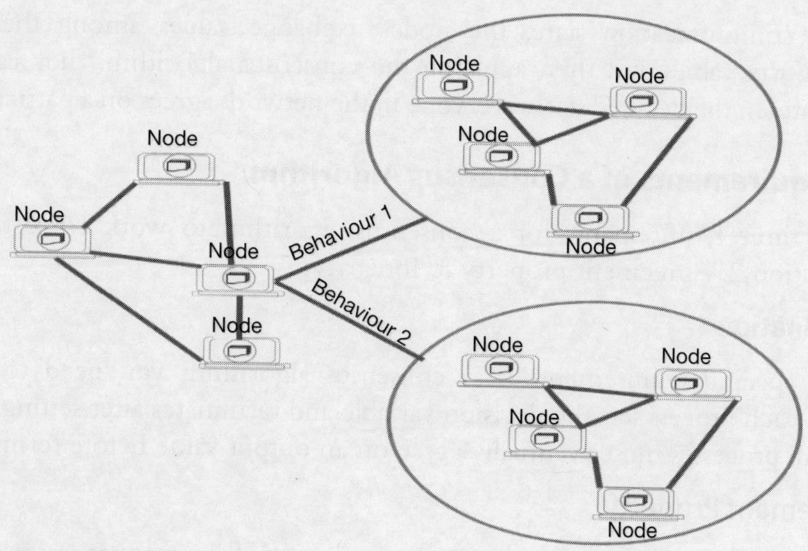

Figure 6.19: Byzantine fault in a distributed environment

In byzantine fault, for specific views, the node behaves in one way, whereas for other views, it behaves in a completely different way.

6.9.6 Consensus for Three Processes

We do not need a consensus in a single-node process. In the case of two nodes, if there is any fault in the network or if there is a network crash, or if the node behaves maliciously, no consensus can be reached. Therefore, to reach a consensus, more than two nodes are required. There is a need for consensus whenever specific decisions are to be taken conventionally by multiple decision-makers. ,

Consider a three-node process. Before consensus is reached among the three processes (P1, P2, and P3), each process can be in one of the three states: 1. Undecided State 2. Communication State 3. Decided State.

In the undecided state, the process at the node has proposed a specific value from a set of possible values. Every node has proposed some value, and they are in an undecided state. Undecided State: Proposed value (V_1, V_2, V_3....V_n) (refer Fig. 6.20).

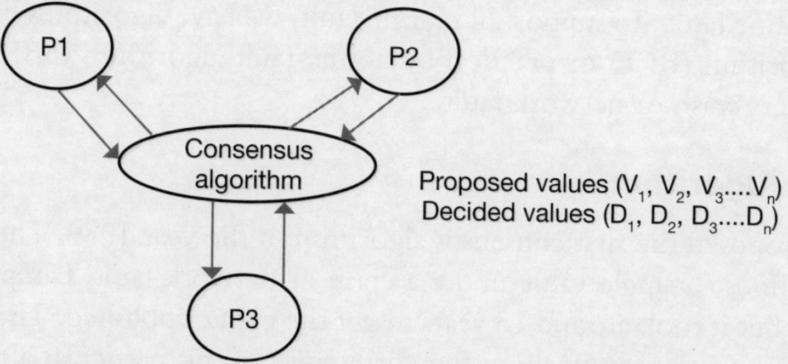

Figure 6.20: State machine consensus algorithm

In the communication state, the nodes exchange values among themselves. By exchanging the values and then applying the consensus algorithm, they can reach the decision state. In the decided state, everyone in the network agrees on a particular variable.

6.9.7 Requirements of a Consensus Algorithm

There are three requirements for a consensus algorithm to work correctly. They are 1. Termination 2. Agreement property 3. Integrity

1. Termination

There are specific requirements of a consensus algorithm; you need to ensure that eventually, each process sets the decision variable and terminates after setting the decision variable. All processes must eventually agree on an output value before termination.

2. Agreement Property

The agreement property says that the decision value for all correct processes should be the same; every correct node should reach a collective agreement.

3. Integrity

The third property is integrity. If the correct processes propose the same value, then all processes in the decision state should choose that value. That is, if all correct processes propose one value X, at the decision state, they should agree on the same value X.

6.10 Different Algorithms of Permissioned Blockchain

Bitcoin and Ethereum rely on the proof-of-work consensus algorithm. However, proof-of-work provides no privacy on transaction details, by design. Hence the nodes are made to compete, as it happens in the case of mining. This is a genuinely trustless network because we do not know who is adding nodes to the system.

We have different kinds of algorithms for ensuring consensus in a typical distributed system applied in a permissioned blockchain. These algorithms are based on the state machine replication principle. There are initial algorithms for distributed consensus: PAXOS and RAFT, which support crash or network faults. However, they cannot support the byzantine fault. To support byzantine fault, we have algorithms like Byzantine fault-tolerant algorithm (BFT) or practical byzantine fault algorithms (pBFT), which in addition, also covers crash or network fault.

6.10.1 PAXOS Algorithm

Leslie Lamport proposed the first consensus algorithm in the year 1989. The objective of PAXOS was to choose a single value under a crash or network fault. Lamport proposed PAXOS in 1989, but it took around 13 years to get the paper published. There was much discussion among the community about the correctness of this particular algorithm. The reviewers were not very convinced that the algorithm proposed by Lamport would be fair enough to ensure consensus in a distributed environment. Lamport tried to come up with a formal proof of the PAXOS consensus algorithm. However, it was considered as a complicated proof and people said it was difficult to understand. Hence, it was tough to implement the PAXOS algorithm.

Let us look at the PAXOS algorithm in a simplified way, and then we will look into the details of how PAXOS can be implemented in a real system to ensure consensus. Let us take a straightforward example to explain this concept.

In Bishop Heber College (Trichy, India), the students have broadly two options on where they can go after a class. The first option is Michael Ice Cream Shop, which is just outside the college campus, and the second option is College Canteen. After class, the students can collectively decide whether all of them want to go to Michael Ice Cream Shop or the College Canteen. It is interesting to see how they decide whether they want to go to "College Canteen" or "Michael Ice Cream Shop" since all of them want to go together. Otherwise, the gathering would not be of much interest. However, there is no central leader. Everyone will propose specific values, and from that value proposal, they will try to come to a consensus.

Let us assume that I am also part of the student community. If none of my friends are proposing any value, then I initiate a proposal. Let us say I propose to go to "Michael Ice Cream Shop." I will check if my friends are accepting that value or not. Consensus is arrived at based on the majority decision. If the majority of the nodes either proposes or agrees to go to "Michael Ice Cream Shop," then everyone goes to "Michael Ice Cream Shop". On the other hand, if the majority of students propose to go to "College Canteen," then all students go to the canteen. This majority selection is the broad idea behind PAXOS.

So let us look into the algorithm of how it works. You have individual proposers, who are proposing the values. The consensus algorithm should choose the proposed values.

Here the proposers can propose a value that chooses either go to "Michael Ice Cream Shop" or go to "College Canteen." Then you have specific acceptors who arrive at a consensus and accept the value. The acceptors, on hearing a specific proposal from the proposers, either accept it or reject it. It may happen that if I want to go to "Michael Ice Cream Shop," but the proposal that comes to me is to go to the canteen, I will reject it. Instead, if I have a friend who proposes to go to "Michael Ice Cream Shop," then I may agree to that. Thus, a majority of the nodes is either proposing or accepting Michael Ice Cream Shop or the Canteen. Then, they collectively accept that particular value. This is the broad idea behind the PAXOS algorithm.

Therefore, there are three types of nodes (refer Fig. 6.21).

1. The proposer
2. The acceptor and
3. The learner

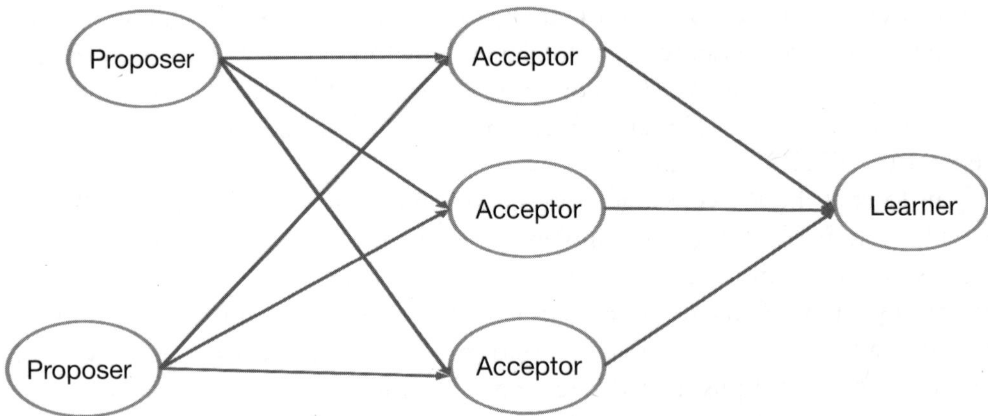

Figure 6.21: PAXOS: Three types of nodes

Everyone is a learner in the network, when it learns "what the majority decision" is. The proposer initially proposes (with a proposal number). This proposal number needs to be good enough to be accepted. The proposer prepares the proposal number and sends it to the acceptor. The proposal number is formed based on the timeline, and the biggest number is considered up to date.

Let us say that one proposal is comes from p1 with proposal number say 101, and another proposal from say p4 with proposal number 202, then we will accept the proposal which is coming from p4. If the new proposal number is higher than your current proposal number, then you accept it; otherwise, you decline it. (refer Fig. 6.22).

Based on the proposal number, the acceptor prepares a response message (Accept/Decline, Proposal #, Accepted values). The message has the result, which has the value of either Accept or Decline. Proposal number is the biggest number the acceptor has seen until now. It also has the values that have been accepted from the proposal (refer Figs 6.23 a and b).

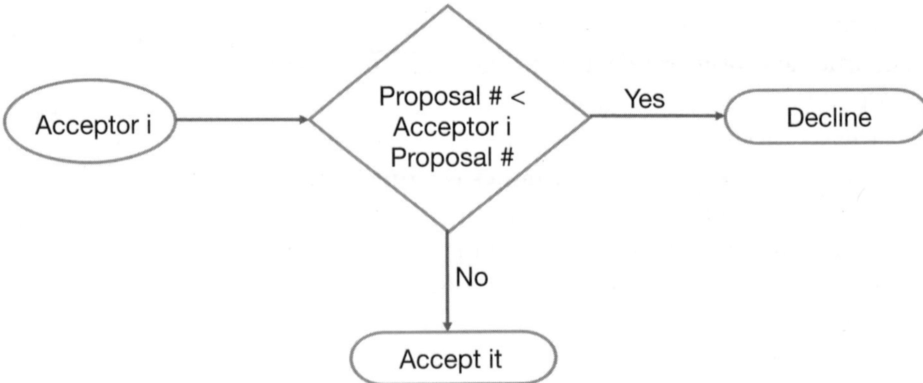

Figure 6.22: PAXOS: Proposal number check for acceptance

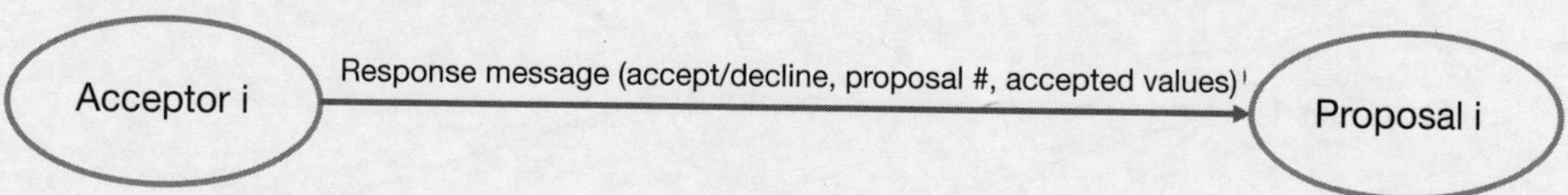

Figure 6.23 a: PAXOS: Response message from the acceptor

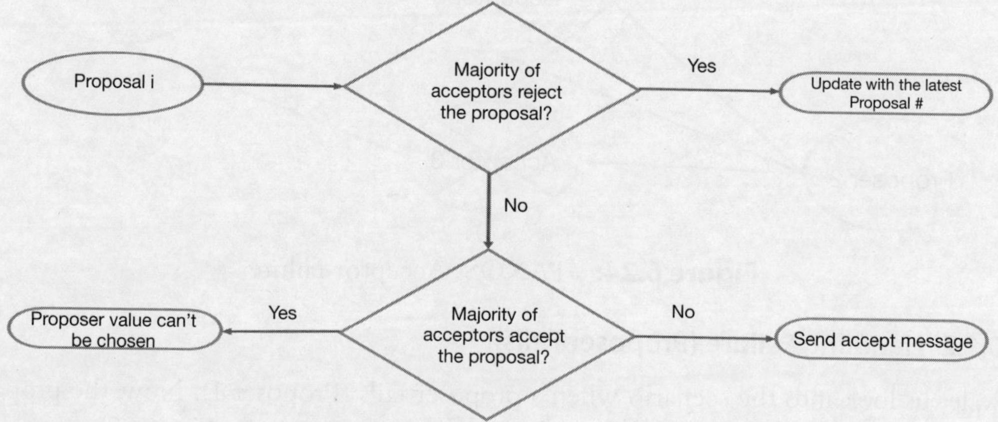

Figure 6.23 b: PAXOS: Majority check

Now, we take a vote based on the majority decision. The proposer looks at whether the majority of the acceptors reject the proposal; if they have rejected the proposal, then the message is updated with the latest proposal number. If the proposal has not been rejected, then you check whether the majority of the acceptors have accepted the value. If the majority of the acceptors have accepted the value, then that particular proposal cannot be chosen. If it is not, then you send the accept message.

The idea is that when a majority of the people have accepted a proposal, they send the accepted values and the value is thus shared becomes a consensus. If you have proposed for Michael Ice Cream Shop, and a majority of your friends have accepted Michael Ice Cream Shop, from that vote all of you can go to Michael Ice Cream Shop. The learner learns from the majority. If you have a single proposal in the system, then that system is straightforward with a single proposal, and every acceptor will accept that proposal because that is going to be the biggest. Therefore, the proposal is not rejected.

6.10.1.1 Handling Failure (Acceptor Fails)

Assume any acceptor (say Acceptor 3) fails. If the acceptor fails during the prepare phase, then there is no issue. There are other acceptors, who can hear the proposal, and they can vote either for the proposal or against it (refer Fig. 6.24).

An acceptor may fail during the accept States. However, this may not cause a problem because there are other acceptors who have already voted for the proposal. The only thing to be ensured is that there are at least N/2 number of acceptors. Since the majority voting principle is followed, in a synchronous environment, a majority of the nodes will have to be correct. If more than N/2 number of acceptors fail, no consensus can be reached.

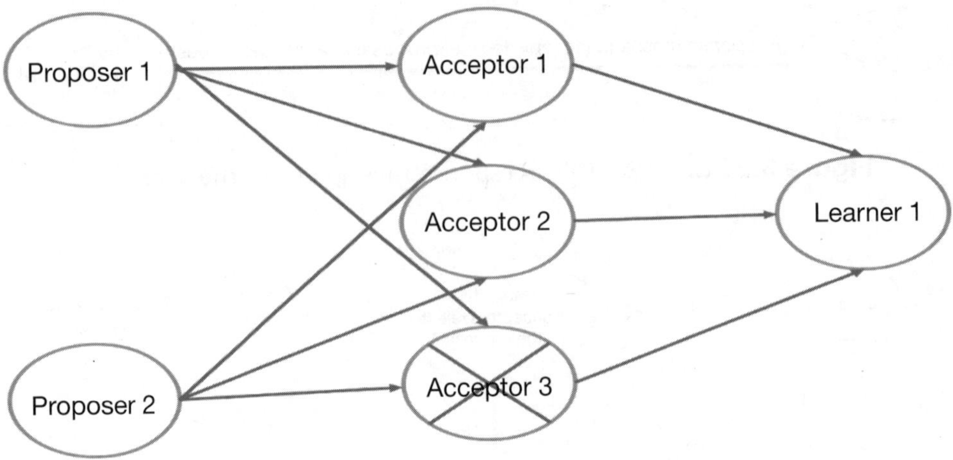

Figure 6.24: PAXOS: Acceptor failure

6.10.1.2 Handling Failure (Proposer Fails)

Now, let us look into the scenario when a proposer fails (Proposer 1). Now, the proposal can fail during the preparation phase or the acceptance phase (refer Fig. 6.25).

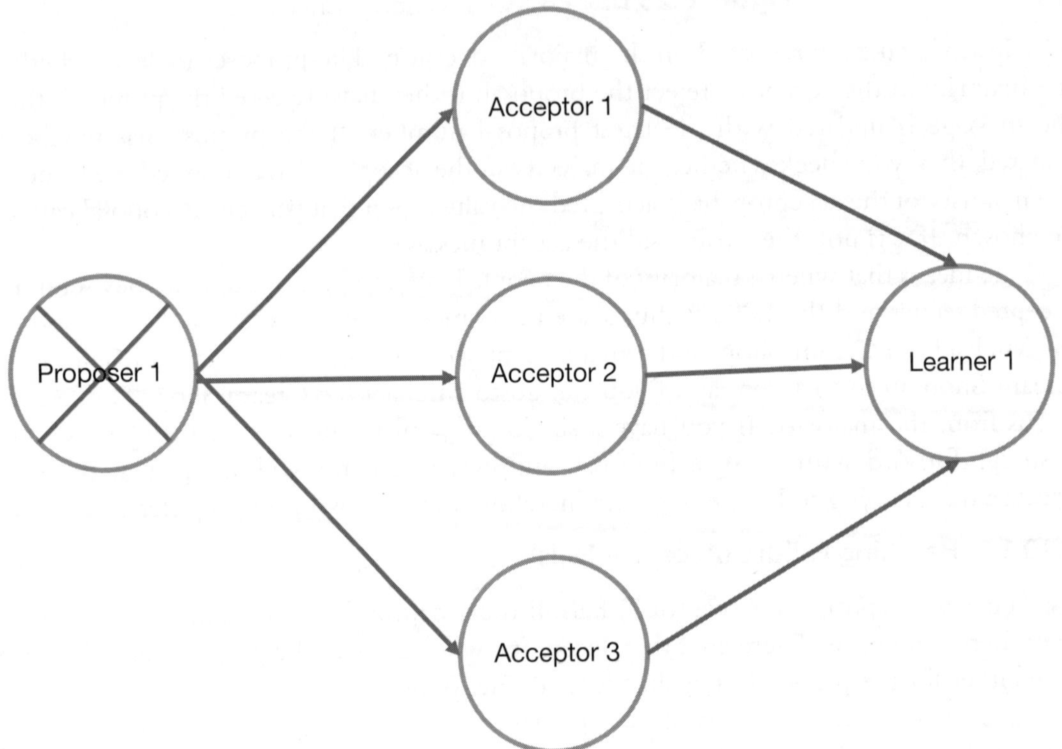

Figure 6.25: PAXOS: Proposer failure

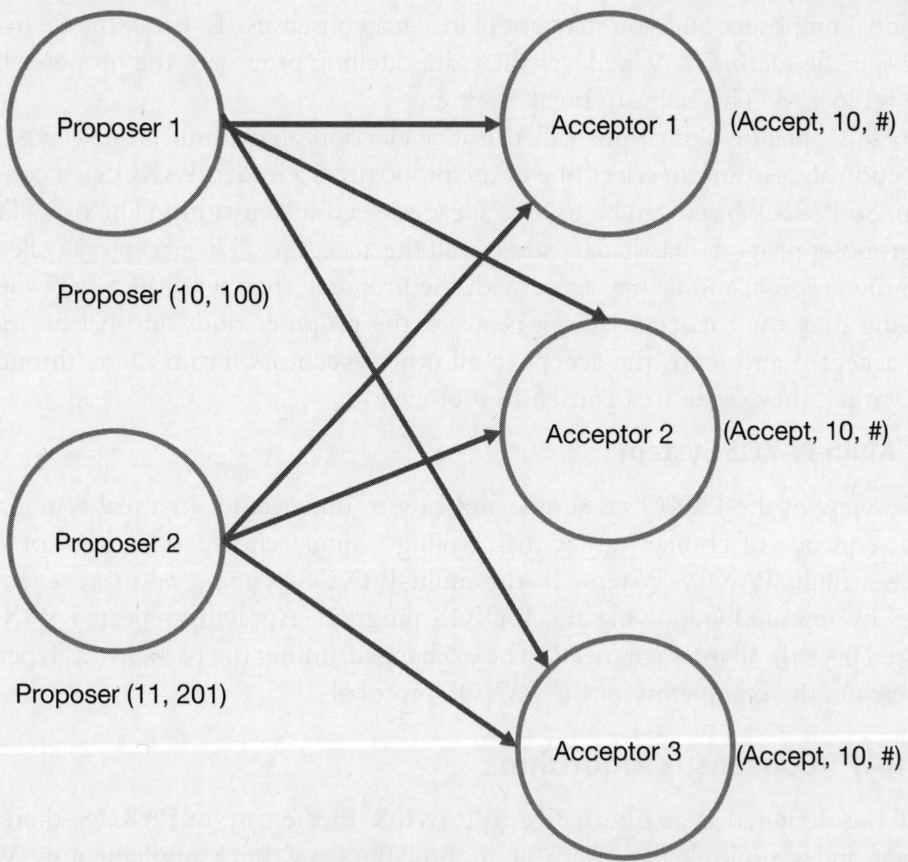

Figure 6.26: PAXOS: Duel proposers

If the proposer fails during the prepare phase, it is just that no one is proposing any value. None of your friends is proposing either to go to Michael Ice Cream Shop or to go to Canteen. Therefore, you wait to get a value from your friend. Moreover, if you are not getting any value, you become a proposer. Similarly, the acceptors wait for some time, and if they do not hear for any proposal, one of the acceptors becomes the proposer and proposes a new value.

However, if the proposer fails during the accept phase, it means the acceptors have already acted upon whether to choose the proposal or not based on the majority vote. Therefore, they have shared a majority vote among themselves. Hence, they can find out whether the proposal has been accepted or not.

Now, there can be a new attack here, which we call as the duelling proposer. There can be two proposers: the Proposer 1 sends specific proposals to all the acceptors. Proposer 2 sends another proposal with a higher proposal number (refer Fig. 6.26).

Now assume that Proposer 1 (attacker) sees that there is a proposal with a high proposal number, and sends another proposal with a higher proposal number. That is, Proposer 1 (Attacker) hears that Proposer 2 has sent a proposal with a proposal number, say 201, and sends another proposal with the higher number, say 301. There can be duelling among the individual proposers, and you may not be reach a consensus. To break the tie here, we use other specific identities. Whenever there are duelling proposals, the proposer having lower ID is blocked. This helps to break the tie.

To do this, specific algorithms called leader election algorithms are executed. This leader election algorithm can select one of the proposals as a leader. PAXOS is a consensus algorithm; So PAXOS itself can be used as a leader election algorithm. The overall idea is that the proposer proposes particular views to all the acceptors. The acceptors collectively look into the proposal and if they agree with the proposal, they send the accept message. By receiving back the particular accept message, the proposer finds out that his message has been accepted and sends the accept to all other acceptors. From there, through the majority voting, they come to a consensus protocol.

6.10.1.3 Multi-PAXOS System

The above view of the PAXOS is simple and easy to understand. In a real system, there may be a sequence of choices rather than having a single choice. This kind of system is called as a multi-PAXOS system. In the multi-PAXOS system, we make a sequence of choices by repeatedly applying the PAXOS program. Applying repeated PAXOS is complicated because all messages need to be exchanged among the participants repeatedly, which increases the complexity of the PAXOS protocol.

6.10.2 RAFT Consensus Algorithm

The RAFT is designed as an alternative to PAXOS. In the case of PAXOS, there is no leader node, and so multiple proposers can propose the same thing simultaneously. Whenever multiple proposers are proposing the same thing simultaneously, it becomes a little complex. The acceptors have to accept one of the proposals and use the highest proposal number used as a tie-breaking mechanism. We can embed a specific algorithm inside PAXOS to ensure that every proposal that is coming from different proposers is unique.

RAFT will first elect a leader, and once it has elected a leader, then the task of the leader is to propose something. There would be a single proposer (leader), and all the accepters are followers of the leader. When the leader is proposes, the acceptors (followers) may or may not follow the leader.

In our example, whether we want to go to Michael Ice Cream or to Canteen, we need to choose a leader among our team. Let the leader of the group propose whether to go to Michael Ice Cream or Canteen, then we start voting. Now, for example, the leader says that he wants to go to Michael Ice Cream, and asks for a show of hands to indicate support. If the majority raises their hands, indicating they also agree with the leader, and then everyone goes to Michael ice Cream. Otherwise, if the majority of them decline to go to Michael Ice Cream, you have to choose the alternative option of going to College Canteen.

The basic idea behind the RAFT consensus protocol (refer Figs 6.27 a and b) is that the nodes collectively select a leader, and the leader is responsible for state transition.

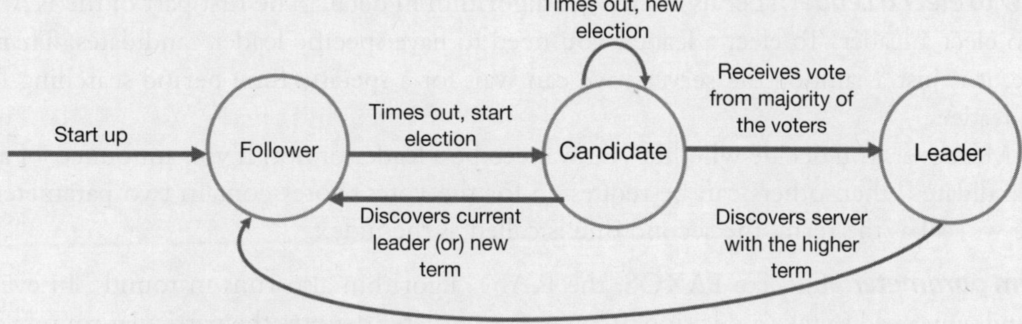

Figure 6.27a: RAFT: Follower, Candidate, and Leader

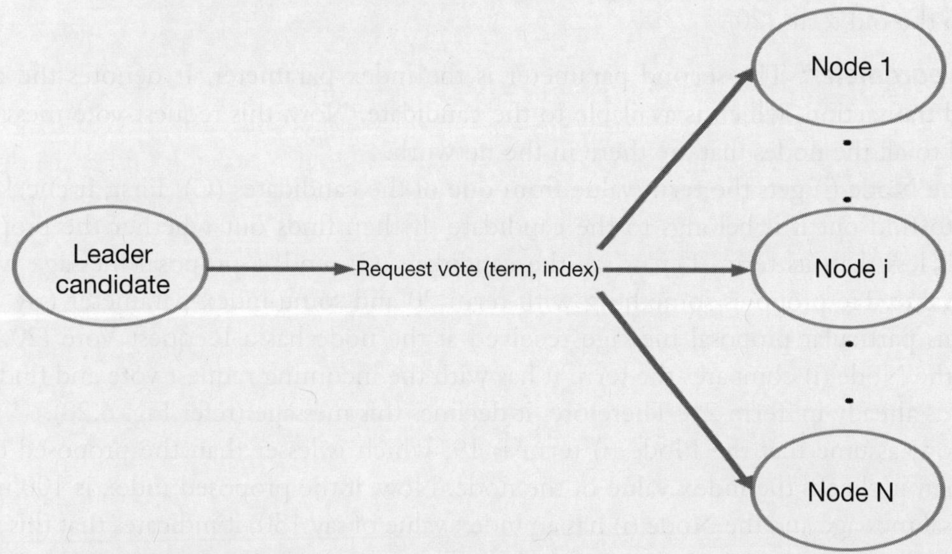

Figure 6.27b: RAFT: Leader Candidate request for vote to nodes

Whenever the system starts up, it automatically has a set of follower nodes. The follower nodes check whether there is already a leader or not. If the node times out, that means there is no leader in the system, and you start the election. In the election, you choose the candidates from among a group of volunteers who want to be a leader. Servers select the leader through voting. Whoever wins the majority of the votes becomes the leader. The leader finds out the proposed value. Then either the followers can vote for the proposal coming from the leader, or they can choose to go against the leader.

6.10.2.1 RAFT Consensus in a Transaction

Let us say that we want to build a consensus among multiple replicated servers. Therefore, whenever there are transactions coming up from the clients, we shall want the replicated servers, say S1, S2, S3, and S4, to make a decision about the consensus. Moreover, they shall also have to decide whether to commit the transactions or not.

How to elect a Leader: Let us discuss the algorithm in detail. The first part of the RAFT is to elect a leader. To elect a leader, you need to have specific leader candidates. Therefore, it is just as among the servers; we can wait for a specific time period searching for the leader.

Moreover, you decide whether you want to be a leader or not. If you announce, "I am a candidate", then others can be requested for the votes. Votes contain two parameters, one we call as the term, the second one is called as the index.

Term parameter: Just like PAXOS, the RAFT algorithm also runs in rounds. In every round, you need to take a decision. The term parameter denotes the particular round you are currently in.

Assuming the old term is numbered as (20), the new term (21) is calculated by adding one to the old term (20).

Index parameter: The second parameter is the index parameter. It denotes the committed transaction, which is available to the candidate. Now, this request vote message is passed to all the nodes that are there in the network.

The Node (i) gets the term value from one of the candidates (C). First, it checks the term to find out if it belongs to the candidate. It then finds out whether the proposed term is less than its term. Let us say the candidate (C) sends a proposal message with a term value. He proposes something with term 20 and some index parameter (say 100); and this particular proposal message received at the node has a Request Vote (20,100). Now the Node (i) compares the term it has with the incoming request vote and finds out that it is already in Term 21. Therefore, it declines this message (refer Fig. 6.28).

Now, assume that the Node (i) term is 19, which is lesser than the proposed term, and then it checks the index value of the node. Now, if the proposed index is 100 in the proposal message and the Node (i) has an index value of say 150, it indicates that this node has already committed up to transaction 150. Therefore, it declines that particular node's request to be a leader (refer Fig. 6.29).

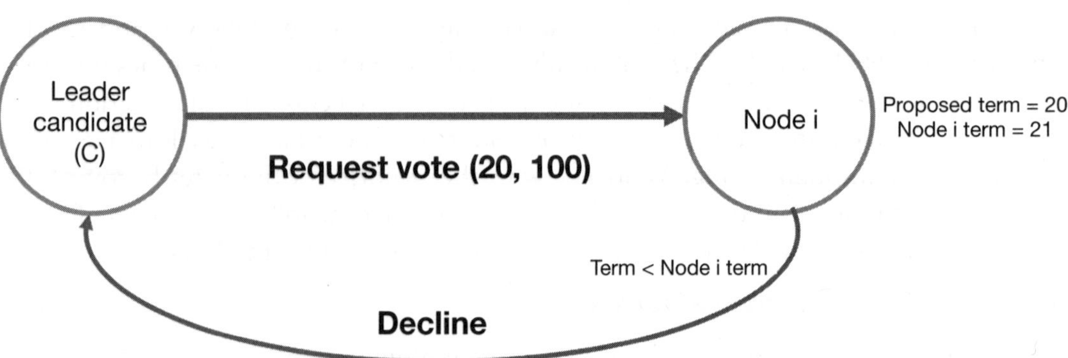

Figure 6.28: RAFT: Leader candidate request for vote to nodes: Term check

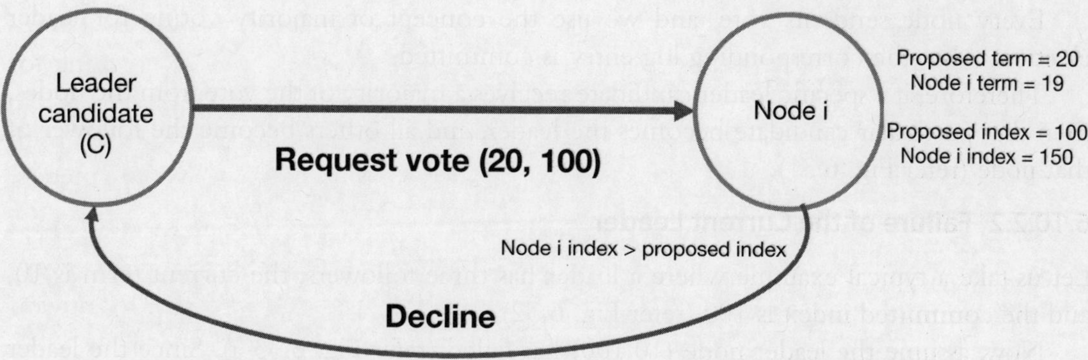

Figure 6.29: RAFT: Leader candidate request for vote to nodes: Term and index check

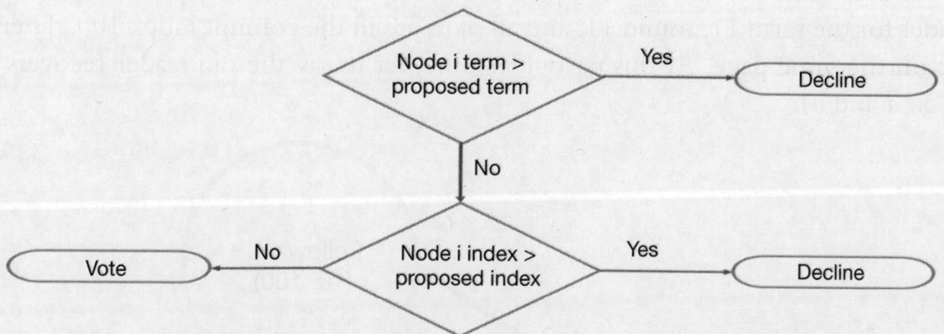

Figure 6.30: RAFT: Accept and Decline check

If on the other hand the index value is lesser than 100, you make a vote in favor of the candidate. This way, you select a leader in the RAFT consensus algorithm. (refer Fig. 6.30).

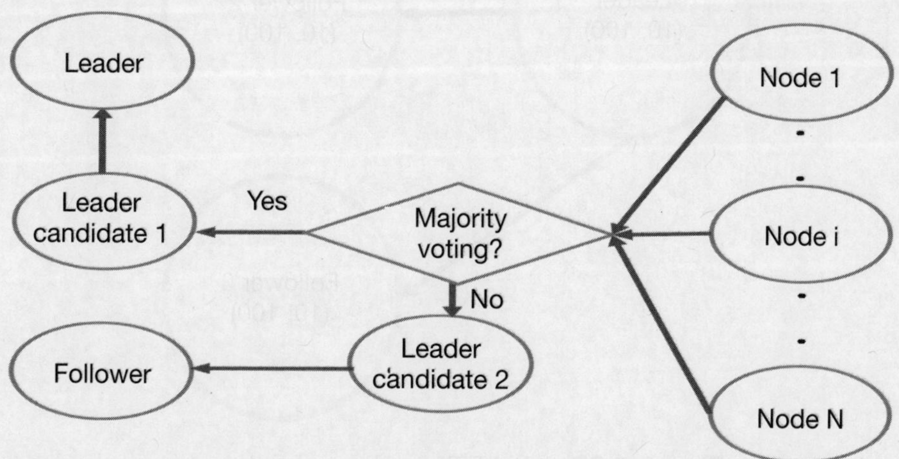

Figure 6.31: RAFT: Majority voting check

Every node sends its vote, and we use the concept of majority voting for leader election. Also, the corresponding log entry is committed.

Therefore, if a specific leader candidate receives a majority of the vote from the nodes, then that particular candidate becomes the leader, and all others become the follower of that node (refer Fig. 6.31).

6.10.2.2 Failure of the Current Leader

Let us take a typical example where a leader has three followers, the current term is 10, and the committed index is 100 (refer Fig. 6.32).

Now, assume the leader node (10,100) has failed (refer Fig. 6.33 a). Since the leader node has failed, a new leader needs to be elected with the new term 11, which is the old term number plus one; that means you need to move from Round 10 to Round 11. A new leader is elected among these three followers. Assume Follower 1 is elected as a new leader for the Term 11. Moreover, let us say the new leader sends the message that he is the leader for the term 11, round 11, and all of us are in the commit index 100. Therefore, we are on the same page. At this particular time, let us say the old leader recovers (refer Figs 6.33 a and b).

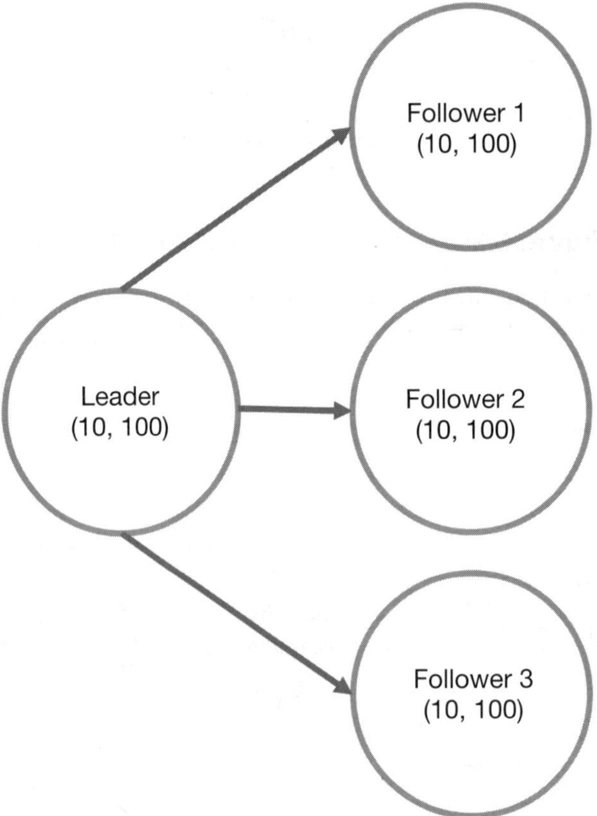

Figure 6.32: RAFT: Leader and follower

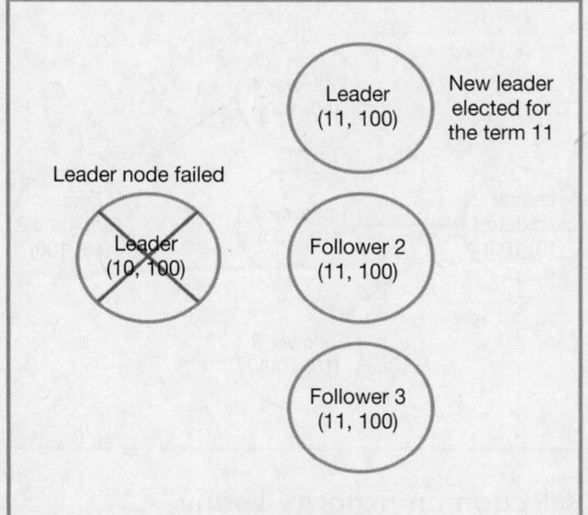

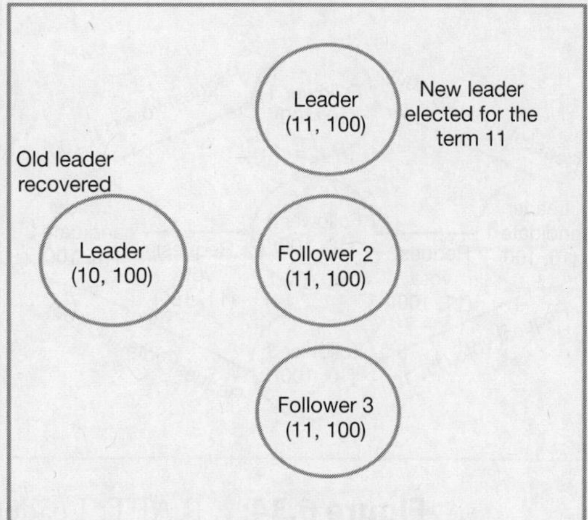

Figure 6.33 a: RAFT: Leader fails and recovers

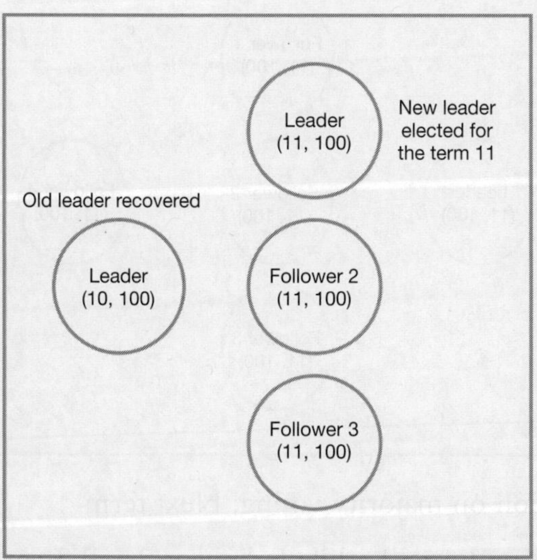

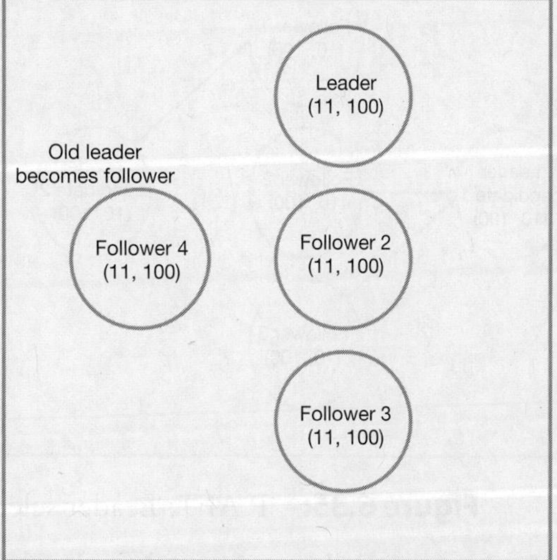

Figure 6.33b: RAFT: Leader fails and recovers, becomes the follower

Case 1

The old leader (10,100) recovers and the new leader's (11, 100) message reaches the old leader. Therefore, the old leader finds out that he was the leader at Round 10, and now there is a new leader in the system, which is a Term 11. Therefore, the Term 11 or Round 11 has already started in the system. Therefore, the old leader (of Round 10) falls from being a leader to a follower. This is one way of handling a new leader when an old leader fails, by utilizing the term parameter.

Case 2

Another case arises when two nodes (Leader Candidate 1 and Leader Candidate 2) send the request for vote message at the same time. The entire system currently stays in Term 10. Therefore, at the Term 10, there are two leader candidates who send a request vote for round 11. Voting happens, and they look for the majority voting.

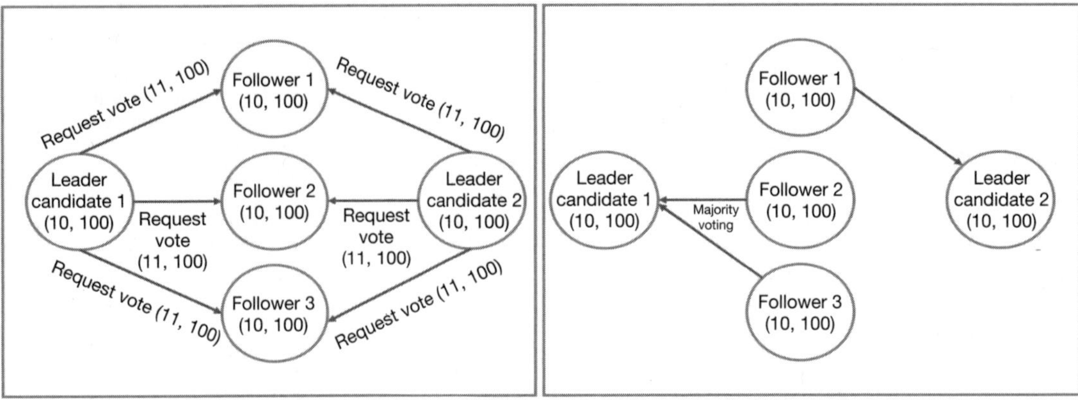

Figure 6.34: RAFT: Leader selection on majority voting

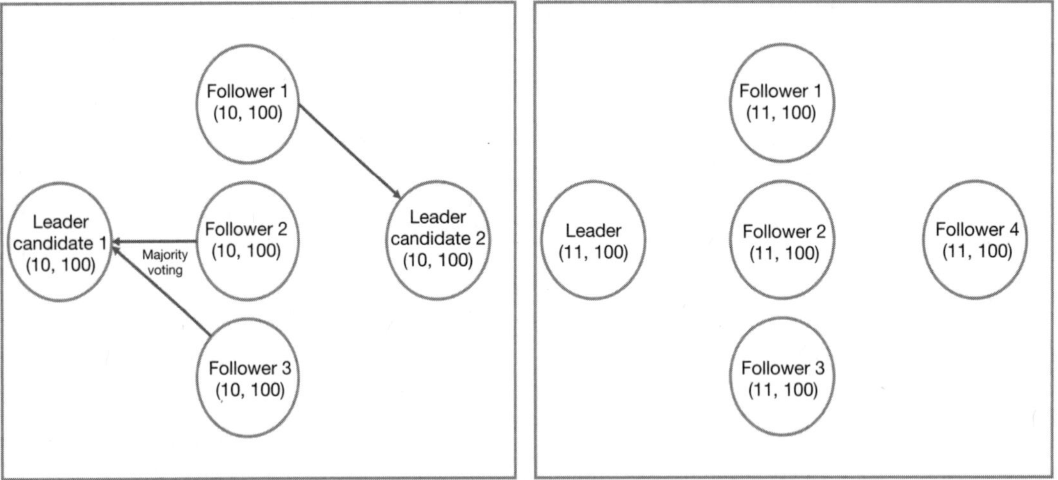

Figure 6.35: RAFT: Leader selection on majority voting: Next term

In this case, Follower 1 voted for Leader Candidate 2 while Followers 2 and 3 voted for Leader Candidate 1. Leader Candidate 1 gets the majority of the voting (refer Fig. 6.34).

Moreover, the node that gets the majority voting sends a special message called a heartbeat message, in which the leader says, "I have received the majority of the voting." The other leader candidate (Leader Candidate 2 in this case) hears the heartbeat message from the winner (Leader Candidate 1) and falls back to a follower, and a Round 11 or Term 11 gets started (refer Fig. 6.35).

6.10.2.3 Committing Entry Log

Once the leader is selected, the second part deals with the index term. It determines how an index log or a transaction is committed as agreed upon by all the followers. Once the leader election is done, the leader has the task to propose a new transaction.

Every node is now at Round 11; that means Term 11 and the last committed transaction index is 100. Now if the leader wants to propose a new transaction, it acts as an entry log with Term 11, the current term where everyone is, and the new transaction index has index 101. Therefore, this particular entry is done in the index log of the leader. Then the leader sends a message called append entries (11,101) to all the followers (refer Fig. 6.36).

These append entries contain the proposal coming from the leader. In a particular proposal, if the leader says that now we are at Index 11, it means we are at Round 11. It also means that the leader is proposing for the Transaction 101. So, once this particular transaction log is broadcast to all the followers, the followers collectively vote either for the transaction or against the transaction (refer Fig. 6.37).

If the leader receives majority votes for the transaction 101, it means that a majority of the nodes are okay with committing that particular log (101). The leader sends an accept message based on the majority voting to all the individual followers. Therefore, the followers update the committed index to 101.

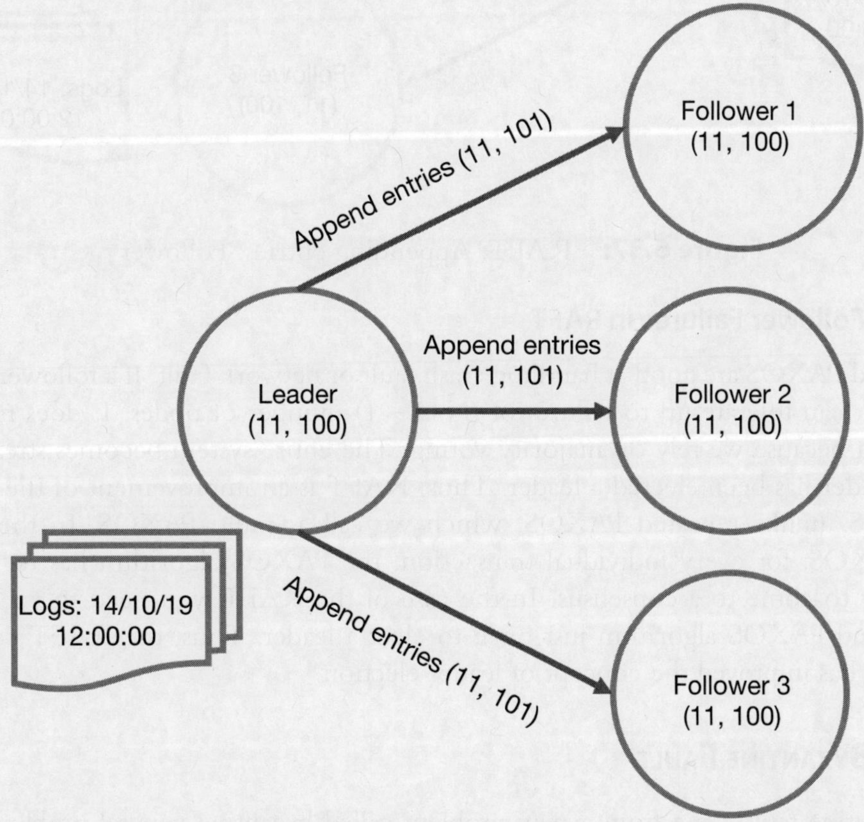

Figure 6.36: RAFT: Appending entries: Leader

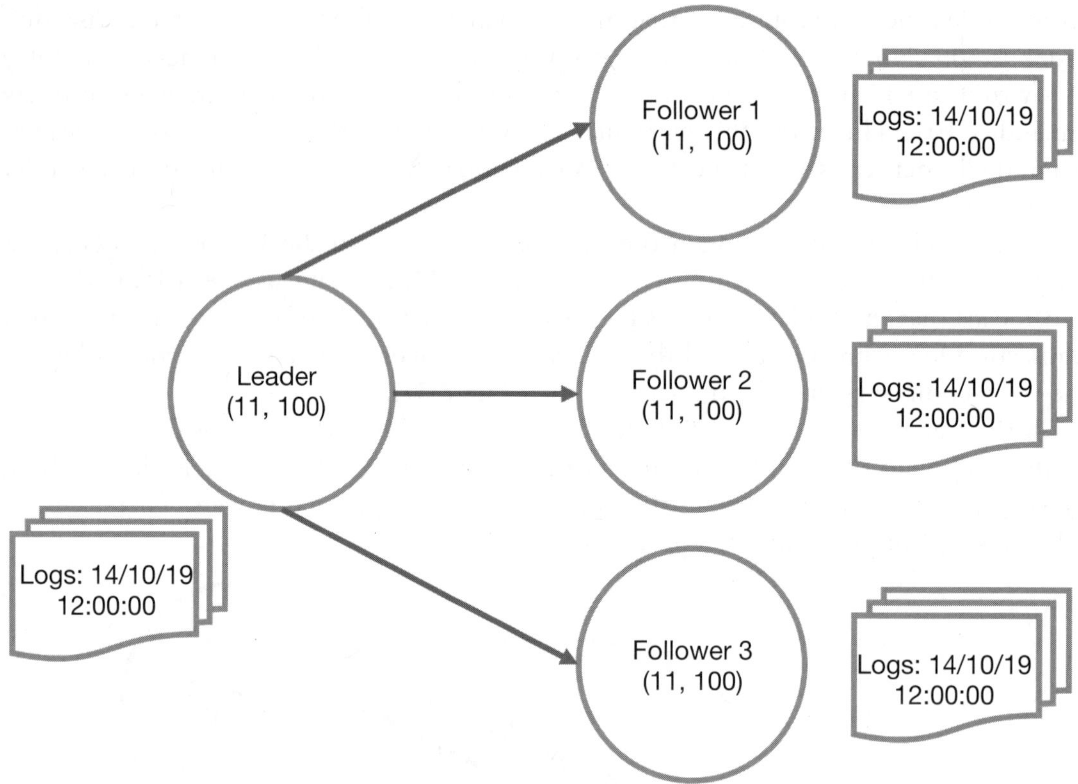

Figure 6.37: RAFT: Appending entries: Follower

6.10.2.4 Follower Failures in RAFT

RAFT and PAXOS are good at handling crash fault or network fault. If a follower crashes, the system can tolerate up to failures of (N/2 – 1) number of nodes. It does not affect the system because we rely on majority voting. The entire system becomes streamlined, once a leader has been elected a leader. Thus, RAFT is an improvement of the concept of PAXOS, unlike repeated PAXOS, which we call as multi-PAXOS. In the case of multi-PAXOS, for every individual transaction, the PAXOS algorithm has to be done repeatedly to come to a consensus. In the case of the RAFT, you need to execute the complicated PAXOS algorithm just once to elect a leader. Thus, the RAFT consensus algorithm has improved the concept of leader election.

6.11 Byzantine Fault

The byzantine fault came from a new problem called byzantine general problem. If the leader behaves in a byzantine way (strange way), that is, if the leader proposes a transaction only to a selective number of nodes among all followers, the PAXOS and the RAFT protocol may not help. In that case, we need to have a strict class of consensus algorithm that can tolerate the byzantine fault, and that consensus algorithm is called as byzantine fault consensus algorithm. We will look into byzantine fault-tolerant protocols in detail.

In byzantine fault, a node (leader) starts behaving maliciously; it sends different messages to different peers (followers). The general (leader) sends an attack message to one troop, whereas he sends a retreat message to another troop. If the general (leader) becomes faulty, then it becomes difficult for a system to determine what to do. Therefore, we will try to develop a fault-tolerant architecture, where the system will be able to tolerate this kind of byzantine faults.

6.11.1 Three Byzantine Problem

Let us examine the byzantine general problem, where you have three generals. In this design, you have one commander who can send message to the lieutenants, and two different lieutenants, who can send the messaging among themselves. Moreover, in this particular case, we will try to design a problem and arrive at a solution for this byzantine fault-tolerant system.

Three Byzantine General: Lieutenant Is Faulty

If a lieutenant is malicious (faulty), then the lieutenant may send different messages. In this case, the commander (non-faulty) sends a retreat message to both the lieutenants. A particular lieutenant (Lieutenant 2) is a faulty lieutenant. This faulty lieutenant does not obey the message from the commander; instead, the malicious (faulty) lieutenant sends a different message (attack message) to the other lieutenant. Lieutenant 2 (faulty) also receives the retreat message from the Lieutenant 1 (non-faulty), which he (Lieutenant 1) received from the commander (refer Fig. 6.38).

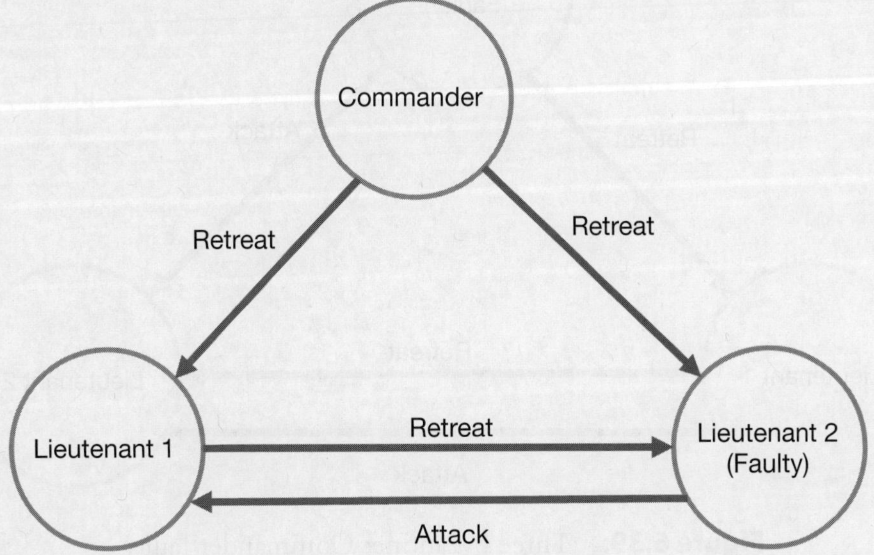

Figure 6.38: Three byzantine: One lieutenant faulty

Let us see whether we can reach a consensus in this scenario. Lieutenant 1 receives different messages. He receives a retreat message from the commander, and an attack message from the Lieutenant 2.

Now, in this case, the lieutenant typically obeys the commander. Moreover, if the commander is non-faulty, then the entire system works correctly. But what if the commander is faulty?

Three Byzantine General: Commander Is Faulty

This is an extreme case (refer Fig. 6.39).

In this case, the commander (faulty) sends Retreat message to one lieutenant (Lieutenant 1), and sends Attack message to the second lieutenant (Lieutenant 2). Lieutenant 1 receives the Retreat instruction from the commander and echoes this Retreat message to the other lieutenant (Lieutenant 2). Lieutenant 2 receives an attack instruction from the commander, and echoes this Attack message to the other lieutenant (Lieutenant 1).

Because Lieutenant 1 has different messages from different generals, he may follow the principle of integrity condition and make a retreat (based on the commander's instruction).However, Lieutenant 2 has received the attack message (the command form the commander) and if he attacks, then, the entire army will get defeated.

When a lieutenant gets equal voting for attack and retreat, he cannot decide what to do. Thus, we will not be able to solve the byzantine general problem with three generals, where we have one commander and two different lieutenants.

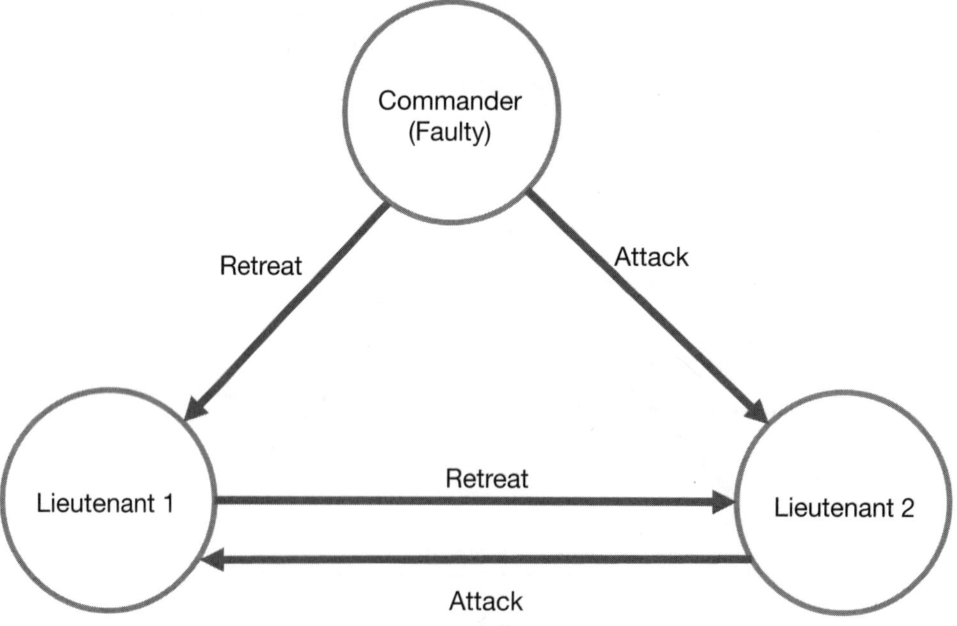

Figure 6.39: Three byzantine: Commander faulty

6.11.2 Four Byzantine Problem

Let us look into another use case where we have four different byzantine generals. With these four byzantine generals, we have three lieutenants (refer Fig. 6.40).

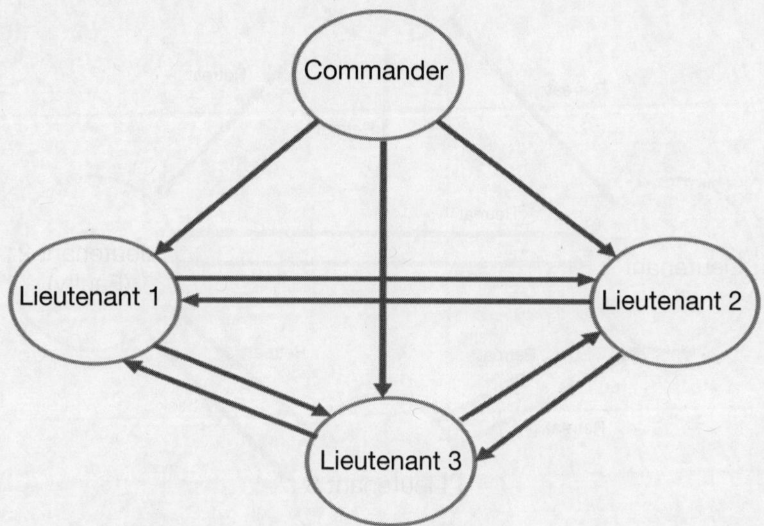

Figure 6.40: Four byzantine structure

Every individual lieutenant based on the message passing principle that we have discussed earlier, talks with each other. The commander (leader) sends the message to all the lieutenants (followers). The lieutenants share the message the commander has shared with them, with each other.

Four Byzantine General: One Lieutenant Is Faulty

There are three lieutenants in total and any of them can be faulty.

Case 1: The faulty node sends the same message to all.

In this case, we assume that Lieutenant 2 is faulty. This faulty lieutenant sends a message, which differs from the message that was sent by the Commander (retreat).

Lieutenant 1 and Lieutenant 3 correctly echo the message (Retreat) to other lieutenants, but Lieutenant 2 differs (refer Fig. 6.41), and sends an attack message to Lieutenant 1 and Lieutenant 3, whereas, the commander has sent a Retreat message. We see that in this case, Lieutenant 1 receives two Retreat messages: one retreat message from the Commander and the second retreat message from Lieutenant 3. Lieutenant 1 can therefore decode the commander's instruction correctly.

Lieutenant 3 (correct) also receives two Retreat messages. The first retreat message is received from the Commander and the next one form Lieutenant 1. He receives two Retreat messages and one Attack message (Lieutenant 2). So he decides to retreat. Lieutenant 2 is faulty; we do not consider the behaviour of the faulty lieutenants. Thus, we observe that even if Lieutenant 2 is a byzantine node, with majority voting, Lieutenant 1 and Lieutenant 3 are able to decode the message correctly.

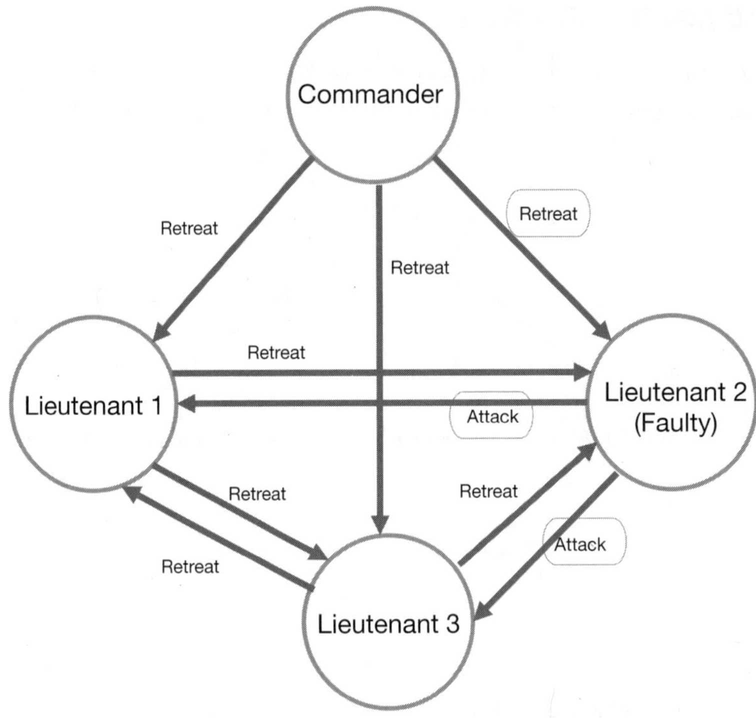

Figure 6.41: Four byzantine structure: One lieutenant faulty: Case 1

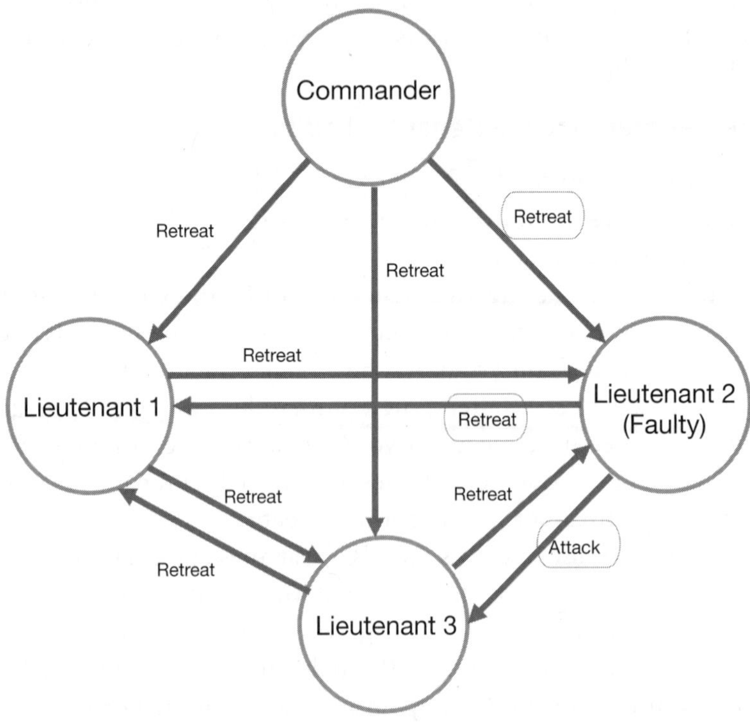

Figure 6.42: Four byzantine structure: One lieutenant faulty: Case 2

Case 2: Faulty node sends a different message to different nodes

In the above case, the faulty node (Lieutenant 2) sent the same "retreat" message to both of the other lieutenants. Now, consider a scenario where the malicious or byzantine behaviour of lieutenant 2 sends different messages to different nodes (refer Fig. 6.42).

Lieutenant 2 sends Attack message to one node (Lieutenant 3) and Retreat message to another node (Lieutenant 1). With the majority voting principle, Lieutenant 1 (Retreat, Retreat, Retreat) can correctly decode the message. Lieutenant 3 (Retreat, Retreat, Attack) can also correctly decode the message. So, when one lieutenant is faulty, we can correctly decode the message.

Four Byzantine General: Two Lieutenants Are Faulty

Here, we assume that Lieutenant 1 and Lieutenant 2 are faulty. Instead of the correct Retreat message, both lieutenants send Attack message at this stage (refer Fig. 6.43).

Lieutenant 3 directly gets one retreat message from the commander, but two attack messages, one each from the faulty Lieutenants 1 and 2. So, Lieutenant 3 will not be able to make a correct decision based on majority voting. Therefore, here we observe that if two out of three lieutenants are faulty, then we will not be able to decode the message correctly.

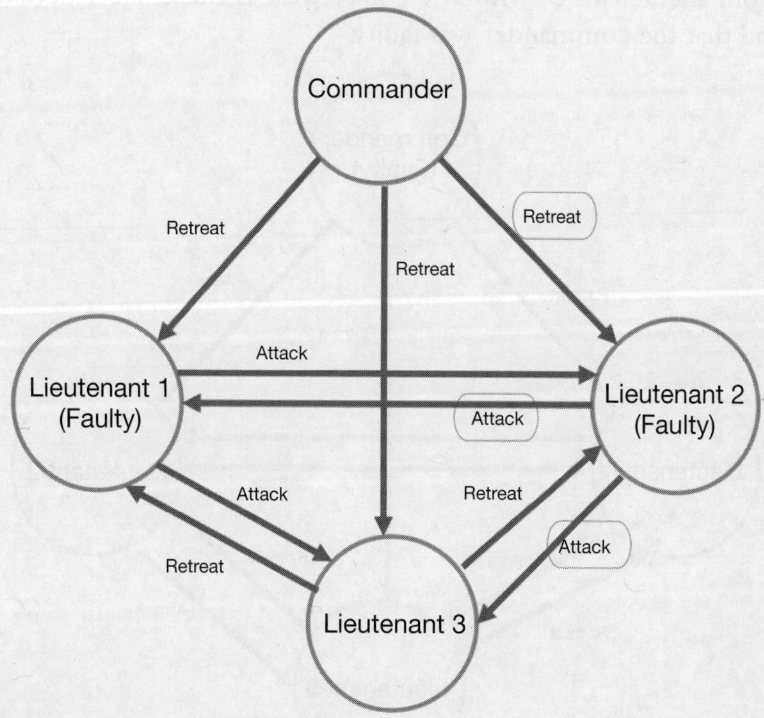

Figure 6.43: Four byzantine structure: Two lieutenants faulty

Four Byzantine General: Commander Is Faulty

Now, let us look into another case of the problem with four byzantine generals. Now, in this case, the commander is faulty and sends Retreat message to Lieutenant 1 and Lieutenant 3 (refer Fig. 6.44).

However, he sends Attack message to Lieutenant 2. Here, the commander starts behaving maliciously, but the lieutenants are behaving correctly. So, in this case, all the lieutenants are correct. Lieutenant 1 hears a Retreat message from the commander and correctly echoes the Retreat message to the other two lieutenants. Lieutenant 2 is again a correct lieutenant, but he has hears an Attack message. So Lieutenant 2 sends Attack message to the other two lieutenants. Lieutenant 3 also is a correct lieutenant. Lieutenant 3 hears a Retreat message. Therefore, he correctly broadcasts the Retreat message to the other two lieutenants.

Let us first look into the case of Lieutenant 1(retreat, retreat, attack). It has received one retreat message from the Commander, one retreat message from Lieutenant 3 and one attack message from Lieutenant 2. Therefore, it has received two retreat votes and one attack vote. Because he has received two retreat votes and one attack vote, he can currently decide that the decision is a retreat.

Lieutenant 2 (attack, retreat, retreat) has received one attack message from the Commander, but it has received one retreat message from Lieutenant 1 and another retreat vote from Lieutenant 3. Moreover, if Retreat is the correct decision, it can be correctly found that the commander was faulty.

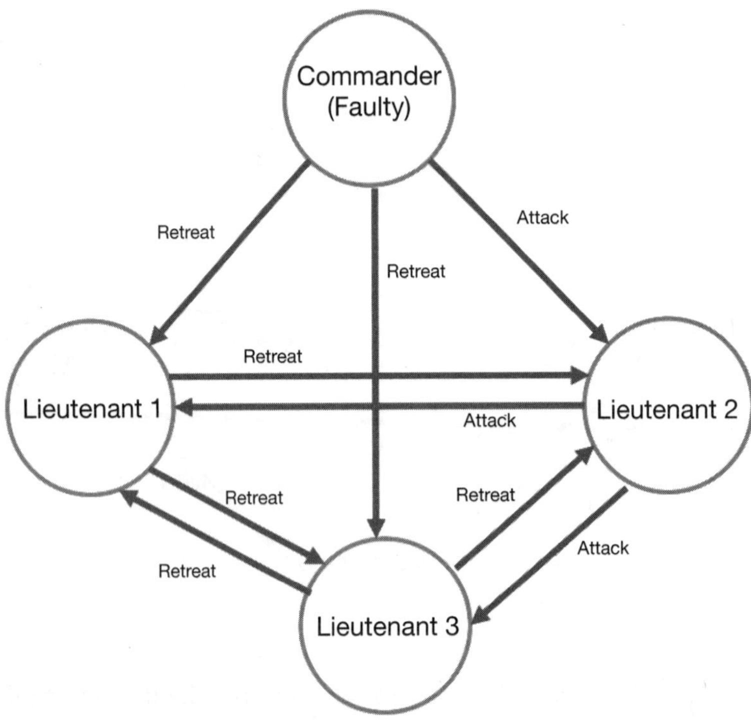

Figure 6.44: Four byzantine structure: Commander faulty

Lieutenant 3 receives Retreat message from the commander followed by one Retreat message from Lieutenant 1 and one Attack message from Lieutenant 2. Therefore, again, from the majority voting principle, he can decide that the correct decision is Retreat.

Again, in this case, we observe that if there are three different lieutenants (all correct), and one commander (malicious), the majority voting principle will give the correct results.

6.11.3 Byzantine Generals Model

If we look into the usual byzantine general problem with an f-number of faulty nodes, then we need to ensure that there is a minimum of $2f + 1$ number of correct lieutenants in total in the system to correctly apply the majority voting principle and find out the byzantine nodes in the system.

Moreover, if the commander is faulty, the consensus is decided based on the majority value proposed by the lieutenants using the four byzantine generals model.

If there are N number of processes out of which at most f number of the processes can be faulty, it has to be ensured that we have $2f + 1$ number of lieutenants in the system. The receiver always knows the identity of the sender. This is a closed model and in the context of blockchain technology, this model helps us to design an algorithm for the permissioned blockchain environment. The system is fully connected, the communication is reliable, and the system is synchronous.

Lamport–Shostak–Pease Algorithm

In the Lamport–Shostak–Pease algorithm, at pulse one (the initial pulse), the commander sends the message to all the lieutenants.

We first consider a base condition for the commander. Broadcast (N = number of process, t (individual rounds) = 0). Therefore, $t = 0$ indicates that you are in Pulse 0 when the commander is ready to send the message to all the lieutenants. So, the commander decides the value, whether to retreat or an attack. Here $N = 3$ because we have three different processes. This is the base condition for the commander (refer Fig. 6.45).

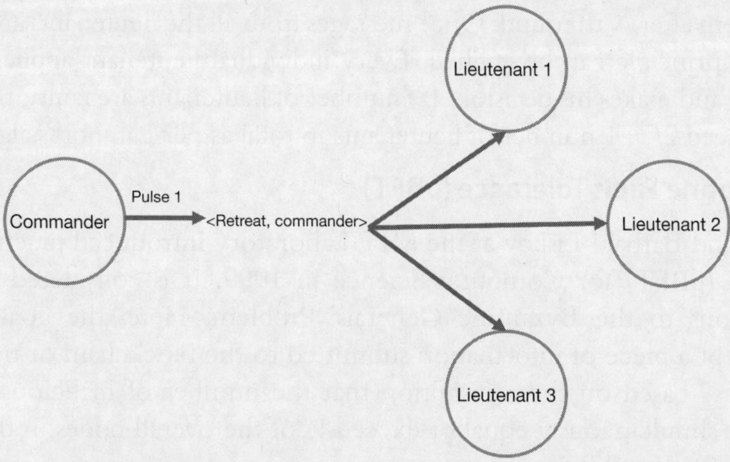

Figure 6.45: Lamport–Shostak–Pease algorithm

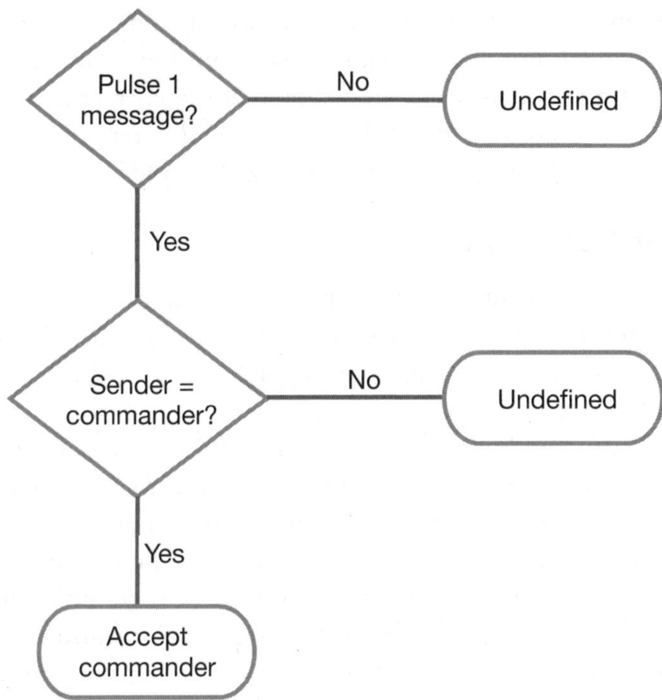

Figure 6.46: Lamport-Shostak-Pease algorithm: Pulse check

The lieutenant receives the message from the commander. Whenever he receives the message from the commander, he first checks whether it is a pulse one message or not. If it is a pulse one message, it means that the message is coming from the commander. If it is not, then he cannot take any decision because he does not know the source of the message (refer Fig. 6.46).

Broadcast is the commander's message to all other processes in the system, and that is the initiation.

If the system is synchronous, you will get the messages from all other lieutenants who are there in the system after N th round. Once messages from all the lieutenants are received, the majority voting principle can be applied. Every individual lieutenant applies the majority voting principle and makes his decision. If f number of lieutenants are faulty, then to achieve consensus, we need $2f + 1$ number of lieutenants in total as per Lamport's algorithm.

Practical Byzantine Fault Tolerance (pBFT)

Miguel Castro and Barbara Liskov at the MIT Laboratory introduced practical Byzantine Fault Tolerance (pBFT) for Computer Science in 1999. It is considered as one of the potential solutions to the Byzantine Generals' Problem. Here, the goal is to decide whether to accept a piece of information submitted to the blockchain or not. It tolerates "byzantine faults" based on the assumption that the number of malicious nodes in the network cannot simultaneously equal or exceed ⅓ of the overall nodes in the system in a given window of vulnerability.

It works on the format of the Byzantine Generals' Problem, where all "generals" (nodes) are considered equal and take their work instruction from the "leader" node. The leader node is the primary node, and all other nodes are called secondary or backup nodes. The leader is selected at random in a round-robin fashion. A node "client" sends a request for a change or transaction to the leader. The leader, in turn, broadcasts it to all the backup nodes. The leader and backup nodes will use the message with their internal state to run computation and transmit the decision result to all the client nodes. The final decision of acceptance or rejection of the transaction is arrived at based on the agreement of the majority.

A high hash rate is not required as pBFT relies on the minimum number of backup nodes to confirm trust, namely $(f + 1)$, where f represents the maximum number of faulty nodes. Hence it is not computationally intensive, and therefor results in substantial energy saving.

Some disadvantages of the pBFT protocol are:

1) Small group sizes – Due to the amount of communication required between all the nodes, this model works best with small group networks for better response times.

2) Sybil attacks – A single party can assume some identities or nodes and manipulate the network. This is mitigated with larger network sizes, but scalability and throughput will be compromised. Stellar, Ripple, and Hyperledger are some blockchains that use variants of the pBFTconsensus mechanism algorithm.

pBFT algorithm is applied for many real applications; it is widely applied for the permissioned blockchain models. In this system, there is a client, and the client submits the request to the commander. Clients are the individual blockchain clients who are submitting the transaction. The commanders are specific nodes in the system that are responsible for ensuring consensus in the system. The system consists of multiple nodes. Once the system comes to a consensus, it sends a response back to the client on whether the system has committed based on the consensus algorithm or not. Here, the entire system works as a synchronous distributed system. Additionally, the pBFT algorithm also supports privacy over the system; it ensures that the messages are tamper-proof by applying a hashing technique.

6.12 Multichain

Individuals/private organizations wanting to start with Blockchain can start with a private blockchain, like Multichain. Multichain offers a platform for the creation of private blockchain. This platform is available in popular operating systems such as Win, Unix/Linux, and Mac Server. The Multichain platform offers users an approach similar to bitcoin but in a private environment.

Multichain also offers API and serves as a platform for the financial domain. It is forked from Bitcoin and has added features of privacy and control

Drawing a parallel between the internet in the late 1980s and Blockchain today, the internet was then about to explode as the communications backbone. Enterprises wanted to leverage the new phenomenon; however, they were skeptical about the privacy, reliability and capacity of the new technology. Hence, many of them chose to start with a scaled-down version of the internet, "the intranet." Similar comparisons are drawn between blockchain and DLT; enterprises are keen to exploit the advantages of the blockchain, although reluctant to invest heavily in the technology. Enterprises need the scale, real-time, or near real-time data points, without comprising the privacy of data. Hence, there is a need for a miniature version of blockchain such as a private blockchain where enterprises could control the people, processes, and machines.

Multichain helps enterprises as a miniature private blockchain while relying on the robust platform of bitcoin. The core aims of Multichain, as per Dr Gideon Greenspan, are threefold:

(a) To ensure that the blockchain's activity is only visible to chosen participants,
(b) To introduce controls over which transactions are permitted, and
(c) To enable mining to take place securely without proof-of-work and its associated costs.

6.12.1 Streams in Multichain

Streams in multichain serve the purpose of ensuring that only authorized or approved participants can see the activity within the block. Each blockchain has an optional 'root or core' stream, which exists from the creation of the chain. Thus, blockchain can be used immediately for storing and retrieving data. As the blockchain progresses, the administrator can create open or closed streams; in an open stream, anybody can write data, while closed stream allows only permissioned nodes to write or read from a stream. The stream has another effective layer of security, which ensures that data, i.e., all transaction data in a blockchain, are encrypted. Streams provide a combination of encryption mechanisms – public key encryption and symmetric encryption to achieve encrypted data transfer.

Step 1: Participants can distribute their public keys for any public-key cryptography scheme using one stream.

Step 2: A second stream is used to publish data. Each piece of data is encrypted (secured) using symmetric cryptography with a unique key.

Step 3: A third stream provides access to data to specific participants.

6.12.2 Control Over Transactions

Multichain uses a permission-based mechanism (refer Table 6.4) to perform data operations into ledgers. However, only two operations are allowed to be made in ledgers – a user is permitted to write data or a user is permitted to read the data. In multichain, the following permissions are granted:

Table 6.4 Commands in multichain

S.No	Command	Remarks
1	Connect	To connect to other nodes and see blockchain's contents
2	Send	To send funds, i.e., sign inputs of transactions.
3	Receive	To receive funds, i.e., appear in the outputs of transactions.
4	Issue	To issue assets, i.e., sign inputs of transactions which create new native assets.
5	Create	To create streams, i.e., sign inputs of transactions which create new streams.
6	Mine	To create blocks, i.e., to sign the metadata of coin base transactions.
7	Activate	To change connect, send and receive permissions for other users.
8	Admin	To change all permissions for other users, including issue, mine, activate and admin.

All permissions are granted on a per-address (IP Address) basis. The creator of a chain's first genesis block automatically receives all regular (non-custom) permissions. This administrator can then grant permission to other addresses (refer Fig. 6.47). We will discuss multichain in detail in another chapter.

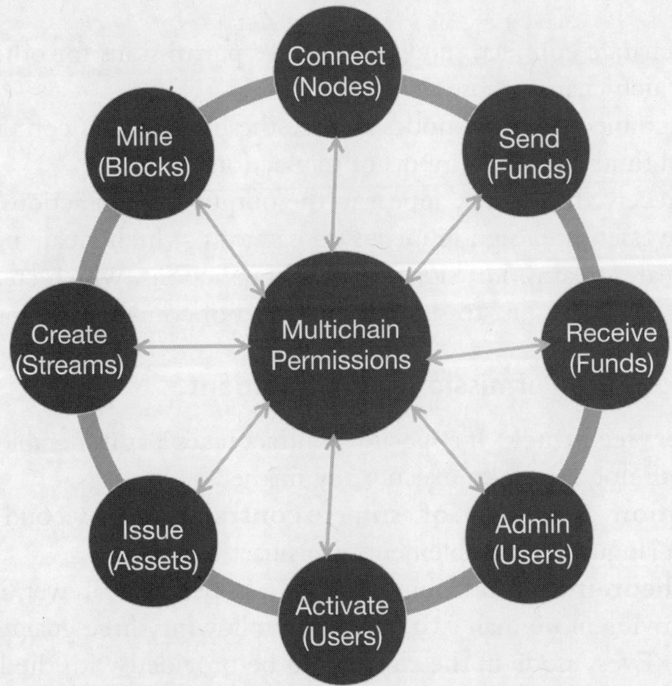

Figure 6.47: Multichain permissions

Summary

Private blockchains are private, as all permissions for the blockchain are kept centralized. Many banks around the world have starting experimenting with the private blockchain model for their banking facilities. Another name of a private blockchain is permissioned or centralized blockchain. These are closed ecosystems where all participants are well-defined. Only pre-approved entities can run nodes. The network is not open, and it is not transparent to all.

Private blockchains are secured by a distributed consensus mechanism such as PAXOS, RAFT, BFT, and pBFT. They have better scalability because of their high transaction speed and lack of need for high storage or computational power.

A private blockchain is different from permissioned blockchain.

A permissioned blockchain network needs a proper and documented mechanism to determine who can do what; an authentication mechanism is needed to identify the node in the network. Hyperledger fabric is an example of a private permissioned blockchain.

We need to use private permissioned blockchain when the state (and date) needs to be stored in the database, if there are multiple writers for transaction, when we cannot use online TTP, when all the writers are known, when all the users cannot be trusted and when public verifiability is not required.

Various commands in private blockchain:

- Admin – to change all permissions for other users, including issue, mine, activate, and admin.
- Activate – to change connect, send, and receive permissions for other users, i.e., sign transactions which change those permissions.
- Connect – to connect to other nodes and see the blockchain's contents.
- Send – to send funds, i.e., sign inputs of transactions.
- Receive – to receive funds, i.e., appear in the outputs of transactions.
- Issue – to issue assets, i.e., sign inputs of transactions, which create new native assets.
- Create – to create streams, i.e., sign inputs of transactions, which create new streams.
- Mine – to create blocks, i.e., to sign the metadata of coin-based transactions.

Design Limitations in a Permissioned Environment

Denial-of-service attack: If a specific contract takes a considerable time to execute, other contracts will not be able to execute any further.

Implementation language of smart contract: The second problem is the identification of a language for implementing a smart contract.

The CAP Theorem: For distributed systems, as per Eric Brewer, it is impossible to simultaneously provide more than two out of the following three guarantees:

Consistency: Every node in the cluster will be provided with the latest and correct data set.

Availability: All requests to a node in the cluster will be provided with a response.

Partition tolerance: The network will continue to operate even if messages are lost due to communication failure between nodes.

The BASE theory: BASE stands for Basically Available Soft State, Eventual Consistency.

- Basically Available—Information is available with all nodes.
- Soft State—Data is retained in the ledger; there is no delete for data.
- Eventual Consistency—Data consistency is ensured among all nodes even if one of the nodes goes temporarily off the network and is connected again

State machine: A state machine can be characterized by a set of parameters based on the system design. There is a specific consensus mechanism to ensure that the states broadcast by multiple state machines are on the same page and correct.

Applications of State Machine Replication

State machine replication is used in flight control systems. It is also used for funds transfer in a distributed environment.

Three Types of Fault in a Distributed Environment

In a distributed environment, you can have primarily three types of fault, namely, crash fault, partition fault and byzantine fault.

We do not need a consensus in a single-node process. In the case of two nodes, if there is any fault – network fault, crash fault, or if the node behaves maliciously – consensus cannot be reached. Consensus is needed when there are multiple decision-makers.

There are three requirements for a consensus algorithm to work correctly. They are: 1. Termination 2. Agreement Property 3.Integrity.

We have different kinds of algorithms for ensuring consensus in a typical distributed system, for permissioned blockchain. These algorithms are based on this state machine replication principle. Initial algorithms for distributed consensus such as PAXOS and RAFT support crash or network faults. However, they cannot support Byzantine fault.

To support byzantine fault, we have algorithms like byzantine fault-tolerant algorithm (BFT) or practical byzantine fault algorithms (pBFT), which, in addition, also support crash and network faults.

The objective of PAXOS was to choose a single value under a crash or network fault. There are three types of nodes: the proposer, the acceptor and the learner. In PAXOS, everyone is a learner in the network, which learns based on what the majority decision is. The proposer initially proposes (with a proposal number). The acceptor, based on that, prepares a response message (Accept/decline, Proposal #, Accepted values). The message has the result, which has the value of either Accept or Decline.

Handling Failure (Acceptor fails): The acceptor may fail during the Accept states. However, if the acceptor fails during the Accept stage, other acceptors would still be able to vote for the proposal, provided there are at least N/2 number of acceptors.

Handling failure (Proposer fails): If the proposer fails during the prepare phase, it is just that no one is proposing any value. Now the proposer may fail during the accept phase, but if the proposer fails during the Accept phase, the acceptors have already acted upon whether or not to choose the proposal based on the majority vote. In addition, they can find out whether the proposal has been accepted or not. When there are duelling proposals, the proposer who is having lower ID is blocked as a means of tie-breaking. To do this, a specific algorithm called the leader election algorithm is executed.

In the multi-PAXOS system, you make a sequence of choices by applying the PAXOS program repeatedly. This is a complicated process since the messages need to be exchanged among the participants repeatedly.

RAFT is designed as an alternative to PAXOS. RAFT will first elect a leader, whose task is to propose something. There would be a single proposer (leader), and all accepters are followers of the leader.

Term parameter: The RAFT algorithm also runs in rounds. A decision is taken in every round. The term parameter denotes in which particular round, you are currently in.

Index parameter: The index parameters say about the committed transaction, which is available to the candidate. Now, this request vote message is passed to all the nodes which are there in the network.

Failure of the current leader: Whenever a leader fails, a new leader can be voted by utilizing the term parameter.

Follower failures: RAFT and PAXOS are good at handling crash fault or network fault. If a follower has crashed, the system can tolerate up to failures of ($N/2 - 1$) number of nodes. It does not affect the system because we rely on majority voting. The entire system becomes streamlined, once you have elected a leader.

Byzantine fault: If there is byzantine behaviour in the network, where the leader node starts behaving maliciously (fault) and proposes a transaction only to a selective number of followers, then the PAXOS and the RAFT protocol may not help. In that case, the byzantine fault consensus algorithm is used.

Three byzantine general problem: In this design, you have one commander who can send message to the lieutenants, and two different lieutenants, who can send the messaging between themselves. There are two possible scenarios: the lieutenant is faulty or the commander is faulty. In a typical scenario, the lieutenant obeys the commander and if the commander is non-faulty, then the entire system works correctly. However, if the commander is faulty and the lieutenant gets equal voting for attack and retreat, he cannot decide what to do. Therefore, we will not be able to solve the byzantine general problem with three generals, where we have one commander and two different lieutenants.

Four byzantine problem: In this model, we have one commander and three lieutenants. Every individual lieutenant talks with each other. The commander (leader) sends the message to all the lieutenants (followers). The lieutenants share the commander's message with each other. Here, the possible scenarios are: One lieutenant is faulty, Two lieutenants are faulty, Commander is faulty. When one lieutenant is faulty we can correctly decode the message. If two lieutenants are faulty, then we will not be able to decode the message correctly. If the commander is faulty, the majority voting principle will give the correct results.

Byzantine Generals Model: In the usual byzantine general problem with an f number of faulty nodes, we need to ensure that there is a minimum of $2f + 1$ number of correct lieutenants in total in the system to correctly apply the majority voting principle to find out the byzantine nodes in the system. Moreover, if the commander is faulty, consensus is decided based on the majority value proposed by the commander.

In the Lamport–Shostak–Pease algorithm, at pulse one (the initial pulse), the commander sends the message to all the lieutenants. The lieutenant receives the message from the commander and first checks whether it is a pulse one message or not. If it is a pulse one message, it means the message is coming from the commander. If it is not, then no decision can be taken because the source of the message is unknown.

Practical Byzantine Fault Tolerance (pBFT): It works on the format of the byzantine general's problem, where all "generals" (nodes) are considered equal and take their work instruction from the "leader" node. The leader node is the primary node, and all other nodes are called secondary or backup nodes. A high hash rate is not required as pBFT relies on the minimum number of backup nodes to confirm trust, namely $(f + 1)$, where f represents the maximum number of faulty nodes. Hence, it is not computationally intensive and results in substantial energy saving.

Some Disadvantages of the pBFT Protocol

1) Small group sizes – Due to the amount of communication required between all the nodes, this model works best with small group networks for better response times.
2) Sybil attacks – A single party can assume some identities or nodes and manipulate the network. This is mitigated with larger network sizes, but scalability and throughput will be compromised. Stellar, Ripple, and Hyperledger are some blockchains that use variants of the pBFT consensus mechanism algorithm.

Multichain: Individuals or private organizations wanting to start using blockchain can start with a private blockchain like multichain. This platform is available in popular operating systems such as Win, Unix/Linux, Mac Server.

Streams in multichain: It serves the purpose of ensuring that only authorized or approved participants can see the activity within the block. Each blockchain has an option of 'root or core' stream, which exists from the creation of the chain. Thus, blockchain can be used for storing and retrieving data. As blockchain progresses, the administrator can create open or closed streams; in an open stream, anybody can write data, while closed stream allows only permissioned nodes to write or read from a stream. The stream also has another layer of security, which ensures that all transaction data in a blockchain are encrypted.

Control over transactions: Multichain uses a permission-based mechanism to perform data operations into ledgers. However, only two operations are allowed to be made in ledgers, a user is permitted to write data or a user is permitted to read the data.

EXERCISES

Multiple Choice Questions

1. **Public blockchains are a great place to start in a business-to-business (b2b) model.**
   A. True
   B. False
   Answer: A
   Explanation: Private blockchains are a great place to start in a business-to-business (b2b) model. In the private blockchain, the degrees of decentralization (or centralization) and transparency are entirely up to the company or companies running and configuring the blockchain. There is no anonymity. It does not require mining to validate transactions or execute smart contracts. It does not require a crypto-economic incentive or tokens for those running nodes (miners).

2. **The private blockchain is a closed environment consisting of a set of nodes that know each other; but they may not trust each other.**
   A. True
   B. False
   Answer: A
   Explanation: The private blockchain is a closed environment consisting of a set of nodes that know each other; but they may not trust each other. Before joining the blockchain network, the nodes have to go through some authentication or pre-authorization mechanism through which they can validate themselves.

3. **Which of the following is not a characteristic of private blockchain?**
   A. Details of the users are known.
   B. Details of the users are not known.
   C. The network is not open for all.
   D. The network type is centralized.
   Answer: B
   Explanation: In private blockchain, the users are trusted participants whose details are known. The network is not open, and it is not transparent to all. The network type is centralized and it is the single point of failure. Only pre-approved participants can read and/or initiate transactions in a private blockchain. There is no miner, and so there is no incentivisation.

4. **There is no miner in a private blockchain.**
   A. True
   B. False
   Answer: A
   Explanation: Same as for Question 3.

5. **Which of the following is not a characteristic/condition common to public permissioned blockchain and private permissioned blockchain?**
   A. When the state (and date) needs to be stored in the database
   B. If there are multiple writers for transaction
   C. When all the writers are known
   D. When public verifiability is required.
   Answer: D
   Explanation: We need to use public permissioned blockchain when the state (and date) needs to be stored in the database, if there are multiple writers for transaction, when we can't always use online TTP (Trusted Third Party), when all the writers are known, when all the users cannot be trusted and when public verifiability is required. We need to use private permissioned blockchain when the state (and date) needs to be stored in the database, if there are multiple writers for transaction, when we cannot always use online TTP (Trusted Third Party), when all the writers are known, when all the users cannot be trusted and when public verifiability is not required.
6. **Which of the following is not a purpose of using private blockchain?**
   A. Organizations prefer to control the overall blockchain.
   B. Organizations would like to control who can use the system.
   C. No audit for the system ensuring privacy.
   D. Users are added via defined, authorized process.
   Answer: C
   Explanation: Organizations prefer using a private blockchain and thereby control who can use the system, who can write to the system and who can read the system. Besides, organizations need a mechanism to ensure users are added via process, and user rights are created, changed, or deleted by an authorized user. In addition, the solution had to ensure faster transactions, proper audit trail, and interactions with the organizations' existing IT systems. These needs were the reason for the birth of private blockchain.
7. **If we consider public blockchain as the internet, then private blockchain can be considered as the intranet.**
   A. True
   B. False
   Answer: A
   Explanation: If we consider public blockchain as the internet, then private blockchain can be considered as the intranet.
8. **Which of the following statements about private blockchain is not True?**
   A. Private Blockchain is centralized to one organization.
   B. It is faster because there are fewer people on it.
   C. Permissioned blockchain uses the multi-signature method.
   D. Private blockchain has high energy consumption due to faster transactions.

Answer: D

Explanation: Private blockchains need read/write permissions and it is centralized to one organization. Permissioned blockchain is faster because there are fewer people on it. Permissioned blockchain uses the multi-signature method. Private Blockchain has low energy consumption; it also has fewer people transacting on the network. The transaction approval frequency is short.

9. **If a specific contract takes a considerable time to execute, other contracts will not be able to execute any further. This is known as:**
   A. 51% Attack
   B. Denial-of-service attack
   C. Vector attack
   D. Sybil attack
   Answer: B
   Explanation: The attacker can introduce a contract that takes a long time to execute. If a specific contract takes considerable time to execute, other contracts will not be able to execute any further. This is because new contracts can be executed only after consensus has been reached for the previous contract. Thus, the transactions are maintained in a serial order preventing execution of the next contract until the previous contract is executed. If malicious content is introduced in the system, it will take considerable time for execution and result in a denial-of-service attack.

10. **As per the CAP Theorem, it is impossible to simultaneously provide more than two out of the following three guarantees:**
   A. Consistency, availability, partition tolerance
   B. Capability, availability, porting principle
   C. Consistency, authority, power principle
   D. Capacity, availability, portability
   Answer: A
   Explanation: As per CAP Theorem for distributed systems, it is impossible to simultaneously provide more than two out of the following three guarantees: Consistency, availability and partition tolerance.

11. **Every node in the cluster will be provided with the latest and correct data set. All nodes see the exact data at precisely the same moment. This is called as**
   A. Capability
   B. Consistency
   C. Availability
   D. Partition tolerance
   Answer: B
   Explanation: Every node in the cluster will be provided with the latest and correct data set. All nodes see the exact data at precisely the same moment. In other words,

performing a read operation will yield the result of the most recent write operation inducing all nodes to get precisely the exact data.

12. **All requests to a node in the cluster will be provided with a response, although it does not mean this is the latest data write to the data set. This is referred as:**
    A. Capability
    B. Consistency
    C. Availability
    D. Partition tolerance
    Answer: C
    Explanation: All requests to a node in the cluster will be provided with a response, although this does not mean this is the latest data write to the data set. This state ensures that every petition receives a response on success/failure. For accessibility in a distributed network, the machine has to remain operational 100 percent of their time. The end-user sees only consistency and availability. This view is formed when connectivity between nodes fails leading to a loss of multiple transactions. Thus, we choose either of consistency or availability.

13. **The network will continue to operate even if messages are lost due to communication failure between the nodes failed. It is known as:**
    A. Capability
    B. Consistency
    C. Availability
    D. Partition tolerance
    Answer: D
    Explanation: The network will continue to operate, even if messages are lost due to communication failure between the nodes. A system that is partition-tolerant can withstand any network failure that does not fail the entire network. Data records are sufficiently replicated across combinations of nodes and networks to keep the system up to date.

14. **Which of the following is not one of the requirements for a consensus algorithm to work correctly?**
    A. Termination
    B. Agreement property
    C. Integrity
    D. State machine replication
    Answer: D
    Explanation: There are three requirements for a consensus algorithm to work correctly. They are 1. Termination 2. Agreement property 3.Integrity.

15. **As per CAP Theorem, there are specific requirements of a consensus algorithm. It has to be ensured that eventually, each process sets the**

decision variable and terminates after setting the decision variable. All processes must eventually agree on an output value before termination. This is known as:

A. Termination

B. Agreement property

C. Integrity

D. State machine replication

Answer: A

Explanation: There are specific requirements of a consensus algorithm; it has to be ensured that eventually, each process sets the decision variable and terminates after setting the decision variable. All processes must eventually agree on an output value before termination.

16. **As per CAP Theorem, the decision value for all correct processes should be the same. This is referred as:**

A. Termination

B. Agreement property

C. Integrity

D. State machine replication

Answer: B

Explanation: The agreement property says that the decision value for all correct processes should be the same; every correct node should arrive at a collective agreement.

17. **As per CAP Theorem, if the correct processes propose the same value, then any exact process in the decision state has chosen that value. This is referred to as**

A. Termination

B. Agreement property

C. Integrity

D. State machine replication

Answer: C

Explanation: If the correct processes propose the same value, then any exact process in the decision state has chosen that value. So if every correct process proposes one value X, at the decision state, they should decide on that same value X.

18. **As per BASE Theorem, letter E stands for:**

A. Enhancement and consistency

B. Event

C. Enablement

D. Eventual consistency

Answer: D

Explanation: BASE – Basically Available Soft State, Eventual Consistency. Basically Available — Information is available with all nodes, so that availability for at all nodes is ensured. Soft state – Data is retained in the ledger; there is no delete of data. Eventual consistency – Assuming a node goes off, data would be consistent once the node is up and running; the ledger would be built to ensure that consistency with other nodes is maintained.

19. As per BASE Theorem, letter S stands for:

A. Sync
B. Soft coin
C. Soft state
D. Software

Answer: C

Explanation: Same as for Question 18

20. The mechanism used in a flight control system is

A. State machine replication
B. PAXOS
C. PoW
D. POI

Answer: A

Explanation: One typical application of the state machine replication is the flight control system. When there are multiple flights and their positions have to be coordinated, the state machine replication technique is used. For fund transferring systems in a distributed environment, a state machine replication-based algorithm can be applied for consensus. In an open environment, we have a challenge-response based consensus algorithm to achieve consensus.

21. In a distributed environment, malicious behaviour among the nodes, which may be due to hardware faults or software faults is a type of:

A. Crash fault
B. Byzantine fault
C. Network fault
D. Software fault

Answer: A

Explanation: In a distributed environment, a crash fault is where a node crashes and is not able to send any message. It may recover after some time and start sending messages again. So this called as fail-stop behaviour.

22. In a distributed environment, the message from one partition cannot be propagated to another partition. This is a type of:

A. Crash fault
B. Byzantine fault
C. Network fault
D. Software fault

Answer: C

Explanation: In a distributed environment, when a network fault occurs, the network is partitioned and the message from one partition cannot be propagated to another partition.

23. **Which fault is difficult to handle in a distributed environment in general?**
A. Crash fault
B. Byzantine fault
C. Network fault
D. Software fault
Answer: B
Explanation: Byzantine fault is a type of fault which is difficult to handle in a distributed environment. It is characterized by malicious behaviour among the nodes, which may be due to hardware faults or software faults.

24. **Leslie Lamport proposed this algorithm and the objective was to choose a single value under a crash or network fault.**
A. RAFT
B. POI
C. PoW
D. PAXOS
Answer: D
Explanation: The objective of PAXOS was to choose a single value under a crash or network fault. Leslie Lamport proposed PAXOS in 1989, but it took around 13 years to get the paper published. There was much discussion among the community about the correctness of this particular algorithm.

25. **In a real system, when a sequence of choices is made rather than having a single choice, this kind of system is called as:**
A. Multi-PAXOS system.
B. Sequence PAXOS system
C. Complicated PAXOS system
D. RAFT PAXOS system
Answer: A
Explanation: In a real system, a sequence of choices may be made rather than having a single choice. This kind of system is a multi PAXOS system. In multi-PAXOS system, a sequence of choices is made by applying the PAXOS program repeatedly. Applying repeated PAXOS is complicated because messages need to be exchanged among all participants repeatedly.

26. **In the case of PAXOS, there is no leader node; so multiple proposers can propose the same thing simultaneously.**
A. True
B. False

Answer: A

Explanation: In the case of PAXOS, there is no leader node; so multiple proposers can propose the same thing simultaneously. Whenever this happens, it becomes a little complex since the acceptors have to accept one of the proposals from the proposer and use the highest proposal number in a tie-breaking mechanism. We can embed a specific algorithm inside PAXOS to ensure that every proposal that is coming from different proposers is unique.

27. In RAFT, there is a leader node. There would be a single proposer (leader), and all the accepters are now the followers of the leader.

A. True

B. False

Answer: A

Explanation: RAFT will first elect a leader. Then the task of the leader is to propose something. There would be a single proposer (leader), and all the acceptors are now followers of the leader. However, the acceptors (followers) can follow the leader's proposal, or they may choose not to follow it.

28. In the Lamport–Shostak–Pease algorithm, at ____________ pulse, the commander sends the message to all the lieutenants.

A. Second pulse

B. First pulse

C. Every alternate pulse

D. Every fifth pulse

Answer: B

Explanation: In the Lamport–Shostak–Pease algorithm, at pulse one (the initial pulse), the commander sends the message to all the lieutenants.

29. Which of the following is not one the core aims of multichain?

A. To ensure that the blockchain's activity is only visible to chosen participants

B. To introduce controls over the permitted transactions

C. To enable mining to take place securely without proof-of-work and its associated costs

D. To ensure that the blockchain's activity is transparent, visible to all participants.

Answer: D

Explanation: Multichain helps enterprises as a miniature private blockchain while relying on the robust platform of bitcoin. The core aims of multichain, as per Dr Gideon Greenspan, is threefold:

(a) To ensure that the blockchain's activity is only visible to chosen participants,

(b) To introduce controls over which transactions are permitted, and

(c) To enable mining to take place securely without proof-of-work and its associated costs.

30. **In multichain, this acts as another layer of security to ensure that the data are encrypted.**
   A. Stream
   B. Channel
   C. Commands
   D. Application layer
   Answer : A
   Explanation: The stream has another layer of security, which ensures that data, i.e., all transaction data in a blockchain, are encrypted. Streams provide a combination of encryptions mechanisms – public key encryption and symmetric encryption to achieve encrypted data transfer.

Short-answer Questions

1. **Define private blockchain.**
   Private blockchains are private, as all permissions for the blockchain are kept centralized. Another name of a private blockchain is permissioned or centralized blockchain. They are closed ecosystems where all participants are well-defined. Only pre-approved entities can run nodes.

2. **What are the key characteristics of private blockchain?**
   In the private blockchain, the degrees of decentralization and transparency are entirely up to the company or companies running and configuring the blockchain. There is no anonymity. It does not require mining to validate transactions or execute smart contracts. It does not require a crypto-economic incentive or tokens for those who run nodes (miners). It can scale much better than public blockchain. Before joining the blockchain network, the nodes have to go through an authentication or pre-authorization mechanism through which they can validate themselves.

3. **List some of the consensus algorithms of private blockchain.**
   Private blockchains are secured by distributed consensus mechanisms such as PAXOS, RAFT, BFT, and pBFT.

4. **What is the need for a private blockchain?**
   Private Blockchains are used by individual hobbyists, private enterprises or organizations for a specific purpose (e.g., an NGO may like to keep a record of money spent in various schools). Organizations prefer using a private blockchain, since it helps them to control
   - Who can use the system
   - Who can write to the system
   - Who can read the system.

5. **What is denial-of-service attack?**
   In denial-of-service attack, the attacker can introduce a contract which will take a long time to execute. If a specific contract takes a considerable time to execute,

other contracts will be held up. This is because new contracts can be executed only after consensus has been reached for the previous contract.

6. **How does identification of suitable language affect blockchain implementation?**

   To implement a smart contract, we need to use a language that gives more power. Bitcoin script is not a "Turing complete" language. For example, it does not support loops, and it has certain limitations in execution. Therefore, developers use other languages such as "`GoLang`," which is widely used in the contract execution platform.

7. **How does `GoLang` help in blockchain implementation?**

   `GoLang` has a construct called a map. The map is a data structure, where there are specific keys and absolute values. Every value is associated with one key. The key can be used to store the hash, which is unique in a block. Value can be used to store the entire block corresponding to the key (hash).

8. **What is state synchronization in smart contract?**

   Generally, we execute the smart contract at all nodes and propagate the state to others to reach a consensus. To ensure consensus, one has to ensure that there is sufficient number of cross-state nodes to validate the execution of smart contracts. If the number of trusted nodes is less than the number of malicious nodes, then the malicious nodes may take control over the entire environment. However, we can prevent that by using proof-of-work-based consensus mechanism.

9. **Explain C in CAP theorem.**

   Consistency: Every node in the cluster will be provided with the latest and correct data set. All nodes see the exact data at precisely the same moment. In other words, performing a read operation will yield the result of the most recent write operation inducing all nodes to get precisely the exact data.

10. **Explain A in CAP theorem.**

    Availability: All requests to a node in the cluster will be provided with a response, although this does not mean that it is the latest data write to the data set. It indicates that every petition receives a response on success/failure. For accessibility in a distributed network, the machine has to remain operational 100 percent of the time.

11. **Explain P in CAP theorem.**

    Partition tolerance: The network will continue to operate, even if messages are lost due to failure of node(s) or communication between nodes. A system that is partition-tolerant can withstand any amount of network failure that does not fail the entire network. Data records are sufficiently replicated across combinations of nodes and networks to keep the system up to date.

12. What is BASE theory?

BASE - Basically Available, Soft State, Eventually Consistency.

- Basically Available – Information is available with all nodes.
- Soft state – Data is retained in the ledger; there is no delete for data.
- Eventual consistency – Assuming a node goes off, data would be consistent once the node is up and running; the ledger would be built to ensure that consistency with other nodes is maintained.

13. What is crash fault in a distributed environment?

In a distributed environment, primarily three types of faults are possible. The first type is the crash fault, where a node crashes and is not able to send any message. It may recover after some time, and it may start sending the message again. So this recall is a fail-stop behaviour.

14. What is network partition fault in a distributed environment?

In a distributed environment, primarily three types of faults are possible. In the second type, because of a network fault, the network is partitioned, and the message from one partition cannot be propagated to another partition.

15. What is byzantine fault?

In a distributed environment, you can have primarily three types of fault. The third type of fault, which is difficult to handle in a distributed environment, is a byzantine fault. In this case, there is malicious behaviour among the nodes, which may be due to hardware faults or software faults.

16. What are the three requirements of a consensus algorithm?

There are three requirements for a consensus algorithm to work correctly are 1. Termination 2. Agreement property 3. Integrity

17. What is termination requirement of a consensus algorithm?

There are specific requirements of a consensus algorithm; it has to be ensured that eventually, each process sets the decision variable and terminates after setting the decision variable. All processes must agree on an output value before termination.

18. What is agreement property of a consensus algorithm?

The agreement property says that the decision value for all correct processes should be the same; every correct node should reach a collective agreement.

19. What is integrity property of a consensus algorithm?

If the correct processes propose the same value, then any exact process in the decision state should also choose that value. That is, if every correct process proposes one value X, at the decision state, they should decide on the same value X.

20. What are the three types of nodes in PAXOS algorithm?

1. The proposer
2. The acceptor and
3. The learner

21. What is duelling proposer problem in Paxos?

There can be an attack in PAXOS, which we call as the duelling proposer. There are two proposers here; Proposer 1 sends specific proposals to all the acceptors. Proposer 2 sends another proposal with a higher proposal number.

22. What is Multi-PAXOS System?

In a real system, a sequence of selections may be made rather than having a single choice. This kind of system is called as a multi-PAXOS system. In the multi-PAXOS system, a sequence of choices is made by applying the PAXOS program repeatedly. Applying repeated PAXOS is complicated because messages need to be exchanged among all participants repeatedly.

23. Write notes on RAFT consensus algorithm.

RAFT is designed as an alternative to PAXOS. In the case of PAXOS, there is no leader node and so multiple proposers can propose the same thing simultaneously. Because of this multiple proposition, it becomes a little complex. The acceptors have to accept one of the proposals from the proposer and use the highest proposal number used, as a tie-breaking mechanism. We can embed a specific algorithm inside PAXOS to ensure that every proposal that comes from different proposers is unique.

24. Explain RAFT term parameter.

Term is one of the two parameters used when requesting for votes in the RAFT algorithm.

Term parameter: Just like PAXOS, the RAFT algorithm also runs in rounds. In every round, you need to take a decision. The term parameter denotes the particular round you are currently in.

25. Explain RAFT index parameter.

Index is one of the two parameters used when requesting for votes in the RAFT algorithm.

Index parameter: The index parameters say about the committed transaction, which is available to the candidate. This request vote message is passed to all the nodes which are there in the network.

26. What is the three byzantine general problem?

In this design, there are three generals – one commander and two lieutenants. The commander can send message to the lieutenants, and the two lieutenants can send messages between themselves. This case is used to design a solution for a byzantine fault-tolerant system.

27. What is the base condition of the commander in Lamport–Shostak–Pease algorithm?

In the Lamport–Shostak–Pease algorithm, at pulse one (the initial pulse), the commander sends the message to all the lieutenants. We first consider a base condition for the commander. Broadcast (N = number of process, t (individual rounds) = 0).

Therefore, $t = 0$ indicates that you are in Pulse 0 when the commander is ready to send the message to all the lieutenants. So, the commander decides the value, whether to retreat or attack.

28. Define multichain in blockchain.

Multichain offers a platform for the creation of private blockchain, this platform is available in popular operating systems such as Windows, Unix/Linux and Mac Server. The multichain platform offers users an approach similar to bitcoin but in a private environment. It also offers API and serves as a platform for the financial domain. It is forked from bitcoin and has the added features of privacy and control.

29. What are the core aims of multichain?

Multichain helps enterprises as a miniature private blockchain while relying on the robust platform of bitcoin. The core aims of multichain, as per Dr Gideon Greenspan, are threefold:

(a) To ensure that the blockchain's activity is only visible to chosen participants
(b) To introduce controls over which transactions are permitted, and
(c) To enable mining to take place securely without proof-of-work and its associated costs.

30. What is stream in multichain?

Stream is another layer of security in multichain, which ensures that data, i.e., all transaction data in a blockchain, are encrypted. Streams provide a combination of encryptions mechanisms – public key encryption and symmetric encryption to achieve encrypted data transfer.

31. What is the purpose of three streams in multichain?

Step 1: Participants distribute their public keys for any public-key cryptography scheme using one stream.

Step 2: A second stream is used to publish data. Each piece of data is encrypted (secured) using symmetric cryptography with a unique key.

Step 3: A third stream provides access to data to specific participants.

Essay-type Questions

1. What is private blockchain? Discuss few of the characteristics of blockchain in detail.
2. Differentiate permissioned and non-permissioned blockchains.
3. Give examples for permissioned blockchains.
4. Write an essay on private blockchain in open source.
5. What are the problems in centralized servers? Explain in detail.
6. How permissioned blockchain concepts can be explained using an e-commerce scenario?
7. Explain the various commands that can be used in e-commerce blockchain.
8. What is denial-of-service attack? Explain in detail.
9. Explain the various design limitations in a permissioned environment.

10. Explain CAP theorem of distributed environment.
11. Explain state machine with three states with the help of a diagram.
12. Explain state machine using the oven as an example.
13. Explain smart contract in a state machine for the below scenario:

    In a crowd-funding platform, you have a set of people (Joe and Ann), who have certain funds available, which they have agreed to donate if specific jobs are done. The proposers propose certain jobs.
14. What are the applications of state machine replication?
15. Explain the three types of faults in a distributed environment.
16. Explain the three requirements of a consensus algorithm in detail.
17. Explain PAXOS algorithm in detail.
18. Explain PAXOS algorithm in the below scenario:

    In Bishop Heber College (Trichy, India), the students have broadly two options where they can go after a class. The first option is Michael Ice Cream Shop, which is just outside the college campus, and the second option is College Canteen.
19. How PAXOS handles acceptor failure and proposer failure? Explain using a diagram.
20. Explain PAXOS duelling proposers problem in detail.
21. Explain RAFT consensus algorithm in detail.
22. Explain RAFT: Follower, Candidate, and Leader interaction in detail.
23. How is failure of the Current Leader handled in RAFT? Explain with a diagram.
24. How Committing Entry log is maintained in RAFT?
25. How failure of the follower is handled in RAFT?
26. How is the Three Byzantine General problem handled when the Lieutenant is faulty?
27. How is the Three Byzantine General problem handled when the Commander is faulty?
28. How is the Four Byzantine General problem handled when one Lieutenant is faulty?
29. How is the Four Byzantine General problem handled when two Lieutenants are faulty?
30. How is the Four Byzantine General problem handled when the Commander is faulty?
31. Explain Lamport–Shostak–Pease algorithm in detail.
32. Explain practical byzantine fault tolerance (pBFT).
33. What are the various controls over transactions in multichain?

CHAPTER 7

Consortium Blockchain

LEARNING OBJECTIVES

Consortium blockchain is a hybrid scheme where the permission to read the block can be public or private depending on the needs of the organizations involved. This can be considered as an extension of public blockchain. However, the consensus mechanism is achieved and controlled by a pre-selected group of nodes rather than a single node (as in the case of private blockchain).

Let us discuss the details of consortium blockchain in this chapter.

7.1 Introduction

In the previous chapters, we discussed the details of public blockchain and private blockchain systems. In a public blockchain, anyone with reasonable computational power and a suitable software such as a CC miner can get on the network. Anybody in the world can download these blockchains and primarily read the data. Private blockchains are private, as all permissions for the blockchain are kept centralized. A single node manages the consensus mechanism. Consortium blockchain is a hybrid scheme where the permission to read the block can be public or private depending on the needs of the organizations involved. This can be considered as an extension of public blockchain; however, the consensus mechanism is achieved and controlled by a pre-selected group of nodes.

7.2 Key Characteristics of Consortium Blockchain

Consortium blockchain is hybrid in nature. In a consortium blockchain, the read and write actions may be restricted. Only certain nodes will verify the transactions. Let us say we have 20 different banks who have decided that they will use the blockchain technology. Out of these 20 banks, say 11 or 12 will have to sign a block for verification (refer Fig. 7.1) to put any new block on the blockchain.

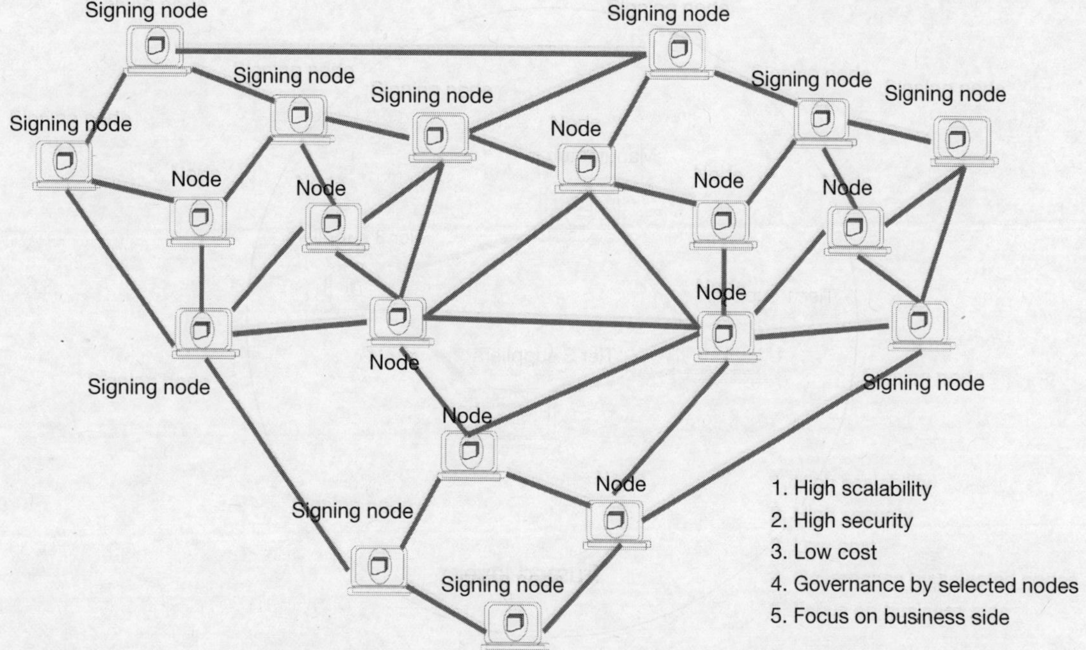

Figure 7.1: Consortium blockchain network

Thus, their read and write actions will be restricted. Further, the game rules are decided only among the members (20 nodes). Examples of consortium blockchain are Hyperledger, Ripple, Corda and Quorum. There are certain limited nodes that take part in the consensus process. Therefore, scalability is the first advantage. Security is the second advantage. We can restrict data access in a consortium, which is not the case in the public blockchain. The third advantage is it is cheaper as the transaction process involves very less nodes.

In a consortium blockchain, we can make changes after the code has been pushed. Any enterprise member of a consortium that has a node is able to see the full contents of the blockchain. Also, in most enterprise networks today, governance is left as an out-of-band process, where the members must agree on changes to the blockchain network rather than have a blockchain governance built right into the network itself. This gives rise to scalability, confidentiality, and built-in configurable governance that are essential for enterprises to achieve credible business results. The hardest part about building these alliances is the creation of trust in business partners to make them do business in new ways or even in existing ways. Consortium blockchain can be used to build trust among the participants of business alliances. Focusing on the business side of blockchain allows one to create these trust profiles.

Example: Let us think of a large manufacturing company like an automotive manufacturer, with tier-one and tier-two suppliers. These suppliers would typically be reluctant to disclose their entire production data to their clients. With blockchain, they can continue to retain their production systems (refer Fig. 7.2), by creating a Trust layer with the client.

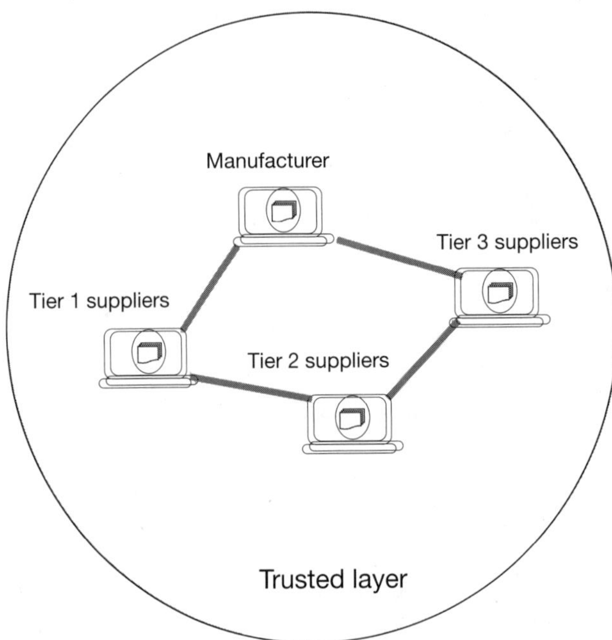

Figure 7.2: Blockchain with trusted layer

This layer allows the supplier to share only selective data, that is, the shared data is limited to only what the client needs to know in order to produce the business results.

7.3 Why We Need Consortium Blockchain

Consortium blockchains are used by private enterprises or organizations with a specific purpose. Organizations prefer to use a consortium blockchain, as they prefer to control the overall blockchain, with focus on the following aspects:

- Who can use the system
- Who can write into the system
- Who can read from the system.

7.4 Hyperledger Platform

Solutions for open source are great for hobbyists, and for small businesses, they often deliver the required speed. However, when enterprises come to work with other enterprises, their requirement from the software more or less remains the same but come with stringent rules about performance, scalability, and auditing. Blockchains in open-source model have been available for the last 5–8 years after the bitcoin became popular. Bitcoin proved that large distributed systems in an open environment (not controlled by a company or consortium) are technically feasible. Large businesses also need blockchain technology; however, with blockchain also came to need for higher performance – fast throughput, proper authentication, and auditing.

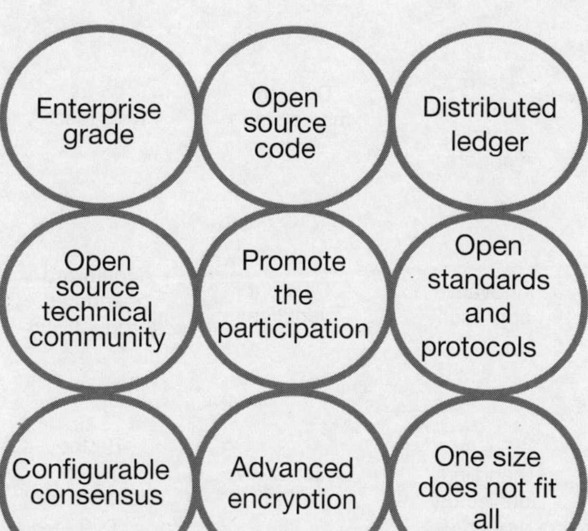

Figure 7.3: Hyperledger characteristics

The Linux Foundation took up the challenge for an open-source enterprise-grade distributed ledger technology and announced the Hyperledger Project in December 2015. The approach was to ensure that the best practices of computer science related to distributed computing are used in blockchain for enterprise solutions. We have seen earlier that bitcoin, launched in 2009, had demonstrated that massive decentralized computing is possible. However, bitcoin is an open blockchain and an unregulated environment. Hyperledger, on the other hand, has stated that they will not be issuing cryptocurrency.

Hyperledger brings into one place multiple efforts in computer science and open standards, solution-centric protocols. The hyperledger project allows multiple alternative solutions. For example, the implementation of Hyperledger can choose the consensus needed for the project. It uses existing computing knowledge, such as Advanced Encryption Standard (AES) and SHA1-256, both of which are encryption standards (refer Fig. 7.3).

As per hyperledger.org, the mission of Hyperledger Project (HLP) is to:

1. Create an enterprise-grade, open-source distributed ledger framework and codebase
2. Create an open-source, technical community to benefit the ecosystem of the HLP (Hyper Ledger Project) solution
3. Promote the participation of leading members of the ecosystem, including developers, service and solution providers and end-users;

Linux Foundation declared the creation of the Hyperledger Project (open source) in December 2015. In early 2016, Hyperledger accepted the code from digital assets, block stream, and IBM; this has evolved into an enterprise-grade solution: Hyperledger Fabric. Again, in 2016, Intel incubated its project; now it is called Hyperledger Sawtooth.

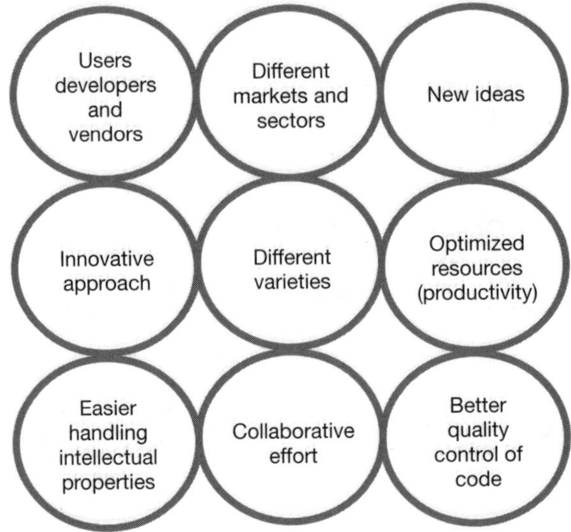

Figure 7.4: Hyperledger greenhouse properties

As per hyperledger white paper, Hyperledger Greenhouse is a place that brings together users, developers, and vendors from many different sectors and market spaces. Now, it is important to note that each enterprise has its own need/expectation from the blockchain. In addition, blockchain does not believe in a "one size fits all" approach. Hyperledger provides a greenhouse structure that can incubate new ideas, support each one with essential resources, and distribute results widely. A greenhouse structure can support many different varieties while consuming far fewer resources.

The greenhouse also brings additional advantage (refer Fig. 7.4):

- Help keeping up with developments
- Better productivity through specialization
- Collaboration to avoid duplicate efforts
- Better quality control of code
- Easier handling of intellectual property.

7.4.1 Fabric

Hyperledger Fabric is by far the most popular project with Hyperledger Greenhouse (refer Fig. 7.5). Fabric is a DLT and smart contract engine. Fabric's status is active, and the first production-ready version was made available in June 2017.

Hyperledger fabric is a permissioned blockchain infrastructure (refer Fig. 7.6) and is famous for its modular, scalable, secure architecture. Hyperledger fabric delivers a uniquely flexible and extensible architecture. The solution is known for its high degrees of confidentiality, resiliency, flexibility, and scalability. Nodes in hyperledger are classified as three types of peers:

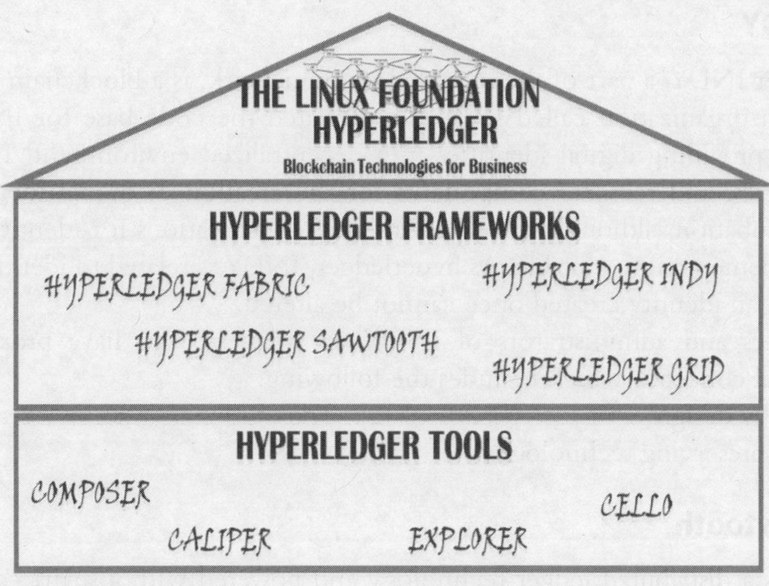

Figure 7.5: Hyperledger greenhouse family

- Endorser
- Committer
- Ledger Peer

Hyperledger, being a permissioned ledger system, is used to authenticate a node by using a mechanism called Membership Service Provider. One of the critical use cases for industry needs is information sharing on ledgers – while some selected information on ledgers may need to be shared with some nodes, all information on the ledger may have to be shared with other nodes. This has been a persistent challenge faced by IT systems for the past three decades. Hyperledger solves this problem by using a channel. Fabric, as part of the open-source movement, has already shared use cases in supply chain management, health care and financial services.

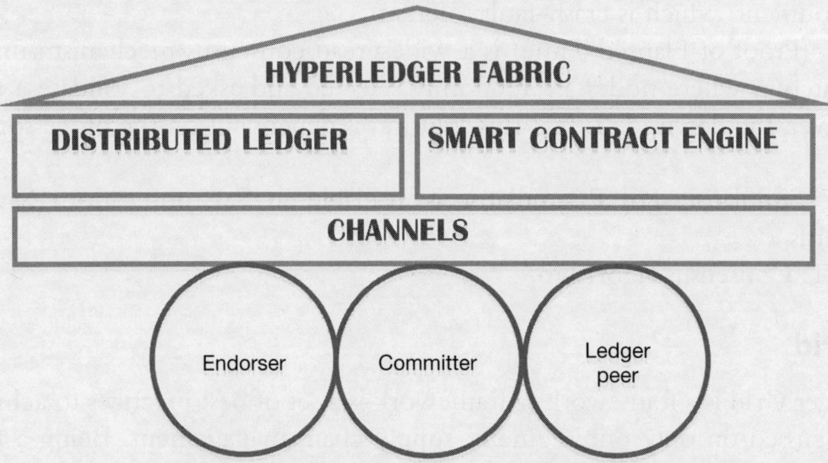

Figure 7.6: Hyperledger fabric system

7.4.2 INDY

Hyperledger INDY, a part of the hyperledger framework, is a blockchain tool for digital identity. An organization called sovrin.org donated the code base for INDY. INDY is known for providing digital identities in a decentralized environment. INDY provides tools, libraries, and reusable components for the creation of digital identities. INDY's status is incubation, although it has documented specifications for identity along with a sample implementation available. As hyperledger, INDY is related to identity, and being a blockchain, an identity created once cannot be altered.

Designers and administrators of INDY are requested to have proper training in foundational concepts, which includes the following:

- Privacy by design
- Privacy-preserving technologies

7.4.3 Sawtooth

Sawtooth is a distributed ledger technology and powered with a smart contract engine. The sawtooth project started with a contribution via Intel. Sawtooth offers hot swapping consensus algorithms and works flawlessly on the execution environment provided by Intel's SGX. Sawtooth offers a robust runtime environment, even allowing change of consensus approach in run time. Sawtooth, being a permissioned layer, brings restrictions via Access Control Lists and nodes are put into these restrictions:

- Who can connect to the network?
- Who can send consensus messages?
- Who can submit transactions to the network?

Sawtooth is the only project within Hyperledger Project that uses Ethereum and smart contracts are written via solidity, as it is written in Ethereum. These smart contracts can be deployed on the fly to the sawtooth network.

For Consensus, sawtooth provides the following implementations:

1. dev_Mode: a simple logic to select a random leader; this is suited for a development environment, which is crash-fault tolerant.
2. PoET (Proof of Elapsed Time) is a widespread consensus mechanism made famous via bitcoin, where nodes are randomly selected and asked to validate a transaction; however, the downside is that this solution is dependent on the processor created by Intel.
3. PoET Simulator: PoET simulator is available in Sawtooth, from Sawtooth 1.1 onwards.
4. RAFT: Consensus algorithm.

7.4.4 Grid

Hyperledger Grid is a framework (a framework is a set of best practices to achieve a task). Grid is focussed on only one segment: supply chain management. Being a framework,

Hyperledger grid does not contain rules; instead, it is an ecosystem detailing a blockchain for supply chain management. It includes data sets and frameworks that work together, letting application developers choose the best-suited technology or methodology as per company's requirement. Supply chain is among the highly rated use case for blockchain implementation. Hyperledger Grid strives to display real and practical ways for hyperledger implementation in the supply chain area.

7.4.5 Hyperledger Tools

Hyperledger framework has a set of collaboration tools for building blockchain business networks (refer Fig. 7.7).

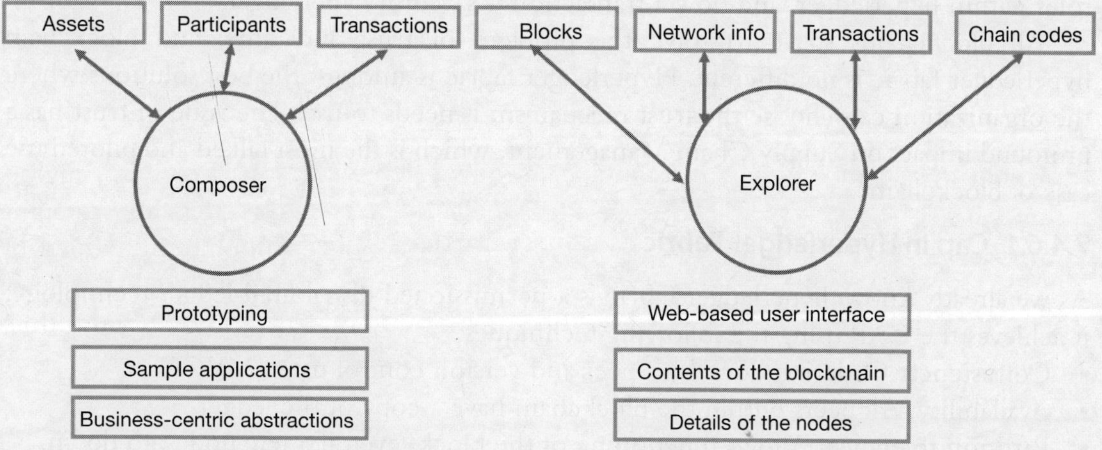

Figure 7.7: Hyperledger tools

7.4.5.1 Composer

Composer offers business-centric abstractions as well as sample apps, which are used to test or replicate business problems. Composer is handy when you have to build an application as part of Proof-of-Concepts (POC), and you have a concise timeline. Composer operates as a rapid prototyping tool for user-facing solutions. Hyperledger fabric is the underlying mechanism to operate a blockchain for the composer. Composer allows for creation/modification of the following:

- Assets
- Participants
- Transactions

7.4.5.2 Explorer

Explorer is another tool hosted in the Hyperledger greenhouse, which provides a web-based user interface. Hyperledger Explorer allows the user to view contents in the blockchain, and list the nodes Hyperledger Explorer can view, invoke, deploy or query. These nodes include the following:

- Blocks
- Transactions
- Network information (name, status, list of nodes)
- Chain codes.

7.4.6 Hyperledger Fabric

In this section, we will explore Hyperledger Fabric (refer Fig. 7.6), which is the most popular of hyperledger project. Let us explore the building blocks of hyperledger, the roles within hyperledger, and how a transaction is saved in hyperledger.

All blockchains strive to solve the problem of trust and time, and blockchain hyperledger fabric is no different. Hyperledger fabric is amongst the best solutions where the organization can choose the trust mechanism it needs to use. The issue of trust has a profound impact on Supply Chain Management, which is the most talked and piloted use case of blockchain.

7.4.6.1 Cap in Hyperledger Fabric

As we already know, hyperledger fabric is a permissioned distributed ledger technology; it achieves the CAP using the following techniques:

- Consistency: Ordering done by a peer and version control using MVCC
- Availability: All peers within the blockchain have a copy of the ledger
- Partition tolerance: Allows functioning of the block, even if a few nodes go down.

Like all blockchains, availability is often allowed in the IT systems; this approach could work in most ledgers: however, stringent consistency requirements are challenging, as companies are trying to reduce lead-time. For enterprise solutions, not having the latest data consistent among its various nodes could create a deficiency, as seen in the following examples.

Supply chain management example: Recall the e-commerce solution discussed in Section 6.7, where the customer places an order, the order is shipped, and the product is installed. A blockchain solution which is unable to give correct data (the latest data) would create more problems than it is trying to address, such as if, for example, the product is dispatched to the wrong address or the installation team reaches out to the customer before the product is dispatched.

Fin-tech example: The presence of a computer in the financial sector has allowed many companies to exploit and churn data and make wise decisions. Assuming you have a trading company which purchases and sells shares in the stock market, the absence of the latest data could mean that the company purchases/sells the wrong company's share (or the wrong number of shares).

Hyperledger fabric tackles this consistency problem using the following techniques:

- Execute, Order, Validate.
- Order works on a total block
- Multi-version concurrency control
- Peers to have ordered ledgers.

Execute, Order and Validate

Hyperledger fabric follows an approach where peers validate transactions submitted by the end-user. The validated transaction is submitted to the orderer; the orderer would further validate the order, especially to avoid double-spending. Orderer orders transactions for a block and approves the block creation, following which peers on the blockchain are notified via gossip protocol. All peers validate the order and make a new block in the chain (refer Fig. 7.8).

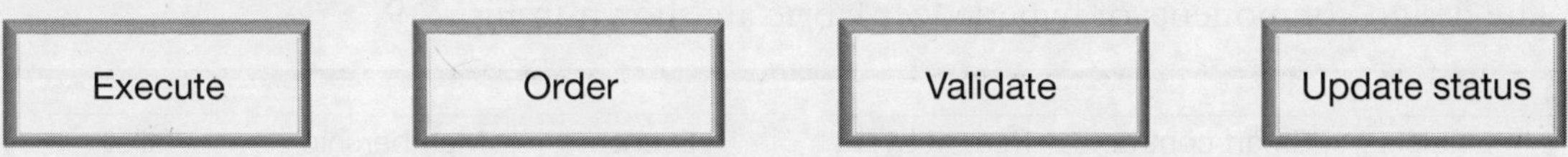

Figure 7.8: Hyperledger fabric approach

Peer validation ensures that service order transactions into blocks are guaranteed to be consistent with all nodes in the blockchain — an ordered block transmitted by the orderer via broadcast to peers. Peers would validate the order with their copy of the existing ledger and make changes to the ledger. If transactions did not pass (fail) the validation, they are marked as invalid in the ledger (they are retained in the ledger but not dropped).

Multi-version Concurrency Control

Multi-version concurrency control (MVCC) is a concurrency control mechanism used by database management system (DBMS), and in programming languages to provide transactional memory. Traditional DBMS handles concurrency problems by applying the lock. Locks can solve the problem only in centralized servers; applying lock in a distributed system may not be feasible.

From the blockchain perspective, concurrency control aims to control two common problems of distributed computing - double-spending and key clashing (key collision). Double-spending is the problem where participants bring transactions to use or modify the same asset, e.g., spending the same money or resource multiple times. Key collision is another problem where two resources try to change key/value simultaneously; the probability of this key collision is higher in a distributed computing system, where clients attempt to change key/value via parallel clients.

Hyperledger fabric uses versioning for keys stored in the ledger to avoid double-spending. The purpose of this logic is that between ordering and writing to the ledger, the key-value pair must not be changed. During the process of validation, that data for the key/value and version of the key is checked before writing to the ledger. If a transaction does not pass the validation, that record is invalidated in the ledger (though not dropped from the ledger). While the given approach solves the double-spending problem, transaction queuing helps to overcome the problem of key collision.

Hyperledger fabric uses Couch DB at all nodes, for keeping a copy of the ledger. Couch DB, with MVCC enabled in its environment ensures that DB is run at full speed, even at high load. Transactions are called Documents in Couch DB, because of its key/value nature. Documents in Couch DB are versioned. In hyperledger fabric, there is no delete or update. Therefore, if a user wants to change a value in a document, DB would create an entirely new version of that document. After doing this, there will be two copies of the same document, with different versions (the old version of the document and the newly created version of the same document.

Components of a Hyperledger Fabric

The design components of hyperledger fabric are shown in Fig. 7.9.

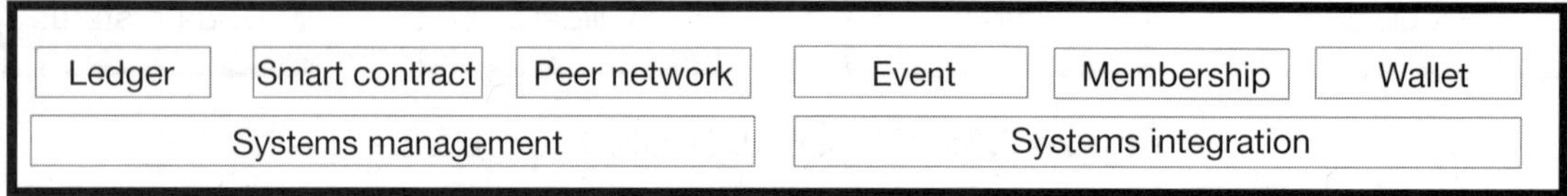

Figure 7.9: Hyperledger fabric components

Being a distributed blockchain, the fabric needs to maintain the ledger. The fabric uses couch DB, for retaining values as key/value pairs. A smart contract allows the creation of assets and allows the transfer of assets based on rules. Peers are the computing resources where ledgers are maintained; also, the peers in the fabric are of three types: **Endorsers, Orders, and Ledgers**. It needs to be noted that all peers keep ledgers; some peers would keep endorse and orders apart from being a ledger. Membership services are crucial for hyperledger fabric, as it is a permissioned network. Blockchain is essentially a state replication machine, where events are generated by the machine when the transaction is requested or processed. An event, by itself, does not bring any change to the system; however, events are used by blockchain to run algorithms. Membership authenticates, authorizes, and manages identities on a permissioned blockchain. The central place a user's identity and related activity is handled is called Wallet. Systems management manages various blockchain components. Blockchain by itself does not exist in the IT system landscape; it co-exists with other IT systems, e.g., with LDAP for user directory, for ERP for resource notification, etc. Systems Integration handles fabric bidirectional integration with other IT systems.

Roles Within Hyperledger Fabric

Hyperledger fabric has two roles, namely, **Clients and Peers**. Clients would submit a transaction to the blockchain network. Peers within a blockchain are of two types (refer Fig. 7.6).

- **Endorser:** Endorser peers will run the smart contract against the transaction and send the result to order.
- **Orderer (Committer):** Orderer will run the consensus and verify if the transaction is valid or not. The orderer adds transaction to a block, in a sequential manner.

Transaction Within Hyperledger Fabric

Transactions with hyperledger fabric are executed as follows (refer Fig. 7.10).

- A transaction request is submitted to the blockchain by a client
- Endorsers and peers validate the request
- Endorsers submit the validation result to orderer peer
- Orderer runs the consensus mechanism on the results from validators
- Orderer decides transaction as valid or invalid
- Orderer adds transaction to block
- Orderer submits a block to peers via a gossip protocol
- Peers add the block to the chain.

Further, when the private network grows, more than one orderer is needed to publish the next block for blockchain. Hyperledger fabric uses Apache Kafka to handle do and put transactions in a sequence and submit the sequence transaction to the orderer. The output is reconciled, and result is published to other peers to update the block (i.e., to add a new block). Hyperledger fabric uses pBFT to collate the transaction result and to arrive at a decision whether the transaction is approved or not.

There are three phases:

- Endorsement phase
- Ordering phase
- Validation Phase

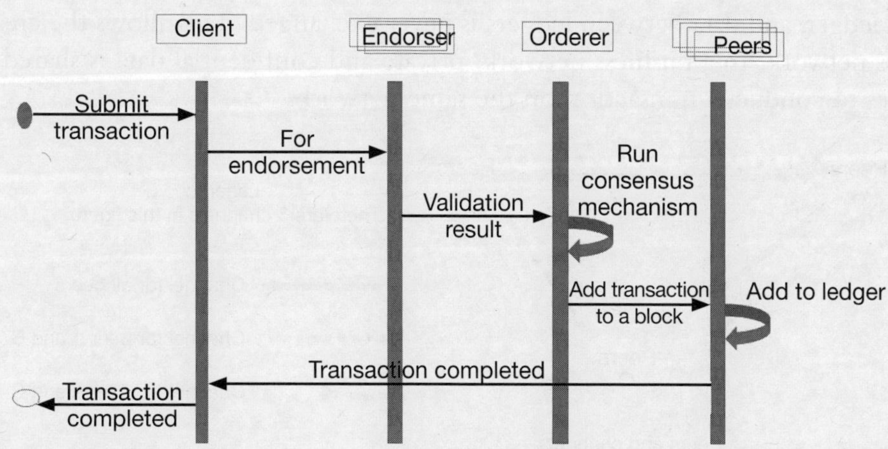

Figure 7.10: Hyperledger fabric transactions

Empirical evidence proves that ordering service (ordering phase) is the most considerable time- and resource-consuming activity within Fabric. This is in spite of the advantage of the fabric design that enables hyperledger fabric to run within the docker. Hence, it can manage RAM efficiently, allowing Fabric to run queries faster. However, Fabric interacts with DB, for efficient working of complex contracts (complex contracts involve working with diverse data inputs and arriving at solutions iteratively), which needs to keep transient data in the DB, for which rocksDB is handy. Modern IT systems use databases and IT architects expect similar data structures from blockchain as well. Fabric, though, prefers key-value approach to solve complex problems. But often, this is not sufficient as composite keys and querying over a range is needed; both these features are supported in Fabric.

During the validation phase, MVCC checks the version of keys read by the transaction during endorsement and ensures that the current state of key is the same as that of the endorsement phase; else the transaction is marked invalid.

Channels in Hyperledger Fabric

Channel is the most powerful tool provided by hyperledger fabric; perhaps there are no other blockchains that have anything similar to hyperledger fabric channels (refer Fig. 7.11). Channel is similar to a subnet within a network (Inside a subnet, only a few allowed computers are allowed to send and receive data). As with subnet of two or more specific network members, channels are the mechanism used by hyperledger fabric for conducting private and confidential transmission of data. Its members, the shared ledger, and the orderer nodes define a channel. Every transaction on the network is executed on a channel. To execute on the channel, each member needs to be authenticated and authorized. Each member is authorized based on the access policy list.

One peer can be part of one or more channels; this ensures that for each channel, there is a ledger, and data between ledgers is never exchanged. This allows the creation of a business network. In a business network, private and confidential data is shared among companies to conduct a transaction on the same network.

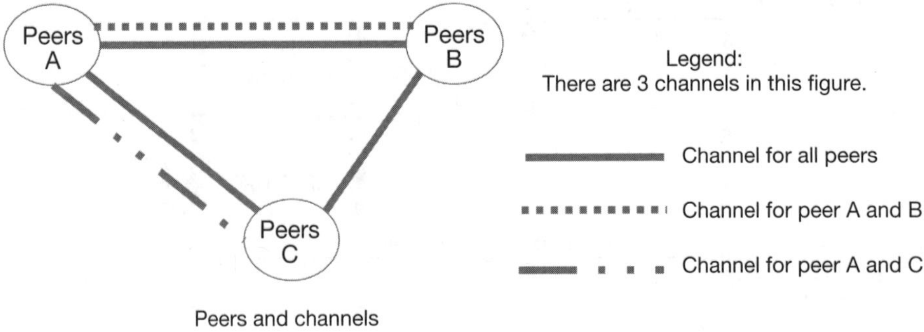

Figure 7.11: Hyperledger fabric channels

7.5 Overview of Ripple

Ripple is an open-payment network for digital currency. It is also a holding company. Ripple is a privately held cash flow positive company, which aims to create and enable a global network of financial institutions and banks to use ripple software, to lower the cost of international payments. It is also cost-efficient and real-time enabled. Ripple calls this global network using ripple software products. The ripple (XRP) ledger is an open-source product created by Ripple. It was created to reduce major point of friction in international payments. XRP can be used by banks to control liquidity on demand in real time. XRP pays providers to expand into new markets, provide faster payment settlement and lower foreign exchange cost.

Unlike Bitcoin, Ripple aims to work with the current financial world; the equivalent of roughly 155 trillions of dollars move across the board every year. For example, let us assume that everyone used PayPal, which charges 3% fees for each transaction; this would be way too much for paying fees, but we are all probably familiar with PayPal taking cut of our payment at some point or another. This means that about 5 trillions of those global payment fees would go straight to PayPal. These payments would likely even take days to process. Today's payment networks are slow, error prone and costly. Ripple net provides internet of value. Ripple was initially launched in 2012. It is already a revenue producing company with over 100 financial situations and banks on its blockchain network, including JPMorgan and Bank of America. In order to understand how Ripple functions, it is useful to know about RTGS. For example, when you send money via Bitcoin, the value is settled in the real time, that is, by real time gross settlement system or RTGS.

Ripple uses what it calls gateways, which are similar to the global ledger and made up of something similar to private blockchain. This is essentially a digital portal that government companies and financial institutions use to join the ripple network. This is called a ripple transaction protocol or ripple net. Ripple is essentially made for any type of entity that regularly moves large amount of money across the boards. For example, companies such as Apple and Amazon, already spending billions across the boards. It is important to note that the ripple network also functions as a currency exchange for all types of fiat currency. However, in order to do this, it needs to have currency liquidity. This is where XRP comes in. XRP is a digital coin built for providing source of liquidity to payment providers, market makers, and banks.

Users of ripple network can be divided into two groups, namely, ripple network users and network members. The primary users of ripple network are corporates, SMEs, small banks and payment providers. They only send payments. A second set of users is called as Network members and it includes banks and payment providers. They serve as the foundation of the network as they process payments and liquidity.

While ripple system makes sense for large corporations, it does not land much to an average consumer. XRP does not provide the average person much of benefits. So, ripple is not likely to change how a normal person receives and sends money. It is looking to revolutionize how money moves around the world in a large scale. Ripple has shown that it can handle tremendous amount of transactions per second. BTC handles about seven transactions per second, whereas the Extra pillager (XRP) can handle approximately 1,500 transactions per second. XRP has already been created with the total supply of 100 billion. There are currently around 38.6 billion ripple XRP tokens in circulation, with the rest held by the ripple labs. In May 2017, Ripple released their decentralization strategy and announced their plan to diversify the validators on the XRP ledger and expand them to 55 validators notes.

Network members process payments through a product called as "xCurrent". It provides pre-validation of transactions and rich data attachments for repayment. It also provides payment certainty through pre-disclosure of information to eliminate failures. It also enables atomic settlement and updates pass/fail status across all intermediaries. Network members can also source on-demand liquidity through the product "xRapid", which provides access-on-demand liquidity through digital assets. It provides lower liquidity costs through on-demand liquidity sourcing. It reduces the need for nostro accounts to make global payments. The product titled "xVia" is an API based on standard interface to access the Ripple Network. With this API, global payments can be made with certainty as it can access across networks to make on-demand and real-time payments. It also enables rich data attachments along with end-to-end visibility.

7.6 Overview of Corda

Corda is another open source distributed ledger technology (DLT) and was developed by R3, which is a consortium of more than 200 companies like Amazon, Microsoft, and Intel. R3 did the research and developed the Corda application. It has built-in app programs for various domains and industries (DApp) like banking (around 150 banks are part of it). It is developed based on blockchain technology. It works on the principle of need-to-know basis, and is a shared ledger, not a gossip network. Based on this principle, information is not shared indiscriminately; only necessary information is shared with the necessary stakeholders.

Corda is freely available on the internet and anybody can use it.

Following are the 4 characteristics of Corda:

- Need to know
- Interoperability
- Shared ledger
- Smart contracts

Need to know: Data is shared only with the intended users, for privacy. It provides assured identity. KYC (Know your Customer) enables single-user profile. Corda is a blockchain flavor and is different from Ethereum and Hyperledger. Its interoperability is the main advantage. It allows using legacy system data and can interact with the legacy systems. Corda has collaborated with Oracle for this. It works on peer-to-peer network with no central database or central provider. A Corda network is a fully connected graph. There is no global broadcast. Communication occurs on a point-to-point basis, similar to email. Let us consider the following example: A and B have some shared contract (refer Fig. 7.12). This contract and related communications will not be shared with other entities (C, D and E).

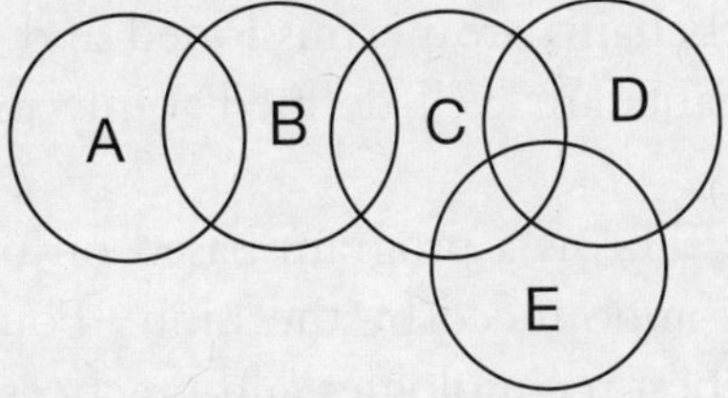

Figure 7.12: Corda example of contracts

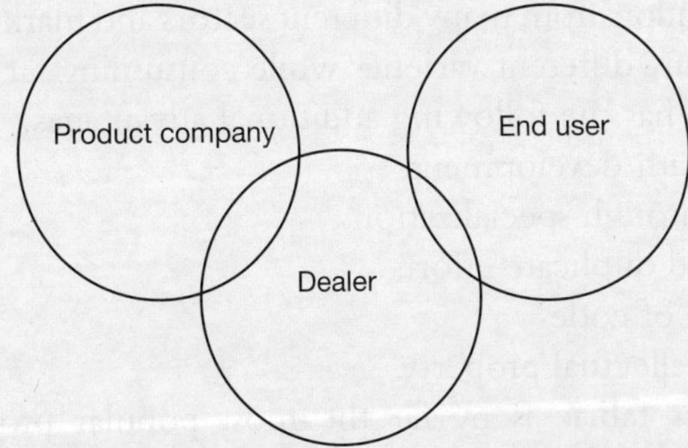

Figure 7.13: Corda example of contracts: Product company

Similarly, there are common contracts among C, D and E. This information will be shared only with the corresponding entities. That means, A and B will have no information about these contracts. Corda is similar to shared ledger between two or more parties. Smart contracts in Corda application can be built for any type of transaction. It can be rental agreement; it can be IPOs. Everything is a contract in Corda. Another example is the contract between a product company, dealer and the end user.

The product company will have a contract with the dealer (refer Fig. 7.13). The dealer will have a contract with the end user. Having one ledger with shared facts will help. Corda is scalable and supports billions of transactions daily across industries.

Longevity: Different versions of Corda will be able to coexist on the same network and applications will continue to run on later versions.

Interoperability: The platform is designed to allow multiple applications to coexist and interoperate on the same network. Oracle provides information services across the network. Consensus pools provide consistent, transparent and resilient uniqueness and consensus services.

Corda allows business and individuals to transact privately between legally identifiable counterparties on a single network that is meant to be highly scalable and follows network wide standards.

Corda can be downloaded from: https://www.corda.net/demobench/

Summary

Private blockchains are private, as all permissions for the blockchain are kept centralized. Many banks and business entities are now testing the use of private blockchains in their respective fields. Private blockchains are permissioned and centralized. They are closed ecosystems where all participants are well-defined. Only pre-approved entities can run nodes.

Hyperledger platform: This is a program based on open-source enterprise-grade distributed ledger technology announced by the Linux Foundation in December 2015. The approach was to use the best technologies and practices of computer science related to distributed computing for building a blockchain to be used for enterprise solutions.

As per hyperledger white paper, hyperledger greenhouse is a place that brings together users, developers, and vendors from many different sectors and market spaces. A greenhouse structure can support many different varieties while consuming far fewer resources.

The greenhouse also has the following additional advantages:

- It helps keeping up with developments
- Better productivity through specialization
- Collaboration to avoid duplicate efforts
- Better quality control of code
- Easier handling of intellectual property.

Fabric: Hyperledger fabric is by far the most popular project with hyperledger greenhouse. Fabric is a DLT and smart contract engine. Nodes in hyperledger fabric are classified as three types of peers: Endorser, Committer and Ledger Peer

INDY: Hyperledger INDY, a part of the hyperledger framework, is a blockchain tool for digital identity. An organization called sovrin.org donated a code base for INDY. As hyperledger, INDY is related to identity, and being a blockchain, an identity created once cannot be altered. Designers and administrators of this tool need to have proper training in the foundation concepts which include privacy by design and privacy-preserving technologies.

Sawtooth: Sawtooth is a distributed ledger technology and powered with a smart contract engine. It offers hot swapping consensus algorithms and works flawlessly in the execution environment provided by Intel's SGX. Sawtooth offers a robust runtime environment, even allowing change of consensus approach in run time. Sawtooth, being a permissioned layer, bring restrictions via Access Control Lists and puts a control on nodes that can connect to the network, send consensus messages and submit transactions.

Grid: Hyperledger grid is a framework (a set of best practices to achieve a task). Grid is focussed on only one segment: supply chain management.

Hyperledger tools: Hyperledger composer has a set of collaboration tools for building blockchain business networks.

Composer tool: Composer offers business-centric abstractions as well as sample apps, which are easy to test or replicate business problems. It is a rapid prototyping tool that comes in handy to build an application as part of Proof-of-Concepts, especially when there is a concise timeline.

Explorer tool: Explorer is another tool hosted in the hyperledger greenhouse, which provides a web-based user interface. Hyperledger explorer allows the user to view contents in the blockchain, and list the nodes Hyperledger Explorer can view, invoke, deploy or query. It includes the following: blocks, transactions, network information (name, status, list of nodes) and chain codes.

Importance of MVCC in consistency: MVCC is a concurrency control mechanism used by database management system (DBMS) and in programming languages to provide transactional memory.

Roles within hyperledger fabric: Hyperledger Fabric has two roles, namely, **Clients** and **Peers**. Clients would submit a transaction to the blockchain network. Peers within the blockchain are of two types: **Endorser:** Endorser peers will run the smart contract against the transaction and send the result to the orderer. **Orderer:** Orderer will run the consensus and decide whether the transaction is valid or not. The orderer adds transaction to a block in a sequential manner.

Transaction within hyperledger fabric: Transactions with hyperledger fabric are executed as follows:

- A transaction request is submitted to the blockchain by a client
- Endorsers and peers validate the request
- Endorsers submit the validation result to orderer peer
- Orderer runs the consensus mechanism on the results from validators
- Orderer decides on the transaction as valid or invalid
- Orderer adds transaction to block
- Orderer submits a block to peers via a gossip protocol
- Peers add the block to the chain.

Channels in hyperledger fabric: Channels are the mechanism by hyperledger fabric used for conducting private and confidential transmission of data. Its members, the shared ledger, and the orderer nodes define a channel. Every transaction on the network is executed on a channel. To execute on the channel, each member needs to be authenticated and authorized. Each member is authorized based on the access policy list.

EXERCISES

Multiple Choice Questions

1. **In this type of blockchain, the consensus mechanism is achieved and controlled by a pre-selected group of nodes.**
   A. Consortium blockchain
   B. Public blockchain
   C. Private blockchain
   D. Permissionless blockchain

Answer: A

Explanation: Consortium blockchain is a hybrid scheme where the permission to read the block can be public or private depending on the needs of the organizations involved. This can be considered as an extension of public blockchain; however, the consensus mechanism is achieved and controlled by a pre-selected group of nodes.

2. **Which of the following is not a characteristic of Consortium Blockchain?**
   A. High scalability
   B. High security
   C. Governance by one organization
   D. Low-priced
   Answer: C
   Explanation: High scalability, high security, low price, governance by selected nodes, and focus on business sides are the characteristics of consortium blockchain. There are certain limited nodes that take part in the consensus process. Therefore, scalability is the first advantage. Security is the second advantage as we can restrict what is accessible to the members, which is not the case in the public blockchain. The third advantage is that it is cheaper as the transaction process involves very less nodes.

3. **Which of the following is not an example of consortium blockchain?**
   A. Ethereum
   B. Hyperledger
   C. Ripple
   D. Corda
   Answer: A
   Explanation: Examples of consortium blockchain are hyperledger, ripple, Corda and Quorum.

4. **___________ declared the creation of the Hyperledger Project (open source) in December 2015.**
   A. Consortium Foundation
   B. Linux Foundation
   C. OpenSource Foundation
   D. Blockchain Foundation
   Answer: B
   Explanation: Linux Foundation declared the creation of the Hyperledger Project (open source) in December 2015. In early 2016, Hyperledger accepted code from Digital assets, Block stream and IBM; this has evolved into an enterprise-grade solution: Hyperledger Fabric. Again, in 2016, Intel incubated its project, and called it Hyperledger Sawtooth.

5. **___________ is a distributed ledger technology and powered with a smart contract engine. It started with a contribution via Intel.**

A. Sawtooth
B. Hyperledger
C. Hyperledger fabric
D. Ethereum
Answer: A

Explanation: Sawtooth is a distributed ledger technology and powered with a smart contract engine. The sawtooth project started with a contribution via Intel. Sawtooth offers hot swapping consensus algorithms and works flawlessly on the execution environment provided by Intel's SGX. Sawtooth offers a robust run-time environment, even allowing change of consensus approach in run-time.

6. As per hyperledger white paper, this is a place that brings together users, developers, and vendors from many different sectors and market spaces.
A. Hyperledger white house
B. Hyperledger greenhouse
C. Hyperledger sawtooth
D. Hyperledger GetHub
Answer: B

Explanation: As per Hyperledger white paper, hyperledger greenhouse is a place that brings together users, developers, and vendors from many different sectors and market spaces.

7. Which of the following is not a characteristic of hyperledger greenhouse?
A. Innovative approach
B. Collaborative effort
C. Optimized resources
D. Similar markets and sectors
Answer: D

Explanation: The greenhouse brings several advantages: It helps keeping up with developments, better productivity through specialization, collaboration to avoid duplicate efforts, better quality control of code and easier handling of intellectual property.

8. This hyperledger framework is the most popular project with hyperledger greenhouse. The first production-ready version available in June 2017 was
A. Hyperledger INDY
B. Hyperledger Fabric
C. Hyperledger Sawtooth
D. Hyperledger Grid
Answer: B

Explanation: Hyperledger Fabric is by far the most popular project with Hyperledger Greenhouse. Fabric is DLT and Smart contract engine. Fabric's status is active and the first production-ready version was made available in June 2017.

9. Which of the following is not the name of a hyperledger fabric?
A. Miner
B. Endorser
C. Committer
D. Ledger Peer
Answer: A
Explanation: Nodes in hyperledger are classified as three types of peers: Endorser, Committer and Ledger Peers.

10. This part of the hyperledger framework acts as a blockchain tool for digital identity. An organization called sovrin.org donated the code base for this:
A. Hyperledger INDY
B. Hyperledger Fabric
C. Hyperledger Sawtooth
D. Hyperledger Grid
Answer: A
Explanation: Hyperledger INDY, a part of the hyperledger framework is a blockchain tool for digital identity. An organization called sovrin.org donated the code base for INDY. INDY is known for providing digital identities in a decentralized environment. It also provides tools, libraries, and reusable components for the creation of digital identities.

11. This is the only project within Hyperledger Project that uses Ethereum:
A. Hyperledger INDY
B. Hyperledger Fabric
C. Hyperledger Sawtooth
D. Hyperledger Grid
Answer: C
Explanation: Sawtooth is the only project within hyperledger project that uses Ethereum. Smart contracts on Sawtooth are written via solidity, as it is written in Ethereum. These smart contracts can be deployed on the fly to the sawtooth network.

12. This is a Hyperledger framework focussed on supply chain management segment:
A. Hyperledger INDY
B. Hyperledger Fabric
C. Hyperledger Sawtooth
D. Hyperledger Grid
Answer: D
Explanation: Hyperledger Grid is a framework (a framework is a set of best practices to achieve a task). Grid is focussed on only one segment: supply chain management. Being a framework, hyperledger grid does not contain rules; instead it is an ecosystem detailing a blockchain for supply chain management. It includes data

sets and frameworks that work together, letting application developers choose the best-suited technology or methodology as per the company's requirement.

13. This hyperledger tool offers business-centric abstraction as well as sample apps.

A. Composer
B. Explorer
C. Sawtooth
D. Grid

Answer: A

Explanation: Composer offers business-centric abstractions as well as sample apps, which are easy to test or replicate business problems. Composer is handy to build an application as part of proof-of-concepts (POC) within concise timelines. Composer operates as a rapid prototyping tool for user-facing solutions. Hyperledger fabric is the underlying mechanism to operate a blockchain for the composer. Composer allows for creation/modification of Assets, Participants and Transactions.

14. Hperledger tool called Composer allows for creation/modification of the following, except:

A. Assets
B. Participants
C. Transactions
D. Blocks

Answer: D

Explanation: Composer operates as a rapid prototyping tool for user-facing solutions. Hyperledger fabric is the underlying mechanism to operate a blockchain for the composer. Composer allows for creation/modification of Assets, Participants and Transactions.

15. This hyperledger tool operates as a rapid prototyping tool for user-facing solutions

A. Composer
B. Explorer
C. Sawtooth
D. Grid

Answer: A

Explanation: Same as that for Question 14.

16. Hyperledger Explorer can view, invoke, deploy or query, all of the following, except:

A. Blocks
B. Transactions
C. Chain Codes

D. Oracles

Answer: D

Explanation: Explorer is another tool hosted in the Hyperledger greenhouse, which provides a web-based user interface. Hyperledger Explorer allows the user to view contents in the blockchain, and list the nodes which Hyperledger Explorer can view, invoke, deploy or query. These nodes include Blocks, Transactions, Network information (name, status, list of nodes), and Chain codes.

17. MVCC stands for:

A. Multi-version concurrency control

B. Multi-version Currency control

C. Multi-version Cost Control

D. Multi-version Consensus Control

Answer : A

Explanation: MVCC (Multi-version concurrency control): MVCC is a concurrency control mechanism used by database management system (DBMS), and in programming languages to provide transactional memory. Traditional DBMS have handled concurrency problems by applying the lock. Locks can solve the problem only in centralized servers; however, applying lock in a distributed system may not be feasible.

18. From the blockchain perspective, concurrency control aims to control two common problems of distributed computing. They are:

A. Double spending and key collision

B. Privacy and security

C. Identity and vulnerability

D. Controlability and accessibility

Answer: A

Explanation: From the blockchain perspective, concurrency control aims to control two common problems of distributed computing – double-spending and key clashing. Double-spending happens when participants bring transactions to use or modify the same asset, e.g., by spending the same money or resource multiple times. Key collision is another problem where two resources try to change key/value simultaneously. The probability of key collision is higher in a distributed computing system, where clients attempt to change key/value via parallel clients.

19. Hyperledger fabric uses versioning for keys stored in the ledger to avoid double-spending.

A. True

B. False

Answer: A

Explanation: Hyperledger fabric uses versioning for keys stored in the ledger to avoid double-spending. The purpose of this logic is that between ordering and writing to the ledger, the key-value pair must not be changed. During the process of validation, that data for the key/value and version of the key is checked before writing to the ledger. If a transaction does not pass the validation, that record is invalidated in the ledger (not dropped from the ledger). While the given approach solves the double-spending problem, transaction queuing helps to overcome the problem of key collision.

20. Hyperledger fabric uses Couch DB at all nodes. Transactions are called ____________ in Couch DB:

A. Records
B. Documents
C. Labels
D. Blocks

Answer: B

Explanation: Hyperledger fabric uses Couch DB at all nodes for keeping a copy of the ledger. Couch DB with MVCC enabled in its environment ensures that DB is run at full speed, even at high load. Transactions are called Documents in Couch DB, because of its key/value nature.

21. Hyperledger fabric uses ____________ to collate the transaction result and to arrive at a decision whether the transaction is approved or not.

A. pBFT
B. BFT
C. PAXOS
D. RAFT

Answer: A

Explanation: Hyperledger fabric uses Apache Kafka to handle or arrange transactions in a sequence and submit the sequence transaction to the orderer. The output is reconciled and the result is published to other peers to update the block (i.e., to add a new block). Hyperledger fabric uses pBFT to collate the transaction result and to arrive at a decision whether the transaction is approved or not.

22. This is similar to subnet, within a network in hyperledger fabric:

A. Fork
B. Channel
C. Documents
D. Nodes

Answer: B

Explanation: Channel is the most powerful tool provided by hyperledger fabric. Perhaps there is no other blockchain having anything similar to hyperledger fabric channels. Channel is similar to subnet, within a network (Inside a subnet, only a few computers are allowed to send and receive data).

23. **This is the mechanism used by hyperledger fabric for conducting private and confidential transmission of data.**
    A. Fork
    B. Channel
    C. Documents
    D. Nodes
    Answer: B
    Explanation: Channels are the mechanism used by hyperledger fabric for conducting private and confidential transmission of data. Its members, the shared ledger, and the orderer nodes define a channel. Every transaction on the network is executed on a channel. To execute on the channel, each member needs to be authenticated and authorized. The member authorization is done based on the access policy list.

24. **Channels are the mechanism used by hyperledger fabric for conducting private and confidential transmission of data. One peer can be part of only one channel for security purpose.**
    A. True
    B. False
    Answer: B
    Explanation: One peer can be part of one or more channels; this ensures that for each channel, there is a ledger, and data between ledgers is never exchanged. This allows the creation of a business network. In a business network, private and confidential data is shared among companies to conduct a transaction on the same network.

25. **In Ripple, the network members process payments through a product called:**
    A. xCurrent
    B. xRapid
    C. xVia
    D. xAPI
    Answer: A
    Explanation: Network members process payments through a product called as "xCurrent". It provides pre-validation of transactions and rich data attachments for repayment. It also provides payment certainty through pre-disclosure of information to eliminate failures. It also enables atomic settlement.

26. **In Ripple, the network members can also source on-demand liquidity through the product**
    A. xCurrent
    B. xRapid
    C. xVia

D. xAPI

Answer: B

Explanation: Network members can also source on-demand liquidity through the product "xRapid", which provides access to on-demand liquidity through digital assets. It provides lower liquidity costs through on-demand liquidity sourcing. It reduces the need for nostro accounts to make global payments.

27. The product titled ____________ is a standard API-based one-standard interface to access the Ripple Network.

A. xCurrent

B. xRapid

C. xVia

D. xAPI

Answer: C

Explanation: The product titled "xVia" is a standard API-based one-standard interface to access the Ripple Network. With this interface, global payments can be made with certainty as it can access across networks to make on-demand and real-time payment. It also enables rich data attachments along with end-to-end visibility.

28. Corda is another open-source distributed ledger technology (DLT) and was developed by R3, which is a consortium of more than __________ number of companies.

A. 10

B. 20

C. 200

D. 1000

Answer: C

Explanation: Corda is another open-source distributed ledger technology (DLT) and was developed by R3, which is a consortium of more than 200 companies like Amazon, Microsoft, and Intel.

29. Which of the following is not a characteristic of Corda

A. Assured KYC (Know Your Customer)

B. Interoperability

C. Shared ledger

D. Anonymity of users

Answer: D

Explanation: The 4 characteristics of Corda are: 1. Need to know. 2. Interoperability. 3. Shared ledger. 4. Smart contacts.

30. In hyperledger, which of the following phase has the most time and resource consuming activity within Fabric?

A. Ordering phase

B. Endorsement phase

C. Validation phase
D. Final phase
Answer: A
Explanation: There are three phases in Fabric: Endorsement phase, Ordering phase and Validation phase.
Empirical evidence proves that ordering service (ordering phase) is the most considerable time- and resource-consuming activity within Fabric.

Short-answer Questions

1. **What is the mission of Hyperledger Project (HLP)?**
   As per hyperledger.org, the mission of Hyperledger Project (HLP) is to:
   a. create an enterprise-grade, open-source distributed ledger framework and codebase
   b. create an open-source, technical community to benefit the ecosystem of the HLP solution
   c. Promote the participation of leading members of the ecosystem, including developers, service and solution providers and end-users.

2. **What are the advantages of greenhouse?**
   The greenhouse has the following advantages:
   - Help keeping up with developments
   - Better productivity through specialization
   - Collaboration to avoid duplicate efforts
   - Better quality control of code
   - Easier handling of intellectual property.

3. **Write notes on hyperledger fabric.**
   Hyperledger fabric is by far the most popular project with hyperledger greenhouse. Fabric is a DLT and smart contract engine. Fabric's first production-ready version was available in June 2017. Hyperledger fabric is a permissioned blockchain infrastructure and is famous for its modular, scalable, secure architecture. Hyperledger fabric delivers a uniquely flexible and extensible architecture. The solution is known for its high degree of confidentiality, resiliency, flexibility, and scalability.

4. **What are the types of nodes in hyperledger?**
   Nodes in hyperledger are classified as three types of peers:
   - Endorser
   - Committer
   - Ledger Peer

5. **Write notes on Hyperledger INDY.**
   Hyperledger INDY, a part of the Hyperledger framework, is a blockchain tool for digital identity. An organization called sovrin.org donated a code base for INDY.

INDY is known for providing digital identities in a decentralized environment. INDY provides tools, libraries, and reusable components for the creation of digital identities. INDY's status is incubation, although it has documented specifications for identity along with those for sample implementation.

6. What is sawtooth in hyperledger?

Sawtooth is a distributed ledger technology and powered with a smart contract engine. The sawtooth project started with a contribution via Intel. Sawtooth offers hot swapping consensus algorithms and works flawlessly on the execution environment provided by Intel's SGX. Sawtooth offers a robust runtime environment, even allowing change of consensus approach in run time.

7. Write notes on hyperledger grid.

Hyperledger grid is a framework (a framework is a set of best practices to achieve a task). Grid is focussed on only one segment: supply chain management. Being a framework, hyperledger grid does not contain rules. Instead, it is an ecosystem detailing a blockchain for supply chain management and includes data sets and frameworks that work together. It lets application developers choose the best-suited technology or methodology as per the company's requirement.

8. Write notes on Composer.

Composer offers business-centric abstractions as well as sample apps, which help to easily test or replicate business problems. Composer is handy to build an application as part of proof-of-concepts within a concise timeline. Composer operates as a rapid prototyping tool for user-facing solutions. Composer has hyperledger fabric as the underlying mechanism to operate the blockchain. Composer allows for creation/modification of the following:

- Assets
- Participants
- Transactions

9. Write notes on Explorer.

Explorer is another tool hosted in the hyperledger greenhouse, which provides a web-based user interface. Hyperledger Explorer allows the user to view contents in the blockchain and list the nodes Hyperledger Explorer can view, invoke, deploy or query. Hyperledger Explorer also allows the user to view the following:

- Blocks
- Transactions
- Network information (name, status, list of nodes)
- Chain codes.

10. What are the roles within hyperledger fabric?

Hyperledger Fabric has two roles, namely: Client and Peers. Clients would submit a transaction to the blockchain network. Peers within blockchain are of two types:

- Endorser: Endorser peers run the smart contract against the transaction and send the result to the orderer.
- Orderer: The orderer runs the consensus and verifies if the transaction is valid or not. The order adds transaction to a block in a sequential manner.

11. What are the three phases of transactions in hyperledger fabric?

There are three phases of transactions:

- Endorsement phase
- Ordering phase
- Validation phase

12. What are the four characteristics of Corda?

Following are the four characteristics of Corda:

1. Need to know
2. Interoperability
3. Shared ledger
4. Smart contacts

13. What is Ripple?

Ripple is an open payment network for digital currency. It is also a holding company. Ripple is a privately held cash-flow positive company, which aims to create and enable a global network of financial institutions and banks to use ripple software, to lower cost of international payments. It is also a cost-efficient and real-time enablement.

Essay-type Questions

1. Explain the hyperledger fabric framework in detail.
2. Explain the various hyperledger fabric tools in detail.
3. Explain how CAP in hyperledger fabric is taken care of.
4. Explain the importance of MVCC in consistency.
5. Write an essay on the various components of a hyperledger fabric
6. Demonstrate transaction within hyperledger fabric.
7. Explain the concept of channels in hyperledger fabric.
8. Write an overview of Ripple.
9. Write an overview of Corda.

CHAPTER **8**

Initial Coin Offering (ICO)

LEARNING OBJECTIVES

Rapid urbanization and technological breakthroughs like blockchain have opened up an avenue of innovative ideas. However, bringing an idea to reality has a whole new set of hurdles from research, design, experimentation, build, promotion and implementation. This requires time and money. Perhaps funding is the most important challenge that new undertakings face. Venture capitalists are reluctant to invest in unproven undertakings having little or no collateral. This is where the new crowd-funding mechanism of Initial Coin Offerings comes in, where a start-up attempts to woo investors by educating them on the future potential and long-term benefits of investing in their project idea.

This chapter aims to educate readers on the subject of Initial Coin Offering, its characteristics, its life-cycle, its distinction from the traditional stock offerings, and the reasons for its popularity as a fundraising instrument for both blockchain entrepreneurs and enthusiasts alike.

8.1 Introduction

Initial Coin Offering, abbreviated ICO, is the method of raising funds for blockchain projects that is similar to the traditional stock market launch or the IPO. In Initial Coin Offering, instead of stocks or shares, tokens or coins are offered to the investors before it is listed on the exchange. ICO took the world by storm in 2017–2018 when a blockchain startup Block.One raised a record-breaking 4.1 billion dollars in 1 year. ICOs are among the biggest financial innovations to spring from Blockchain Technology for raising business venture capital.

According to ICO data, 875 ICOs raised over 6 billion dollars in 2017 and 1253 ICOs raised 7.8 billion dollars in 2018. This was largely driven by investor hype around cryptocurrency. According to the ICO market analysis report published by ICO bench, of the 2517 ICOs that were completed in 2018, only 1012 were able to raise funds.

However, the year 2018/2019 has seen blockchain-based businesses raising more capital via ICOs than through traditional venture capital. The popularity of the ICO is mainly attributed to its ease of setup and usage. In this chapter, we will understand the basic Initial Coin Offering or the ICO.

Most of the ICOs today run on Ethereum blockchain through a smart contract that collects Ethereum tokens and automatically exchanges these for a new token created by the start-up company that is issuing the ICO. This chapter will explore the novel phenomenon of the Initial Coin Offering, and examine how to launch an ICO and invest in it. The chapter also discusses the advantages and disadvantages from both the company and the investor perspective.

8.2 Blockchain Fundraising Methods

8.2.1 Initial Coin Offering

Initial Coin Offering or ICO is a method of crowd-funding projects or new ventures in the crypto-industry. However, the concept in ICO is slightly different from crowd-funding. While in crowd-funding, the project initiator is not expecting a substantial return on investment, the backer of an ICO is motivated by profit. ICO fundraising is known by the term "crowd-sale" so as not to confuse it with crowd-funding.

Did you know?

Crowd-funding is the practice of funding an undertaking or venture by raising small amounts of money from a large number of people interested in investing money in different companies and projects they want to support, typically via the Internet. As per the global statistic portal Statista, the transaction value in crowd-funding segment amounted to USD 6.9 billion in 2019 and is expected to reach nearly USD 12 billion by 2023.

One of the most popular crowd-funding platforms is Kickstarter that has nearly 159,000 successfully funded projects. Started in 2009, Kickstarter currently has over USD 4 billion pledged to projects that span across films, arts, publishing, technology and other fields.

Popular traditional crowd-funding platforms for startups are StartSomeGood, Indiegogo, Patreon and Republic.

ICOs are the cryptocurrency version of crowd-funding. They are popular among technology companies, specifically blockchain start-ups, looking to raise capital by issuing cryptocurrency or tokens that are tied to the value or services of that company. Put in another way, we can say that Initial Coin Offering (ICO) is to cryptocurrency world as Initial Public Offering (IPO) is to the investment world. It enables start-ups or individuals to raise funds for their projects and people to invest in the projects that they see value in. However the key differentiator between the two is that while IPO investors get company shares for their investment, blockchain companies provide the investor with "tokens" that are used to access the technology or any form of compensation.

A high-level basic process of ICO fundraising mechanism is represented in Fig. 8.1. The key stakeholders are the founders who organize the ICO and the investors. The process may vary slightly based on the ICO project.

The ICO model can be simplified to three basic workflows:

1. **Fund Flow**: A person interested in investing in the ICO buys popular cryptocurrency like Bitcoin or Ether with USD/EUR or any other accepted fiat currency. The investor sends the bitcoin/ether either to a smart contract or special wallet that stores the funds typically till the end of crowd-sale. Finally, the company organizing the ICO will convert the bitcoin/ether to USD/fiat via Exchange to pay for the project development, salary payments, marketing, etc.
2. **Token Issuance**: After the crowd-sale, the company organizing the ICO will distribute an equivalent value of new tokens/coins to the investors based on their individual fund contribution. Investors can utilize the tokens to participate in the project ecosystem.

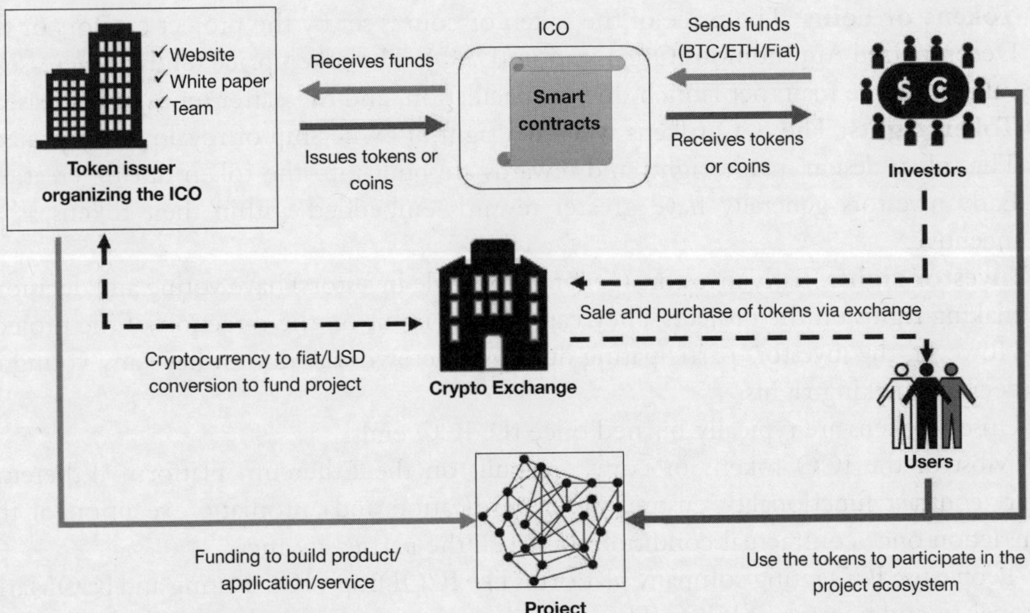

Figure 8.1: The ICO high-level process

3. **Crypto Exchange**: Once the ICO ends, the company lists the token on a cryptocurrency exchange. This would require standard information like the coin name, symbol, coin price, total quantity, etc. Once it is listed, tokens can be traded. Investors can sell their tokens for money or BTC/ETH. Public users who wish to access or use the project platform or service can purchase the necessary tokens via the exchange.

Other key features of the ICO include:

- ICOs give investors coins or tokens in return for their investment at a later point in time, usually as soon as the ICO ends. The crypto coin is an asset native to its own blockchain and can be used as money in the cryptocurrency world. The crypto token provides investors with a right of ownership or royalties to the project or DAO.
- **Online platform**: They are fully online, operating via a website that is built to inform and convince the public of the potential value of the project. The website typically has a catchy headline announcing the project idea/concept to instill a sense of urgency, in addition to project description, the white paper, details of the team and their role, the roadmap, community links like news blogs, medium, telegram, etc., and the legal terms of service.
- **ICO lifespan**: There may be multiple rounds of fundraising. The crowd-sale may span from between a few days to a couple of months. Some ICOs stay open for more than a year incorporating multiple rounds of sale. ICO concludes once the coins or tokens are tradable in the open market.
- **Tokens or coins**: The price of the token or coin is set by the project creators or the Decentralized Autonomous Organization (DAO). The token price tends to increase in value until the main net launch due to speculation, and thereafter tends to decrease.
- **Token rights**: The ICO tokens will sell a right of ownership or royalties to a project. The token design, usage rights and rewards are built into the token during creation. Early investors generally have greater rewards embedded within their tokens as an incentive.
- **Investor rights**: Early investors (generally pre-sale investors) have voting and decision-making rights on the project. They can provide input on the direction of the project. However, the investors participating during the crowd-sale do not have any voting or decision-making rights.
- Unsold tokens are typically burned once the ICO ends.

Most of the ICO tokens or coins are built on the Ethereum platform. Ethereum smart contract functionality ensures faster verification and automatic execution of the transaction once contractual conditions between the parties are met.

People use ICO rating company platforms like ICODrops, ICORating and ICOMarks for analysis and research of ICOs. They provide an overview of the current and upcoming ICOs including a rating of the best ICOs to invest in. Investors can use the services of the rating company to get an independent detailed analysis report of the projects they plan to invest in.

8.2.2 ICO and the Traditional IPO

Both ICO (Initial Coin Offering) and IPO (Initial Public Offering) serve as a means for businesses to raise capital to fund a specific project or venture. Though they appear similar in nature, there are key distinctions between the two offerings especially with respect to processes, working platform, and legal and regulatory aspects.

8.2.2.1 ICO vs. IPO – Key Differences

In an ICO, investors are given the opportunity to purchase a part of the blockchain start-up or project's total coin/token supply before the mining process begins, with the promise of potential rise in the market value of the coin. However in an IPO, the company aims to raise funds for its operations and projects by providing investors a share in the company whereby investors can earn dividends from company profits.

Some of the differences between the two models are:

- Aim of an ICO company is to promote adoption of the company and its associated platform or services while the aim of the IPO is to collect dividends.
- ICOs receive contributions in both fiat currency and cryptocurrency, usually Bitcoin and Ethereum, while IPOs deal only in fiat currency.
- ICOs generally fund projects that are conceptual and exist only on paper, while IPOs are launched by well-established companies.
- ICOs projects are generally open-source and decision-making is decentralized depending on the project type. It gives investors decision-making and approval rights that can direct the project. However, the decision-making in the IPO companies are centralized through the CEO or the Board of Directors.
- ICOs are made to online investors who can be from anywhere in the world.
- ICOs give investors only governing rights to the project platform and specific utility rights and rewards designed in the tokens they invested in, while IPOs give ownership of the company based on the number of shares held by the investor.
- ICOs can have multiple rounds of fundraising and pricing may vary, while an IPO is a one-time sale where the share price is pre-determined with the help of intermediaries like banks, underwriters and rating agencies.
- ICOs listed on the cryptocurrency exchanges are largely not regulated, while the stock exchanges and IPO companies are heavily regulated.

8.2.2.2 ICOs vs. IPOs – Pros and Cons

Advantages of an ICO are:

- Anyone globally who owns cryptocurrency can participate in the ICO, while the investor base of IPOs is limited as they are mainly adapted towards established investors like banks.
- ICOs are easier-to set up. All that is required is a white paper and smart contract to get started, compared to IPOs which are more complex due to pre-requisite legal formalities.

- ICOs have shorter duration and raise capital more quickly, with the actual crowd-sale lasting no more than a month to reach its small cap or hard cap, while an IPO can take anywhere from 6 months to 2 years.
- ICOs are held entirely online thus benefiting from reduced cost of marketing, payments and settlements, unlike the IPOs that are fundamentally paper-based.

Fundraising with an ICO also has a few disadvantages:

- ICO is used to raise funds to launch a project and enter the market hoping for market appreciation compared to an IPO where the company is financially stable and fundraising is for future expansions.
- Lack of regulatory oversight: A start-up can launch an ICO with a white paper that is not standardized and is discretionary. However, IPOs are heavily regulated and can be launched only with a government approved legal document called prospectus that includes key verifiable information about the company.
- ICO start-ups are not fully trustworthy and prone to fail as they do not have any certifiable track record of accomplishments, while a company needs to present an attested track record of earnings and other accomplishments to list its shares via an IPO.
- ICOs hold a poor market reputation with high failure rate, scams and fraudulent activities while IPOs are more difficult to attack.
- ICOs are fairly recent with the first ICO launch of Mastercoin in 2013, while IPOs are quite mature having been around for centuries.

8.2.3 Important ICO Terms

There are some important terms inherent in an ICO that anyone planning on launching an ICO or anyone interested in investing in an ICO should be aware of aside from the blockchain technology. These include, but are not limited to the following:

1) ICO Token

A token is a unit of value that is created by the company to govern the ICO ecosystem, enabling users to interact with the product or platform and facilitate the distribution of rewards and benefits to all of its stakeholders, e.g. utility token, security token, etc.

2) Decentralized Autonomous Organization

Decentralized autonomous organization or DAO, as the name suggests, is a decentralized organization that is run autonomously without the need for managerial supervision, by embedding the organization bylaws into the system code through protocols and smart contracts.

The first organization, the Ethereum powered DOA, failed due to a glitch in the initial coding. However, later DOAs such as Dash, Digix Global, and BitShares were successful. Many others are in the proposal stage.

3) Token Sale

Token sale is another term for Initial Coin Offering or ICO. It is a limited period sale wherein tokens are issued to the public in exchange for cryptocurrency, typically Bitcoin and Ether.

4) White Paper

A white paper is an official document, prepared by the blockchain project company or entrepreneurs to enable the reader to make an informed decision on whether to participate in the ICO. It details everything that a potential investor would need to know about the blockchain project, including but not limited to the company details, vision and objective of the project, description of the project and its core team, product description, technological aspects, market analysis, business plan, token and its distribution, and guides the investor on how to participate in the ICO, providing timelines and terms and conditions.

5) Whitelist

Whitelist refers to a listing of investors who are registered to participate in the token sale. Whitelisted or pre-registered investors are typically given the first cut of the token sale ahead of the crowd-sale with added benefits, bonuses and/or rewards.

6) Private Sale

Private sale is an early round of token sale, typically held prior to public crowd-sale announcement, for the benefit of strategic investors with considerable investment capacity.

7) Maximum, Circulating and Total Supply

Maximum supply refers to the maximum number of coins or tokens that will ever be created for the given cryptocurrency. For example, the maximum number of Bitcoins is 21 million units.

Circulating supply refers to the coins or tokens circulating in the real world for day-to-day transactions. There are over 17.7 million BTC in circulation as of mid-2019

Total supply refers to the total number of coins or tokens that are either circulating or yet to hit the market. In short, it is the sum of the circulating supply and tokens under lock-up.

8) Token Lock-up

Token lock-up refers to a pre-set time period following the ICO sale, typically 1–2 years, within which the tokens cannot be transacted or traded. The lock-up process is controlled with smart contracts and its strategic purpose is to strengthen the market value of the coin.

9) Hard Cap and Soft Cap

Hard cap is the maximum amount of funds that the ICO plans to raise from investors during the crowd-sale.

Soft cap is the minimum amount of funds the ICO plans to raise from investors to term the venture as success. If the minimum amount is not collected, the funds will be refunded to the investors.

10) Airdrop

Airdrop is distribution of ICO tokens into a wallet for free. It is a marketing strategy to spread the word of the ICO project to the relevant audience in hope of acquiring potential investors. Airdrop could translate to free money if the ICO project succeeds even if no investment is made in the ICO. This is because, the token value of successful ICO projects increases automatically and people who receive free tokens (at the initial stage) can sell the tokens at a later point for a higher value and earn profit.

11) Alpha and Beta (Release)

Alpha is the very first version of the product or platform that is released by the development team to an internal team or small group of users to internally test the system while the developers continue to enhance features and fix issues.

Beta is released after the Alpha release. It is the advanced version of the product, released to the user community to test and obtain feedback.

12) Decentralized Exchange

Decentralized exchange or DEX is an exchange where cryptocurrencies are traded directly from peer to peer using their own wallets, without a company holding the funds or crypto.

13) Market Cap

Market cap or market capitalization refers to market value of the cryptocurrency and is calculated by multiplying the total supply of coins by the latest price.

14) HODL, FOMO and FUD

These are jargon now in common use in the cryptocurrency investment world.

HODL, a typo of the investment term 'hold' originating from bitcointalk, refers to maintaining ownership or holding on to the coins without selling despite fluctuating prices. A person holding on to his coins is referred to as a HODLer. HODL is now used as abbreviation for 'Hold on to Dear Life'.

FOMO is an acronym for 'Fear of Missing Out', and refers to an investor's or customer's fear that he or she may be missing out on a profitable opportunity. This normally occurs when the value of a coin that you do not own starts appreciating.

FUD is an acronym for 'Fear, Uncertainty and Doubt'. It is a marketing strategy to spread fear and insecurity among customers and traders

15) Flipping

Flipping refers to the strategy used by investors to make a fast profit wherein they buy the coins or tokens during the pre-sale or token sale and sell it for a profit and exit within a short time of ICO being listed on the exchange.

8.3 Launching an ICO

Blockchain start-ups look to initial coin offerings as the best financing option mainly because there are limited legal and regulatory hurdles to cross to fund a cryptocurrency business/project. The structure of an ICO can vary based on the type and size of the business, but there are typically five stages to an ICO life-cycle as shown in Fig. 8.2.

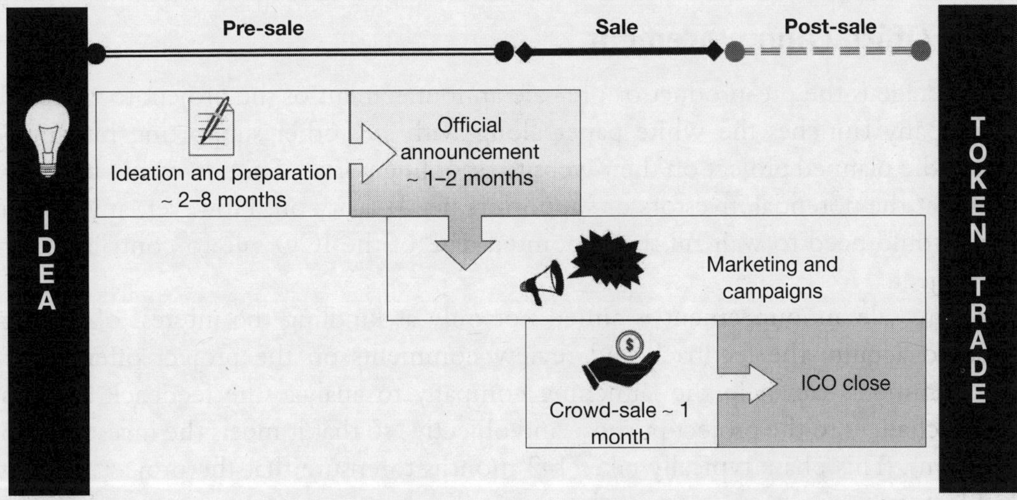

Figure 8.2: Basic stages of an ICO

Stage I: Ideation and Preparation

This stage can be considered as the planning phase of the ICO. It starts with a company or group of individuals or individual, also referred to as **founder**, who has an innovative idea for a technology platform or service that revolves round the blockchain technology and desires to make it a reality. The idea results in a **white paper** detailing the company's strategic plan, vision, business model and expected future growth. The paper is aimed at aiding the "investor" or "supporter" to make a decision towards contributing funds to the project. This includes but is not limited to important information like:

a) Description of the project and its uniqueness
b) The problem or need that the project aims to address
c) Benefits of implementing the project and advantage over any similar existing project
d) Targeted capital – minimum and maximum amount
e) The number of coins/tokens available for sale
f) Form of payment accepted, i.e., whether in fiat, cryptocurrency or any other
g) Percentage of token or crypto that the investor or supporter can benefit
h) Duration of the ICO campaign.

Ideally, at the end of this stage, the company should have a well-written and thought-out white paper, a business model, a qualified product and development team, pre-sale and public sale timeline, and a marketing strategy in place. There is no specific timeline to this stage as it will depend on the readiness of the founder or founding team. But in general, it could take anywhere from 2 and 8 months.

In some cases, the ICO conducts a **private sale** at the end of this phase. It is open to early investors and generally not announced or opened to the public. The expected contributions are higher than the main sale, but the bonus incentives are higher as well.

Stage II: Official Announcement

The next stage is the pre-product or pre-sale announcement of the project to the public. The company launches the white paper along with any other supporting information related to the planned project on their website or online portal. The site would be open to queries that the potential investors or supporters need clarity on. However, any potential investor would need to wait till the appointed date of the ICO sale to contribute funds to the project.

The pre-sale announcement is aimed not only at kindling the interest of investors, but also to acquire the feedback and review comments on the project offering. The review comments can help the launching company to analyze the feedback and make necessary changes to the project purpose and objective so that it meets the investor/public expectations. This phase typically takes 1-2 months to ensure that the project objectives are satisfactory to everyone concerned.

Stage III: Marketing and Campaigns

The key proposition is that if the company takes off and grows large and successful, the tokens will gain value too. The primary focus of the campaigning is to spread the word and reach angel investors who would be interested in promising business ideas. Hence a PR or digital marketing campaigns play a major role in ICO launching as it markets, attracts, and forecasts the present and future of the ICO project through events, social media, forums, etc.

The pre-sale campaign process may last approximately a month, so as to reach both small and big investors and customers from every category. Ideally at the end of the phase, the investors or customers will be ready for the ICO launch and partake in the buying and selling of ICO tokens or cryptocurrency.

If required by the ICO, the Whitelist and KYC process work takes place at this time, along with some pre-sale participation. Pre-sale participants can get good discounts on the tokens.

In addition, there could be an **ICO Bounty** which is a reward program where promoters of the ICO are rewarded in tokens. Bounty programs can include different promotional activities like publicizing the ICO project on social media, YouTube channels, translating white papers into different languages, writing articles on ICO listing websites like CoinSchedule, ICOBench, etc.

Stage IV: ICO Sale

On the set date, tokens are sold in return for investor money. Any investor or supporter can buy tokens from the project launching company in exchange for fiat or cryptocurrency. This event where the company raises funds from investors by issuing tokens is called a **token crowd-sale**. It should be noted that a percentage of the pre-mined tokens or coins will be retained by the founders and shareholders of the company and the rest will be made available to the public for crowd-sale.

Investors send cryptocurrency like BTC or Ether to the company's ICO wallet.

Once the company acquires the minimum amount stated in the proposal, the investors receive an allocation of the company's tokens as specified in the white paper.

The company distributes the tokens to the investors either manually or via smart contracts. The distribution of the tokens will depend on the ICO that one has invested into. The ICO details its distribution structure in its roadmap.

The various distribution structures used in ICO sale is as follows:

a) Limited: As the name suggests, there is a limited number of tokens with a pre-set price. In this distribution method, a predetermined number of tokens are sold on a first-come-first-serve basis till all the tokens are sold. The cap on the number of tokens to be sold means there is a cap on the amount raised. There can be a further cap on the number of tokens that an individual can buy.
b) Unlimited: Here, an unlimited number of tokens are sold at a pre-set price over an extended period of time. Buyers can buy as many tokens as they desire.
c) Pledged amount: Buyers pledge a total spend. A fixed number of tokens are sold to the buyers at a fixed price, in proportion to each buyer's pledged total spend.
d) Auction: Buyers bid on the desired price of tokens and the total spend. The tokens are then sold at the lowest successful bid price in proportional to each buyers' pledged total spend.

Another bidding option used is to bid on the desired price and the quantity of tokens. In this instance, a fixed number of tokens are sold to the bidders in descending price order until all tokens are sold.

This stage normally runs for 30 days. It can close before time if the ICO has achieved the target or the maximum limit of the funds. A percentage of the raised funds, around 5–20%, will be retained by the founders for personal use towards the building of the project.

Stage V: ICO Close

This is the final stage where the ICO is closed after it achieves its target. Once the sale is done, the tokens are listed on the cryptocurrency exchange enabling trade and exchange on secondary markets as well.

The minimum funds required to launch the project is generally specified in the white paper. If the total amount of funds or the value of cryptocurrencies collected through the

ICO sale is equal to or greater than the amount specified, the ICO is deemed successful. If the startup is unable to raise the minimum funds required, the sale is declared unsuccessful and the funds collected are refunded to the investors.

In some cases, if the project exceeds its funding goals, the company keeps the contributions and deploys it to fund future developments or innovations. Eventually the tokens issued will be useful on the software platform that the company is building.

Post the closure, the company issues the tokens or coins to the investors or supporters, provides customer support and post-ICO marketing needs. In addition, the company commences building the platform or development milestones as per the roadmap promised in the white paper.

The value of coins, after successful ICO, is calculated by dividing the total amount of funds raised by the number of coins offered in the project. Once the coins or tokens are issued, they can be traded via public platforms like cryptocurrency exchanges. The coin value can increase or decrease based on the future value of the product or service offered by the company. This enables people to make money by selling coins that were initially purchased at low cost.

Thus ICOs help both in capital fundraising to initialize blockchain projects as well as establishing a whole economy of the cryptocurrency.

8.4 Investing in an ICO

8.4.1 Why Invest in an ICO

It is widely published that over 90% of the ICOs launched have failed to achieve their milestones and failed to make a profit for the investors. This begs the question on why one should invest in ICOs.

The answer lies in the fact that ICOs are investment opportunities just like investing in stocks, government bonds, real-estate, etc., except they are comparatively more risky. However Bitcoin and Ethereum are still strong in the market and many tokens like Ripple, LTC and others have followed suit.

Some main reasons behind investing in an ICO are:

1) ***The profit potential:*** ICOs are high-risk high-reward investments. Early bitcoin investors are millionaires today. If the ICO project is successful, the value of the token increases. With the "buy low, sell high" principle, many ICO investors have stood to gain high return on their investments. Smart investors familiarize themselves with the digital currency market via research site and forums like Bitcoin forum, ICO Drops, ICOStats, Coin Schedule as well as digital exchange sites.

   But just like any wise investment decision, a person should invest only what they can afford to lose. The digital currency market is volatile and ICO/cryptocurrency holds very less value until the token actually hits the market and gains some popularity. Many ICO investment failures can be attributed to inferior product, poor customer base/market demand, or diminished revenue, aside from scams and frauds.

2) ***Solution to a real world issue:*** Some people's interest in backing the ICO may not be purely profit motivated, but rather a belief in the vision of the ICO. Examples include ventures that address real-world issues such as Grid+ for smart energy distribution or BitDegree for building a service-oriented platform that makes quality education accessible to the global community.

3) ***Token rights for usage:*** Tokens provide rights to use the product or platform. People invest in the ICO if they believe in the proposition. The token offers multiple rights to the user like access rights to use the platform, payment rights where the token is the only form of payment within the blockchain network, fee rights where the profit is passed to token holders, governance rights where the token holder can influence the project direction and features, and many others. There can be a combination of rights as well. The token rights should be clearly specified in the white paper.

A 2018 report published in Fortune's Technology newsletter stated that 59% of ICOs launched in 2017 either fully or partially failed. However, around 75% of all startups that are launched with traditional venture funding also fail. Hence, the key question that serious investors ask is not why invest in an ICO, rather which ICO to invest in and what is the right way to invest in them.

8.4.2 How to Invest in an ICO

To participate in an ICO, one needs to have a basic understanding of how to use cryptocurrency wallets and exchanges. There is no guarantee of winning by investing in an ICO even if you have done your homework. This is a risk that every investor has to take. However by taking the necessary precautions, falling prey to a scam or poorly organized ICO can be avoided.

The process may vary from ICO to ICO, but the basic steps to participate in a token sale are as follows:

Step 1 - ICO Identification

Information on upcoming ICOs is gathered from websites that list upcoming and ongoing ICOs like TopICOList, ICOWatchlist, ICOAlert, CoinSchedule and others. This will help to research and plan well in advance. Next, a close study and query on the ICOs are made on community forums like BitcoinTalk, Reddit, and CryptocurrencyTalk to identify the ICO(s) that one would want to invest in. Choosing a reputable company with an established online presence reduces the risk.

It has to be checked if there is a whitelist for the ICO. A whitelist is a listing of investors of the ICO that is guaranteed a token sale. This means that you have to register in advance to participate. There is an advantage to being part of the whitelist as the investor would then have early access to the tokens. In some ICOs, the registration requires the potential investor to go through a KYC process.

Step 2 - Due Diligence

Once the ICO is identified, perform due diligence to ensure the validity of the ICO. Analyze the ICO white paper to learn more about the project and potential for benefits realization. Join the company official blog and other related forums to follow the updates and interaction between the developers and the community. This space can be used to clarify one's own queries.

Sites like CrushCrypto can be used for an independent fundamental analysis of the ICO.

Step 3 – Sign up to an Exchange

Once you have decided on the ICO, you need to purchase digital currency in order to participate in the ICO as ICOs typically do not accept fiat currencies and accept only popular cryptocurrency like BTC and ETH. Since all exchanges require account verification before depositing funds, you must first open a domestic cryptocurrency exchange account that accepts fiat currency and convert your money to cryptocurrency like BTC or ETH.

In instances where the domestic exchange will not accept various coin deposits, you need to open another account in a crypto exchange that supports the ICO coin enabling coin transfer between the two exchanges. You now possess the funds and means to participate in the ICO.

Step 4 – Set up Cryptocurrency Wallet

The next step is to set up your cryptocurrency wallet. Note that it is not safe to store coins on the exchange as the risk of being hacked is high. The best practice is to set up one's own software and hardware wallet. Software wallets like Coinomi and MyEtherWallet (MEW) are convenient if the base currency is Bitcoin or Ether or if the tokens are ERC20 compliant. However, to avoid the risk of hackers, it is advisable to eventually store the coins offline in the hardware wallet. Ledger Nano S and Trezor are popular hardware wallets.

Step 5 – Purchase ICO Tokens

The ICO will generally provide a detailed guide with screenshots on how to participate in the ICO. To participate, transfer the funds from your wallet to the address specified in the ICO website by following the instructions. To ensure that the address has not been tampered with, it is important to follow the news on the ICO on the company website, social media and community sites to remain up to date on the ICO sale and news on attacks. Remember that if the funds have left your wallet to the wrong address, it is lost forever.

Once the token sale ends or within the timeline specified in the white paper, you can collect your tokens from the ICO wallet or the company may send the tokens directly to your wallet.

Step 6 – Secure Your Token

Finally, secure your tokens in your hardware wallet or cold wallet. You must budget for the fees to the miners for wallet-to-wallet transactions and hence retain additional funds for the purpose.

Ideally, the ICO token should get listed in the exchange and appreciate in value. Coins can be held on to for medium to long term until it reaches your target price or it can be flipped for a fast profit as soon as the ICO gets listed on the exchange.

8.4.3 Understanding the White Paper

The white paper is the most important piece of any ICO. It is the single most critical instrument for the potential investor to arrive at an informed decision on whether to fund the project. The document is prepared by the company or project team to give people an understanding of what the project is about and how the team plans on executing it. It is the one-stop shop that should ideally address all the queries related to the ICO project like the product to be developed, the team behind the enterprise, the roadmap, etc. If analyzed correctly, the white paper, in combination with external research, can help minimize the risk of ill-judged investments.

Following are the main aspects of the ICO that the investor must evaluate thoroughly to make an informed decision.

1) Quality of the White Paper

A well-written white paper should be clear, concise and leave no room for conjecture. The content should be precise enough to sustain reader interest. Each topic or section should be to-the-point for the reader to understand quickly and easily rather than being spread out across various sections in the document. The language should be professional with correct spelling and grammar, and at the same time simple enough for a layman to understand. If there are too many technical jargons that seem complex, it should raise a red flag that the project proposition is highly risky or it could be a scam.

In addition, any facts, figures and statistical information provided should be referenced by citing the source or the study done for obtaining the information.

2) Company Information

An ICO backed by a well-established company that is a known leader in the industry will have a better value proposition as compared to a start-up or new entrepreneurial venture. Part of the ICO risk is mitigated if a person can glean information on the company with respect to its ownership, location, subsidiaries, products, years of operation, and other related factors that can be tracked online.

However, the paper should be more issue focused, rather than company focused. A bad idea from a good company could spell disaster for the investor, while a great idea from a new enterprise could easily reap long-term benefits. It helps if the project is endorsed by well-known and reputable organizations or people.

3) Problem, Proposed Solution and Benefits

The problem, the proposed solution and its underlying benefits are the basic building blocks of any white paper. Any paper that does not clearly explain the problem case should be avoided. There should be enough information for the diligent investor to understand whether the company is trying to solve a real-world problem that has a market demand or whether it is just a trumped-up problem that does not have a user case. To leverage the "Blockchain" hype, there are many ICOs that are just building a "decentralized" solution to an existing problem that already has well established centralized solutions which are more feasible or cost effective. The onus is on the investor to not get pulled into projects that have zero to low shelf life.

4) Product or Service Being Offered

Every white paper will have a section on the product that is being offered. However, it is important to analyze whether the product or service being proposed is feasible and clearly addresses the problem at hand. The investor has to check if there is an alpha version or prototype available and if yes, should check on the proposed availability of its results. If no such prototype is in the offing, the timeline, technology used/proposed, target users, benefits to the users, ease of adoption, etc. are all information relevant to the investor. This section should be described clearly and in depth without any ambiguity.

The technical details should be explanatory enough but not too technical as to digress from its intended purpose. The investor needs to keep abreast of the blockchain technology to understand the solution being offered; like whether the product is an entirely new blockchain or built on an established platform like Ethereum. Other technological aspects to be considered are the consensus mechanism, built-in stakeholder incentives, ERC20 standards, open source code, security, etc.

5) Market Competition

It helps to make an informed decision on funding the ICO if one has a clear understanding of the market competition. If there are other solutions, blockchain-based or otherwise, addressing or proposing to address the same/similar problem, studying the aspects that set apart the ICO solution from the rest of the crowd can direct your decision.

6) The ICO Team

Another key aspect of the white paper that the smart investor should study is the core team behind the ICO. Sufficient research must be done to ascertain whether they have the required qualifications and expertise necessary for the project. In addition, an ICO run by a strong leadership or management team can drive the project successfully to completion. The short-term and long-term motivation of the team have to be understood. There should be a good blockchain advisory and marketing team backing the ICO team. Having a strong advisory board or advisor(s) as part of the core team can advance and grow the ICO in the right direction.

A viable ICO must have team members with prior experience in their respective role set in the ICO project. If the team looks in any way incomplete or lacking the required skill-sets, it raises a red flag on whether the team can deliver.

7) Tokenization

The investor should understand whether the token being offered serves as a utility for using the product or service that the company is offering or whether it represents a stake in the company. The investor needs to know the role of the new token in the ICO blockchain ecosystem, i.e., whether the token is a utility, equity, security, combo, etc. Also, questions such as whether the new token is really required or it is just a fundraising tactic, what rights do the token give the holder, details of the token liquidity and its unique edge in the market, growth prospects of the token and the company's plans on sustaining the intrinsic value of the token have to be studied carefully.

8) Token Sale and Distribution

The investor should check the total number of tokens that will be created and released during the various stages of the ICO, i.e., pre-sale and crowd-sale and if there is a whitelist. It is important to check the token distribution among the various players like the founders, developers, early investors, and the public. If the tokens reserved for the core team is too high, a demand for the tokens may inflate the value of the token. The investor needs to check if soft cap and hard cap exists, for liquidity factors, existence of any lockdown period, bonus and incentive distribution and analyze the ICO's overall practicality as an investment option.

9) Development Roadmap

Another aspect to be checked is whether there is a clearly defined roadmap with specific timelines. The prospective investor should be able to assess if the timeline is realistic with achievable milestones. Are token lockdown periods specified and in line with the product delivery? If the development timeframe of the product is too long, it may not be a viable investment.

10) Legal Aspects

It is critical to look for the fine print and study the legal terms and conditions set for the ICO. Understand the country of operation of the ICO and whether there are any country-specific regulations like KYC and AML that the ICO may not have considered. Note that though the ICO is unregulated in most places, it is banned in countries like China, South Korea, Nepal and Bangladesh. However, the overall acceptance of ICOs among regulators is increasing with many countries developing a set of legislative rules around ICOs.

As ICOs, for the most part, fall outside the regulatory purview, it is up to an investor to ensure the legitimacy of the ICO. The investor must exercise a high degree of caution and diligence when researching and analyzing ICOs in order to capitalize on its potential. An ICO having strong innovative idea with a strong engaged community backing it, makes for a strong investment opportunity.

8.5 Pros and Cons of Initial Coin Offering

Just like any fund-raising model, there are advantages and disadvantages in ICOs for both the entrepreneurs and founders as well as for the venture capitalists or investors.

8.5.1 Advantages of an ICO

Following are the positives for the entrepreneurs:

a) Funding Promising Projects

According to ICODATA, ICOs raised over 7.8 billion dollars of capital in 2018. For most entrepreneurs and start-ups, getting the initial funding is the principal issue. Venture capitalists are scarce and may not be willing to pour in money into an undertaking or scheme that is at best a vision of a future success story.

Hence, fundraising via ICOs is quite popular with blockchain start-ups as it gives the companies an opportunity to reach investors who cannot be reached through the traditional route. ICOs act as a platform offering potential investors a deeper understanding of the project, its benefits, roadmap and future growth potential.

Blockchain companies have raised huge capital via ICOs. In 2017, Bancor raised 153 million dollars while Tezos start-up raised over 230 million dollars for sourcing their projects. Ethereum project was sourced via an ICO and is now currently the second largest cryptocurrency and the number one go-to platform for DApp developers. Refer Section 8.6.2 for examples of successful ICOs.

Thus, ICOs are beneficial for startup companies to acquire all the required funding upfront. This enables them to focus all the energy on developing their product or platform to increase in value over time.

b) Faster Processing Speed

The traditional fundraising methods like IPOs, stocks, crowd-funding and other conventional assets require various regulatory documentation, filings and approvals that take time and energy. Time to market is essential for any start-up to achieve its goal and viable ROI. Setting up a company is usually a lengthy process involving huge amount of time and money that a start-up founder cannot afford to spend. More often than not, lack of a specific paperwork can mean the rejection of the company or project.

ICOs offer speed because all it takes for the start-up founder(s) is to issue a white paper to get things started. The white paper released online contains all the vital project details for the potential investors and partners to make an investment decision. Lack of strict government regulations and bureaucratic red tape surrounding ICOs allows for faster execution of projects. In addition, unlike traditional methods that require time-consuming paperwork, the start-ups use blockchain technology where transactions can be updated in mere seconds.

c) Insight on Market Demand

As there is no working model of the product or platform, ICOs can provide a way for founders to access the market demand for their product or service. It helps to evaluate the product's sustainability in the pre-product sale. Based on the market demand and feedback, the founders can also make any project changes if required. In addition, public confidence in the digital end-product of the startup and the team support can drive up the market price of the token.

d) Global Investment Base

ICOs allow for a global investment base. This is an advantage for founders and entrepreneurs from poorly developed venture capital ecosystems that help them reach investors from all over the world. The right marketing of the project can popularize the ICO and expose the project at a global level. The more the exposure gained, the more the number of potential investors in the project. The use of cryptocurrency as investment tokens can attract investors from all economic levels.

Following are the positives for the investors:

a) Open to Public

ICOs allow for anyone in the world to invest, irrespective their socio-economic background or geographical location. Any interested investor can find information on the ICO online. There are dedicated sites that offer information on the various current and upcoming ICOs that allow investors to compare the offerings and opportunities thereof. This gives an opening for anyone to invest in potentially valuable tokens early enough, when they are low-priced or discounted.

Interested parties in a prospective ICO can study the white paper in detail. The white paper is largely non-technical and easy to understand by anyone. As all the information is assessable online, every aspect of information can be easily traced and investigated. Background checks and further research can be done on the project team and company profile through LinkedIn and blockchain community forums.

Thus ICOs have opened the doors to investors from all walks of life who can find the right investment that fits their budget.

b) High Liquidity

Liquidity is the degree to which an asset can be bought or sold in the market without significant impact on its value. In other words, it is the ease at which it can be converted to cash. Cryptocurrencies are relatively more liquid compared to other financial and tangible assets like real-estate, equities, bonds, etc. ICO investments become liquid as soon as the token is listed and can be easily exchanged on a secondary market. As tokens are digital, it does not require a physical form to be exchanged and can be moved faster.

This is attractive to investors as it affords them a high-degree liquidity, provided there is sufficient market demand for the token. The investors can monitor the performance of the company in the secondary market and token price real time.

c) Potential for High ROI

Early blockchain start-ups have provided investors with very high ROI. Some of the most successful ICOs of all time started from a relatively low value and steadily increased their market value. For example, Ether started at $0.311 and now stands over $163. The success of these ICOs can be attributed to the value delivery that included innovative and exciting new features compared to other ICOs.

8.5.2 Drawbacks of an ICO

Though there are many advantages in an ICO, one cannot underestimate its underlying risks. The drawbacks of an ICO that one should be aware of before investing or buying into the concept are as follows:

a) Lack of Transparency and Governance

ICOs are not constrained by the need for regulatory compliance, banks or stock exchanges like traditional fund raising models, thus making it easy for entrepreneurs to initiate an ICO with just a white paper. There are no standards for the white paper. Hence, the information provided in the white paper is left to the discretion of the founder or developer. This could spell downfall as there is no concrete transparency around the company or initiative. The global investment base for ICO would mean that the start-up company and its investors may be in different countries or continents. This would make monitoring near-to-impossible and the investor would have to trust the information or news published by the company. Hence the investor is at a disadvantage if the start-up is fraudulent or fails to meet its project milestones, bringing down the price value of its token.

b) Lack of Accountability

Unlike regular corporations, many start-ups offering ICOs lack true leadership that can steer the company to success. Once the start-ups receive the funding, there is no guarantee that the founders can deliver on their promises. This negatively impacts the investors, who expect a return on their investment.

Lack of accountability is one of the key reasons that ICOs either fail, i.e., they raise the required funding but the project is abandoned and investors refunded; or they go dead, i.e., the required funding is raised but does not get listed on an exchange.

c) Speculative

The main reason for investing in an ICO is the potential future value of the token. However, it is important to understand that putting money into an ICO is not really an "investment", but a "speculation". The investor is betting on the possible future success of an idea or concept. The product, platform or service is yet to be built, unlike say, in IPO where the product or business is already established. Hence fund-raising via ICO is a huge risk as there are no strict regulations or disclosure requirements similar to the IPO business that protects the investors.

Though investors have profited from high-risk high-return investments in the financial market, the digital market is quite volatile. Most ICO investors are not market savvy or blockchain technology savvy and the hype surrounding the ICOs may blind them to the potential risks. A person stands to lose all the money invested if he or she, with due diligence, does not understand the white paper, cryptocurrency trade and inherent risks thereof.

d) ICO Whales

Whales are people or enterprises that hold large amounts of digital currencies and can manipulate the cryptocurrency market as it remains largely unregulated. During an ICO, the "whales" who have the money and resources can buy up most of the tokens. The whales are in it for the gains and will sell their tokens at a premium to make a profit. The whaling monopoly puts other potential investors and participants at a disadvantage in the ICO sale. For example, the BAT ICO that raised 35 million dollars was basically a whale hijack where 50% of the BAT tokens were bought up by just five accounts.

e) Incentive Misalignment

There are many participants in an ICO and include founders, advisors, employees, early investors, ICO investors, future investors, and end-users, among others. It is important to understand the token allocation between the participants, its price over time, incentive for keeping developers motivated, handling of excess tokens, token split pre-sale and post-sale, and fund usage across the project milestones or roadmap. Ideally all key factors related to token economics like price, token allocation, fees, details related to pre-sale, and crowd-sale relevant for investors to calculate the market cap should be listed in the white paper. Unfortunately, that may not be the case with all ICOs.

ICO start-ups usually go through several rounds of fund raising. When the required funding is raised upfront in full and sometimes even in excess, it can potentially cause overspending, lack of motivation and an overall incentive misalignment between the founder and the investor. A well thought-out white paper could be the solution that could address this issue.

f) Frauds and Scams

The very model that makes ICOs popular also makes them susceptible to fraud and malicious intent. Unlike IPOs and other traditional assets that need to undergo strict regulatory processes, the ICO can be initiated with a simple white paper. Due to the hype surrounding the ICOs and cryptocurrency, many people fall easy prey to scammers who take advantage of them with an attractive white paper and professional looking websites. Not every investor is diligent or informed well enough to differentiate between a good ICO from a fraudulent one.

A study paper published by Statis Group in July 2018 showed that around 78% of ICO's raised in 2017 were identified scams. It is easy for a scammer to create a bogus white paper and con the people off their money. Many projects fail or go dead as the founders purposely leave out critical information from the white paper to make it seem more appealing to the investors. This is nothing short of defrauding the public.

The large amount of money involved along with the short time-frame make the ICO a prime target for cybercriminals. Though a blockchain may be tough to hack, websites and digital exchanges are not. According to positive.com, a research and consulting company for cyber security, there are five vulnerabilities common to ICO projects, around 47% of which was identified having medium- to high-severity levels.

The vulnerability attacks identified by positive's anti-fraud team are:

1) ***Against the ICO organizers:*** This involves hacking the organizer's email account, resetting passwords of the ICO domain, replacing wallet address, etc. CoinDash had lost over 7 million dollars in an ICO attack when hackers swapped the CoinDash token sale address with a fraudulent one that was under their control. The whole hacking event took place within a 15-minute time period prior to the public ICO.

2) ***Against the ICO investors:*** 28 % of projects tested by Positive were vulnerable to attacks against the investors. Hackers manipulate the investor psyche using social engineering attack techniques to trick the investor into revealing sensitive information or breaking security measures that will allow an attacker to gain access to the network. In 2018, an attacker successfully stole 150K dollars equivalent from participants of the Experty ICO via breach and phishing attack.

   A very common phishing scam of hackers is using domain names similar to the ICO domain and copying the website design, thus tricking investors to route the investors to the fraudulent wallet.

3) ***In smart contracts:*** This vulnerability occurs with poorly coded smart contracts either due to programmer inefficiency or lack of sufficient source code testing. Once an ICO starts, it is open to all and the contract cannot be amended, making it an easy prey for cyber hackers. The famous 2016 DOA hack allowed an attacker to steal 3.6 million ETH (equivalent to 70 million dollars at the time). The DOA was built as a smart contract on the Ethereum blockchain. The hack was not due to a problem in the Ethereum blockchain, but due to poorly coded smart contract.

4) ***In web applications:*** Half the projects tested contained vulnerabilities in the application itself. This includes security of the blockchain and its backend implementations, code injection, sensitive data exposure, insecure data transfer, and arbitrary file reading among others.

5) ***In mobile applications:*** The ICO mobile apps become vulnerable when investor usage convenience takes precedence over security. Common faults detected were insecure data transfer, user data storage in back-ups and session ID disclosure.

As can be inferred from the pros and cons, the risk impact of a failed ICO is more on the investor than on the founder. Investors who look at ICOs as a formula for a get-rich-quick scheme have more probability of losing than winning. Lack of governance and security structure make ICOs more vulnerable to attack from malicious people.

The biggest consequence of these scams is the decreased faith of the public in blockchain technology when the inherent technology is quite sound and the negative impact is due to surrounding factors. However, many countries and financial institutions are now moving towards identifying ways to regulate and legalize ICOs.

8.6 Successful Initial Coin Offerings

We have seen in the previous section that a majority of ICOs fail to achieve their project objectives or goals post-sale mainly due to overpricing of tokens, overselling or overestimating the utility value and, above all, their vulnerability to scammers and security attacks.

It is estimated that 50% of ICOs fail. However, with innovative ideas and due diligence, the failure rate can be substantially reduced.

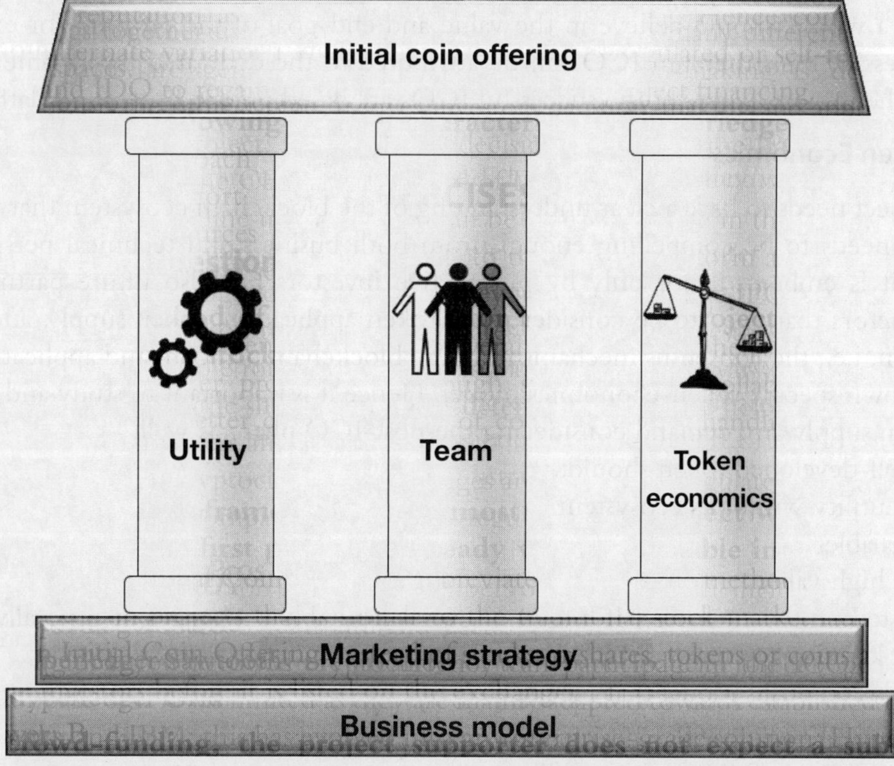

Figure 8.3: ICO success pillars

8.6.1 Success Pillars of ICO

Many factors contribute towards the success of ICOs as illustrated in Fig. 8.3. However, there are three pillars of success, namely Utility, Team and Economics that are essential for any good ICO.

1) Utility

The utility value is perhaps the must-have attribute that can ensure the success of an ICO. It should be the solution to a tangible problem that has a real case or global application. ICOs that have an intrinsic utility attached to their token have a higher rate of success than their counterparts. For example, Bitcoin and Ethereum have continued to increase in value. The value of a token is directly linked to the success of the company. The ICO startup that creates a utility that has sustainable demand ensures not only success in ICO but also in the token market value.

2) Team

The project needs to be supported by a qualified and capable team. Investors will believe in a project when it is backed by an experienced team that is passionate about the new venture and has the requisite skill set and prior knowledge of similar projects or technology. The team, as a whole, should believe in the value and end-goal of the project. For example, the success of the Ethereum ICO can be attributed to the dedicated development team behind the project. Many of the successful ICOs are based off the Ethereum platform.

3) Token Economics

The project needs to have a clear understanding of the blockchain ecosystem that is being built. It needs to be compelling enough from both business and technical perspectives so that it is embraced, not only by prospective investors but also future partners and users. Factors that need to be considered are token application, token supply, and token validation, i.e., the consensus mechanism. Every blockchain platform and application will have its own specific token-economics model. Hence it is important to study and analyze the token supply and demand considering the post-ICO phase as well.

A well-developed token should:

- Have utility within its ecosystem
- Be scalable
- Have high value
- Be listed on the exchange. If listed on several reputable exchanges, it generally means the token is recognized and has market value
- Have a consensus model that incentivizes
- Be resistant to inflation
- Have potential to increase in usage.

Hence, it is important to study and analyse the token supply and demand considering the post-ICO phase as well. There also has to be a fair token distribution model across all stakeholders including but not limited to the founders, administrators, advisors and partners.

Fair financial incentive distribution ensures a well-balanced ecosystem where the participants can build the blockchain and public/users can derive utility on the network. Token pricing, hard and soft caps, token supply, allocation levels, ICO discounts, all need to be diligently considered in token economics.

The pillars with a secure, well thought-out business model, backed by a well-planned and organized marketing strategy can potentially sustain even after the end of the ICO. The right amount of marketing via social media, blockchain forums and campaigns can peak investor interest and website traffic, bringing in the funding required for the ICO.

8.6.2 Examples of Successful ICOs

Given below are examples of few of the successful initial coin offerings in terms of funds raised during the ICO period:

1) NXT

ICO Period: 28 Sept 2013 to 18 November 2013
Coin/Token: NXT
ICO price per token: USD 0.0000168
Funds raised: 21 BTC worth approximately USD 16.8K at the time
Token distribution: 1 billion coins

Nxt is one of the earliest ICOs with a total distribution of 1 billion Nxt tokens. It is an open-source blockchain project launched by an anonymous software developer called *BCNext*. The first blockchain that utilized the proof-of-stake consensus protocol, it was designed as a flexible payment platform for applications and services catering to the financial sector.

From a performance perspective, at its peak Nxt tokens have given investors a return on investment (ROI) of over 1477000%. According to ICO Stats, as of the third quarter 2019, the ROI of Nxt sits at 184000%.

2) Ethereum

ICO Period: 20 July 2014 to 2 September 2014
Coin/Token: ETH
ICO price per token: USD 0.311
Funds raised: USD +15.5 million
Token distribution: 50 million tokens

Ethereum, the distributed blockchain computing platform continues to grow in utility as it functions both as a digital currency, called "Ether", as well as the foundation for decentralized applications by making use of smart contracts. The Ethereum platform, enabled by smart contract technology, has found business project applications across critical sectors such as finance, healthcare, supply chain, energy, government and many more.

Ether is the second-largest digital currency after Bitcoin by market cap in 2018. It is a utility token that sold at $ 0.31 at the time of ICO and valued at over $180 in April 2020. At its peak, the token had an ROI of 200000%. According to ICO Stats, the ROI since ICO as of Q3 2019 is +54500%.

3) NEO (Formerly Antshares)

ICO Period: 01 October 2015 to 10 October 2015
Coin/Token: NEO
ICO Price per Token: USD 0.032
Funds raised: 2100 BTC worth around USD 556.5 K
Token distribution: 17.5 million coins

Nicknamed "China's Ethereum", NEO is considered one of the most successful ICOs. While Phase 1 of the ICO ended in October 2015 earning over half a million dollars, the second phase of the ICO continued till September 2016 raising nearly USD 4.5 million. This was, in part, due to the massive support of the Chinese government and major companies like Microsoft, Alibaba and others.

NEO is a Chinese open-source blockchain supporting C++ and JavaScript. Similar to Ethereum platform, NEO supports the development of decentralized apps (DApps). It utilizes Smart Contract technology that supports common programming languages like C#, JavaScript and other mainstream languages. It includes exciting features like decentralized forms of commerce, digitization of various assets and secure digital identification like facial recognition, fingerprint identification, voice recognition and others.

The NEO ecosystem has two tokens:

i. ***NEO tokens:*** The NEO token gives the owner voting and decision-making rights in the network. There is a maximum supply of 100 million tokens in circulation. NEO is indivisible, i.e., its smallest unit is '1' and cannot be divided into smaller fractions like many other cryptocurrencies. One needs to hold NEO in the NeoWallet to be rewarded with GAS tokens.

ii. ***GAS tokens:*** Similar to Ether of Ethereum, GAS is the fuel that allows one to use the NEO blockchain. It also has a maximum supply of 100 million tokens. GAS is required to run transactions and smart contract operations on the NEO blockchain applications. Unlike NEO, GAS is divisible and its smallest unit is 0.00000001.

NEO token reached an all-time high of 196.85 USD in Jan 2018. According to ICO Stats, NEO's return-on-investment since ICO was +28077% as of Q3 2019.

4) Lisk

ICO Period: 22 February 2016 to 21 March 2016
Coin/Token: LSK
ICO price per token: USD 0.076
Funds raised: USD +6.5 million
Token distribution: 85 million coins

Lisk is a sidechain development platform that enables developers to build and deploy decentralized applications in Javascript. The Lisk App SDK enables the creation of a sidechain that is a fully customizable blockchain including a new custom token if the developer does not want to use LSK tokens from the mainchain.

Lisk is also a cryptocurrency and its token is LSK. It uses dPoS (Delegated Proof of Stake) as its consensus mechanism.

Lisk is also integrated into Microsoft Azure Blockchain creating a BaaS (Blockchain-as-a-Service) solution wherein developers can build and deploy Lisk blockchain applications using Microsoft's Azure cloud computing platform and infrastructure.

The ICO price of the LSK was $0.076 and reached an all-time high of $39.28 in January 2018. According to ICO Stats, Lisk's ROI was +2091% as of Q3 2019.

5) Stratis

ICO Period: 21 June 2016 to 26 July 2016
Coin/Token: STRAT
ICO price per token: USD 0.007
Funds raised: 915 BTC worth approximately USD 611 K at the time
Token distribution: 84 million coins

Stratis, built by a UK based company, is a BaaS platform that allows developers to build and deploy blockchain-based custom applications without the overhead of infrastructure and complex programming. These private blockchains (sidechains) can be assessed by APIs and lite clients. Similar to NEO and Lisk, the platform supports smart contract technology and is compatible with common programming languages like C#, .NET and Java. Stratis is also integrated with Microsoft Azure cloud service.

Stratis is looked at as the go-to solution for many financial institutions. It is also finding user cases in medical research, identity management and others. According to ICO Stats, the ROI of Stratis since ICO is +10714% as of Q3 2019.

Stratis is the obvious choice for financial institutions looking to make a neat profit.

The UK-based startup has craftily created a platform that is compatible with .NET and C# and, as a result, the product appeals to veteran users of Microsoft products. The Stratis platform offers businesses the ability to create, test and deploy custom applications without the need to set up or maintain their own infrastructure; so in a way, it is blockchain for (corporate) dummies.

Few companies lend as much legitimacy as Microsoft, so their ICO had every reason to see success. Raising 915 BTC in 5 weeks, those who cashed in on the low investment of $0.01 per token have seen a titanic ROI of 56,000%.

Aside from the success of the ICOs in terms of funds raised, the common key factor that links their performance is their utility or value. To sustain performance and remain relevant, the cryptocurrencies need to either bring something new or innovate/improve on an existing model. Launching such an ICO where an investor can see potential value and growth is a sure step to its success.

8.7 Evolution of ICO

Ever since first token sale held by MasterCoin in 2013, Initial Coin Offerings have become the most popular alternative to traditional fundraising methods. ICOs enabled entrepreneurs to raise capital directly from the public without being impacted by their personal assets, or lack thereof.

However, the popularity started to wane as it became the target of frauds and other malicious actors. People took to ICOs as a get-rich-quick investment mechanism. Companies took advantage of the people psyche with embellished white papers promising unrealistic returns to investors.

As the percentage of failed ICOs began to rise, it brought with it public awareness and, in turn, a maturity within the blockchain community. Companies started releasing alternative versions of the ICOs that aimed to regain public trust on the fundraising model. This resulted in variations of the ICO model, like STO, IEO and the latest IDO.

8.7.1 ICO Variants

8.7.1.1 Security Token Offering (STO)

Security Token Offering or STO is the most trusted investment tool in cryptocurrency world as it complies with government rules. If ICOs are meant to operate in an unregulated environment, the STO was launched with governance in mind. The ICOs circumvent the need for compliance with regulatory bodies as the token issued is flagged as a utility for usage within the blockchain ecosystem. In the case of STO, the issued security tokens are tied to a real-world asset like stocks, bonds, REIT, etc. In short, it can be considered as a hybrid between an ICO and IPO.

In STO, the ownership information of the asset is recorded on the blockchain and issued to the investor as a security token. Due to the strict regulatory laws, only accredited investors can participate in the token sale. Since they are legally compliant, they are a low-risk investment opportunity for serious investors and can promote blockchain technology firms.

So far, STOs are dominated by the finance and banking sector, e.g. Swarm Fund (2017, raised 5.5 million), Nexo (2018, raised $52 million) and Harbor (2018, raised $38 million).

Advantages of STO

- STO tokens enable fractional ownership of an asset
- Regulated offerings ensure investor security
- More secure than traditional ICOs as token sale participation is limited to only accredited investors
- Less speculative and lower chance of market manipulation
- Project launched by STOs are generally more trustworthy with higher chances of successful completion
- More cost effective than IPOs
- Increased liquidity compared to ICOs and IPOs with 24/7 global access for trading.

Disadvantages of STO

- High cost of launching due to the regulatory pre-requisites and approvals
- Reduced investor pool as only recognized investors can participate
- Process timeline is longer as all transactions needs to adhere to strict KYC (know your customer) and AML (anti-money laundering) requirements
- Liquidity constraints due to tight government regulations.
- High administrative burden brought on by need for regulatory compliance.

8.7.1.2 Initial Exchange Offering (IEO)

Initial Exchange Offering or IEO is the answer to the high-risk environment of the ICO, where participants are at risk of being scammed by a fraudulent company or Ponzi schemes. The fundraising method is the same for both IEO and ICO, i.e., tokens are sold to raise money. The difference lies in the platform. In a traditional ICO, the start-up company is responsible for conducting the ICO sale through their website and, as standard practice, fund transfer is executed via smart contracts. However, in an IEO scenario, a crypto exchange company conducts the token sale on behalf of the start-up via the exchange platform.

The company needs to pay the exchange a listing fee along with a percentage of the tokens sold. The company, in turn, benefits as the exchange takes care of the marketing and the token listing once the IEO is closed. It is easy for the participants as well, as they do not have to deal with multiple wallets on different blockchains.

Binance was the first to launch the IEO platform, Binance Launchpad. Huobi Prime and Coineal Launchpad are other popular IEO platforms. Examples of some IEOs in 2019 are BitTorrent ($7.2M on Binance LaunchPad), Matic ($5M on Binance LaunchPad) and VeriBlock ($7M on Bittrex International).

Advantages of IEO

- Simplified process for start-ups as the launch is executed by the exchange, including most of the marketing. The start-up can also benefit from the existing user base on the exchange
- High degree of trust as the IEO is backed by the exchange as a trusted third-party offering a certain level of credibility and quality assurance to the IEO
- Reduced risk of fraud on the platform due to the regulatory pre-requisites
- Mandatory KYC/AML verification ensures high security to investors
- More flexibility and transparency for customers/traders as there are options for multiple coin selection, including fiat currency.

Disadvantages of IEO

- High cost compared to ICO
- Relatively complex and time consuming to set-up
- Lower liquidity compared to ICOs and STOs
- Limited privacy as the investor has to provide personal and sometimes sensitive information to the exchange
- Exchanges are at a higher risk of hacker attacks.

8.7.1.3 Initial Decentralized Offering (IDO)

IDO or Initial Decentralized Offering or Initial DEX Offering started in 2019 and is practically same as the IEO. The difference between them is that while the IEO is launched on a centralized exchange, the IDO is launched on a decentralized exchange (DEX). DEX allows for direct P2P cryptocurrency transactions between traders on the blockchain without third-party involvement.

Both centralized and decentralized offerings have their advantages and disadvantages. As the centralized exchange is run by a company, privacy is limited. There is risk of hackers and identity theft compared to a DEX where fund or customer information is not stored. DEX only acts as a matching and routing layer for trade orders. This ensures a high level of privacy as transactions are anonymous.

However, this architecture also means that IDO has scalability issues, limited speed and low liquidity compared to IEO. In addition, they face the same concerns raised for ICOs like lack of transparency and lack of legal regulatory control.

In June 2019, Ravencoin IDO was launched on Binance DEX. The IDO successfully raised $ 500K capital for the company.

8.7.2 Regulatory Aspects

Though 2017 saw a huge spike in ICOs, their popularity waned by mid-2018. This was mainly due to increase in people's awareness about blockchain technology, the ICO and the overall hype surrounding both.

ICO advisory firm Statis Group identified abound 80% of the 2017 ICOs to be scams. The year 2018 saw Pincoin ICO scam where the founders Modern Tech raised

$660M from the ICO and disappeared leaving the investors high and dry. Other well-known scams are Plexcoin and Bitcard were investors were cheated out of $15M and $5M respectively. Though ICOs are not regulated, the SEC (the Securities and Exchange Commission) has jurisdiction in the US to intervene if they suspect illegal activities. SEC halted the Plexcoin and froze their accounts. In Oct 2019, SEC obtained a temporary restraining order against Telegram Group Inc. for alleged illegal activity in their ICO.

To date there are limited legal, regulatory and disclosure requirements for launching an ICO. More often than not, the product/platform is only a concept during the fundraising stage. Founders can utilize the contribution funds with minimal to no impact to their personal wealth, while investors have no guarantee on how the project will eventually turn out. In short, ICOs are mostly speculative and their quality varies from project to project.

But with the world-wide benefits of blockchain technology, user cases and digital currency adoption, countries and governments are beginning to accept the intrinsic value of ICOs, provided they can build regulatory frameworks to mitigate risk of frauds like Ponzi schemes, pump-and-dump schemes, etc.

One of the main issues faced by regulators is whether to treat ICOs as a security or a commodity. The ICO token can be treated as a security if it can be used as currency or a store of value that can be used to conduct transactions. However the token can be considered as a commodity if it is just a utility token that is used purely for the purpose of access and participation in the platform. The confusion arises as in initial stages of the ICO, the token acts as a security token while post-ICO, it becomes a utility token.

Since 2017, there have been discussions and strategies within governments and financial institutions on regulating and legalizing ICOs.

- In 2017, U.S. Securities and Exchange Commission (SEC) started applying federal security laws to ICOs on a case-to-case basis. The Commodity Futures Trading Commission (CFTC) also considered defining Bitcoin and other crypto assets as commodities.
- ICOs are not regulated in countries like India, but they are not strictly considered illegal either. RBI, Securities Exchange Board of India (SEBI), Income Tax authorities and other institutions are considering a regulatory regime governing ICOs. Similarly, governments of Pakistan and Morocco too discourage the use of ICOs.
- China has embraced blockchain technology. However ICOs are banned in China as well as in countries like South Korea, Nepal, Bangladesh, Macedonia, Algeria and Bolivia.
- France, Russia, Switzerland and Singapore have allowed usage with strict regulations while countries like UK, Canada, UAE, Australia, and New Zealand are working on regulating them.

Until the government standards come into play, a few trustworthy companies follow self-imposed standards or guidelines in order to boost investor confidence. Some of the guidelines followed are:

a) Make KYC and AML procedures mandatory
b) Limit the total tokens produced to 100 million for healthy economy
c) Sell tokens in batches to enable many investors to participate as against token monopoly by limited investors
d) Use ERC20 standards in token creation
e) Reserve < 20% tokens for founders and 50–60% for investors
f) Reserve 20–25% of the balance tokens for legal fees, platform maintenance, security and other essential needs to mitigate risk of malicious actors. Investor security is generally prioritized over other segments.
g) Special benefits like investor involvement in the project development process and investment refund if the progress is unsatisfactory.

As organizations continue to raise hundreds and millions of dollars in ICOs, it is increasingly important for industry leaders and policy makers to understand the economic landscape of cryptocurrency and ICO and form regulatory governance around the ICO space.

8.8 ICO Platforms

8.8.1 ICO Launching Platforms

Developing your own ICO is a complicated process. An ICO launch platform is a beneficial tool that can help founders and entrepreneurs with every aspect of their ICO launch. These platforms significantly reduce the time and costs associated with an ICO launch. ICO service providers like Blockchain App Factory, IBC Group and Bacancy Technology, among others, provide tailor-made solutions and expertise to start-ups or entrepreneurs to develop and launch compliant ICO/STO tokens. A few blockchain companies now provide end-to-end turnkey solutions where start-ups can launch their own ICOs without the need for coding or advanced programming knowledge. They generally provide a simple user-friendly interface and features for setting up crypto-assets, creating smart contracts, pre/post sale processes and campaigns.

8.8.1.1 CoinFactory

Developed by Accubits Technologies, the CoinFactory platform launched in 2016 has a proven track record of conducting several successful ICOs, having raised over $340 million with 27 ICOs as of Jan 2020. CoinFactory platform is a self-hosted, end-to-end platform, compliant with SEC and FCA, to issue, manage and trade utility and security token offerings. It supports a wide range of essential and customizable options for launching ICOs and STOs The platform provides three packages for hosting, namely, self-hosted (application can be hosted on your own server for greater control and security), snippet (run the crowdsale on your website by adding CoinFactory JavaScript snippet), and finally the API model (where CoinFactory provides APIs for the user's own token transactions) that the entrepreneur or start-up can choose from. The user interface is intuitive and friendly to investors to enable them peruse and contribute to an ICO of choice and also provides a platform for post ICO investor management.

A key feature is the Milestone Contract where milestones are created and, when published, investors get to see their contributions held by an intermediary smart contract. The controlled distribution of funds based on the milestones is transparent to the investors, thus providing protection against scams and failed ICOs. The CoinFactory platform has built-in marketing tools like Referral programs, Bonus Programs, and Airdrop programs. It supports over 14 languages and has built-in features to support regulatory compliance, integrated KYC, two-factor authentication, user management, pre-defined milestone templates, whitelisting, community management, admin dashboard, investor dashboard, custom smart contracts and many other features making it the best-in-class ICO launch platform.

The Platform accepts over 10+ cryptocurrencies including fiat and supports Ethereum, Stellar, EOS and Tezos blockchains.

8.8.1.2 TokenGet

Founded in 2017, TokenGet is a secure ICO and STO turnkey platform. It has helped launch 50+ projects, and helped to raise over $900 million. Its key focus is on STO issuance platform with global regulatory compliance integration, 24/7 global market access, and smart contracts for issuer and investor rights. TokenGet offers smart contract generator, and marketing tools like referral program, bounty program, automated airdrop and a marketing module where the ROI of the marketing campaigns can be monitored. Other features include:

- ICO management module where all the stages of the ICO/STO can be set up from pre-sale, crowd-sale and post-sale.
- Token sale configuration
- Integrated system for KYC and AML checks
- Customizable investor dashboard
- Multiple token support on Ethereum, EOS, NEO and Stellar
- PCI DSS compliance, real-time monitoring, security against Distributed Denial of Service (DDoS) and other security features.

8.8.1.3 BlockStarter

BlockStarter (ticker ZBS; Total supply 100,000,000 ZBS) is a new ICO launch platform for Ethereum-based token sales. The aim of BlockStarter is to automatize the entire ICO process. BlockStarter boasts of an advanced UI that is simple and intuitive that enables entrepreneurs with basic technical knowledge to list their campaign and launch a crowd-sale. The solution consists of Ethereum smart contracts and a customizable white-label webapp called Tokensale Dashboard.

Its features include a campaign builder, ICO database, smart contract generator and wallet. The main smart contracts enable customized ERC20 token, token sale contract, whitelist manager and bonus rule setup. Blockstarter has a built-in token calculator and accepts multiple cryptocurrencies like ETH, BTC, LTC, BCH, DASH, DOGE. Blockstarter does not store the private keys to the addresses for the contributions; thus adding a level of security from hackers.

It is currently being tested with live customers and the platform is expected to launch in 2020. CoinLaunch, Coral and ICOBox are other ICO launch platforms in the making.

8.8.2 ICO Listing Platforms

With ICOs being typically unregulated, investors need a trustworthy place where they can study and analyze the ICO market. ICO listing websites address the need by offering an overview of the ICO landscape including details of each individual ICO/STO project, thus providing a one-shop-stop to track all token offerings. The listing sites do not provide investment advice. The onus is on the investor to analyze the information provided and research further to ascertain the accuracy of the details and decide on the investment opportunity. Start-ups can woo potential investors by listing their projects on the listing websites. Some of the popular listing platforms are provided below.

8.8.2.1 ICO Drops

ICO Drops is an independent ICO review and rating website that provides a calendar list of upcoming, active and ended ICOs. ICO Drops measures the quality of an ICO by an interest level rating system for investors based on three factors, namely, hype rate (analysis based on the size of the genuinely interested audience in chats, forums, social network, etc.), risk level (based on the % token to raise funds, the hard and soft cap, pre-sale outcomes, and other related factors) and ROI level (factor relies on the ICO performance and conditions of similar projects in retrospect). According to the outcome of these measurements, ICO Drops rate the interest of the ICOs from Low to Very high.

The webpage design is simple and easy to read with minimal and relevant information. The main page displays all three lists of Active, Upcoming and Ended ICOs. It also shows the category of the ICO, i.e., whether it is a blockchain service, an exchange, a blockchain platform, etc., whether it is rated or not, as also the date the ICO and the funding goal. For active and ended ICOs, the webpage also shows the funds raised in dollars and the percentage of hard cap amount reached.

Links are provided for each individual project, where brief information of the project is displayed. Links to the ICO project website, token sale period, the ICO fundraising goal, ICO token price, the white paper, the KYC form and Whitelist required if applicable, currencies accepted, terms and conditions and social links among others are also provided.

There are links to Twitter and Telegram page for the latest ICO news, updates and alerts. ICO calendar can be downloaded to the user's device to mark the dates of upcoming ICO events. Subscribers get weekly ICO plans sent to their email address.

The percentage of available tokens for sale and current fundraising numbers are displayed making it easy to determine whether the project is about to end or not.

8.8.2.2 ICOBench

Founded in 2107, ICOBench offers listings for ICOs, IEOs and STO. The projects are categorized by industry types such as retail, real-estate, tourism, etc. as well as by technology e.g. big data, platform, cryptocurrency, smart contract, etc.

Each individual ICO/STO/IEO provides information on the project, the team, milestones, KYC requirements, financials (token price, bonus, currency accepted, etc.), ICOBench ratings and white paper.

ICOBench is popular for its ratings and review on each ICO. Ratings are done on a daily basis and each factor is rated on a scale of 1 to 5 with 5 being the best. There are three kinds of ratings done:

1) Through an assessment algorithm done by a bot 'Benchy' based on 20 different criteria. Benchy evaluates the ICO profile by analyzing key factors like the team, ICO information, product presentation (white paper, milestones, goals, etc.) and marketing and social media presence.
2) By the community who are experts in their specific field. The experts rate the ICO based on given guidelines across three factors – the team, the vision, the product.
3) In some instances, the ICO project is evaluated by an ICOBench of legal experts who provide a legal review.

ICOBench provides information on the start and end date of each ICO/IEO/STO, the ratings, with filtering options on token types (utility, security, payment), bounty available, bonus available, ICO KYC passed, free tokens, expert reviews conducted, etc. This makes it an excellent tool for investors who are looking for specific investment opportunities that are rated by experts.

8.8.2.3 CoinSchedule

CoinSchedule is another popular ICO listing website launched in 2016. It has all the basic requirements of a good listing site and includes a list of all upcoming and ongoing ICOs, ICO event calendar, information on each project related to the milestones, white paper, ICO website, sale start and end period, team information, token offerings, statistics of funds raised, tokens sold, etc.

CoinSchedule categorizes its ratings across Exchanges, Coins, ICOs, IEOs, Stablecoins and Airdrops. CoinSchedule has two types of rating:

1) **CSRating**: A proprietary algorithm that rates Exchanges based on several variables that include but are not limited to web visitors, audience volume consistency, origin, and other factors.
2) **Crowdscore**: Score given by CoinSchedule users.

There is a TrustScore given to each ICO project. TrustScore is a proprietary algorithm, immune to manipulation, which uses artificial intelligence to combine a lot of data points like company verification, team member(s) verification, white paper, interviews and many other data points to arrive at the score. The scores are on a scale of A (Excellent) to E (Low). Details of the score are provided to subscribers only.

ICOAlert, ICORating, TokenMarket, Top ICO List, and ICO Watchlist are other popular ICO listing sites.

Summary

Initial Coin Offering or ICO is a popular fundraising practice for blockchain projects. It has similarities to the traditional fundraising methods for new ventures like IPOs and crowd-funding. The key difference between IPOs and ICOs is that while investors in the former practice are offered company shares for their investment, in ICOs tokens or altcoins are offered before they are listed on the market exchange. The specific fundraising process in the ICO is referred to as crowd-sale.

Investors are keen to participate in ICOs as they see it as an investment opportunity to buy cryptocurrencies while they are at their lowest price value and sell high. However, cryptocurrency is a volatile market and there is no guarantee that the token or coin will appreciate in value. Many cryptocurrencies have traded lower than their ICO price once they hit the market.

There can be multiple rounds of fundraising that occur during a single ICO. In some instances, select investors are offered cryptocurrencies prior to the public crowd-sale. This practice is called pre-sale. The pre-sale is conducted to buy-in support from influential and wealthy investors. The pre-sale comes with additional bonus or benefits.

ICOs are launched by blockchain startups, typically for raising funds to build a blockchain product or platform that is still in concept stage. The investor needs to research the people and team leading and executing the project, their past experience and knowledge and feasibility of the concept for mass adoption to make an informed decision. This type of ICO forms the highest risk to the investor as there is very little guarantee that the project can deliver on its promises. When the team behind the project has a successful track record of delivering similar projects, the risk is reduced.

The advantage of ICOs is that they are not regulated and hence very easy to set up and execute. The duration of most ICOs is typically 6 months to less than a year. However, some ICOs span across years as the blockchain company continues to raise funds to build the blockchain product further.

An ICO consists of three main processes: the pre-sale, the actual sale (crowd-sale) and post-sale. The activities within each process can vary based on the company, product being built and other factors. However, the basic activities are common. An ICO starts with an idea or concept of a potential blockchain product or application taking shape. This is formalized as a white paper detailing the company's strategic plan, vision, the user case, details of the technology that will be used, the team, the sale duration, roadmap and business model including the expected future growth of the investment. The white paper aims to attract potential investors and supporters. The white paper is followed by the official announcement of the crowd-sale date. Publicity and marketing campaigns are conducted to create hype around the ICO. Once the crowd-sale is successfully completed, the tokens are distributed to the investors as promised. The new cryptocurrency is listed in the cryptocurrency exchange for open trade.

ICO can be regarded as successful once its soft cap is reached. Soft cap is the predetermined minimal amount required for the project to move forward. Soft cap is determined during the planning stage, along with the hard cap which is the maximum amount expected. If an ICO fails to reach its soft cap, the ICO is considered failed and all contributions are returned to investors.

An investor needs to perform due diligence on the white paper and conduct in-depth research and background checks on all information provided on the website, blockchain forums, social media platforms and other community sites to minimize the risk of unwise investments and exposure to phishing sites and other potential scams.

The act of participating in an ICO is relatively simple. The investor sends funds to the public address defined when the sale starts and receives tokens in return after the sale ends. The investor has to be aware of the steps required for participating as defined by the particular ICO. The investor also needs to get information on pre-requisites, if any, like KYC or whitelist, the wallet and currencies accepted and any other relevant details as cryptocurrency transactions are irreversible, and any money sent to the wrong address is lost forever.

As ICOs are unregulated, they are more prone to scams and frauds, bringing a bad name to an otherwise cutting-edge technology that has potential for immense growth. It has a bad reputation as a get-rich-quick investment tool. Hence companies have introduced alternate variations of the ICO that are more regulated or self-regulated like STO, IEO and IDO to regain public trust in blockchain project financing.

EXERCISES

Multiple Choice Questions

1. **This is the method of raising funds for blockchain projects that is similar to the traditional stock market.**

   A. KPO

   B. ICO

   C. BPO

   D. Crowd funding

   Answer: B

   Explanation: Initial Coin Offering, abbreviated ICO, is the method of raising funds for blockchain projects that is similar to the traditional stock market launch or the IPO. In Initial Coin Offering, instead of stocks or shares, tokens or coins are offered to the investors before it is listed on the exchange.

2. **In crowd-funding, the project supporter does not expect a substantial return on investment.**

   A. True

B. False

Answer: A

Explanation: Initial Coin Offering or ICO is a method of crowd-funding projects or new ventures in the crypto industry. However, the concept in ICO is slightly different from crowd-funding. While in crowd-funding the backer of the project does not expect a substantial return on investment, the backer of an ICO however, is motivated by profit. ICO fundraising is known by the term "crowd-sale" so as not to confuse it with crowd-funding.

3. **This is the practice of funding an undertaking or venture by raising small amounts of money from a large number of people interested in investing money in different companies and projects:**

   A. KPO

   B. ICO

   C. BPO

   D. Crowd-funding

   Answer: D

   Explanation: Crowd-funding is the practice of funding an undertaking or venture by raising small amounts of money from a large number of people interested in investing money in different companies and projects they want to support, typically via the Internet. As per the global statistic portal Statista, the transaction value in crowd-funding segment came to USD 6.9 billion in 2019 and is expected to reach nearly USD 12 billion by 2023.

4. **This is the most popular crowd-funding platform:**

   A. Google map

   B. AWS

   C. Kickstarter

   D. Amazon

   Answer: C

   Explanation: One of the most popular crowd-funding platforms is Kickstarter that has nearly 159,000 successfully funded projects. Started in 2009, it currently has over USD 4 billion pledged to kickstarter projects that span across films, arts, publishing, technology and others.

5. **Which of the following is not a crowd-funding platform?**

   A. Patreon

   B. Google map

   C. Kickstarter

   D. Republic.

   Answer: B

   Explanation: One of the most popular crowd-funding platforms is Kickstarter that has nearly 159,000 successfully funded projects. In addition, popular crowd-funding platforms for startups are StartSomeGood, Indiegogo, Patreon and Republic.

6. **Which of the following is not part of the workflow of ICO?**
   A. Fund flow
   B. Token issuance
   C. Crypto exchange
   D. Fiat currency issuance
   Answer: D
   Explanation: The ICO model can be simplified to three basic workflows, namely, 1. Fund flow 2. Token issuance 3. Crypto exchange.
7. **In this workflow, the company organizing the ICO will convert the bitcoin/ether to USD/fiat via Exchange to pay for the project development, salary payments, marketing, etc.**
   A. Fund flow
   B. Token issuance
   C. Crypto exchange
   D. Fiat currency issuance
   Answer: A
   Explanation: In fund flow, a person interested in investing in the ICO buys popular cryptocurrency like Bitcoin or Ether with USD/EUR or any other accepted fiat currency. The investor sends the bitcoin/ether either to a smart contract or special wallet that stores the funds typically until the end of crowd-sale. Finally, the company organizing the ICO will convert the bitcoin/ether to USD/fiat via Exchange to pay for the project development, salary payments, marketing, etc.
8. **In this step, the company organizing the ICO will distribute an equivalent value of new tokens/coins to the investors based on their individual fund contribution.**
   A. Fund flow
   B. Token issuance
   C. Crypto exchange
   D. Fiat currency issuance
   Answer: B
   Explanation: After the crowd-sale, the next step for the company organizing the ICO is Token Issuance, in which the company will distribute an equivalent value of new tokens/coins to the investors based on their individual fund contribution. Investors can utilize the tokens to participate in the project ecosystem.
9. **In this step, investors can sell their tokens for money or BTC/ETH. Public users who wish to access or use the project platform or service can purchase the necessary tokens via the exchange.**
   A. Fund flow
   B. Token issuance
   C. Crypto exchange
   D. Fiat currency issuance

Answer: C

Explanation: Crypto exchange happens once the ICO ends. The company lists the token on a cryptocurrency exchange. This would require standard information like the coin name, symbol, coin price, total quantity, etc. Once it is listed, tokens can be traded. Investors can sell their tokens for money or BTC/ETH. Public users who wish to access or use the project platform or service can purchase the necessary tokens via the exchange.

10. **The ICO tokens will sell a right of ownership or royalties to a project. The token design, usage rights and rewards are built into the token during creation.**

    A. True

    B. False

    Answer: A

    Explanation: The ICO tokens will sell a right of ownership or royalties to a project. The token design, usage rights and rewards are built into the token during creation. Early investors generally have greater rewards embedded within their tokens as an incentive.

11. **This is the name of an official document, prepared by blockchain project company or entrepreneurs to enable the reader to make an informed decision on whether to participate in the ICO.**

    A. Profit and loss Statement

    B. A white paper

    C. Company declaration

    D. Bank statement

    Answer: B

    Explanation: A white paper is an official document prepared by blockchain project company or entrepreneurs to enable the reader to make an informed decision on whether to participate in the ICO. It details everything that a potential investor would need to know about the blockchain project, including but not limited to the company details, and vision, objective and description of the project.

12. **The maximum number of Bitcoins that will be supplied is 21 million units and it will not exceed this number. This is an example of:**

    A. Maximum supply

    B. Total supply

    C. Market supply

    D. Cap limit supply

    Answer: A

    Explanation: Maximum supply refers to the maximum number of coins or tokens that will ever be created for the given cryptocurrency. E.g. the maximum number of Bitcoins is 21 million units.

13. There are over 17.7 million BTC in circulation as of mid-2019. This is an example of:

A. Circulating supply
B. Total supply
C. Market supply
D. Cap limit supply
Answer: A

Explanation: Circulating supply refers to the coins or tokens circulating in the real world for day-to-day transactions. There are over 17.7 million BTC in circulation as of mid-2019.

14. This is the minimum amount of funds the ICO plans to raise from investors to term it as a success; else the funds will be refunded to the investors.

A. Mid cap
B. Min cap
C. Soft cap
D. Hard cap
Answer: C

Explanation: Soft cap is the minimum amount of funds the ICO plans to raise from investors to term it as a success; else the funds will be refunded to the investors.

15. This is the maximum amount of funds that the ICO plans to raise from investors during the crowd-sale.

A. Max cap
B. Total cap
C. Soft cap
D. Hard cap
Answer: D

Explanation: Hard cap is the maximum amount of funds that the ICO plans to raise from investors during the crowd-sale.

16. This is a marketing strategy, to spread the word of the ICO project to the relevant audience in hope of acquiring potential investors. It is a distribution of ICO tokens into a wallet for free.

A. Airdrop
B. Marketdrop
C. Wallet drop
D. Free drop
Answer: A

Explanation: Airdrop is distribution of ICO tokens into a wallet for free. It is a marketing strategy to spread the word of the ICO project to the relevant audience in hope of acquiring potential investors. Airdrop could translate to free money if the ICO project succeeds even, if you do not invest in the ICO.

17. **This refers to maintaining ownership or holding on to the coins without selling despite fluctuating prices.**
    A. FOMO
    B. HODL
    C. FUD
    D. Airdrop
    Answer: B
    Explanation: HODL is a jargon in common use in the cryptocurrency investment world. It is a typo of the investment term 'hold' originating from bitcointalk and refers to maintaining ownership or holding on to the coins without selling despite fluctuating prices. A person holding on to his coins is referred to as a HODLer. HODL is now used as abbreviation for "Hold on to Dear Life".

18. **This refers to an investor's or customer's fear that he or she may be missing out on a profitable opportunity. This is normally occurs when the value of a coin that you don't own starts appreciating.**
    A. FOMO
    B. FODL
    C. FUD
    D. Airdrop
    Answer: A
    Explanation: FOMO is an acronym for "Fear of Missing Out" and refers to an investor's or customer's fear that he or she may be missing out on a profitable opportunity. This normally occurs when the value of a coin that you don't own starts appreciating.

19. **This is a marketing strategy to spread fear and insecurity among customers and traders.**
    A. FOMO
    B. FODL
    C. FUD
    D. Airdrop
    Answer: C
    Explanation: FUD, an acronym for "Fear, Uncertainty and Doubt", is a marketing strategy to spread fear and insecurity among customers and traders.

20. **This refers to the strategy used by investors to make a fast profit wherein they buy the coins or tokens during the pre-sale or token sale and sell it for a profit and exit out within a short time of ICO being listed on the exchange.**
    A. FOMO
    B. FODL
    C. FUD

D. Flipping
Answer: D
Explanation: Flipping refers to the strategy used by investors to make a fast profit wherein they buy the coins or tokens during the pre-sale or token sale and sell it for a profit and exit within a short time of ICO being listed on the exchange.

21. In this distribution method, a pre-determined number of tokens are sold on a first-come-first-serve basis until all the tokens are sold.
A. FOMO
B. Limited
C. FUD
D. Flipping
Answer: B
Explanation: In limited distribution method, there is a limited number of tokens with a pre-set price. A predetermined number of tokens are sold on a first-come-first-serve basis until all the tokens are sold. The cap on the number of tokens to be sold means there is a cap on the amount raised. There can be a further cap on the number of tokens that an individual can buy.

22. These are people or enterprises who hold large amounts of digital currencies and can manipulate the cryptocurrency market as it remains largely unregulated. During an ICO, they buy up most of the tokens.
A. ICO Whales
B. ICO tigers
C. ICO Fish
D. ICO Giant
Answer: A
Explanation: ICO Whales are people or enterprises who hold large amounts of digital currencies and can manipulate the cryptocurrency market as it remains largely unregulated. During an ICO, the "whales" who have the money and resources can buy up most of the tokens. The whales are in it for the gains and will sell their tokens at a premium to make a profit. The whaling monopoly puts other potential investors and participants at a disadvantage in the ICO sale. For example, the BAT ICO that raised 35 million dollars was basically a whale hijack where 50% of the BAT tokens were bought up by just 5 accounts.

23. This the most trusted investment tool in cryptocurrency world as it complies with government rules. This was launched with governance in mind.
A. ICO
B. IPO
C. STO
D. IEO

Answer: C

Explanation: Security Token Offering or STO is the most trusted investment tool in cryptocurrency world as it complies with government rules. If ICOs are meant to operate in an unregulated environment, the STO was launched with governance in mind. The ICOs circumvent the need for compliance with regulatory bodies as the token issued is flagged as a utility for usage within the blockchain ecosystem. In the case of STO, security tokens are issued that are tied to real-world assets like stocks, bonds, REIT, etc. In short, it can be considered as a hybrid between an ICO and IPO.

24. **This is the answer to the high-risk environment of the ICO, where participants are at risk of being scammed by a fraudulent company or Ponzi schemes. A crypto exchange company conducts the token sale on behalf of the start-up via the exchange platform.**
A. RICO
B. IPO
C. STO
D. IEO
Answer: D
Explanation: Initial Exchange Offering or IEO is the answer to the high-risk environment of the ICO, where participants are at risk of being scammed by a fraudulent company or Ponzi schemes. The fundraising method is the same for both IEO and ICO, i.e., tokens are sold to raise money. The difference lies in the platform. In a traditional ICO, the start-up company is responsible for conducting the ICO sale through their website and as standard practice, fund transfer is executed via smart contracts. However in an IEO scenario, a crypto exchange company conducts the token sale on behalf of the start-up via the exchange platform.

25. **This is launched on a decentralized exchange (DEX). DEX allows for direct P2P cryptocurrency transactions between traders on the blockchain without third-party involvement.**
A. IDO
B. IPO
C. STO
D. IEO
Answer: A
Explanation: IDO or Initial Decentralized Offering or Initial DEX Offering started in 2019 and is practically same as the IEO. The difference between them is that while the IEO is launched on a centralized exchange, the IDO is launched on a decentralized exchange (DEX). DEX allows for direct P2P cryptocurrency transactions between traders on the blockchain without third-party involvement.

Short-answer Questions

1. **What is ICO?**

   Initial Coin Offering, abbreviated ICO, is the method of raising funds for blockchain projects that is similar to the traditional stock market launch or the IPO. In Initial Coin Offering, instead of stocks or shares, tokens or coins are offered to the investors before it is listed on the exchange. ICO took the world by storm in 2017–2018 when a blockchain startup Block.One raised a record-breaking 4.1 billion dollars in 1 year. ICOs are among the biggest financial innovations to spring from Blockchain Technology for raising business venture capital.

2. **Distinguish between crowd-funding and ICO.**

   Initial Coin Offering or ICO is a method of crowd-funding projects or new ventures in the crypto industry. However, the concept in ICO is slightly different from crowd-funding. While in crowd-funding the backer of the project is not expecting a substantial return on investment, the backer of an ICO, however, is motivated by profit. ICO fundraising is known by the term "crowd-sale" so as not to confuse it with crowd-funding.

3. **What is crowd-funding?**

   Crowd-funding is the practice of funding an undertaking or venture by raising small amounts of money from a large number of people interested in investing money to different companies and projects they want to support, typically via the internet. As per the global statistic portal Statista, the transaction value in crowd-funding segment amounted to USD 6.9 billion in 2019 and is expected to reach nearly USD 12 billion by 2023.

4. **Write notes on Kickstarter project.**

   One of the most popular crowd-funding platforms is Kickstarter that has nearly 159,000 successfully funded projects. Started in 2009, Kickstarter currently has over USD 4 billion pledged to projects that span across films, arts, publishing, technology and others.

5. **Name a few popular crowd-funding platforms for startups.**

   Popular crowd-funding platforms for startups are StartSomeGood, Indiegogo, Patreon and Republic.

6. **What are the three basic workflows of ICO model?**

   The ICO model can be simplified to three basic workflows: They are 1. Fund flow 2. Token issuance and 3. Crypto exchange.

7. **Write notes on fund flow of ICO model.**

   Fund flow: A person interested in investing in the ICO buys popular cryptocurrency like Bitcoin or Ether with USD/EUR or any other accepted fiat currency. The investor sends the bitcoin/ether either to a smart contract or special wallet that stores the funds typically till the end of crowdsale. Finally, the company organizing the ICO will convert the bitcoin/ether to USD/fiat via Exchange to pay for the project development, salary payments, marketing, etc.

8. Write notes on Token Issuance workflow of ICO Model.

Token issuance: After the crowdsale, the company organizing the ICO will distribute an equivalent value of new tokens/coins to the investors based on their individual fund contribution. Investors can utilize the tokens to participate in the project ecosystem.

9. Write notes on Crypto Exchange workflow of ICO Model.

Crypto exchange: Once the ICO ends, the company lists the token on a cryptocurrency exchange. This would require standard information like the coin name, symbol, coin price, total quantity, etc. Once it is listed, tokens can be traded. Investors can sell their tokens for money or BTC/ETH. Public users who wish to access or use the project platform or service can purchase the necessary tokens via the exchange.

10. What are Token Rights in ICO?

Token rights: The ICO tokens will provide users the right of ownership or royalties to a project. The token design, usage rights and rewards are built into the token during creation. Early investors generally have greater rewards embedded within their tokens as an incentive.

11. What is Investor Rights in ICO?

Investor rights: Early investors (generally pre-sale investors) have voting and decision-making rights on the project. They can provide inputs on the direction of the project. However the investors participating during the crowd-sale do not have any voting or decision-making rights.

12. What is ICO Token?

A token is a unit of value that is created by the company to govern the ICO ecosystem, enabling users to interact with the product or platform and facilitating the distribution of rewards and benefits to all of its stakeholders. Utility token, security token, etc., are examples.

13. What is white paper in ICO?

A white paper is an official document prepared by blockchain project company or entrepreneurs to enable the reader to make an informed decision on whether to participate in the ICO. It details everything that a potential investor would need to know about the blockchain project, including but not limited to the company details, vision and objective of the project, description of the project and its core team, product description, technological aspects, market analysis, business plan, token and its distribution, and guides the investor on how to participate in the ICO providing timelines and terms and conditions.

14. Write notes on Maximum, Circulating and Total Supply in ICO.

Maximum Supply refers to the maximum number of coins or tokens that will ever be created for the given cryptocurrency. For example, the maximum number of Bitcoins is 21 million units.

Circulating Supply refers to the coins or tokens circulating in the real world for day-to-day transactions. There are over 17.7 million BTC in circulation as of mid-2019

Total supply refers to the total number of coins or tokens that are either circulating or yet to hit the market. In short it is the sum of the Circulating Supply and tokens under lock-up.

15. What is Token Lock-up?

Token Lock-up refers to a pre-set time period following the ICO sale, typically 1–2 years, within which the tokens cannot be transacted or traded. The lock-up process is controlled with smart contracts with a strategic purpose of strengthening the market value of the coin.

16. What is Hard Cap and Soft Cap in ICO?

Hard Cap is the maximum amount of funds that the ICO plans to raise from investors during the crowdsale.

Soft Cap is the minimum amount of funds the ICO plans to raise from investors to term it as a success; else the funds will be refunded to the investors.

17. What is Airdrop?

Airdrop is distribution of ICO tokens into a wallet for free. It is a marketing strategy to spread the word of the ICO project to the relevant audience in hope of acquiring potential investors. Airdrop could translate to free money if the ICO project succeeds even if you do not invest in the ICO.

18. What is HODL, FOMO and FUD in Cryptocurrency?

These are jargons now in common use in the cryptocurrency investment world.

HODL, a typo of the investment term 'hold' originating from bitcointalk, refers to maintaining ownership or holding on to the coins despite fluctuating prices. A person holding on to his coins is referred to as a HODLer. HODL is now used as abbreviation for "Hold on to Dear Life".

FOMO is an acronym for "Fear of Missing Out". It refers to an investor's or customer's fear that he or she may be missing out on a profitable opportunity. This normally occurs when the value of a coin that the investor does not own starts appreciating.

FUD is an acronym for "Fear, Uncertainty and Doubt" and is a marketing strategy to spread fear and insecurity among customers and traders.

19. What is Flipping in ICO?

Flipping refers to the strategy used by investors to make a fast profit wherein they buy the coins or tokens during the pre-sale or token sale and sell it for a profit and exit within a short time of ICO being listed on the exchange.

20. What is Limited Structure in ICO Sale?

As the name suggests, in limited structure, there is a limited number of tokens with a pre-set price. In this distribution method, a predetermined number of tokens are sold on a first-come-first-serve basis until all the tokens are sold. The cap on the number of tokens to be sold means there is a cap on the amount raised. There can be a further cap on the number of tokens that an individual can buy.

21. Write notes on the profit potential in ICO.

ICOs are high-risk high-reward investments. Early bitcoin investors are millionaires today. If the ICO project is successful, the value of the token increases. With the "buy low, sell high" principle, many ICO investors have stood to gain high return on their investments. Smart investors familiarize themselves with the digital currency market via research site and forums like Bitcoin forum, ICO Drops, ICOStats, Coin Schedule as well as digital exchange sites before buying the offers.

22. How is ICO a solution to a real-world issue?

Some people's interest in backing the ICO may not be purely profit motivated, but based on their belief in the vision of the ICO on a real-world issue that needs to be addressed, e.g., Grid+ for smart energy distribution or for building a service-oriented platform like BitDegree that makes quality education accessible to the global community.

23. What do you mean by the Quality of the White Paper?

A well-written white paper should be clear, concise and leave no room for conjecture. The content should be precise enough to sustain reader interest. Each topic or section should be to-the-point for the reader to understand quickly and easily rather than being spread out across various sections in the document. The language should be professional with correct spelling and grammar, and at the same time simple enough for a layman to understand. If there are too many technical jargons that seem complex, it should raise a red flag that the project proposition is highly risky or it could be a scam.

24. What is tokenization?

The investor should understand whether the token being offered serves as a product utility or service that the company is offering or whether it represents a stake in the company. The investor has to understand the role of the new token in the ICO blockchain ecosystem, that is, whether it is a utility, equity, security, combo, etc. He or she should be able to judge if the new token is really required or just a fundraising tactic. The investor also has to understand the rights given by the token to the holder and its liquidity. Also, an indication of the token's edge in the market in terms of growth prospects and the company's plan to sustain the token's intrinsic value would be of added advantage.

25. Write notes on lack of transparency and governance in ICO offerings.

ICOs are not constrained by the need for regulatory compliance, banks or stock exchanges like traditional fund raising models, thus making it easy for entrepreneurs to initiate an ICO with just a white paper. There are no standards for the white paper. Hence, the information provided in the white paper is left to the discretion of the founder or developer. This could spell downfall as there is no concrete transparency around the company or initiative. The global investment base for ICO would mean that the start-up company and its investors may be in different countries or continents. This would make monitoring near-to-impossible and the investor would have to trust the information or news published by the company. Hence, the investor is at a disadvantage if the start-up is fraudulent or fails to meet its project milestones, bringing down the price value of its token.

26. Write notes on lack of accountability in ICO.

Unlike regular corporations, many start-ups offering ICOs lack true leadership that can steer the company to success. Once the start-ups receive the funding, there is no guarantee that the founders can deliver on their promises. This negatively impacts the investors who expect a return on their investment. Lack of accountability is one of the key reasons that ICOs either fail, i.e., they raise the required funding but the project is abandoned and investors refunded; or they go dead, i.e., the required funding is raised but does not get listed on an exchange.

27. Write notes on speculation in ICO Offerings.

The main reason for investing in an ICO is the potential future value of the token. However, it is important to understand that putting money into an ICO is not really an "investment", but a "speculation". The investor is betting on the possible future success of an idea or concept. The product, platform or service is yet to be built, unlike say, in IPO where the product or business is already established. Hence fund-raising via ICO is a huge risk as there are no strict regulations or disclosure requirements similar to the IPO business that protects the investors.

28. Who are ICO whales?

Whales are people or enterprises that hold large amounts of digital currencies and can manipulate the cryptocurrency market as it remains largely unregulated. During an ICO, the "whales" who have the money and resources can buy up most of the tokens. The whales are in it for the gains and will sell their tokens at a premium to make a profit. The whaling monopoly puts other potential investors and participants at a disadvantage in the ICO sale. For example, the BAT ICO that raised 35 million dollars was basically a whale hijack where 50% of the BAT tokens were bought up by just 5 accounts.

29. Write notes on incentive misalignment in ICO.

There are many participants in an ICO, including founders, advisors, employees, early investors, ICO investors, future investors, and end-users among others. It is important to understand the token allocation between the participants, its price over time, incentive for keeping developers motivated, handling of excess tokens, token split pre-sale and post-sale, and fund usage across the project milestones or roadmap. Ideally all key factors related to token economics like price, token allocation, fees, and details related to pre-sale and crowd-sale relevant for investors to calculate the market cap should be listed in the white paper. Unfortunately that may not be the case with all ICOs.

30. How do utility values contribute to the success of ICO?

The utility value is perhaps the must-have attribute that can ensure the success of an ICO. It should be the solution to a tangible problem that has a real-case or global application. ICOs that have an intrinsic utility attached to their token have a higher rate of success than their counterparts. For example, Bitcoin and Ethereum have continued to increase in value. The value of a token is directly linked to the success of the company. The ICO startup that creates a utility that has sustainable demand ensures not only success in ICO but also in the token market value.

31. How does token economics contribute to the success of ICO?

The project needs to provide a clear understanding of the blockchain ecosystem that is being built. It needs to be compelling enough from both business and technical perspective that it is embraced by not only prospective investors but also future partners and users. Factors that need to be considered are token application, token supply, and token validation, i.e., the consensus mechanism. Every blockchain platform and application will have its own specific token-economics model. Hence, it is important to study and analyze the token supply and demand considering the post-ICO phase as well.

32. What are the characteristics of a well-developed token?

A well-developed token should:

- Have utility within its ecosystem
- Be scalable
- Have high value
- Be listed on the exchange. If listed on several reputable exchanges, it generally means the token is recognized and has market value
- Have a consensus model that incentivizes
- Be resistant to inflation
- Have potential to increase in usage.

33. What is STO?

Security Token Offering or STO is the most trusted investment tool in cryptocurrency world as it complies with government rules. If ICOs are meant to operate in an unregulated environment, the STO was launched with governance in mind. The ICOs circumvent the need for compliance with regulatory bodies as the token issued is flagged as a utility for usage within the blockchain ecosystem. In the case of STO, the security tokens issued are tied to real-world assets like stocks, bonds, REIT, etc. In short, it can be considered as a hybrid between an ICO and IPO.

34. What is IEO?

Initial Exchange Offering or IEO is the answer to the high-risk environment of the ICO, where participants are at risk of being scammed by a fraudulent company or Ponzi schemes. The fundraising method is the same for both IEO and ICO, i.e., tokens are sold to raise money. The difference lies in the platform. In a traditional ICO, the start-up company is responsible for conducting the ICO sale through their website and, as standard practice, fund transfer is executed via smart contracts. However in an IEO scenario, a crypto exchange company conducts the token sale on behalf of the start-up via the exchange platform.

35. What is IDO?

IDO or Initial Decentralized Offering or Initial DEX Offering started in 2019 and is practically same as the IEO. The difference between them is that while the IEO is launched on a centralized exchange, the IDO is launched on a decentralized exchange (DEX). DEX allows for direct P2P cryptocurrency transactions between traders on the blockchain without third-party involvement.

Essay-type Questions

1. Explain the basic process of ICO fundraising mechanism through a diagram along with its key features.
2. What are the key differences between ICO and IPO?
3. What are the pros and cons of ICO and IPO?
4. Explain various steps for launching an ICO.
5. Explain the various distribution structures used in ICO sale.
6. What are the main reasons behind investing in an ICO?
7. What are the basic steps to participate in a token sale?
8. What are the aspects of the ICO that the investor must evaluate thoroughly through white papers to make an informed decision?
9. What are the pros and cons of ICO?
10. Write the details of vulnerability attacks identified by Positive's anti-fraud team on ICO offerings.
11. Write an essay on the various success pillars of ICO.
12. Write an essay on various examples of successful ICOs.
13. Write an essay on STO in detail.
14. Write an essay on IEO in detail.
15. Write an essay on IDO in detail.
16. Write an essay on the various regulatory aspects of ICO.

CHAPTER 9

Security in Blockchain

LEARNING OBJECTIVES

Data breach in companies can cause considerable hardships to consumers and companies. Just as security is a concern for all individuals like us, we also believe that the right to remain anonymous is equally essential and applicable for blockchain also. This chapter aims to educate readers on the subject of security aspects in blockchain, which include security and privacy challenges, performance and scalability, identity management and authentication, regulatory compliance and assurance apart from security aspects in Ethereum and Hyperledger.

9.1 Introduction

Individual security is important in our day-to-day lives. People want to feel secure; hence, they make efforts to ensure that their lives, assets and reputation are not compromised. We spend huge amounts on hiring security guards and installing anti-theft devices in our automobiles or buildings. We install a firewall to secure the computer while connecting networks. Today, as we write this book, big data has started to play prominent roles in the lives of ordinary people. People are generating more data, and large corporations are making use of those data. While significant data-related activities are legitimate and aimed at enhancing the quality of life for humankind (for example, data maps collected during floods in Kerala (India) that helped in rescues), there can also be data breaches that can cause considerable hardships to consumers and companies.

Blockchain has created many buzzes in the academia and industry. There are a few blockchain projects such as interconnecting settlements in the financial sector and telecom that are a huge success. However, other uses for blockchain, like general public access to the network, like electronic voting, are still a distant dream due to various concerns related to security and privacy aspects, along with its scalability and performance issues. Blockchain security focuses on protecting and safeguarding computer systems (connected to blockchain networks), blockchain networks, blockchain programs (code), and data from unintended or unauthorized access.

In a traditional security model, a central authority manages access to control and privileges. The transaction contains the user's private identifiers, e.g., social security numbers, credit card numbers, aadhar numbers, etc., which are susceptible to attacks. The identity of parties (like social security numbers, credit card numbers, aadhar numbers, etc.) is embedded as part of the stored data. The rest of the data are susceptible to tampering if a hacker gains access to it. On the contrary, in a blockchain security model, access control and privileges are either decentralized or pushed to the end-user by entrusting them with the management of private/public keys. Transactions authorize just a particular key value from a specific user, which cannot be modified or forged. The key value will not disclose any personal information, like the identities of those parties, and cannot be utilized to authorize additional transactions.

Blockchain, though promises to be secure, could be challenged by smart cryptographers. Let us discuss some of the security aspects of blockchain in this chapter.

a. Security aspects in bitcoin
b. Security and privacy challenges in blockchain
c. Performance and scalability in blockchain
d. Identity management and authentication in blockchain
e. Regulatory compliance and assurance in blockchain
f. Security aspects in ethereum
g. Security aspects in hyperledger.

9.2 Security Aspects in Bitcoin

Bitcoin, the predecessor for blockchain technology, had a long unbeaten run in the last decade. However, experts concede that as hardware and computing power increases, its security mechanism could be under stress. Bitcoin is facing challenges via cartelization, where miners form a pool. It is worthwhile to note that bitcoin was devised to be an anonymous, trustless mode (refer Fig. 9.1). However, this feature is under stress. Bitcoin, though it raises a challenge to the banking systems worldwide, does not have a secrecy feature. The banking software has better secure settings and policies than bitcoin. Bitcoin is a public network, ensures all transactions are available for all users; preferably, transactions are recorded in every node of the bitcoin network. This is unsafe compared to banking applications. As bitcoin transactions are visible to all members, it is possible that anonymity may be compromised when people choose to exchange bitcoin into local currency.

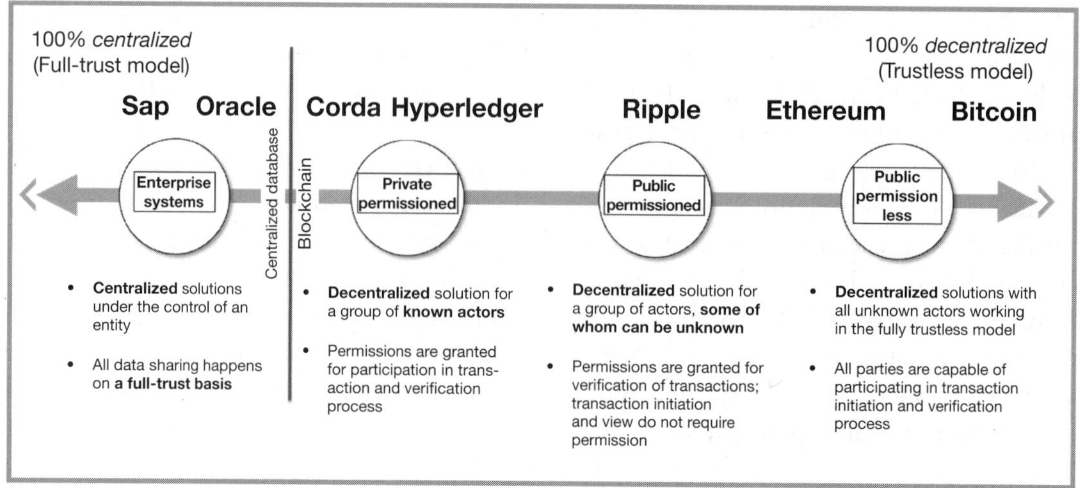

Figure 9.1: Security aspects in bitcoin vs. others

9.2.1 Key and Signature Mechanism in Bitcoin

However, security is enabled in bitcoin through its key and signature mechanisms. When a user joins a bitcoin network, a private key is generated (a user creates a private key). A public key is made from the private key, using cryptographic algorithms. This is because cryptographically generated public keys cannot be used to trace the user back to his private key. The public key for the user is available throughout the network (refer Fig. 9.2). Sometimes, unauthorized people may try to find the private key by attacking the public key, and the anonymity function may not always be followed. A signature is a number that proves that signing operations took place. A signature needs two inputs – private key and public key. A signature can be compared to a hash.

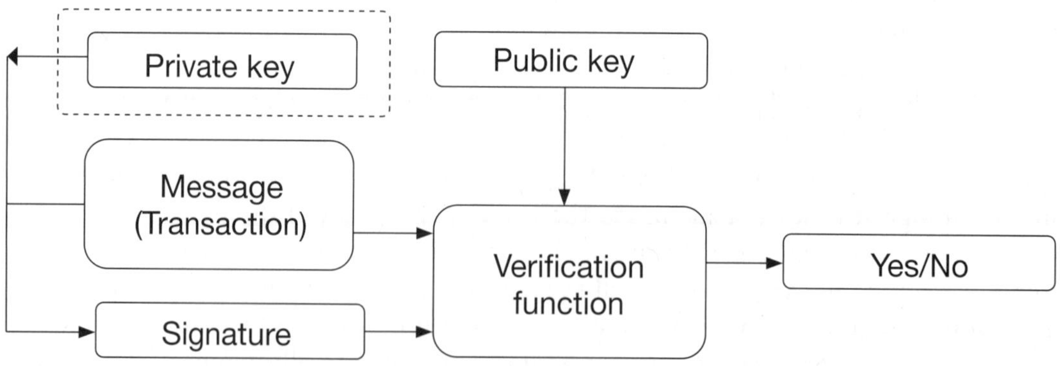

Figure 9.2: Private and public key verification

Digital signature algorithm (DSA) is an asymmetric key cryptography system, where a private key is used to sign messages digitally, and the corresponding public key may be used to verify those messages. The elliptic curve variant of DSA called as ECDSA (elliptic curve digital signature algorithm), is commonly used in blockchain implementations. This offers several advantages, such as reduced key size and faster computation as compared to other discrete logarithm-based algorithms and factoring modulus algorithms. Bitcoin currently uses ECDSA. It is embedded into the bitcoin software. However, with the advancement of computing power, it is possible that this cryptographic algorithm could be broken in a few years.

Today, smart hackers are busy trying to find out the private key holder by reverse-engineering the public key or signature. Also, it is worth noting that IP address of the user-entered network can leave a trail towards identifying a person. Also, an exchange account for exchanging coins earned/gifted in local currencies compromises anonymity. However, for an attacker to be able to successfully interfere with the bitcoin network, block, and reverse transactions, he should have the support of the majority (51%) of the node's support. The probability of rolling back a transaction decreases exponentially with the number of confirmations it has received. Bitcoin also uses SHA 256 encryption for both its proof-of-work (PoW) system and transaction verification.

9.3 Security and Privacy Challenges of Blockchain in General

Security can be defined as actions that safeguard the data and prevent unauthorized access to those data. We usually put various security controls to restrict access to the stored data/information. Privacy is about the safeguarding (not revealing to outsiders) user identity. Although the meaning of security and privacy overlap with each other, it is possible to have security without privacy, but impossible to have privacy without security. Security needs to ensure that all the transactions are valid before they are committed to the ledger. Once it is committed to the ledger, it is immutable, and no single participant can unilaterally modify the contents of the ledger.

9.3.1 Majority Attack

In majority attack, more than 51% of the nodes get corrupted and start sending wrong signals to other nodes. As per blockchain protocol, majority wins. Some of the assumptions that guarantee the terms of security and privacy are that not more than 50 percent of the computing nodes have been compromised (meaning that at least 50 percent of the nodes are following the defined protocol) ensuring safety and security.

9.3.2 Deterministic Transactions

All transactions running on blockchain have to be deterministic (refer Fig. 9.3). That means, the result should be the same irrespective of who transacts or where the transaction is executed. In ethereum, it is achieved by reducing the set of operations that can be performed.

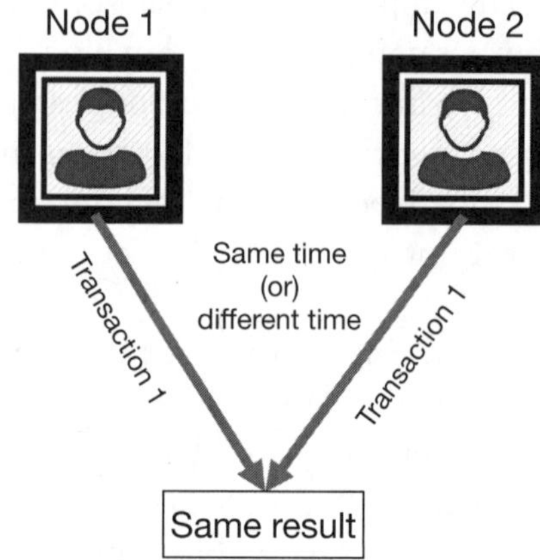

Figure 9.3: Deterministic transaction

These assumptions may not always hold, and there have been many attacks that happened on these open networks. As an example, among the most usual cases is where both user wallets and the decentralized trades have been compromised. In a certain way, secret keys have been stolen to embezzle bitcoins. Other types of cryptocurrencies have also faced a similar type of issue, since the ledger immutability gets compromised if nodes behave in a notably different way. For instance, there is a recent research article, which states that even if just 25% of the network is compromised, it can cause bad things to the network. Ethereum purports that code is the law. There is no single governing body or legal resource that governs cryptocurrency transactions. Similarly, there is no legal framework to fall back on to actually prove the theft, if it occurs. There are hard forks generated in the blockchain. Individuals have to undo transactions on the blockchain, which goes contrary to the fundamentals of the blockchain itself. It is supposed to become immutable and never susceptible to modification.

9.3.3 Inherent Security Attributes of Blockchain (CIAR)

CIAR (refer Fig. 9.4) stands for Confidentiality, Integrity, Availability, and Non-repudiation.

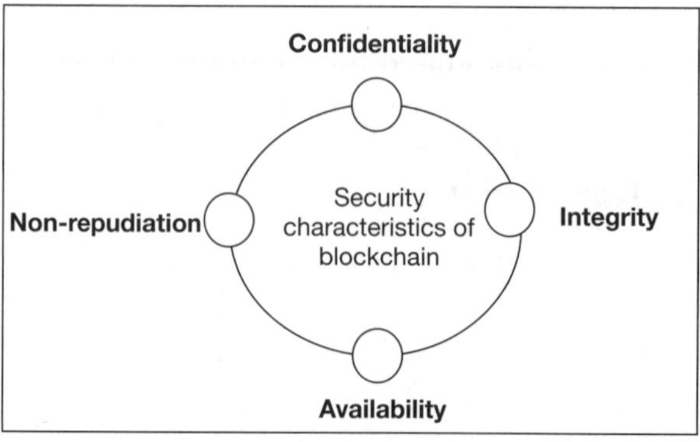

Figure 9.4: CIAR aspect of security

Confidentiality: It gives assurance that information is disclosed only to the relevant authorized parties (refer Fig. 9.5). Confidentiality denotes that any sensitive information should only be shared with a limited number of authorized people.

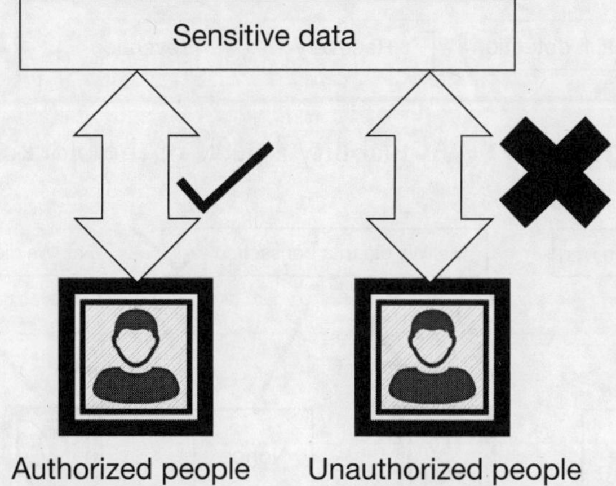

Figure 9.5: Confidentiality of sensitive data

For example, if credit card information and transaction details were shared with a few other users (say hackers), in addition to money theft, your credit rating and reputation could also be compromised very quickly. The transactions executed by a user with a bank are considered secure only if authorized users have access to it; of course, supervisors within the bank would also have some access to it with reasonable restrictions governed by the laws in the country.

Integrity: Assurance for the accuracy and completeness of data (refer Fig. 9.6). Integrity involves keeping information from being altered. It refers to the reliability and trustworthiness of data. It consists of the maintenance and assurance of the accuracy and consistency of data over its entire life cycle. If a transaction detail is updated in a single node, integrity ensures that the same data is updated in all other nodes.

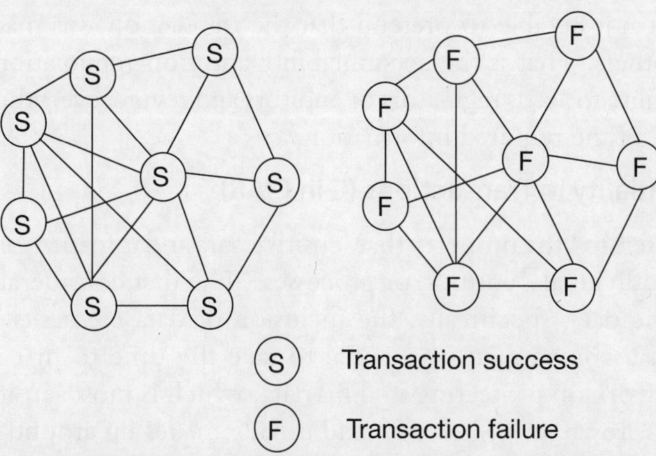

Figure 9.6: Integrity of blockchain nodes

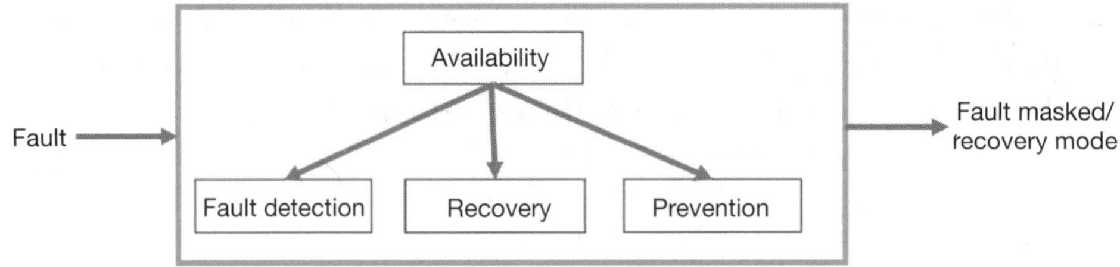

Figure 9.7: Availability aspects of the blockchain

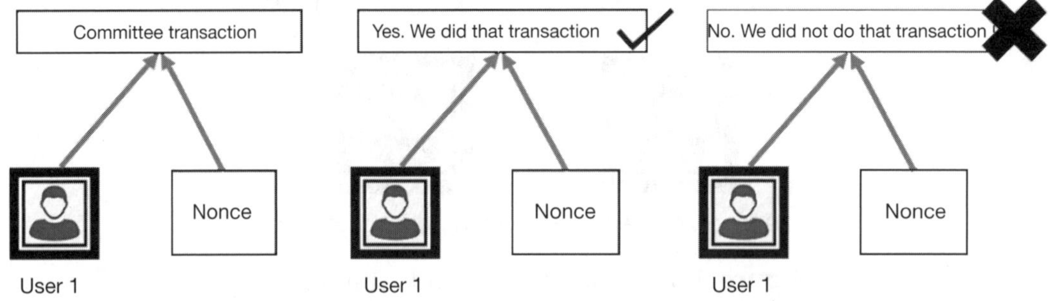

Figure 9.8: Non-repudiation aspects of the blockchain

Availability: Assurance that the system will be available for use when it is required (refer Fig. 9.7). It involves ensuring that those who rely on accurate and truthful information can gain access to the data without hassles. Availability is linked to ethics. However, it can also involve matters such as a cyber-attack, which prevents people from accessing specific servers, or from getting into the blockchain network.

Non-repudiation: Assurance that users who participate in a transaction (refer Fig. 9.8) will not be able to deny their participation at a later point. Once information is present in the system, we need to have the ability to prove that this person and the nonce produced this information.

The user must not be able to pretend that the transaction was made by some other person or many others. That is the accountability and non-repudiation aspect. All users should have the ability to act as regulators or auditors and review everything in the network or a select portion of the required information.

9.3.3.1 Confidentiality in Transactions (C in CIAR)

Confidentiality refers to "the property that sensitive data and information are not disclosed to unauthorized individuals, entities, or processes." The first consideration would be the classification of the data; specifically, the inclusion of data elements that require high confidentiality. As a company, it is essential to take the time to first identify the most valuable data and work on protecting it. This data, which is most sensitive, is commonly referred to as the "crown jewels" of data and usually makes up around 10 percent of the overall data. If this data is compromised, it would cause severe damage to the company.

A subset of crown jewels data would include Personal Identifiable Information (PII). Public and consortium blockchains may not be suitable for sharing data across parties when the data needs to remain confidential to those parties. Consortiums could rely on off-chain compensating controls to address confidentiality issues, as part of consortium governance. Public blockchains currently lack the commands required to provide forward-looking confidentiality requirements. The use of blockchain technology for storing PII is strongly not recommended, as it is not likely to comply with privacy legislation.

After identifying the crown jewels data, procedures have to be set up to not only secure the data but also limit the number of nodes that have access to it. An organization takes inventory of what information every node may or may not have access to. Nodes that need access to sensitive data and those which do not participate in decision-making are identified. The nodes with access to sensitive information are limited to a small, manageable number. Besides, the admins need to determine which type of access each node needs (refer Fig. 9.9). Data privacy defines who has authorized access to permission.

Develop a data security plan/policy: Data protection is all about the mechanism, i.e., the procedures and tools. The data protection system focuses mainly on keeping the information away from hackers. Data protection secures the data from unauthorized access by unknown entities. Data privacy standards are invoked for legal situations. Data privacy regulations or policies govern the use of data shared with other entities. The controls of data privacy are usually given to the end-users. The end-users can control and manage the data shared with other individuals. An important strategy when looking to improve data security is to develop a data security policy. This policy should also be open to edits and changes, as amendments need to be designed to match the developing technological inventions and new company coverages.

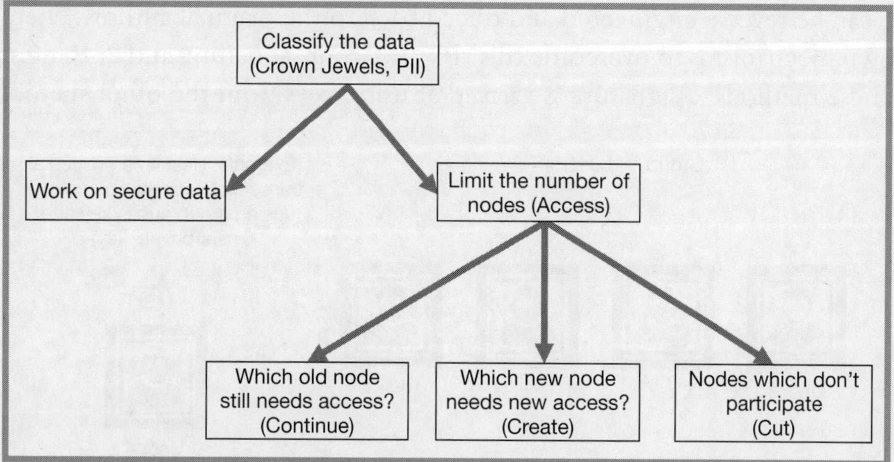

Figure 9.9: Confidentiality aspects of the blockchain

To ensure confidentiality, transactions must not be linked to the user. There are few methods using which this can be achieved:

a. Mixing/bartering transactions
b. Anonymous signatures
c. Advanced encryption techniques.

Mixing/bartering transactions: The blockchain, bitcoin, in particular, allows mixing/bartering coins as a way of ensuring that transactions recorded into the chain are not linked to the user's public key. The system does a mix of coins, or swaps coins or submits a joint transaction. This ensures that the transaction recorded in the block is not linked to the user and reduces the probability of relating transactions to a particular user. Mixing or joining transactions into a single transaction ensures complete anonymity. Mixing brings in anonymity like traditional techniques and maintains a registry to ensure that only valid members of the group are allowed to mix; this is to ensure that members do not take advantage of fellow members in the course of mining or validating transactions.

Meanwhile, joining transactions will have two or more users join hands to make a single transaction; this ensures that outside agencies cannot link an individual user to the committed transaction. CoinJoin, coin for joining, allows the joining of transactions; users are allowed to choose other users to enter a transaction, and negotiate with other users to join a transaction. However, a server within CoinJoin keeps a list of users who have joined to form joint transactions. This solution brings the coin closer to anonymity; however, it also brings in a centralized server, which is against the concept of a distributed ledger. A centralized server adds a risk that it could publish (or hack) the list of users who formed joint transactions.

Anonymous signatures: Digital signatures are the footprint in any electronic system. Bitcoin and other blockchains rely on them heavily; however, it is sometimes found that a signature can be reverse engineered and traced back to the original author. There are two well-known mechanisms to overcome this risk: 1. Group signature, and 2. Ring signature. In both these methods, a signature is arrived at using keys from the other members.

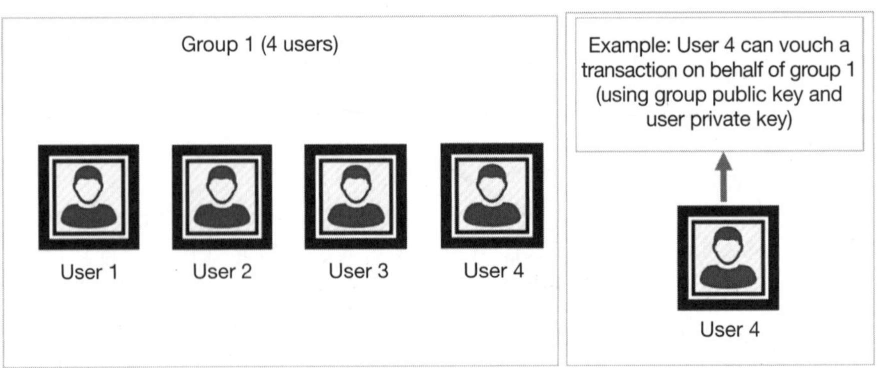

Figure 9.10: Group signature

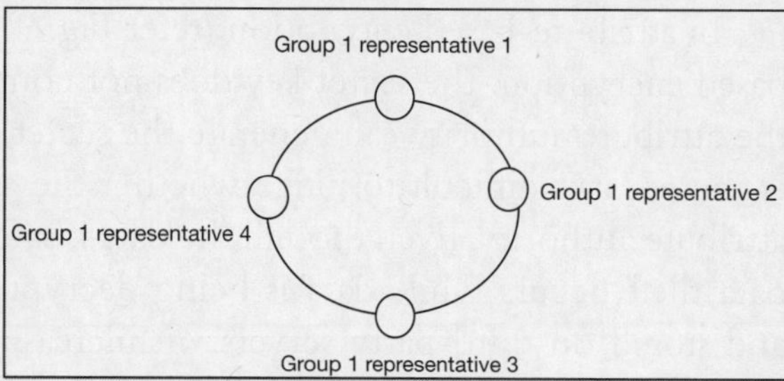

Figure 9.11: Ring signature

In group signature (refer Fig. 9.10), a member can sign using his private key, and using the group's public key. This ensures that a user's identity is not revealed. However, the users' group has to be revealed.

In the case of ring signature (refer Fig. 9.11), signing is done by a group of people and the user's identity is not shown. Only the group to which the signature belongs is exposed.

Advanced encryption techniques: Bitcoin uses SHA 256 encryption for both its proof-of-Work (PoW) system and transaction verification. You can also apply application-level encryption to provide privacy guarantees on who can see the data. Two advanced encryption-based techniques, namely, attribute-based encryption and homomorphic encryption, are used for security purpose.

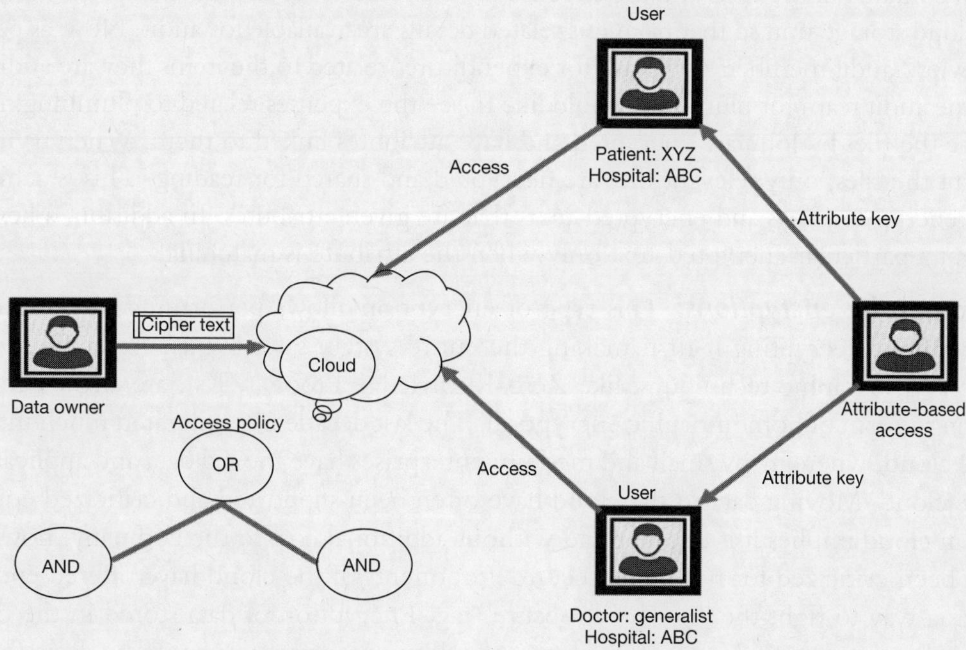

Figure 9.12: Attribute-based encryption

In attribute-based encryption (refer Fig. 9.12), especially cipher text-policy attribute-based encryption, the secret key does not contain the specific information of users. Also, the attribute authority can generate the secret key from any attribute set. If the secret key is abused, it is difficult to judge whether the abused private key comes from users or the attribute authority. Even though the data is safe in its encrypted form, unfortunately, most data theft occurs while data is being decrypted for use. As more information is shared and stored on third-party servers, an increasing number of data protection schemes are now being used. One way of fighting and aiding in data protection is to use encryption. Encryption ensures that data stays secure even when it moves to the wrong hands and intruders cannot make the most of the data. However, the purpose of data is to be read and to read the information you need to decrypt it. Hence, if an intruder (hacker) has access to the keys, then the whole purpose of encryption fails.

Further, in the approach called coarse-grained level approach, sharing of the private key is not an issue; users are always comfortable at doing this. In this approach, fine-grained access to encrypted data is given based on the need. Some parts of the key are shared to find the solution, which allows partial encryption. This is sometimes called key-policy attribute-based encryption. The solution works such that ciphertexts are given a set of features (or attributes) and private keys associated with access structures, so as to control which ciphertexts a user is allowed to decrypt. Let us see an example. Assuming John is the secretary of a housing society. He is responsible for plumbing works, generator set, maintenance of lawn and medical checks in the society. Assume John maintains an account, which contains all the tasks and the expenses of the society. He uploads all the account files to a cloud storage unit so that payment related details are available for audit. Now, as part of the review, audit members could ask for expenditures related to the items they are auditing, e.g., the audit team for plumbing would like to see the expenses related to plumbing. Now assume the files by John are encrypted and have attributes linked to them. When trying to decrypt the files, only relevant files are decrypted and shared for reading. This is achieved using encryption keys and encrypted-texts that are given attributes. In addition, a key can decrypt a particular encrypted-text only when the attribute is matching.

Homomorphic encryptions: This type of encryption allows performing operations on data without decrypting it first, making the entire system safer for use (refer Fig. 9.13). Some cryptographic techniques, like Zero-Knowledge Proofs (ZKs), are already implementing a form of homomorphic encryption. The last decade has generated much interest in the cloud, where many small and medium enterprises have moved data and applications to the cloud. Moving data to the cloud have often been supported and criticized equally. Data in cloud implies it is cheaper and without additional cost to the company; however, it has been criticized for not being secure. Proponents of the cloud have offered encryption as a way to fight the "not being secure" tag. Encryption of data stored in the cloud needs a large amount of computing power; in addition, encryption has to be done for the whole block. This mechanism of encrypting and decrypting an entire block is time-consuming and takes much energy. It has thus been correctly been flagged as inefficient.

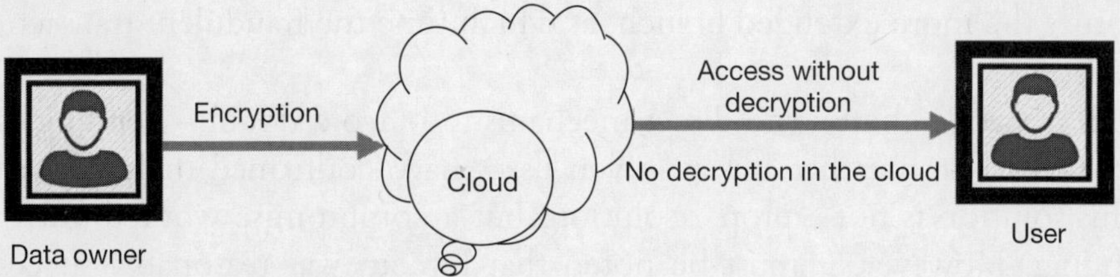

Figure 9.13: Homomorphic encryption

Another critical application of homomorphic encryption is PIR (Private Information Retrieval). When a patient visits the clinic, his records need to be pulled; this is a time-consuming activity. Suppose there is a large set of data related to the patient such as medical reports, which are encrypted and stored in a cloud server. We need a smart system that retrieves documents relevant to the patient only. Homomorphic encryption allows us to pull only pertinent, timely records. Homomorphic encryption is the stepping-stone towards secure multiparty computation. It is a mechanism to verify data, without having access to data.

9.3.3.2 Integrity in Transactions (I in CIAR)

Integrity is described as "guarding against improper information modification or destruction and includes assuring information non-repudiation and credibility," according to NIST cyber security framework. Due to its internal security aspects, blockchain technology would appear to be a perfect fit for data security solutions, where high ethics is demanded. In simple terms, data integrity denotes the reliability and trustworthiness of data. It involves the maintenance of confidence in the accuracy and consistency of data within its whole life cycle. Indeed, almost all firms must maintain data integrity and compliance with all government regulations and rules over it. However, every year in the United States, organizations lose over millions of dollars because of data integrity issues (as well as other issues, including that of data security). Blockchain may be the remedy to improve data integrity to the highest standards.

Ledgers across the nodes should reflect the same values. This is achieved via consensus. A suitable consensus mechanism ensures all nodes reflect the same benefits at any time. PoW (Proof-of-work) is the earliest known network-level consensus mechanism in a public system. It indeed offers a consensus mechanism: however, it is not error-proof and is prone to double-spending. In addition, PoW allows users to add blocks behind a block, which competes with other blocks already in the network.

Bribery attack: Bribery attacks are a type of attack in which an attacker creates a typical transaction to be included in a block. After waiting for some number of confirmations such that the transaction is irreversible, the attacker creates a conflicting fraudulent transaction and introduces it into a new crooked block. The attacker then bribes or rents a significant portion of the mining power in the network to extend the fraudulent branch until it becomes the more extended branch, at which time the fraudulent transaction will become valid.

In bitcoin network, there is no direct mechanism to avoid double-spending; instead, users are advised to ensure that at least seven users have confirmed their transaction as genuine. This solution is not a proper solution, but a compromise worked out to avoid double-spending. However, it must be noted that bitcoin was reasonably successful in warding off the challenges of cryptography and distributed ledgers. Hyperledger also offers solutions to difficulties like consistency: it advocates two forms of consensus mechanism, pBFT and RAFT. pBFT is the consensus mechanism, which allows nodes in the network to have the same data across the network. pBFT allows for consistency even though there could be faulty nodes in the network. RAFT allows a nomination-based model, which is faster than pBFT. Both these mechanisms ensure the consistency (integrity) of data across all nodes in the network.

Transactions are immutable: Blockchain ledgers are immutable, meaning that if data addition or transaction has been made, it cannot be edited or deleted. In addition, blockchains are not only a data structure but also a timekeeping mechanism for the data structure. Therefore, proof of the history of data is easily reportable and updated up to the second. Organizations facing an audit, regulatory compliance requirements, or legal challenges can use blockchain technology to improve data integrity and save millions.

While it may seem that blockchain immutability is an all-around positive attribute, it does create some challenges with data management. Data that has been stored in a blockchain, especially public ones, are not susceptible to data retention policies. Once recorded, the data would remain as is for the lifetime of the blockchain. The immutability security attribute of blockchain technology creates additional challenges for confidentiality, as it makes it virtually impossible to address legacy data protection controls: it can neither be deleted nor be re-encrypted. Blockchain transactions are immutable. They are immutable because they are helped and served by a robust cryptographic technology; this includes hashing and Merkle tree. Immutable transactions cannot be changed. The transactions that go into a blockchain ledger cannot be changed, as it is one of the fundamental features of blockchain databases. Blockchain only allows transactions to be added into the ledger. Once data have been written to a ledger, blockchain ensures that the data is added (distributed) to all other nodes in the network. Nodes in the network must have consistent data. Within a block, data are stored using the Merkle tree mechanism, which allows for fast verification of data.

Hashing and Merkle Tree: Proof-of-work (PoW) uses Hash and Nonce. Mining comprises hashing a block, introducing a nonce to the hashing function and then running the hash all over again. A nonce is a number that can be used just once in the cryptographic communication. Adding a nonce to a transactions' identifier makes it additionally unique, thus reducing the chance of duplicate transactions. The nonce is key to creating a block to a blockchain database. Another aspect of blockchain technology that is particularly important for improved data integrity is Merkle Tree. It ensures the integrity of data in the blockchain. A Merkle root is synonymous with a fingerprint of all the transactions in the block, which is created by hashing together pairs of Transaction IDs. It gives a short and unique fingerprint for all the transactions in a block. A Merkle tree (binary hash tree) involves taking large amounts of transaction data and constructing it in a way that is more convenient to process. This is also a field in a block header. Similarly, as data is saved in all nodes of the network, a hacker cannot go to the back-end and change the records. Hackers would have to change data in all the nodes of the system, which is practically not feasible for individuals.

Now, assuming we have put a team of people who are smart enough to go into the whole network (all nodes in the system), can they make changes in all the nodes? They cannot change it so quickly, because, for each written data, we have a saved hash of the data. Identifying the hash of the data satisfying all the criteria within the time available is difficult. Smart contracts are immutable, and so fixing a bug, if found, is very difficult. What if a code is wrongly written, which opens up the way for hackers?

Collusion attack: Perhaps the most well-known attack is the 51% attack due to its ability to completely subvert the blockchain. In this type of attack, the mining power of over 50% of the network is under the control of a single entity or group (refer Fig. 9.14). Such a group would be able to control the history of the blockchain. For example, if a double-spend happens, the colluding miners may force the acceptance of the fraudulent transaction by including it in a block. Even if the rest of the network disagrees, the entire system will reach consensus when the 51% mining pool outpaces the mining output of the rest of the network until the fraudulent chain becomes longer than the legitimate chain.

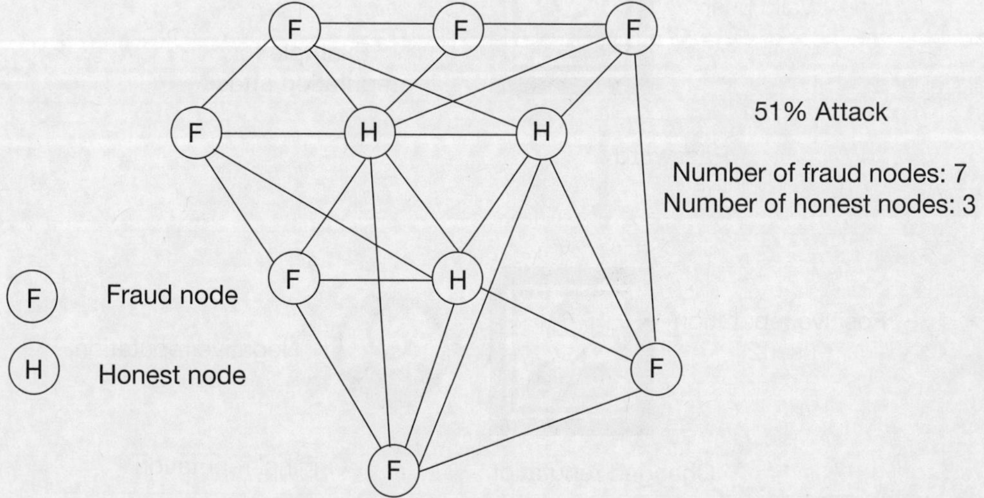

Figure 9.14: Collusion attack

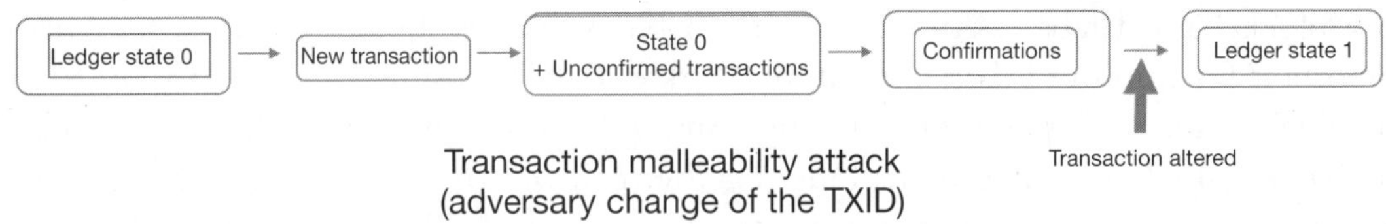

Figure 9.15: Transaction malleability model

Transaction malleability attack: Transaction malleability is a design flaw in bitcoin by which transactions may be altered after being created (refer Fig. 9.15) but before being added to a block. The source and destination addresses and the transaction amount cannot be manipulated, but other portions of the transaction may be altered, which results in a transaction ID (TXID) that differs from the original.

Reputation-based attack: A user in the blockchain can change his reputation from negative to a positive one (refer Fig. 9.16) and can fool the framework. Tampering the user reputation is another big concern for the blockchain community. It can be done mainly in two ways – one is hiding the negative transaction, and another is creating a new account.

The Finney attack: It is a double-spending attack that requires a block to be pre-mined. This block will hold a fraudulent transaction but it will not be broadcast yet. The same coins used in the fraudulent transaction will meanwhile be used as payment to a target node. After receiving some goods or services, the crooked block will be broadcast to the rest of the network, thus invalidating the transaction sent to the target. Waiting for additional confirmation before accepting a transaction, defeats this attack.

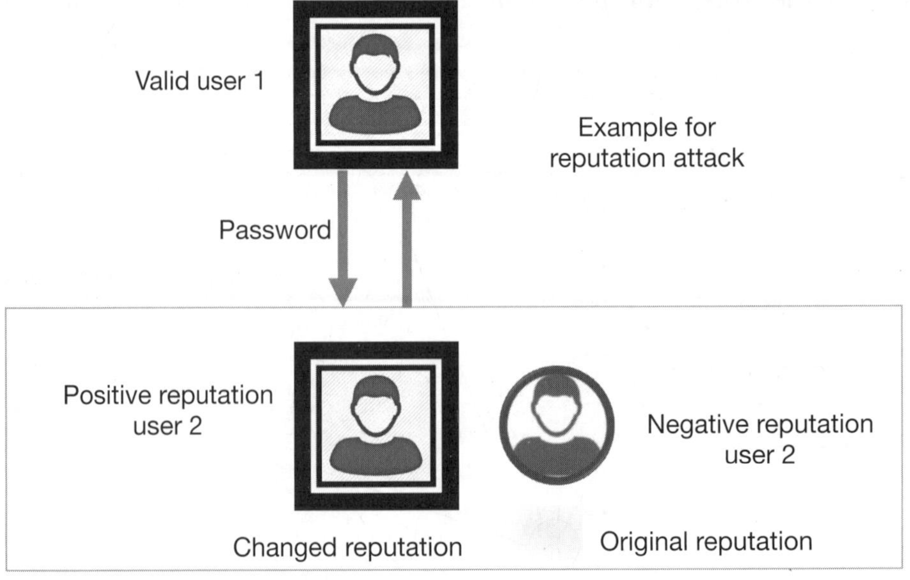

Figure 9.16: Reputation-based attack

The vector76 attack: It is a combination of the race and Finney attacks that target victims that only require a single confirmation. In this case, the attacker establishes a direct connection to the target, such as an e-wallet service, and a mining pool. The attacker creates a significant fraudulent transaction to deposit a substantial amount into the e-wallet, and a small transaction with the same tokens to send to the miners. The attacker mines until a block is found. The fraudulent transaction is included in the block, and both the block and the small transaction are broadcast at the same time. Once the e-wallet service sees the fraudulent transaction, the attacker immediately withdraws a large number of tokens that were credited to his account. Meanwhile, the rest of the network is much more likely to accept the small transaction, thus invalidating the significant fraudulent transaction to the e-wallet service.

9.3.3.3 Availability in Transactions (A in CIAR)

The availability attribute indicates timely and reliable access to and use of information. A transaction or smart contract execution that takes too long a time to execute is susceptible to attack. Denial of service (DoS) or distributed denial of service (DDoS) have worked in favour of hackers and troublemakers ever since computers joined the network of networks: the internet. Troublemakers do not need to create a new virus or attack; they need to raise a request to the website and do it repeatedly, so much that all resources of the website are busy serving the troublemaker's request. This does not do any harm to the underlying systems, or data stored in the file systems; however genuine consumers of service would not gain access to the network. To avoid being detected, troublemakers create a distributed way of attacking the target. Recent history shows that DDoS was successful in the attack on Telegram, during the Hong Kong protest in June 2019 and the attack on Wikipedia during September 06, 2019. While the attack on Telegram seems to have originated from a large organization, later attacks appear to have been the task of a few techies. Nevertheless, DoS or DDoS is a reality and we have to learn to work around it. If attackers attack a node, they could make it unavailable for other (genuine) users. However, a useful blockchain network can never be subjected to a DDoS attack as the network itself is distributed implying that if one node goes down, other nodes continue to work perfectly fine and can still serve customer requests.

Manipulation-based attacks: Various types of routing attacks are possible where one or more nodes in a blockchain network may be partially or fully partitioned (refer Fig. 9.17) from the rest of the network for malicious purposes. Using such attacks, it may be possible to delay block propagation for a significant amount of time, perform DoS attacks, isolate a large portion of the network mining power, or perform other similar attacks.

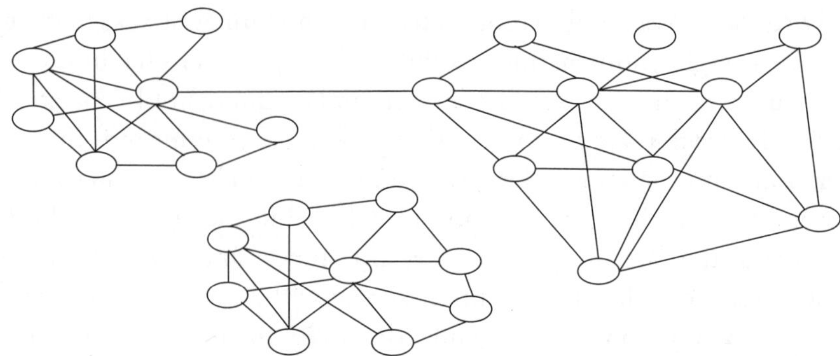

Figure 9.17: Manipulation-based attack

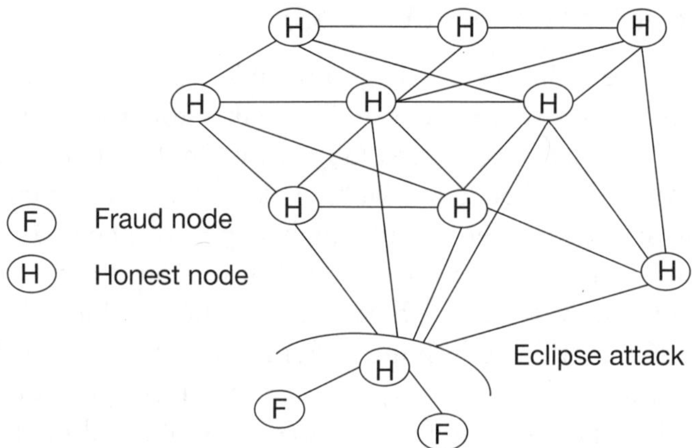

Figure 9.18: Eclipse attack in blockchain

Eclipse attacks: In eclipse attack (refer Fig. 9.18), an attacker attempts to isolate a target from the rest of the network by monopolizing all of the target's incoming and outgoing connections. This allows the attacker to corrupt the view of the target of the blockchain, force it to waste compute power or subvert the target's compute power for nefarious purposes.

No single point of failure: IP-based DDoS is not possible on a blockchain. All nodes maintain a full copy of the blockchain. When a node goes down, data can be picked up from other nodes. The distributed nature of the technology solves the Byzantine General's problem of false consensus.

9.3.3.4 Non-repudiation (R in CIAR)

Non-repudiation is an assurance of parties not being able to deny having participated in a transaction.

Once information is present in the system, we need to have the ability to prove that this person and the nonce produced this information. One person must not be able to make the transaction seem like it came from some other person or many others. That is the accountability and non-repudiation aspect. All users should have the ability to act as regulators or auditors to review everything in the network or a select portion of the required information.

9.4 Performance and Scalability

Scalability and performance are essential concepts. Performance is related to the number of transactions one (node) can handle per second (refer Fig. 9.19), and the associated latency (how quickly data can be transferred from one node to the other). Scalability is the ability to grow with participants; it shows how much the network can grow in size (refer Fig. 9.20).

We do not want to compromise on performance and scalability while providing richer security. Security and privacy are a clear mark of differentiation for cybersecurity fabric compared to many other platforms that are out there.

Blockchain 1	Blockchain 2	Blockchain 3	Blockchain 4	Blockchain 5
Transactions per second	Transactions per second	Transactions per second	Transactions per second	Transactions per second
Time for one transaction	Time for one transaction	Time for one transaction	Time for one transaction	Time for one transaction

Figure 9.19: Performance in blockchain

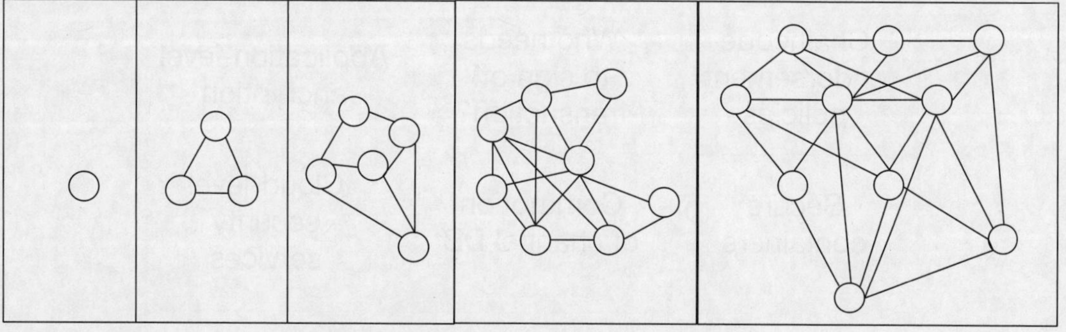

Figure 9.20: Scalability in blockchain

From the beginning, the hyperledger fabric (commonly referred to as 'fabric', refer Chapter 7, Section 7.4.1) has been designed keeping security and privacy as the foremost priority. There is membership and access control in fabric. At every stage of fabric, there can be access control. We can monitor those who are participating in a channel; within the channel, those executing the chain code; within the chain code, the endorsement policies, and those who need to sign off on a particular transaction. We can also apply application-level encryption to provide privacy guarantees on who can see the data. So, unless we have the application layer key, that is, the secret key that was used to encrypt the transaction, we will not be able to decrypt it. There are also hardware capabilities, like trusted chain code execution using inside secure containers. At every stage, there is some level of security and privacy embedded, in terms of membership access control, endorsement policies, and the commotion of channel side DB, etc.

We can also obtain security properties through cloud services or hardware. In the IBM cloud blockchain platform, all components of the network run inside a secure services container (refer Fig. 9.21). This secure services container is isolated, in terms of CPU memory, and in terms of the disk. In other words, everything is isolated. We write pure order chain code; all these codes execute in separate containers, which are all isolated from each other to prevent tampering. A secure services container (refer Fig. 9.22) is a combination of hardware and software. It has a firmware boot loader that is first loaded into memory. Then, it loads a software boot loader from disk. While loading the software, the container does integrity check to make sure that what is read from disk is actually put inside. Thus, we can make sure the drive has not been tampered with.

Membership and access control	Users in channel	Who is executing chain code?
Chaincode endorsement policies	Who needs to sign off transaction?	Application-level encryption
Secure containers	Commotion of channel DB	Cloud-level security services

Figure 9.21: Security and privacy aspects

Isolated secure services container 1	Isolated secure services container 2	Isolated secure services container 3	Isolated secure services container 4
APIs	APIs	APIs	APIs
Application	Application	Application	Application
Integrity check	Integrity check	Integrity check	Integrity check
Software boot loader	Software boot loader	Software boot loader	Software boot loader
Firmware boot loader	Firmware boot loader	Firmware boot loader	Firmware boot loader

Figure 9.22: Isolated secure services container

Software boot loader has keys stored in the disks that enable activation of encrypted drives, and this encryption-decryption is done every time the Applications Program Interface (API) is called. The application is designed to interact with the container only through APIs. There is no admin user. Even IBM will not be able to get into this container and look at what is inside. Thus, the application is designed to run a network function entirely isolated and the user has complete control over the contents provided. The cloud provider will have no access or control over the user data.

In the cloud, when virtual machines are used, it is possible that multiple applications may run on the same operating system or the same infrastructure. To prevent other malicious apps from being able to attack or obtain information from any user application, we have this notion of an enclave. It is a silicon chip plus CPU microcode, which is an isolated user-level code. Enclaves are protected from other processes that are simultaneously running on the system. Their codes can be verified remotely to check if they have been tampered with, or are running the original piece of code. The enclave also has a sandbox to protect a specific piece of code or data in that object. The user can plug in his or her consensus algorithm on the framework.

Enclave also has a persistent store; it maintains the user's ledger. There is a blockchain core, which has a pluggable component. We can write an adapter for our favorite blockchain application and plug it into the cocoa framework. It will manage the consensus and transaction execution and so on. It gives the user permission for identity management. It also uses a trusted execution environment. The user can vote on whether a new organization should join the network or not. Enclaves provide an exciting set of capabilities for the management and governance of the blockchain network. For example, we can use an adapter and bring in any blockchain platform to it.

9.5 Identity Management and Authentication

Permissionless networks do not provide strong identity management facilities and do not have the ability to audit transactions. This is so to ensure that performance does not degrade while such capabilities are provided and make certain that only a permissioned set of entities gain access to the network. Other entities do not even know that this network exists and cannot read what is on the network. The participation in the network is authorized and governed, and that is the strong Identity Management.

Robust identity management is typical in systems where the enterprise knows the particular set of other enterprises it is going to do business with. However, knowing these companies does not mean that you trust them.

Hence, only authorized personnel should have access to information and perform transactions. Only a specific set of entities should be allowed to perform certain types of transactions on the blockchain. This is called as authorized execution. There might be different subsets of entities that need to learn specific pieces of logic, and they are the ones who are execute the logic.

Replay attack: It is a form of attack where the attacker spoofs the communication between two valid parties and gains access (refer Fig. 9.23). Stealing the hash key and re-using it makes the attacker a legitimate user; this is a common threat to the blockchain community. However, using a key pair based exchange protocol is useful in protecting the user from these types of attacks. Some blockchains use a one-time private, crucial public pair to detect such a replay attack, while some use elliptic curve digital signature based encryption to have protection against it.

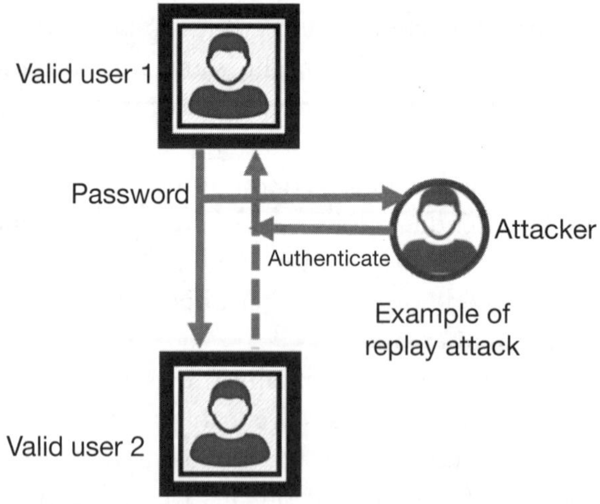

Figure 9.23: Replay attack in blockchain

Impersonation attack: Illegal access can also be gained by impersonating a legitimate user (refer Fig. 9.24). The elliptic curve digital signature algorithm (ECDSA) can create a defense against this type of attack. Some other methods, on the other hand, use the attribute-based signature to validate users.

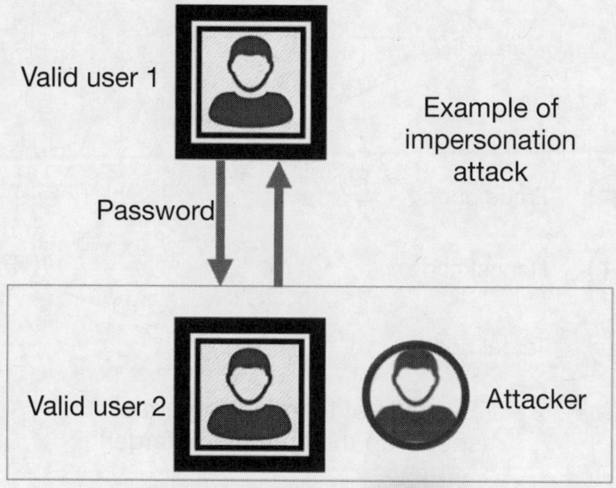

Figure 9.24: Impersonation attack in blockchain

Sybil attack: This is a general type of attack on peer-to-peer networks, where multiple fraudulent identities are created and controlled by a single rogue entity. In blockchain networks, this type of attack is used to isolate a target node from the rest of the honest network (refer Fig. 9.25), which in turn is used to launch different types of attacks.

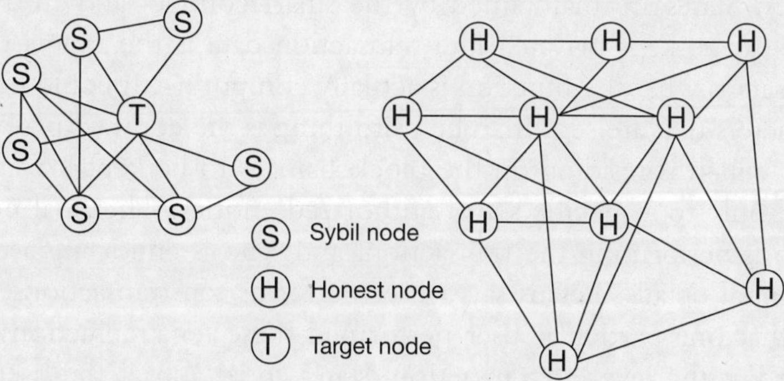

Figure 9.25: Sybil attack in blockchain

Time-jacking: It is an attack (refer Fig. 9.26) that attempts to skew a target node's timestamp by connecting to a target with multiple peers and reporting an incorrect time to the target. A node uses the network time to validate new blocks. By skewing a node's view of the network time, it would reject new blocks with a timestamp more significant than a predetermined duration. Like the Sybil and Eclipse attacks, this would allow the malicious nodes to isolate the target node from the overall network virtually. By isolating a target node, fraudulent transactions could be created and sent to the target.

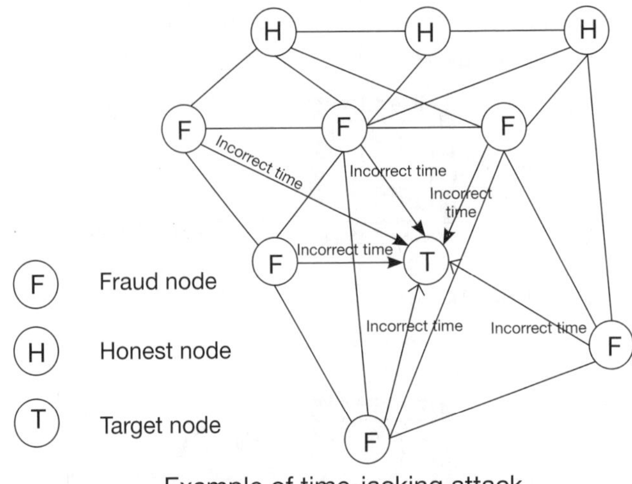

Figure 9.26: Time-jacking attack in blockchain

9.6 Regulatory Compliance and Assurance

A blockchain is about decentralization with multiple peers holding a replica of data; smart contracts are also replicated and decentralized. There is transaction privacy, which allows the inputs of a not contract-based transaction to be kept private. There is also a state data privacy. States are maintained by the smart contract and are recorded on the ledger, which is to be kept private. The transaction data is the inputs to the function, while the state data is what the function is actually computing. It could be more than just the input parameters that are sent in. The third thing is the contract privacy; the logic of the chain code, which is residing on the blockchain, is to be kept private. The relevant code is exposed only to a specific set of authorized entities. The final notion is privacy of the user who is performing the transaction, and who is endorsing a transaction with user-level credential details. Let us say a user performs ten transactions; no one should be able to say that one particular user performed these ten transactions. Privacy needs to be maintained at the level of transaction, state, code, and at the user level. Most of the permissionless blockchain platforms do not support these aspects of privacy. Bitcoin and Ethereum only provide the notion of pseudo-anonymity. As a user, I can assume a pseudonym is performing transactions, but one really cannot link the pseudonym with the original user. It does not provide the privacy notion of transaction data or state data. It is left to the applications to handle these notions of privacy explicitly. These notions of privacy are fundamental for many of the enterprise use cases.

The healthcare records are sensitive. Health records, whether they are kept on the blockchain or otherwise, should be confidential. The information has to remain private and can be authorized for users only on a need-to-know basis. Another example is when banks transact with enterprises, exchanging money or assets, and wish to ensure that user information is kept confidential. Thus, we need the notion of privacy even as the licensed entities are enabled to see our critical details. Some indications of ownership (operational or financial) within a blockchain system coping with real assets usually finish with somebody being at the receiving end of Compliance and Audit. Finally, the perfect method to deal with Audit and Compliance is self-audit and using an independent party creating audit reports.

International laws associated with security: In the UK, the Data Protection Act is utilized to make sure personal data is accessible only to individuals/companies with permissions and offer a remedy to individuals in case of inaccuracies. That is essential to make sure people are treated fairly; for example, for credit checking purposes. The Data Protection Act says that only companies and individuals who have valid and legal reasons can process personal data, and this data cannot be shared easily with others. Data Privacy Day can be a global holiday – it was started by the Council of Europe and is observed on January 28. General Data Protection Regulation (GDPR) of the European Union (EU) became law on May 25, 2018; organizations might face substantial penalties in case they do not obey the regulation. GDPR forces companies to know their data confidentiality risks and choose the right measures to decrease the possibility of unauthorized disclosure of users' private details.

9.7 Safeguarding Blockchain Smart Contract (DApp)

Despite its inherent security features, even properly designed and employed blockchain technology remains vulnerable to additional aspects that may endanger its security. Additionally, the trustless performance of a blockchain does not expand to components residing away from the consensus network. Individuals outside of blocks do not gain from any one of the inherent or emerging blockchain attributes and will be vulnerable to STRIDE risks (Spoofing, Tampering, Repudiation, Information Disclosure, Denial of Service and Elevation of privilege). As an example, Zero coin bug in the code enables the consumer to reevaluate his present legal proofs to create extra Zero coin spend transactions. Ethereum Smart Contract Attack is another example. Decentralized Autonomous Organization (DAO) with no central authority is the idea of running an organization through rules encoded as smart contracts and scripts running on blockchains. It is similar to conducting a company without CEO. "The DAO" is the name of a particular DAO launched in 2016 that was developed and run on ethereum. It had the most significant crowd-funding ever raised, with over $150m generated from more than 11,000 members. However, it soon became a victim of hacking. The attacker was able to empty more than 3.6 million ethers to a "child DAO" that had precisely the same settings as the "parent" DAO. Web security is essential to keeping hackers and cyber-thieves from accessing sensitive information. Without a proactive security strategy, businesses risk malware attacks from other websites, networks, and other IT infrastructures.

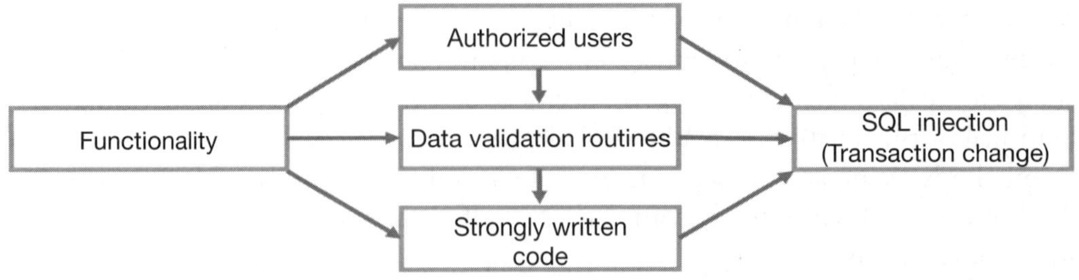

Figure 9.27: Minimizing attack area in blockchain

Minimize attack area: Each feature that is put into an application adds a quantity of risk to the total form. The aim is to reduce exposure. For example (refer Fig. 9.27), let us talk about an internet application having a search functionality. The search functionality might be directly exposed to SQL injection attacks. In case this feature is confined only to authorized users, then the risk of attack is reduced. In case the search function has been gated through data validation routines, the possibility to do SQL injection (transaction change before updating) is further reduced. In the event, the search feature is rewritten (through better user interface, as an example); this nearly removes the attack area, even when the help feature open to the internet is large.

Principle of least privilege: The principle states that accounts should have the smallest number of privilege required to perform their business processes. This encompasses user rights, resource permissions like CPU limits, memory, network, and file system permissions. With a smart contract, in general, a good rule of thumb is to give the minimum amount of privilege to any entity. So, first, try with the private keyword in a function to see if it works; if it does not work, then check progressively, in turn using internal, external, and at the end, public key words (refer Fig. 9.28). However, a typical beginner in coding tries to use public keywords for all the functions, which may create some security and vulnerability issues. Variables can also have some visibility keywords.

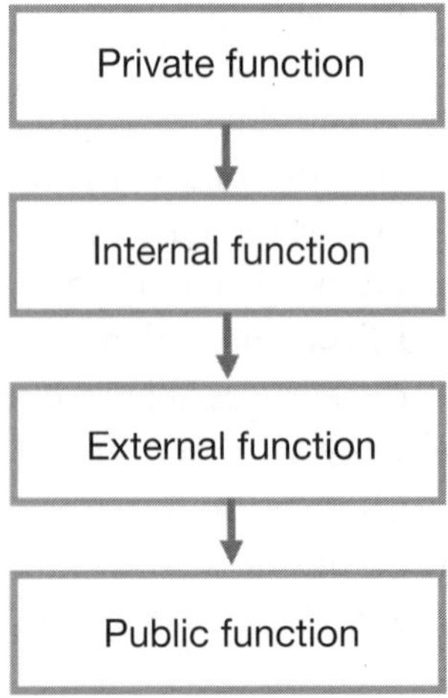

Figure 9.28: Principle of least privilege

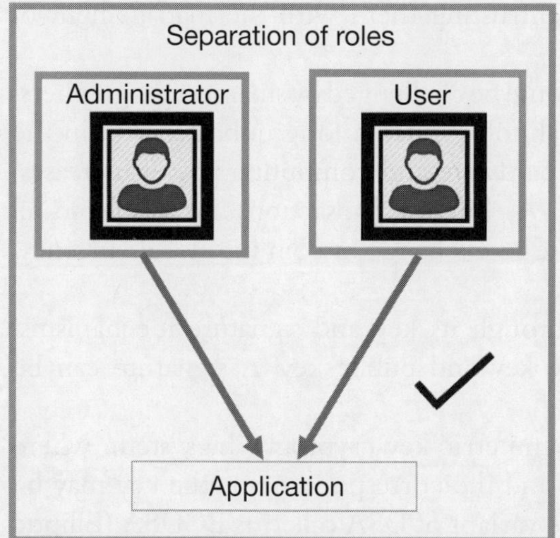

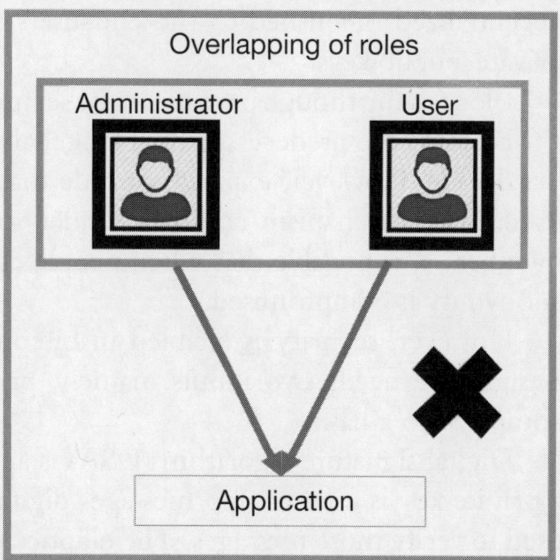

Figure 9.29: Separation of duties

Separation of duties: Particular roles have various degrees of trust compared to ordinary users. Specifically, administrators (refer Fig. 9.29) will vary from normal users. Generally, administrators must perhaps not be users of the application. For example, an administrator will have the ability to show the system as off or on and set password protection; however, he or she really should not have the ability to sign on into the storefront.

9.8 Security Aspects in Hyperledger Fabric

Hyperledger fabric necessitates policies and identities at every point of processing. Therefore, within an organization, you can have admins and members, and they have specific roles. Organizations can run peers, and these peers have identities. The ordering service also has personalities, and organizations can participate and order nodes. Only an admin can deploy or install a chain code. If anyone else tries to establish a particular chain code on a pair, it will fail.

Summary

Individual security affects our day-to-day lives. Just as security is a concern for all individuals like us, we also believe that the right to remain anonymous is equally essential and applicable for blockchain also.

Uses for blockchain, like access of the general public to the network, like electronic voting, etc. are still a distant dream due to various concerns related to security and privacy, along with its scalability and performance issues.

In a traditional security model, a central authority manages access to control and privileges. In a blockchain security model, access control and privileges are either decentralized or pushed to the end-users by entrusting them with the management of private/public keys.

Blockchain, though promises to be secure, could be challenged by smart cryptographers.

Bitcoin, the predecessor for blockchain technology, had a long unbeaten run in the last decade. However, experts concede that as hardware and computing power increases, its security mechanism could be under stress. As bitcoin transactions are visible to all members, it is possible that when people choose to exchange bitcoin into local currency, anonymity is compromised.

However, security is enabled in bitcoin through its key and signature mechanisms. A signature needs two inputs, namely, private key and public key. A signature can be compared to a hash.

Digital signature algorithm (DSA) is an asymmetric key cryptography system, where a private key is used to sign messages digitally, and the corresponding public key may be used to verify those messages. The elliptic curve variant of DSA called as ECDSA (Elliptic Curve Digital Signature Algorithm), is commonly used in blockchain implementations.

For an attacker to be able to successfully interfere with the bitcoin network, block, and reverse transactions, he or she should have the support of the majority (51%) of the node's support. The probability of rolling back a transaction decreases exponentially with the number of confirmations it has received.

Bitcoin also uses SHA 256 encryption for both its proof-of-work (PoW) system and transaction verification.

Security can be defined as actions that safeguard the data and prevent unauthorized access to those data. Privacy is about the safeguarding (not revealing to outsiders) of user identity.

Ethereum purports that, code is the law. There is not any governing body; there is not any legal resource. There is not any legal frame actually to prove the theft.

CIAR stands for Confidentiality, Integrity, Availability, and Non-repudiation.

Confidentiality: It gives assurance that information is disclosed only to the relevant authorized parties.

Integrity: Assurance for the accuracy and completeness of data. Integrity involves keeping information from being altered.

Availability: Assurance that the system will be available for use when it is required.

Non-repudiation: Assurance that users who participate in a transaction will not be able to deny their participation at a later point.

As a company, it is essential to take time to first identify the most valuable data and work on it. This data, which is most sensitive, is commonly referred to as the "crown jewels" of data. A subset of crown jewels data would include personal identifiable information (PII).

In order to ensure that transactions are confidential, it is important they are not linked to the user; there are a few methods by which this can be achieved:

a. Mixing/Barter transactions
b. Anonymous signatures
c. Advanced encryption techniques.

In the attribute-based encryption, especially cipher text-policy attribute-based encryption, the secret key does not contain the specific information of users. In addition, the attribute authority can generate the secret key from any attribute set.

Homomorphic encryptions: It allows performing operations on data without decrypting it first, making the entire system safer for use. Some cryptographic techniques, like Zero-Knowledge Proofs (ZKs), are already implementing a form of homomorphic encryption.

In simple terms, data integrity denotes the reliability and trustworthiness of data. It involves the maintenance of confidence in the accuracy and consistency of data within its whole life-cycle.

A suitable consensus mechanism ensures that all nodes reflect the same benefits at any time. Proof-of-work (PoW) is the earliest known network level consensus mechanism in a public system.

Bribery attack: Bribery attacks are a type of attack in which an attacker creates a typical transaction to be included in a block.

Transactions are immutable: Blockchain ledgers are immutable, meaning that if data addition or transaction has been made, it cannot be edited or deleted. In addition, blockchains are not only a data structure but also a timekeeping mechanism for the data structure.

Hashing and Merkle Tree: Proof-of-work (PoW) uses Hash and Nonce. Mining comprises hashing a block and then introducing a nonce to the hashing function and running the hash all over again. A nonce is a number that can be used just once in the cryptographic communication.

A Merkle tree (binary hash tree) involves taking large amounts of transaction data and constructing it in a way that is more convenient to process. This is also a field in a block header.

Collusion attack: Perhaps the most well-known attack is the 51% attack due to its ability to completely subvert the blockchain. In this type of attack, the mining power of over 50% of the network is under the control of a single entity or group.

Transaction malleability attack: Transaction malleability is a design flaw in bitcoin by which transactions may be altered after being created but before being added to a block.

Reputation-based attack: A user in the blockchain can change his reputation from negative to a positive one and can fool the framework. Tampering the user reputation is a cause of big concern to the blockchain community. It can be done mainly in two ways – one is hiding the negative transaction, and another is creating a new account.

The Finney attack: It is a double-spending attack that requires a block to be pre-mined. This block will hold a fraudulent transaction but will not be broadcast yet. The same coins used in the fraudulent transaction will meanwhile be used as payment to a target node. After receiving some goods or services, the crooked block will then be broadcast to the rest of the network, thus invalidating the transaction sent to the target. Waiting for additional confirmation before accepting a transaction, defeats this attack.

The vector76 attack: It is both the combination of the race and Finney attacks that target victims that only require a single confirmation.

The availability attribute indicates timely and reliable access to and use of information. A transaction or smart contract execution that takes an unusually long time to execute is susceptible to attack. Denial of service (DoS) or Distributed Denial of Service (DDoS) work in favor of hackers and troublemakers.

Manipulation-based attacks: Various types of routing attacks are possible where one or more nodes in a blockchain network may be partially or fully partitioned from the rest of the network for malicious purposes. Using such attacks, it may be possible to delay block propagation for a significant amount of time, perform DoS attacks, isolate a large portion of the network mining power, or perform other similar attacks.

Eclipse attacks: In eclipse attack, an attacker attempts to isolate a target from the rest of the network by monopolizing all of the target's incoming and outgoing connections. This allows the attacker to corrupt the view of the target of the blockchain, force it to waste compute power or subvert the target's compute power for nefarious purposes.

No single point of failure: IP-based DDoS is not possible on a blockchain. All nodes maintain a full copy of the blockchain. When a node goes down, data can be picked up from other nodes. The distributed nature of the technology solves the Byzantine General's problem of false consensus.

Scalability and performance are essential concepts. Scalability is the ability to grow with participants, how much the network can grow in size. Performance is related to how many transactions the system can handle per second and the latency involved. Performance and scalability should not be affected while providing richer security.

Permissionless networks do not give strong identity management and do not have the ability to audit transactions. Robust identity management is typical in enterprise systems because any enterprise knows that it is only going to do business with a particular set of other enterprises.

Replay attack: It is a form of attack where the attacker spoofs the communication between two valid parties and gain access.

Impersonation attack: Impersonating a legitimate user is also used to gain access. Using an ECDSA algorithm can help to defend against it. Some other methods, on the other hand, uses the attribute-based signature to validate users.

Sybil attack: These are a general type of attack on peer-to-peer networks in which multiple fraudulent identities are created and controlled by a single rogue entity. In blockchain networks, this type of attack is used to isolate a target node from the rest of the honest network, which in turn is used to launch different types of attacks.

Time-jacking: It is an attack that attempts to skew a target node's timestamp by connecting to a target with multiple peers and reporting an incorrect time to the target.

Bitcoin and Ethereum only provide the notion of pseudo-anonymity. As a user, I can assume a pseudonym is performing transactions, but one really cannot link the pseudonym with the original user. It does not provide the privacy notion of transaction data or state data. It is left to the applications to handle these notions of privacy explicitly.

International laws associated with security: In the UK, the Data Protection Act is utilized to make sure personal data is accessible only to individuals/companies with permissions and offers a remedy to individuals, in case of inaccuracies. GDPR forces companies to know their data confidentiality risks and choose the right measures to decrease the possibility of unauthorized disclosure of users' private details.

Decentralized Autonomous Organization (DAO) with no central authority is the idea of running an organization through rules encoded as smart contracts and scripts running on blockchains. It is similar to conducting a company without CEO.

Hyperledger fabric necessitates policies and identities at every point of processing. Therefore, within an organization, you can have admins and members, and they have specific roles. Organizations can run peers, and these peers have identities. The ordering service also has personalities, and organizations can participate and order nodes. Only an admin can deploy or install a chain code. If anyone else tries to establish a particular chain code on a pair, it will fail.

EXERCISES

Multiple Choice Questions

1. **In a blockchain security model, access control and privileges are either decentralized or pushed to the end-user by entrusting them with the management of private/public keys.**

   A. True

   B. False

   Answer: A

   Explanation: In a blockchain security model, access control and privileges are either decentralized or pushed to the end-user by entrusting them with the management of private/public keys. Transactions authorize just a particular value to a specific user, which cannot be modified or forged. It will not disclose any personal information, like the identities of the parties, and cannot be utilized to authorize additional transactions.

2. **Security is enabled in bitcoin through its __________ and __________ mechanisms.**

   A. Consensus and mining

   B. Mining and transaction pooling

   C. Consensus and confidentiality

   D. Key and signature

   Answer: D

   Explanation: Security is enabled in bitcoin through its key and signature mechanisms. When a user joins a bitcoin network, a private key is generated (a user creates a private key). A public key is made from the private key, using cryptographic algorithms. This is because, cryptographically generated keys will ensure that the public key cannot be used trace back the owner to his private key. The public key for the user is available throughout the network.

3. **When a user joins a bitcoin network, a public key is generated (a user creates a public key). A private key is made from the public key using cryptographic algorithms.**
   A. True
   B. False
   Answer: B
   Explanation: Same as that for Question 2.
4. **This needs two inputs: private key and public key. This can be compared to a hash.**
   A. Security
   B. Signature
   C. Mining
   D. Consensus mechanism
   Answer: B
   Explanation: Same as that for Question 2.
5. **This is about the safeguarding (not revealing to outsiders) of user identity.**
   A. Security
   B. Privacy
   C. Mining
   D. Hacking
   Answer: B
   Explanation: Security vs. Privacy: Security can be defined as actions that safeguard the data and prevent unauthorized access to those data. We usually put various security controls to restrict who can access the stored data/information. Privacy is about the safeguarding (not revealing to outsiders) of user identity. Although the meaning of security and privacy overlap with each other, it is possible to have security without privacy, but impossible to have privacy without security.
6. **This can be defined as actions that safeguard the data and prevent unauthorized access to those data.**
   A. Security
   B. Privacy
   C. Mining
   D. Hacking
   Answer: A
   Explanation: Same as that for Question 5.

7. **In blockchain security, CIAR stands for:**
   A. Concurrency, Identity, Authorization and Repudiation
   B. Concurrency, Integrity, Authorization and Non-repudiation
   C. Confidentiality, Integrity, Availability and Non-repudiation
   D. Confidentiality, Identity, Authorization and Non-repudiation
   Answer: C
   Explanation: CIAR stands for Confidentiality, Integrity, Availability and Non-repudiation.

8. **It gives assurance that information is disclosed only to the relevant authorized parties. It denotes that any sensitive information should only be shared with a limited number of authorized people.**
   A. Non-repudiation
   B. Confidentiality
   C. Integrity
   D. Availability
   Answer: B
   Explanation: Confidentiality: It gives assurance that information is disclosed only to the relevant authorized parties. Confidentiality denotes that any sensitive information should only be shared with a limited number of authorized people.

9. **This refers to the reliability and trustworthiness of data.**
   A. Non-repudiation
   B. Confidentiality
   C. Integrity
   D. Availability
   Answer: C
   Explanation: Integrity: Assurance for the accuracy and completeness of data. Integrity involves keeping information from being altered. It refers to the reliability and trustworthiness of data. It consists of the maintenance of and the assurance of the accuracy and consistency of data over its entire life cycle. If a transaction detail is updated in a single node, integrity ensures that the same data is updated in all other nodes.

10. **Assurance that the system will be available for use when it is required. It involves ensuring that those who rely on accurate and truthful information can get it.**
   A. Non-repudiation
   B. Confidentiality
   C. Integrity
   D. Availability
   Answer: D
   Explanation: Availability: Assurance that the system will be available for use when it is required. It involves ensuring those who rely on accurate and truthful information

can get it. Availability is linked to ethics. However, it can also involve matters such as a cyberattack, preventing people from accessing specific servers, or from getting into the blockchain network.

11. **Assurance on parties not being able to deny having participated in a transaction. Once information is present in the system, it is the ability to prove that this person and the nonce produced this information.**
    A. Non-repudiation
    B. Confidentiality
    C. Integrity
    D. Availability
    Answer: A
    Explanation: Non-repudiation: Assurance on parties not being able to deny having participated in a transaction. Once information is present in the system, it is the ability to prove that this person and the nonce produced this information. The user must not be able to pretend that the transaction did not take place or make it seem like it came from some other person or many others. This is the accountability and non-repudiation aspect.

12. **As a company, it is essential to take the time to identify the most valuable data and work on protecting it first. This data is commonly referred to as:**
    A. Valuable data
    B. Secret data
    C. Crown jewels of data
    D. High priority data
    Answer: C
    Explanation: Crown jewels of data: As a company, it is essential to take the time to identify the most valuable data and work on protecting it first. This data is commonly referred to as the "crown jewels" of data and it usually makes up around 10 percent of the overall data, and if it were compromised, it would cause severe damage to the company.

13. **An important strategy when looking to improve data security is developing a___________ policy.**
    A. Data security
    B. Data protection
    C. Data improvement
    D. Data identification
    Answer: A
    Explanation: An important strategy when looking to improve data security is developing a data security policy. Data security policies help to keep employees organized. Such policies should be open to edits and changes, as amendments need to be designed to match the developing technological inventions and new company

coverages. When data access rules are strictly enforced, higher data protection can be achieved on an everyday basis.

14. In order to ensure transactions are confidential, it is not linked to the user. Which of the following is not one of the methods to achieve it.

A. Mixing/Barter transactions
B. Anonymous Signatures
C. Advanced encryption techniques
D. Initial coin offering

Answer: D

Explanation: In order to ensure transactions are confidential it is important they be not linked to the user. A few methods to achieve this are (a) Mixing/Barter transactions (b) Anonymous signatures and (c) Advanced encryption techniques.

15. The blockchain, bitcoin in particular, allows this as a way of ensuring that transactions recorded into the chain are not linked to the user's public key.

A. Mixing/Barter transactions
B. Anonymous signatures
C. Advanced encryption techniques
D. Initial coin offering

Answer: A

Explanation: Mixing/bartering transactions: The blockchain, bitcoin, in particular, allows mixing/bartering coins as a way of ensuring that transactions recorded into the chain are not linked to the user's public key. They could do a mix of coins, or swap coins or submit a joint transaction. This would ensure that the transaction recorded in the block is not linked to the user.

16. It is sometimes found that a signature can be reverse engineered to trace back to the original author. To avoid this, this mechanism is used:

A. Mixing/Barter transactions
B. Anonymous signatures
C. Advanced encryption techniques
D. Initial coin offering

Answer: B

Explanation: Anonymous signatures: Digital signatures are the footprint in any electronic system. Bitcoin and other blockchains rely on them heavily; however, it is sometimes found that a signature can be reverse engineered to trace back to the original author. There are two well-known mechanisms to overcome this risk: 1. Group signature, and 2. Ring signature.

17. In this, the secret key does not contain the specific information of users.

A. Attribute-based encryption
B. Homomorphic encryption
C. Briber attack

D. Immutability

Answer: A

Explanation: In the attribute-based encryption, especially cipher text-policy attribute-based encryption, the secret key does not contain the specific information of users. In addition, the attribute authority can generate the secret key from any attribute set. If the secret key is abused, it is difficult to judge whether the abused private key comes from the users or the attribute authority.

18. It allows performing operations on data without decrypting it first, making the entire system safe for use.

A. Attribute based encryption

B. Homomorphic encryption

C. Bribery attack

D. Immutability

Answer: B

Explanation: Homomorphic encryptions allow performing operations on data without decrypting it first, making the entire system safer for use. Some cryptographic techniques like Zero-Knowledge Proofs (ZKs) are already implementing a form of homomorphic encryption.

19. It is a type of attack in which an attacker creates a typical transaction to be included in a block. After waiting for some number of confirmations such that the transaction is irreversible, the attacker creates a conflicting fraudulent transaction and introduces it into a new crooked block.

A. Eclipse-based attack

B. Manipulation-based attack

C. Bribery attack

D. Relay-based attack

Answer: C

Explanation: Bribery attack: Bribery attacks are a type of attack in which an attacker creates a typical transaction to be included in a block. After waiting for some number of confirmations such that the transaction is irreversible, the attacker creates a conflicting fraudulent transaction and introduces it into a new crooked block. The attacker then bribes or rents a significant portion of the mining power in the network to extend the fraudulent branch until it becomes the more extended branch, at which time the fraudulent transaction will become valid.

20. It is perhaps the most well-known attack due to its ability to completely subvert the blockchain.

A. Eclipse-based attack

B. Transaction malleability attack

C. Collision attack

D. Relay-based attack

Answer: C

Explanation: Collusion attack: Perhaps the most well-known attack is the 51% attack due to its ability to completely subvert the blockchain. In this type of attack, the mining power of over 50% of the network is under the control of a single entity or group. Such a group would be able to control the history of the blockchain.

21. This is a design flaw in bitcoin through which transactions may be altered after being created but before being added to a block.

A. Eclipse-based attack
B. Transaction malleability attack
C. Collision attack
D. Relay-based attack

Answer: B

Explanation: Transaction malleability attack: Transaction malleability is a design flaw in bitcoin through which transactions may be altered after being created but before being added to a block. The source and destination addresses and the transaction amount cannot be manipulated, but other portions of the transaction may be altered, which results in a transaction ID (TXID) that differs from the original.

22. It is a big concern for the blockchain community. It can be done mainly two ways – one is hiding the negative transaction and another is creating a new account.

A. Eclipse-based attack
B. Transaction malleability attack
C. Collision attack
D. Reputation-based attack

Answer: D

Explanation: Reputation-based attack: A user in a blockchain can change his reputation from negative to a positive one and fool the framework. This can be done mainly in two ways, namely, by hiding the negative transaction and by creating a new account.

23. It is a double-spending attack that requires a block to be pre-mined. This block will hold a fraudulent transaction but it will not be broadcast yet. The same coins used in the fraudulent transaction will meanwhile be used as payment to a target node.

A. Eclipse-based attack
B. The Finney attack
C. Collision attack
D. Reputation-based attack

Answer: B

Explanation: The Finney attack: It is a double-spending attack that requires a block to be pre-mined. This block will hold a fraudulent transaction but it will not be broadcast yet. The same coins used in the fraudulent transaction will meanwhile be used as payment to a target node. After receiving some goods or services, the crooked block will then be broadcast to the rest of the network, thus invalidating the transaction sent to the target. Waiting for additional confirmation before accepting a transaction, defeats this attack.

24. **Various types of routing attacks are possible where one or more nodes in a blockchain network may be partially or fully partitioned from the rest of the network for mischievous purposes.**
    A. Eclipse-based attack
    B. The Finney attack
    C. Manipulation-based attack
    D. Reputation-based attack
    Answer: C
    Explanation: Manipulation-based attacks: Various types of routing attacks are possible where one or more nodes in a blockchain network may be partially or fully partitioned from the rest of the network for malicious purposes. Using such attacks, it may be possible to delay block propagation for a significant amount of time, perform DoS attacks, isolate a large portion of the network mining power, and perform other similar attacks.

25. **In this attack, the attacker attempts to isolate a target from the rest of the network by monopolizing all of the target's incoming and outgoing connections.**
    A. Eclipse-based attack
    B. The Finney attack
    C. Manipulation-based attacks
    D. Reputation-based attack
    Answer: A
    Explanation: Eclipse attacks: In eclipse attack, an attacker attempts to isolate a target from the rest of the network by monopolizing all of the target's incoming and outgoing connections. This allows the attacker to corrupt the view of the target of the blockchain, force it to waste compute power or subvert the target's compute power for nefarious purposes.

26. **It is a form of attack where the attacker spoofs the communication between two valid parties and gains access. Stealing the hash key and reusing it to makes the attacker a legitimate user, which is a common threat to the blockchain community.**
    A. Eclipse-based attack
    B. The Finney attack
    C. Manipulation-based attacks
    D. Replay-based attack
    Answer : D

Explanation: Replay attack: It is a form of attack where the attacker spoofs the communication between two valid parties and gains access. Stealing hash key and reusing it to make the attacker a legitimate user, which is a common threat to the blockchain community. However, using a key pair based exchange protocol is useful in protecting the user from these types of attacks.

27. **It is an attack that attempts to skew a target node's timestamp by connecting to a target with multiple peers and reporting an incorrect time to the target.**

    A. Eclipse-based Attack
    B. The Finney attack
    C. Time-jacking attack
    D. Timer Attack

    Answer : C

    Explanation: Time-jacking: It is an attack that attempts to skew a target node's timestamp by connecting to a target with multiple peers and reporting an incorrect time to the target. A node uses the network time to validate new blocks. By skewing a node's view of the network time, it would reject new blocks with a timestamp more significant than a predetermined duration. Like the Sybil and eclipse attacks, this would allow the malicious nodes to virtually isolate the target node from the overall network. By isolating a target node, fraudulent transactions could be created and sent to the target.

Short-answer Questions

1. **What is traditional security model?**

   In a traditional security model, a central authority manages to access control and privileges. The transaction contains the user's private identifiers e.g., social security numbers, credit card numbers, aadhar numbers, etc., which are susceptible to attacks. Such private identifiers of parties are embedded as part of the stored data. The rest of the data are susceptible to tampering, if a hacker gains access to it.

2. **What is blockchain security model?**

   In a blockchain security model, access control and privileges are either decentralized or pushed to the end-user by entrusting them with the management of private/public keys. Transactions authorize just a particular value to a specific user, which cannot be modified or forged. It will not disclose any personal information, like the identities of the parties, and cannot be utilized to authorize additional transactions.

3. **Write notes on keying mechanism of bitcoin.**

   Security is enabled in bitcoin through its key and signature mechanisms. When a user joins a bitcoin network, a private key is generated (a user creates a private key).

A public key is made from the private key using cryptographic algorithms. This is because cryptographically generated keys will ensure that the public key cannot be used to trace the owner back to his private key. The public key for the user is available throughout the network.

4. **Write notes on signature mechanism of bitcoin.**

   Security is enabled in bitcoin through its key and signature mechanisms. A signature is a number that proves that signing operations took place. A signature needs two inputs – private key and public key. A signature can be compared to a hash. Digital signature algorithm (DSA) is an asymmetric key cryptography system, where a private key is used to sign messages digitally, and the corresponding public key may be used to verify those messages.

5. **Write notes on Security vs. Privacy.**

   Security can be defined as actions that safeguard the data and prevent unauthorized access to those data. We usually put various security controls to restrict access to the stored data/information. Privacy is about the safeguarding (not revealing to outsiders) of user identity. Although the meaning of security and privacy are overlapping with each other, it is possible to have security without privacy, but impossible to have privacy without security.

6. **What is majority attack?**

   Some of the assumptions to guarantees in terms of security and privacy are that, not more than 50 percent of the computing nodes have been compromised (meaning that at least 50 percent of the nodes are following the defined protocol). A majority attack on a blockchain is one in which a group of miners control more than 50 percent of the network's mining hash rate or computing power and prevent new transactions from gaining confirmations.

7. **What are the components of CIAR?**

   CIAR stands for Confidentiality, Integrity, Availability and Non-repudiation.

8. **What is confidentiality in CIAR?**

   Confidentiality: It gives assurance that information is disclosed only to the relevant authorized parties. Confidentiality denotes that any sensitive information should only be shared with a limited number of authorized people.

9. **What is Integrity in CIAR?**

   Integrity: Assurance for the accuracy and completeness of data. Integrity involves keeping information from being altered. Integrity refers to the reliability and trustworthiness of data. It consists of the maintenance of and the assurance of the accuracy and consistency of data over its entire life cycle. If a transaction detail is updated in a single node, it also ensures updation in all other nodes.

10. **Write notes on availability in CIAR?**

    Availability: Assurance that the system will be available for use when it is required. It involves ensuring that those who rely on accurate and truthful information can get it. Availability is linked to ethics. However, it can also involve matters such as a cyberattack or preventing people's access to specific servers or the blockchain network.

11. Write notes on non-repudiation in CIAR?

Non-repudiation: Assurance on parties not being able to deny having participated in a transaction. Once information is present in the system, we need to have the ability to prove that this person and the nonce produced this information. The user must not be able to make the transaction seem like it came from some other person or many others. That is the accountability and non-repudiation aspect.

12. What is "crown jewels" of data?

As a company, it is essential to take the time to identify the most valuable data and work on protecting it first. Such data is commonly referred to as the "crown jewels" of data. This type of data usually makes up around 10 percent of the overall data, and if it were compromised, it would cause severe damage to the company.

13. Write notes on data security policy.

An important strategy when looking to improve data security is developing a data security policy. Such policies may help to keep employees organized. Data security policy should also be open to edits and changes, as amendments need to be designed to match the developing technological inventions and new company coverages. When data access rules are strictly enforced, higher data protection can be achieved on an everyday basis.

14. What are the mechanisms to ensure that the user and transaction are not linked, for security purpose?

In order to ensure transactions are confidential, it is important they be not linked to the user. There are a few methods using which this can be achieved:

a. Mixing/Barter transactions
b. Anonymous signatures
c. Advanced encryption techniques.

15. What is mixing/bartering transactions?

Mixing/bartering transactions: The blockchain, bitcoin, in particular, allows mixing/bartering of coins as a way of ensuring that transactions recorded into the chain are not linked to the user's public key. They could do a mix of coins, or swap coins or submit a joint transaction. This would ensure that the transaction recorded in the block is not linked to the user.

16. What is anonymous signature?

Anonymous signatures: Digital signatures are the footprint in any electronic system, Bitcoins and other blockchains rely on them heavily; however, it is sometimes found that a signature can be reverse engineered and traced back to the original author. There are two well-known mechanisms to overcome this risk: 1. Group signature, and 2. Ring signature.

17. What is attribute-based encryption?

In attribute-based encryption, especially cipher text-policy attribute-based encryption, the secret key does not contain the specific information of users. In addition, the attribute authority can generate the secret key from any attribute set. If the secret key is abused, it is difficult to judge whether the abused private key comes from users or the attribute authority.

18. What is homomorphic encryption?

Homomorphic encryptions: It allows performing operations on data without decrypting it first, making the entire system safe for use. Some cryptographic techniques like Zero-Knowledge Proofs (ZKs), are already implementing a form of homomorphic encryption.

19. What is bribery attack?

Briber Attack: Bribery attacks are a type of attack in which an attacker creates a typical transaction to be included in a block. After waiting for some number of confirmations such that the transaction is irreversible, the attacker creates a conflicting fraudulent transaction and introduces it into a new crooked block. The attacker then bribes or rents a significant portion of the mining power in the network to extend the fraudulent branch until it becomes the more extended branch, at which time the fraudulent transaction will become valid.

20. What is immutability in blockchain ledger?

Blockchain ledgers are immutable, meaning that if data addition or transaction has been made, it cannot be edited or deleted. In addition, blockchains are not only a data structure but also a timekeeping mechanism for the data structure. Therefore, proof of the history of data is easily reportable and updated up to the second. Organizations facing an audit, regulatory compliance requirements, or legal challenges can use blockchain technology to improve data integrity and save millions.

21. What is collusion attack?

Collusion attack: Perhaps the most well-known attack is the 51% attack due to its ability to completely subvert the blockchain. In this type of attack, the mining power of over 50% of the network is under the control of a single entity or group. Such a group would be able to control the history of the blockchain.

22. Give an example for collusion attack in blockchain.

If a double-spend happens, the colluding miners may force the acceptance of the fraudulent transaction by including it in a block. Even if the rest of the network disagrees, the entire system will reach consensus when the 51% mining pool outpaces the mining output of the rest of the network until the fraudulent chain becomes longer than the legitimate chain.

23. What is transaction malleability attack in blockchain?

Transaction malleability attack: Transaction malleability is a design flaw in bitcoin through which transactions may be altered after being created but before being added to a block. The source and destination addresses and the transaction amount cannot be manipulated, but other portions of the transaction may be altered, which results in a transaction ID (TXID) that differs from the original.

24. What is reputation-based attack in blockchain?

Reputation-based attack: A user in the blockchain can change his reputation from negative to a positive one and can fool the framework. Tampering with the user reputation is a big concern for the blockchain community. It can be done mainly in two ways – one is hiding the negative transaction and another is creating a new account.

25. What is Finney attack in blockchain?

The Finney attack: It is a double-spending attack that requires a block to be pre-mined. This block will hold a fraudulent transaction but will not be broadcast yet. The same coins used in the fraudulent transaction will meanwhile be used as payment to a target node. After receiving some goods or services, the crooked block will be broadcast to the rest of the network, thus invalidating the transaction sent to the target. Waiting for additional confirmation before accepting a transaction, defeats this attack.

26. What is manipulation-based attack in blockchain?

Manipulation-based attacks: Various types of routing attacks are possible where one or more nodes in a blockchain network may be partially or fully partitioned from the rest of the network for malicious purposes. Using such attacks, it may be possible to delay block propagation time for a significant amount of time, perform DoS attacks, isolate a large portion of the network mining power, or perform other similar attacks.

27. What is eclipse attack in blockchain?

Eclipse attacks: In eclipse attack, an attacker attempts to isolate a target from the rest of the network by monopolizing all of the target's incoming and outgoing connections. This allows the attacker to corrupt the view of the target of the blockchain, force it to waste compute power or subvert the target's compute power for nefarious purposes.

28. What is replay attack in blockchain?

Replay attack: It is a form of attack where the attacker spoofs the communication between two valid parties and gains access. Stealing hash key and reusing it makes the attacker appear as a legitimate user, and this is a common threat to the blockchain community. However, using a key pair based exchange protocol is useful in protecting the user from these types of attacks.

29. What is impersonation attack in blockchain?

Impersonation attack: This type of attack results in impersonating a legitimate user to gain access to the blockchain network. Using an ECDSA algorithm can create a defense against it. Other methods such as attribute-based signature to validate users are also used.

30. What is Sybil attack in blockchain?

Sybil attack: This is a general type of attack on peer-to-peer networks in which multiple fraudulent identities are created and controlled by a single rogue entity. In blockchain networks, this type of attack is used to isolate a target node from the rest of the honest network, which is then used to launch different types of attacks.

31. What is time-jacking attack in blockchain?

Time-jacking: It is an attack that attempts to skew a target node's timestamp by connecting to a target with multiple peers and reporting an incorrect time to the target. A node uses the network time to validate new blocks. By skewing a node's view of the network time, it would reject new blocks with a timestamp more significant than a predetermined duration. Like the Sybil and eclipse attacks, this would allow the malicious nodes to virtually isolate the target node from the overall network. By isolating a target node, fraudulent transactions could be created and sent to the target.

32. How is security taken care of in hyperledger fabric?

Hyperledger fabric necessitates policies and identities at every point of processing. Therefore, within an organization, you can have admins, and members and they have specific roles to play. Organizations can run peers, and these peers have identities. The ordering service also has personalities, and organizations can participate and order nodes. Only an admin can deploy or install a chain code. If anyone else tries to establish a particular chain code on a pair, it will fail.

Essay-type Questions

1. Compare traditional security model vs blockchain security model.
2. Explain the security aspects in bitcoin in detail.
3. Explain CIAR in detail.
4. Write an essay on Confidentiality in blockchain transactions (C in CIAR).
5. Write an essay on Integrity in blockchain transactions (I in CIAR).
6. Write an essay on Availability in blockchain transactions (A in CIAR).
7. Write an essay on Mixing/Barter transactions.
8. Write an essay on advanced encryption techniques.
9. Write an essay on the various types of security attacks on blockchain.
10. How does hashing and Merkle tree help in security?
11. How regulatory compliance and assurance is taken care of in blockchain?
12. How can smart contracts be safeguarded from security attacks?

CHAPTER 10

Applications of Blockchain

LEARNING OBJECTIVES

Open-source codes and open platforms like Ripple, Hyperledger, Ethereum, and others allow the world to benefit from blockchain technology and build innovative enterprises and projects that can counteract the common and unique challenges faced by all industry sectors. With blockchain-based smart contract system, most businesses can reduce the risk of fraud and costs of intermediaries apart from saving on time and workforce by the automatic execution of business agreements. The automated, verifiable and trustless nature of smart contracts has added a new aspect of tangible blockchain uses in the real world.

This chapter aims to provide an overview of the challenges faced by various industries and how blockchain technology can potentially meet those challenges.

10.1 Introduction

"If you can imagine it, you can create it."

– *William Arthur Ward*

Blockchain Technology has the potential to revolutionize any business that stores structured data. Blockchain is a shared distributed ledger of data collected through a network that sits in the Internet. The way the information is handled makes it unique. It is decentralized, transparent, permanent and secure. As the first foray of blockchain into the finance sector is considered largely successful with bitcoin and cryptocurrency, the underlying technology of the blockchain has inspired the think-tanks and technocrats to explore the potential applications of this revolutionary technology in various industrial sectors like healthcare, retail, real-estate and others.

Though still in the nascent stage, blockchain technology has already penetrated the vision and mission of the government of some countries. The Estonian government has digitized its services using blockchain technology. The health records of the country's 1.3 million residents are on the blockchain. The Government of UAE launched the

Emirates Blockchain Strategy 2021 that aims to capitalize on the blockchain technology to transform 50 percent of government transactions into the blockchain platform by 2021. Countries like Japan, USA, UK, China, Australia, and Singapore are all investing in blockchain application innovations.

10.2 Blockchain in Banking and Finance

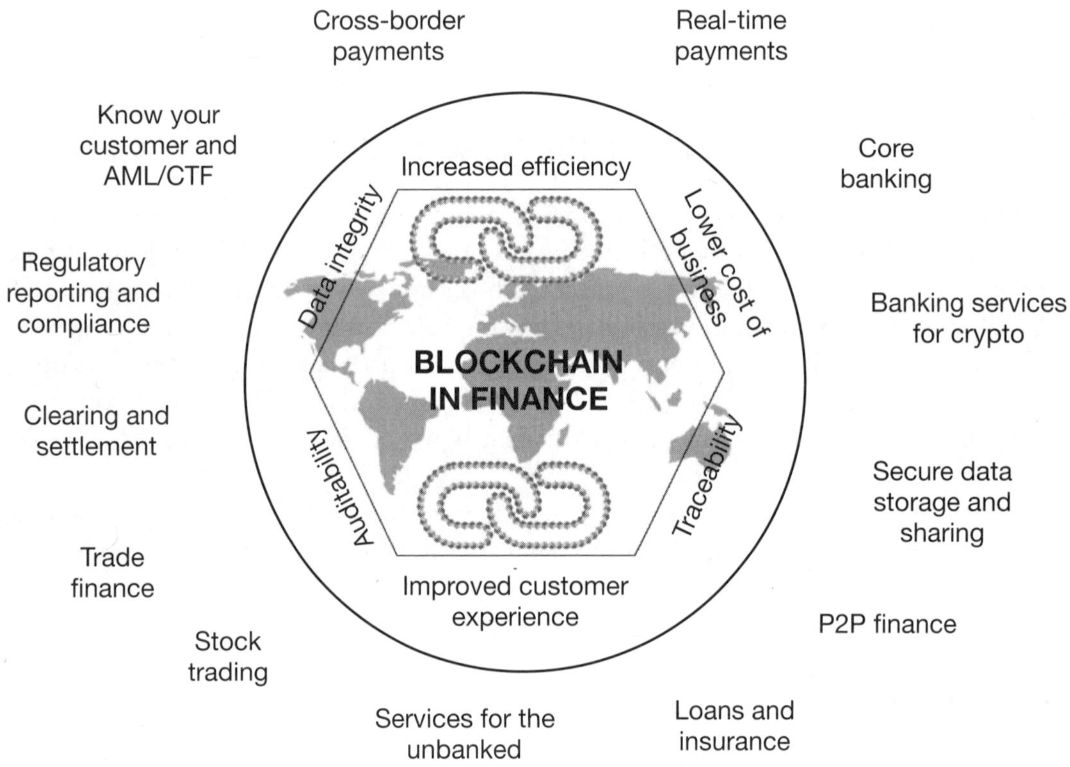

Figure 10.1: Potential blockchain applications in finance

The financial sector consists of banks, investment funds, insurance companies and real-estate. After banks, the second largest industry within the sector is insurance companies.

Most credit and financial institutions function with the help of mediators or intermediaries. In the traditional system of business, the intermediaries bring about the much needed element of trust, making the services of the institutions expensive. Blockchain can help in eliminating the need for mediators, thus enabling banks to provide customers with cheaper services.

Blockchain can pave the way for a new operating model for finance (refer Fig. 10.1). The immutability, transparency, high security, low cost of operations and distributed ledger aspects of blockchain technology make for an up-and-coming and in-demand solution even in the extremely conservative and restricted bank industry. Smart Contracts could introduce another new layer of documenting and meet contractual obligation, reducing reliance on intermediaries for settling disputes. Major pain points linked to cross-border payments and trade finance can be addressed by blockchain. The disintermediation and geographically agnostic nature of blockchain solutions can bring about further savings in capital markets post-trade settlement and regulatory reporting. Some of the finance sector applications that can benefit from blockchain are listed in the sections below.

10.2.1 Challenges in Finance Sector

The finance sector can leverage blockchain for the following benefits:

1) Data Integrity

Blockchain offers the potential to create a single source of truth around customer identity. The immutability and transparency of blockchain solutions can improve data accuracy and security, reduce delays from paperwork and the risk of fraud, and show compliance through an audit trail.

2) Increased Efficiency

Research by KPMG suggests a 40% increase in efficiency by leveraging blockchain for data processing and maintaining a single version of the truth. Blockchain technology can create a ledger that is continuously synchronized throughout the network, thus eliminating the need for reconciliations.

3) Improved Customer Experience

Using blockchain to share permissioned and relevant information with clients and partners may allow financial institutions to serve customers more quickly, fostering trust and even find new sales opportunities. This could boost customer experience and loyalty due to faster processing and use of digital channels.

4) Lower Cost of Business

With blockchain consensus mechanisms and smart contracts, automatic transfer of funds can be triggered when the agreed set of conditions is met. This reduces the amount of time the capital is tied up in a transaction. Blockchain can also, to an extent, eliminate transaction fees by reducing reliance on third parties.

10.2.2 Know Your Customer (KYC)

Know Your Customer (KYC) is a regulatory and legal requirement of banks and financial institutions (FIs) to verify the identity of their customers. According to a report published by Burton Taylor International Consulting, global AML/KYC spending was estimated at about $749 million in 2018 with Asia accounting for $301.1 million.

Credit organizations need to perform KYC when processing applications. Anti-money laundering (AML) regulations, Counter-Terrorism Financing (CTF) and other regulatory pressures to curb money laundering and other intentional or non-intentional criminal activities, have led banks and financial institutions to err on the side of caution. However, the KYC processes are tedious, time-consuming and expensive as the verifications involve massive paper-work and information. Departmental information is not collated, resulting in duplication of work, slow pace of delivery and risk of error, delivering poor customer experience.

A customer identification system based on the distributed ledger technology could be the solution. Blockchain enables user identity information to be incorporated just once and stored securely with access granted to other banks in the system. Once a customer/ client is identified in the blockchain, the information is stored securely with access granted to other banks or FI by the relevant authority. Hence, the KYC process for the customer need not be restarted for new services or applications.

A permissioned blockchain can eliminate the chance of unauthorized access. The KYC data, once stored, is immutable and transparent resulting in greater operational efficiency, increased trust between institutions and reduction of labour-intensive data gathering, processing time and costs. Waiting times for the customer is considerably reduced as it eliminates the need for customer to repeatedly provide the same information to the service providers. This fosters trust and customer satisfaction.

For regulators, this single source of customer data leads the way for better understanding and transparency of customer activity across FIs' and could theoretically bring about massive reduction in financial fraud and crimes in the long term.

Current initiatives at play

- CordaKYC, a KYC utility built on R3's Corda blockchain platform, works on a self-sovereign model, where corporate customers can create and control their own identities, including relevant documentation. The trial encompassed 39 banks, firms and regulators that included the likes of BNP Paribas, Deutsche Bank, ING and Société Générale.
- Belrium, a hybrid blockchain KYC solution from a Malaysian based start-up Belfrics, is already being deployed by the governments of Nigeria, Kenya, Bahrain, Abu Dhabi and Malaysia. A MoU has been signed with the government of the Indian state of Andhra Pradesh.

10.2.3 Cross-border Payments

There is a steady growth in global businesses and international trade over the last two centuries, transforming the global economy and contributing to an increase in intricate cross-border payments. As per McKinsey, cross-border payments account for around 40% of global payments transactional revenues with payment flows of more than $135 trillion during 2016.

However, the current cross-border payment process suffers certain shortcomings. On an average, a cross-border transaction takes 3–5 business days. International payments lack visibility leaving the beneficiary guesstimating on the delivery time of the payment as the exchange rate is not fixed till the arrival of the funds. In addition, the involvement of multiple financial institutions creates a complex web of procedures accompanied by transactional fees and charges at each stage. This makes the process costly, slow and less suitable for today's global e-commerce demands.

To address the current operational weaknesses, financial institutions and banks are adopting blockchain for near-instant cross-border payments with higher security, transparency, reliability and lowering costs by cutting off the middlemen. With blockchain, transactions are tamper-proof, thereby cutting costs associated with investigation of cases and litigation.

Current initiatives at play

- Global banking payments network, SWIFT plans to launch a PoC with blockchain consortium R3 Tech to use its GPI Link payments standard via R3's Corda platform on a trial basis.
- Ripple has over 100 customers globally using their enterprise blockchain network RippleNet for cross-border payments.
- IIN (Interbank Information Network), led by JPMorgan Chase and other banks, uses blockchain technology to transfer US dollars across borders and institutions

10.2.4 Trade Finance

According to World Trade Organization, around 80–90% of world trade relies on trade finance (trade credit and insurance/guarantees). Financial institutions act as the guarantor of payment between seller and buyer. Trade finance is expected to mitigate the many risks involved in trade related to product, manufacturing, transportation and even currency, if we are to consider international trade.

Banks, shippers, importers, exporters, regulatory bodies and the customs officers are just some of many parties involved in a typical trade transaction. There are verification points at each node on the supply chain and any fault at any of these verification points would cause a delay or complete voiding of a transaction. Streamlining the process of obtaining approvals from multiple legal entities like customs, port authorities, trucking or rail transportation firms, and others for the movement of goods across borders is critical.

With Blockchain, the complex processes of trade finance can be simplified where all entities have the required access to the ledger, thus fostering increased trust and accountability among the enterprises, regulators, and consumers. Blockchain technology could help decrease recordkeeping costs and eliminate some intermediaries while simultaneously reducing paper-based systems that cost time and money. The blockchain can not only be used by the legal entities for approvals, but also by all concerned parties to gather information on the approval status. Blockchain ledger can be used as the single source of truth to capture receipt of goods, and record the payment transfer between importer–exporter banks. There is an increased access to capital due to faster settlement times and avoidance of errors and disputes.

Current initiatives at play

- The Institute for Development and Research in Banking Technology (IDRBT), the Reserve Bank of India's research arm, has successfully tested blockchain technology for trade application. The evaluation was carried out by conducting a workshop involving all the stakeholders, including academicians, bankers, regulators and technology partners. A white paper that details the technology, concerns, global experiences and possible areas of adoption in the financial sector in India was released.
- We.Trade (we-trade.com), a consortium powered by the Hyperledger Fabric blockchain, built a cross-border trade finance platform for tracking, managing and protecting trade transactions amongst SMEs. With 12 participating banks (including HSBC, KBC, Natixis, Deutsche Bank, Nordea, Santander, Société Générale, Rabobank, and Unicredit) and 24/7 availability, it connects the seller, buyer, seller's bank, buyer's bank and transit operators on a single network spanning multiple European countries. Since all the KYC/AML checks are in place, the entities are trustworthy and verification points can trigger automatic payments if goods are received in agreed condition.

10.2.5 Stock Trading

Decentralized blockchains and secure ledger platforms give every party a say in the validation of a transaction, speed up the settlement process and allow for greater trade accuracy. Buying and selling stocks and shares involves many parties including middlemen like brokers and the stock exchange itself. Blockchain technology can potentially cut out not only the brokers, but also decentralize the stock exchange.

The trade of securities is governed by security laws and involves intermediaries, such as bankers, portfolio managers, underwriters, stock brokers and registrars, who provide trust and safety. Blockchain with smart contracts can be compliant with the law, promote trust and save on commission fees by cutting out the intermediaries.

Banks are looking at blockchain to improve settlements. With blockchain, transactions can be settled near instantly, removing the need for a sizeable middle- and back-office staff. Blockchain smart contracts can settle trades on the day of the transaction itself efficiently and create self-executing payments to release dividends to shareholders more quickly and cost-effectively.

Tokenization is another aspect that is being targeted. An asset-backed token digitizes an asset, i.e., anything of value both tangible (a house, land, or cash) or intangible (like intellectual property) that then gets recorded and traded with minimized costs and risk for all parties involved. Double-spending can be prevented as the entire blockchain is shared by all participants, and it is easy to check who owns which token at any given time. The benefits of tokenization are numerous: liquidity of assets, their securitization, reduced volatility, and a possibility for intangible assets to be precisely defined.

Current initiatives at play

- Nasdaq is using Nasdaq Linq Blockchain Ledger technology blockchain to record private securities transactions.
- India is in the early stages where National Stock Exchange (NSE) conducted a blockchain trial with collaboration between the country's leading banks (IDFC, Kotak Mahindra, ICICI, IndusInd and RBL) and HDFC Securities using KYC data.

10.2.6 Insurance

Blockchain has a lot of potential user cases in the insurance field. Just like any other financial sector institution, the insurance industry also faces challenges of stringent rules and regulations, need for custom-tailored options meeting customer expectations, complex processes, high costs and fraud risks to list a few.

Today insurers and brokers need to manage loads of paperwork and electronic forms created by parties such as claims assessors, lawyers, and other experts. A blockchain-based claims system would make the process of sharing data faster and more efficient, creating significant operational savings for all parties. Payment of insurance can be automated by programming smart contracts that are executed automatically, enabling people to receive payments immediately. Whenever an insurable event is reported by a trusted source, the insurance policy is automatically triggered, processing the claim as per the policy terms specified in the smart contract, and the customer is paid on time.

For insuring high-value assets like art, jewellery, or antiques, blockchain can provide secure and provenance tracking authenticating the proof of ownership and value of assets. By moving insurance claims onto an immutable ledger, blockchain technology can help eliminate common sources of deception in the insurance industry.

Thus the insurance industry can benefit from blockchain technology in eliminating the cost of insurance claims processing, reducing insurance fraud and improving customer satisfaction.

Current initiatives at play

- Etherisc's DIP (Decentralized Insurance Protocol/ Platform) aims to provide a permissionless, standard mechanism by which risk of any kind can be priced, serviced, and transferred by a group of independent service providers without having to rely on centralized parties, thus making the purchase and sale of insurance more efficient and affordable. It has launched flight delay insurance and worked towards the launch of crop insurance, social insurance, hurricane protection and crypto-wallet and loan protection.
- VouchForMe (vouchforme.co) uses the power of social network to lower the cost of insurance. The platform aims to reduce risk and fraud by social proof endorsements from

family and friends. The system works on the principle that higher the endorsement, lesser the risk to the insurer and lower the premium.

– Toyota Research Institute (TRI) is collaborating with partners and looking into usage-based insurance. Vehicle owners can save money on their insurance rates by allowing the vehicle's sensors to collect driving data pertaining to safe driving habits and store it in a blockchain that can be made transparent to their insurance companies. The increased transparency prevents fraud cases as well.

10.2.7 Mortgages

The mortgage industry is perhaps the most complex home financing market in the world. A typical mortgage lending process is lengthy, heavily paper based, labour intensive and time-consuming. Aside from the debtor (property owner) and creditor (loaner, typically a bank), the process involves multiple parties from surveyors for up-to-date property evaluations, lawyers for advisory and legal documentation, credit agencies to assess the loan eligibility, underwriters for risk assessment, land registry offices to attest and update property ownership, among others. This results in increased cost, high processing times and lack of transparency.

All the entities work on disparate systems, leading to duplication of effort and the need for a large team of administrative staff to deal with the heavy paperwork. Transaction fees at each intermediary level add to the overall cost of the mortgage process. Also, the additional days taken by each intermediary adds to the overall processing time. There is also a lack of transparency as all the different institutions involved like banks, brokers, credit agencies, property dealers, and other third parties have their own independent systems and processes. Hence, communication between the various parties are either physical paper based or via emails. In such a scenario, it is difficult to track the exact state of mortgage application at any given time making it a long-drawn process. Post approval, as soon as the property changes hands, the title deed information needs to be updated in all the systems. The manual process is prone to human error leading to incorrect, incomplete or inconsistent data.

However, if the property deeds and other relevant information were stored on the blockchain, it easily addresses the challenges of time, cost and transparency. In a blockchain enabled system, the property could be tracked directly on the system by the various institutions/parties based on the access rights provided. Property information, digitized documents, valuations and ownership linked to the property can be updated by the relevant parties. Banks can check out the required information and verify the chain of the title without relying on third party like lawyers, surveyors, etc. making for faster processing times and lesser transaction fees. Smart contract features can be implemented to release funds as soon as specific conditions are met. The distributed ledger technology makes for transparency allowing the property owner to check out the exact state of the mortgage. This reduces the need for administrative overhead, improves approval turn-around time, faster settlement times and relieves all-round stress. Post-approval, the mortgage loan repayment status can also be monitored, if required.

Current initiatives at play

- Liquid Mortgage (www.liquidmortgage.io) is a 2018 blockchain-based digital asset platform for both individual and corporate borrowers to manage loan information, payments, and provide direct interaction with lenders. The platform provides cashflow management, transaction validation and portfolio management enabling lenders to benefit from smart contract features, lower costs and real-time transaction data.
- As per a 2018 report, Bank of China Hong Kong (BOCHK) processes 85% of its mortgage-related property valuations that are necessary to calculate monthly mortgage payments, using blockchain technology. Processes between banks and RE appraisers that earlier necessitated communication via emails, faxes and physical documents is now executed in seconds within the blockchain.

  (Source: www.scmp.com/business/banking-finance/article/2142997/bank-china-embracing-blockchain-use-hong-kong)
- Moneycatcha (mcatcha.com) is a startup that offers blockchain-based solution for cheaper and faster loan applications, increased transparency, pre-vetted borrowers, retrieval and verification of data, compliance, risk assessment and streamlined deployment of capital. It has two solutions: Homechain for end-to-end loan origination and RegChain, a regulatory compliance tool for real time risk-monitoring, loan performance assessment and reporting. HSBC Australia is currently using Regchain to accelerate regulatory reporting.

10.3 Blockchain in Education

The internet was a disruptive technology that dramatically changed the traditional classroom education model. It is said that the first instinct of students (over 93%) is to turn to the internet (Google or Bing) when faced with a research problem. Over 32% of students in higher education take at least one online course. In addition, they use social media to improve their network, join study groups, follow topics of interest, and last but not least, for online résumés and job portfolios.

Today, it is not just the students who turn to internet and social media, but also the faculty and administration. With online education programs taking the industry by storm, online applications are now a norm for most universities and colleges. Public colleges and universities experienced considerable growth in online course enrollment in 2015 and 2016. However, some of the challenges that the internet, with its virtual classrooms and one-shop-stop for information, has not been able to address adequately, is the quality of content and lack of accreditation. The internet is an excellent tool for connecting and collaborating, but it is built for storing and passing information and not for creating and monitoring value.

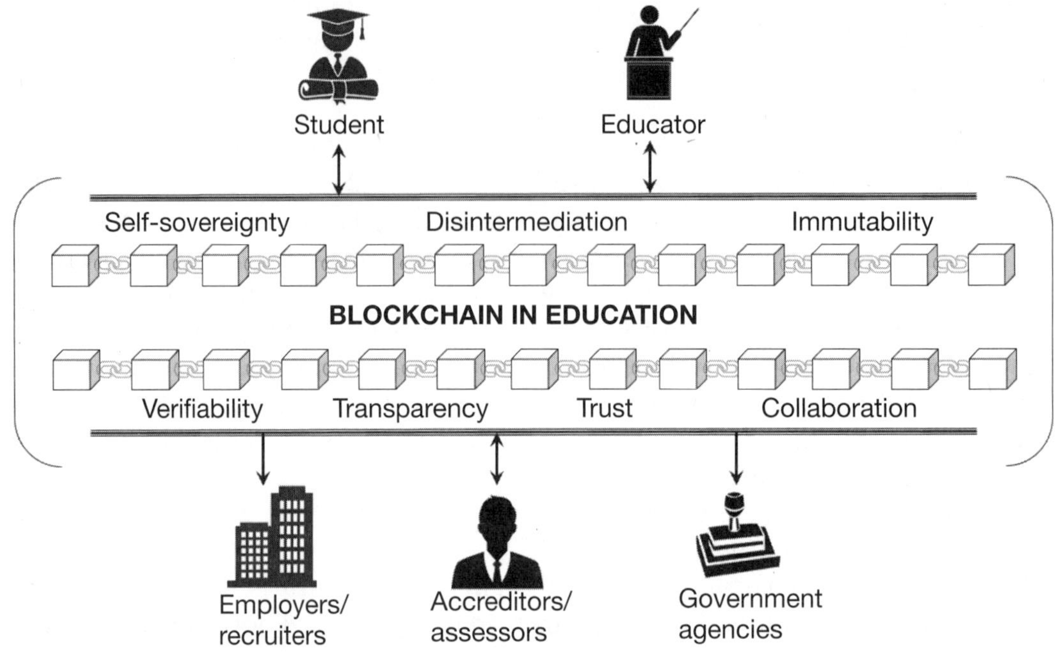

Figure 10.2: Players within the education blockchain

This is where the second disruptive technology steps in – the Blockchain (refer Fig. 10.2).

Currently schools, colleges and institutes store records like student private information and transcripts either as paper records or in centralized servers. Both these methods are not efficient as they are prone to misplacement, destruction or hacking. In addition, with the rise of unauthorized institutions and fraudulent certificates, employers and organizations question the validity of the educational qualifications and skills reported by students and educators. As education becomes more diversified and decentralized, the educational institutions continue to seek ways to maintain reputation, trust in certification, and proof of learning. These are challenges that the blockchain proponents believe can be addressed by blockchain technology.

10.3.1 Challenges in Education

Some of the benefits from blockchain that the education sector expects to harness are:

1) Self-sovereignty

Self-sovereignty is all about owning one's own data and identity. Our identity is our personal property and it is our sovereign right to keep it private and have control over the management of our personal data. Blockchain offers a robust infrastructure to identity attestations. This could give control to the students over their educational portfolio and avoid dependency on educational institutions for the storing of personally identifiable information (PII). A person can create a self-sovereign identity on a blockchain which is fully controlled and maintained personally by the individual. This could address the issue of identity theft prevalent on the traditional identity management system.

A permissioned blockchain-based decentralized registration system can provide an individual with an independent ID that is not under any centralized authority and hence cannot be controlled or interfered with by any third party without the individual's consent.

2) Immutability

Certificates can be faked, credentials can be weak, and education may not be accredited. With blockchain, once transaction records like transcripts, attendance sheets, assessment records and results, certificates, credits and other records of achievement are created, they are permanently stored and cannot be modified or deleted. A block in the Blockchain contains the cryptographic hash of the previous block, as well as a time-stamp and the transaction data. This means that once transaction records like transcripts, attendance sheets, assessment records and results, certificates, credits and other records of achievement are created, they are permanently added to the blockchain in chronological order as a block. These blocks cannot be modified or deleted, once they are added to the blockchain.

Every network participant (node) holds the exact same copy of the blockchain, ensuring traceability. Any modification to academic data can only be created as a new record with the consensus of every network participant/node. This guarantees a real-time view of the student's certifications and qualifications that cannot be faked or hyped.

3) Transparency and Trust

In blockchain, transaction is validated through a consensus process. Data is kept secure and private through cryptographic techniques. Transparency is important in education for a variety of reasons. It gives a wide and in-depth view of the education process involving the students, teachers, colleges, universities and related organizations. The visibility of educational transactions at all levels of the student's education builds confidence and trust in the education system enhancing the authenticity and strength of certificates/degrees issued.

4) Disintermediation and Collaboration

Professor John Dominique, Director at Knowledge Media Institute, an Innovation Lab at Open University in UK, has stated that the centralized model of present-day education is no longer sustainable. Blockchain technology allows for total disintermediation and disaggregation of education. Eliminating the need for a central controlling authority for transaction and record maintenance moderates the number of player interactions, saves time and cost and, best of all, promotes collaboration between students, teachers, and employers. Students can engage with top-quality educators anywhere in the world directly, provided they all exist in the blockchain ecosystem.

10.3.2 Identity and Student Records

Learning is a life-long journey. Learning does not end when one graduates but it continues throughout one's professional life to stay ahead of the competition and even further. Keeping track of all the transcripts, degrees or diplomas and certificates is not an easy task primarily when one refers to it only for a job change or immigration purposes. Currently, the degree and other academic certificates and transcripts are stored at the Universities or Institutes, along with the personal information like date of birth, parents/guardian names, passport or identification number, student loans, and related information. This data is vulnerable to hackers. It is the responsibility of the college or university to ensure the security, privacy and reliability of the student's data. Photocopies of certificates are sent out while applying for a job or higher education and the organizations, in turn, use intermediary services to verify the certificates and qualifications. Academic credential verification remains largely a manual process. It is not only time consuming, but quite expensive and prone to fraud as well. People make fake profiles, exaggerate or falsify resumes and every country has institutions that issue fake certificates as well. So it is not surprising that employers want to validate the original certificates from the issuing universities or institutes. As processing these requests are an overhead, the universities charge a transaction fee for the same.

With blockchain technology, students or applicants can have full control of their personal identity information, enabling them to share their degrees/diplomas/certificates directly with the prospective employer or institution without contacting the university. Blockchain provides a tamper-proof system of record that can host the educational data and securely share it among permissioned organizations. This removes the burden on the educational institutions to store and maintain documents, saving millions of dollars on administrative costs eliminating paperwork and manual labour for degree verification. By cutting down on degree fraud and promoting transparency, the integrity of the institute and its credentials is protected, therby promoting trust in the system of accreditation.

When on the subject of identity and records, blockchain is not limited to just academic learning. Continuing Professional Education (CPE) and talent management trainings delivered by companies are often fragmented and poorly tracked.

Current initiatives at play

- BlockCerts (http://www.blockcerts.org/), founded in 2017, is an open blockchain platform for creating and issuing verifiable certificates.
- The Knowledge Media Institute (KMI) within the Open University is working with APPII, a blockchain career verification platform, to build a platform that can register and verify student academic records.

10.3.3 Student Financing

Education is a costly affair and higher education is quite expensive for average to low-income families. The economic growth of a country depends on a number of factors like its natural resources, capital, technology and most of all, on the quantity and quality of its workforce. Education is the most critical component out of various components of social infrastructure. A well-educated and adequately trained workforce can accelerate the pace of economic development. That is why the student loan system exists in every country, though the operating procedure may vary. However, what is common is the issues related to the student loan debt. In UK, student debts stand at £100.5 billion ($132 billion). In the US, around 44 million borrowers owe over $1.48 trillion in student loan debt. At Rs. 72,000 crore ($10 million), public loans for higher education is considered to be the fastest growing non-performing assets in the banking system in India. These numbers are expected to increase due to rising cost of education, rising student numbers, very slow employment rates and mismatched skills to job requirements. The unemployed students cannot be expected to repay their loans.

Currently, student loans are sanctioned with hopes of repayment by the student post-employment. Instead through smart contracts, financial aid can be targeted toward personal development. Companies that need specific skills could sponsor individual students, identify goal-based targets and payout according to achievement. Another option is tagging loans to employment contracts. Employers can finance the learning program of student/employee, and the wages paid can automatically be adjusted to repay loans attached to the student/employee. This would considerably bring down threats of non-performing assets; and benefit both students and employers alike.

Blockchain technology can improve the system for calculating scholarships for students and provide a transparent and fair mechanism for funding grants and projects.

Current initiatives at play

- EduCoin (edu.one) is an India-based blockchain startup company that aims to facilitate online education at regions where traditional payment is not possible and where skills development is necessary to avoid poverty and professional obsolescence.
- Tutellus (tutellus.io) works on the model where students, instead of paying for their education, are rewarded for completing their courses. In addition, a student's success is directly proportional to the reward for their educators as well. Thus the incentive to bring up the quality of education.

10.3.4 Verification of Academic Credentials

One of the biggest challenges educational institutions, recruiters, immigration or visa processing authorities face is ascertaining the authenticity of the school and degree certificates of the candidates. With the incidence of doctored certificates growing, recruiters and companies have been seeking the help of verification companies. There is no global standard for education. Hence, verification requires diverse approaches that are not too well-equipped to identify fake certificates. Also the verification process is time consuming, sometimes taking weeks.

With blockchain technology, we can get a system that provides tamper-proof verification of education records. A blockchain system storing national or global level digital educational certificates where each block represents a country that has sub-blocks for every accredited university or degree awarding bodies in that country could be the answer.

Did you know?

According to a 2014 survey conducted by CareerBuilder, a global leader in human capital solutions, 57% of job applicants have embellished their skill set, 55% have embellished their job responsibilities, 42% the dates of employment, 34% their job titles, and 33% have lied about their academic degree.

Source: www.careerbuilder.com

Press Release 07-Aug-2014: Fifty-eight Percent of Employers Have Caught a Lie on a Resume, According to a New CareerBuilder Survey

The digital certificates of the students are directed to the blockchain network. They are stored as a one-way hash coding. Whenever an organization or prospective employer requires a copy of the certification, the student can send a public key to the academic institute from where they completed their study. The university or educational institutions sends the academic credentials/documents through blockchain to the student who in turn shares it with the prospective employer.

With the consensus mechanism of blockchain, there cannot be any tampering or manipulations leading to the total eradication of bogus degree and courses, saving time and money for students, immigration authorities and employers. The same system could be used by students to verify authenticity of educational programs offered by universities around the globe.

Current initiatives at play

- RecordsKeeper (recordskeeper.co), as the name suggests, is a public blockchain where users can have their own private infrastructure to store records.
- Blockcerts, BitDegree are all blockchain based solutions for digital educational certificates to make the process of document verification transparent and secure.
- Edgecoin (edgecoin.io) aims to build a platform that contains a person's full education history, qualifications, skills, certifications, and employment history for a swift and fully automated hiring process.

10.3.5 New Pedagogy

Learning is now independent of time and space. We can learn anything, from anywhere at any time. Pedagogy is study of teaching methods, including the aims of education and the ways in which such goals may be achieved. Blockchain paves the way for new teaching and learning models as we move past the traditional broadcast models.

Online or distance education is a platform that can leverage blockchain technology to fix some of its issues related to lack of quality, lack of student motivation to complete the courses and questionable benefits. Online educators and contributors can be paid in cryptocurrency. In addition, blockchain tokens via smart contracts can be used as incentives for creating high-quality educational content and superior teaching courses that can be attested by the students and other participating nodes. This concept is similar to that of YouTube, where more the videos that are watched, more the YouTubers are paid in royalties.

Students can be rewarded with blockchain token to incentivize them to complete their courses. These tokens can be used to fund further studies or courses or converted to cryptocurrency and cashed.

Blockchain-based learning systems allow people to access education around the world, and without the barrier of high costs. We could potentially be looking at a global learning model where students gain access to courses across the globe, attend courses in parallel that are tailored to the global/country requirements and all the while competing with other learners for token points and credits.

Mozilla's Open Badges is an IMS standard used for digitally embedding accomplishments in portable image files. It is used by thousands of organizations in the micro-credential space. The digital badge system aims to give students the freedom to pick the subject and skill areas of their choice, thus allowing them to create their personalized curriculum. However, open badges can be used for just about anything including soft skills, gamification and for achievements not traditionally associated with formal education. Hence, it lacks reliability as a definite proof of credential. In addition, the infrastructure is not secure and prone to manipulation. However, blockchain with the concept of Open Badges can move into the high-stake credential space like degrees, diplomas and certificates. The blockchain technology features of permanence and immutability are used to secure and authenticate formal education qualifications. Also as a hashed digital fingerprint of the document is stored in the blockchain, personal information is never at risk.

Colleges and universities are looking to move to a more interactive, adaptive and incentivized approach to education. A new model proposed for promoting classroom collaboration is by replacing management hierarchy with self-organized student teams. Students agree on what needs to be learned, the roles and responsibilities and rewards. These are then codified and executed as smart contracts. In another concept, model learning can be treated as a currency. Students can get paid for their time in self-driven tasks powered by smart contracts, thus rewarding mastery-based learning and leading students towards skills needed by employers.

Many of the blockchain education models are conceptual, but disruptive technology has a way of revolutionizing the concept.

Current initiatives at play

- OpenBlockchain, an initiative of the Open University, UK, is experimenting on ways to store and issue Open Badges using Ethereum platform. Its aim is not limited to university certification, it presents a more detailed picture of students' learning experiences throughout their life, like participation, official certification, community involvement, and other skills people develop outside of formal credentialing systems.
- Vanywhere (http://www.vanywhere.com/) is a live skill sharing platform where a follower can connect with influencers for 1-on-1 voice or video chat making it more personal than on social media. Privacy and payment will be facilitated by Vanywhere's blockchain integration.

10.4 Blockchain in Energy

As the global economy moves from the industrialization age to the information age, the need for energy has reached unparalleled heights. According to the most recent United Nations estimates, the population of the world as of January 2019 is 7.7 billion. A sizeable number of these people across the world have access to lighting, transportation, communication devices, computers and other electrical and electronic devices and conveniences, which is possible only by utilizing energy. With economic growth and continuous improvement in the standards of living, the world's energy consumption is estimated to increase by almost 50% by 2040. Energy supplies mainly come from fossil fuels like crude oil, coal and gas; as well as nuclear power and renewable sources like solar, hydropower and others. However, fossil fuel is the most important energy source accounting for 70% of electricity generation as published by World Energy Council. To meet the challenges of rising energy demand and global emissions, it is no surprise that the innovators of the energy sector are speculating on blockchain technology as a solution.

10.4.1 Challenges in Energy Sector

The main concerns impacting the Energy Sector that needs to be addressed are:

1) Energy Inefficiencies

Countries mainly rely on centralized power plants that generate electricity which moves through a complex system of substations, transformers and power lines. Power is transmitted over long distances and distributed to the homes of consumers through high-power transmission lines. This complex system is called a grid. The transmission over long distances causes power losses to an extent estimated to be between 8–15%. Transmission and distribution losses vary from country to country as well. In some countries, the loss is between 30–60%, attributed to electricity thieves and other factors. These losses not only hit our electricity bill but also indirectly contribute to carbonization as approximately 40% of global CO_2 emissions are emitted from electricity generations through the combustion of fossil fuels.

In addition, if the energy supply and demand is not balanced in the grid, it can damage the infrastructure and cause safety issues like fire and human causalities. Blockchain is expected to help make enhanced, modern and optimized power grids.

2) Environment and Climate Changes

All methods of electricity generation have an environmental impact on our air, water and land. Fossil fuels like coal, oil, and natural gas cause substantially more environmental damage than renewable energy sources like wind, solar, geothermal, biomass, and hydropower.

The global energy sector is responsible for two-thirds of all the greenhouse gases emitted into the atmosphere as reported by the International Energy Agency. From unstable climate change to poor water and air quality, pollution and disease, our natural habitat is under threat. It is time for our energy system to evolve in response to these environmental challenges.

Blockchain technology enables the production and distribution of electricity more efficiently, thereby reducing both the amount of greenhouse gases emitted and other forms of air pollution.

Energy Inequality

Energy retailers purchase energy (natural gas or electricity) in wholesale markets and sell it to end-users at fixed price. They charge consumers a premium for this service along with a provision for ancillary services, which is reflected in the energy bill. Thus energy retail companies make profit by marking up the price of energy the customer consumes. This system is inefficient and unfair to the consumers.

Blockchain technology can help reduce energy inequality and inefficiency by empowering consumers to buy and sell energy from other consumers or suppliers directly, thus removing the need for intermediaries/middlemen.

10.4.2 Peer-to-Peer (P2P) Trading

In peer-to-peer (P2P) energy trading, people can generate their own energy from renewable sources like solar, wind, biomass, etc. for homes, offices, factories and trade or exchange excess energy with each other. With people becoming prosumers, i.e., both producer and consumer of energy, blockchain has an extensive user case establishing a decentralized energy supply system. A decentralized energy transaction and supply system on the blockchain network would empower consumers to manage their own electricity supply contracts and consumption data. All transactions made between the individual parties are directly executed through a peer-to-peer network facilitated by a smart contract without a central or intermediary third party intervention. This "microgrid" network on Blockchain is more efficient with saved cost, reduced power loss as it is a short distance for energy distribution and low carbon emissions.

The "microgrids" are self-sustainable and can theoretically be enlarged to interact with other P2P grids in a distributed and secure network enabling the buying and selling of renewable energy at cheaper costs. Smart meter data could be shared with participating consumers without compromising on privacy or security, while at the same time offering the opportunity to generate income from sharing. Consumers can also use cryptocurrency for payment of consumer bills.

Another P2P prosumer opportunity is in Electrical Vehicle (EV) charging posts. It uses the electrical energy stored in rechargeable batteries. Electric vehicles are expected to grow from 3 million to 125 million by 2030 according to International Energy Agency forecasts. EV owners can charge their vehicles both at home and at charging posts located at gas stations, retail stores or parking lots. Nearly half the owners have their own chargers. Through the blockchain enabled P2P network with smart contracts, private owners can make available their charging posts to other participating EV owners for charging their vehicle while on the road and earn revenue as well.

Current initiatives at play

- PowerLedger (powerledger.io) is an Australian energy trading platform that uses blockchain technology enabling households to trade their excess rooftop solar power with their neighbours. Their projects are currently running in Australia, Thailand, US and Japan.
- Share&Charge (shareandcharge.com) is a decentralized blockchain protocol for electric vehicle charging, transactions and data sharing.

10.4.3 Smart Grids

The electric grid is a complex interconnected network for electricity generation, transmission, and distribution from producers to consumers. It consists of generating stations producing electrical power, high-voltage transmission lines, and distribution lines that connect individual customers. Because of the physicality and distance from source to consumer, experts largely agree that power loss and occasional failure of the power transmission system are inevitable. Power outages, both planned and unplanned, are a common state in some countries, which cause business disruption, heavy losses and security issues. For an ideal power grid, there has to be a balance between supply and demand. This is practically impossible as there are many stakeholder systems like the energy generators, distributors, traders, regulators, and industrial end users who are equipped with no real-time communication.

However, the Blockchain can help by making smart power grids. With digital technology combining controls, automation, IoT, computers, smart meters and other devices communicating with each other in the power grid and responding digitally to the changing consumer demands, the blockchain can store and track energy data including failures. With intercommunication between power grids, points of failure or outages can be detected faster and directly communicated to the companies, making the repairing process more efficient and effective. It can also take advantage of customer-owned power generators (P2P) to produce power when it is not available from utilities.

The high amount of transaction data connected to the power grid, including past issues stored in the blockchain, can provide authorities information on the need for renovations or new installations. Analysis of the data can predict failures in advance and also suggest options on how to bring the grid back up to normal working condition faster. Smart contracts can be enabled to control energy flows as required to create a balance between supply and demand.

Current initiatives at play

- Grid+ (gridplus.io) uses Smart and Blockchain technologies for consumers to trade and buy directly from the grid. Factories that consume significant energy can now acquire as per requirement with minimal wastage, thus making substantial save in energy bills.

10.4.4 Energy Trading

The gas and energy commodity trading industry, as with any commodity trading, is a complex structure that requires third party intermediaries like brokers, trading agents, exchanges, banks and regulators among others. The current procedure involves massive manual processing involving paper work, multiple reconciliation and communication between various entities to consolidate the information at different sources. This makes the procedure slow and insecure with high risk of documentation fraud. Currently traders buy and sell energy on the exchanges, while banks act as intermediaries to handle the transactions made by the parties involved. With blockchain's distributed ledger technology and smart contracts, generating units can directly trade with a consumer or an energy retail supplier, thus cutting out the role of middlemen and intermediaries like brokers, exchanges and trading agencies. With a digital token, trading can be done directly and securely. The implementation of blockchain technology in commodity trading has the ability to reduce costs associated with not only the build and maintenance of complex trading systems, but also labour costs, data management, data visibility, settlement delays, and inter-system communication.

Current initiatives at play

- WePower (wepower.network) is a blockchain-based green energy trading platform where everyone can buy, trade or invest in tokenized green energy at scale.
- EnergyPremier (energypremier.com) is a blockchain and cloud-based electricity retail bidding platform for both suppliers and consumers, aimed at transparent pricing and direct market access, promising fair and better prices to consumers.

10.5 Blockchain in Healthcare

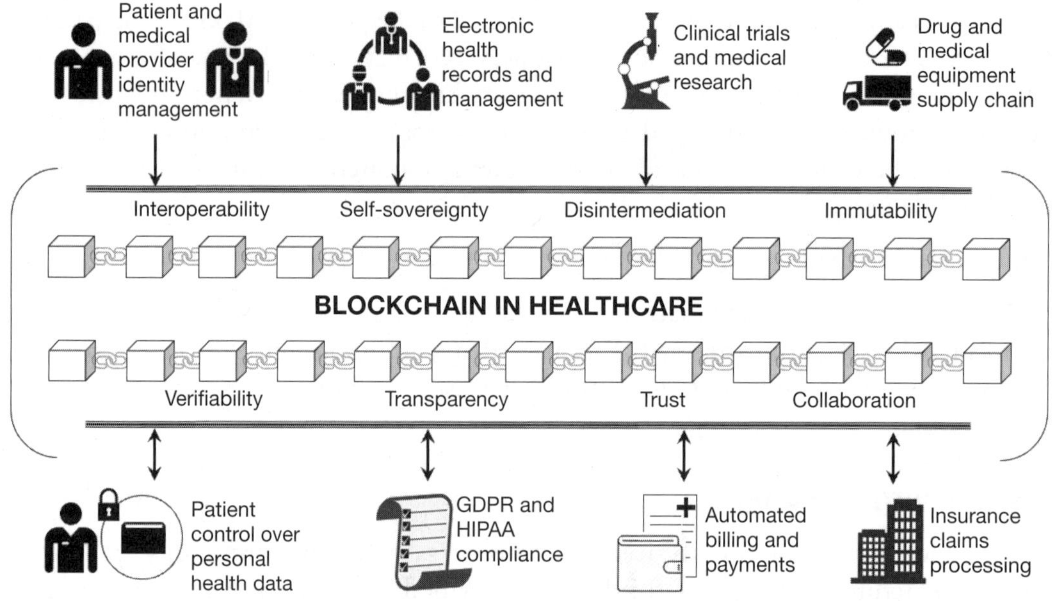

Figure 10.3: Potential blockchain applications in healthcare

Access to healthcare is one of the most important factors that affect the quality of our life. It is not without its challenges considering the changing landscape of the healthcare industry. People are becoming more and more aware of the importance of healthcare and the medical treatment options that are available. Information on medicines, treatments, healthcare providers and practitioners are all just a click away in the internet. With this information overload comes the challenge of ascertaining how accurate and reliable the information is. But this also means that the government, health authorities and medical professionals need to change their strategy on how they provide services as healthcare moves from being a supply-driven industry to a demand driven-one, with people wanting to make more informed choices about their healthcare (refer Fig. 10.3).

10.5.1 Challenges in Healthcare

The main challenges that exist in healthcare today that can potentially be addressed by blockchain are:

1) Interoperability

Interoperability means different things to different people. In healthcare, it is the process of freely exchanging healthcare information among electronic systems to ensure delivery of high-quality, effective, and efficient care to patients. This means that any and all electronic health records (EHR) that are stored in disparate systems should be integrated to make the consolidated information available for the patient at the point of care. To address the issue, the healthcare industry has developed and appropriated standards for both syntactic interoperability based on the structure of the communication, and for semantic interoperability, which stresses on the meaning of the communication. Some messaging standards to facilitate communication between different clinical systems are HL7 V3, DICOM, and IEEE 1073. Some organizations use the Clinical Document Architecture (CDA) standard while others use Fast Healthcare Interoperability Resource (FHIR). Varying data standards across organizations is a key challenge to true interoperability.

A US research study, published by healthaffairs.org in Oct 2017 (https://doi.org/10.1377/hlthaff.2017.0546), revealed that only 29.7% of hospitals were able to meet the four key metrics necessary for true interoperability: data integration, reception, distribution and finding. Results showed that progress towards interoperability is slow with the focus more on information movement between hospitals rather than on managing and maximizing healthcare value. In a country like India where patient health records in many private and public healthcare sectors are still paper-based, this poses a big problem.

Blockchain technology can address the challenge of varying standards by accessing data through application program interfaces (APIs). Data transfer through APIs achieves standardization of data formats. Thus, data can be transmitted irrespective of the capabilities of EHRs to communicate with different HL7 versions (HL 7 is the messaging standard for electronic data exchange in the clinical domain, widely implemented in the healthcare sector).

2) Security

It is a foregone conclusion for financial institutions like banks that it is mandatory to adopt a robust security system for data protection. Two-factor authentication of customer accounts along with multiple firewalls and cybersecurity has become a standard for banks to protect themselves from hackers. Unfortunately, the same cannot be said for healthcare institutions. However, given the sensitive nature of patient records and even the need to share such information with third parties like pharmacies, specialty hospitals, etc., it goes without saying that protected health information (PHI) or personal health information should be secure and protected.

PHI is any demographic information that can identify a patient. It contains such information as patient's name, date of birth, addresses, phone numbers, IDs, card numbers, place of work, insurance information, medical history and others. Misappropriation of such information can lead to a complete identity theft. Hence, it is not surprising that information security breaches in the healthcare industry affected more than 27 million patients in 2016. With cybercriminals putting more time and resources into exploiting and monetizing healthcare data, data and PHI security has become important than ever.

Blockchain is immutable, i.e., nobody, in theory, can tamper with the block data. Data is encrypted, keeping it indecipherable for anyone who does not have access to the decryption keys. In addition, via blockchain, one has a full history of who can and has accessed one's data, what was accessed and when it was assessed.

3) Accessibility

Another component of quality healthcare is universal accessibility. More often than not, a pharmacy or a medical laboratory may need access to patient files for a particular or routine diagnosis. Providing relevant and specific patient information to the healthcare entity can be tricky. In majority of the cases, the whole patient file is shared or access to patient files is not possible and medical tests need to be redone. With blockchain-enabled smart contracts, the patient or authorized doctor can share the patient's electronic health records (EHR), even between organizations, say a medical clinic and an insurance company, that do not have a business relationship with each other.

4) Integrity

Integrity refers to ensuring that the patient health record is reliable, accurate and consistent over its full life-cycle. A security breach can easily hamper the integrity of protected health information (PHI). There are disparate healthcare systems that have similar data but stored differently. Clinical, financial and supply chain systems of healthcare work in silos and are unable to 'talk' to each other and even when they do, they do not speak the same language. Even a hospital staff routinely changing the personal information like a phone number or address can compromise the integrity of the data such that the information is no longer reliable.

Blockchain, through its distributed technology, can address the challenges of synchronizing patient data across multiple, contrasting healthcare information systems. Multiple instances of obsolete and redundant patient data can be replaced with a single source of up-to-date patient information through the distributed ledger while assuring patient data security and privacy.

5) Rising Costs

Healthcare is becoming expensive. There are many factors that are responsible for the rising number of tests. One of the reasons is the incongruent systems as mentioned earlier. This leads to overtreatment and repetitive tests. Medical practitioners are rewarded for each test, visit or procedure rather than the end result or quality of treatment. Increasing life expectancy along with lifestyle sickness leads to inflated medical spending and adds to the cost of healthcare. Patients and doctors demand the latest and most expensive treatments even if there is little or no evidence of its effectiveness. Despite a wealth of information at our fingertips via the internet, there is no way to compare treatment options and costs associated with them. In addition, there is also the cost of maintenance of the healthcare information systems. As data increases exponentially year by year, so does the cost of back-up servers, data recovery and business continuity mechanisms and procedures.

Harnessing the immutable nature of blockchain, meaningful data can be shared across healthcare organizations and still be General Data Protection Regulation (GDPR) or Health Insurance Portability and Accountability Act (HIPAA) compliant. By addressing high transaction cost through blockchain technologies, healthcare costs can decline and could result in savings for taxpayers.

10.5.2 Health Records Management

Healthcare is a complex system that requires patients to share their data and medical records with multiple entities across the healthcare ecosystem. This brings about the prime issue facing the healthcare industry – the lack of interoperability. The issue of interoperability can be addressed by a public distributed ledger of patient data (transactions) that is made available to doctors, nurses, pharmacists, lab technicians and anyone else who is granted access to the said information. Upon patient initiation, smart "healthcare" contracts enable the service provision of individualized health and research/treatment record data in a high-security environment.

Medical records contain patient information like name, address, tests conducted, medications and other vital factors that can help healthcare professionals make informed decisions about patient care. This information, when stored on a distributed ledger or blockchain, can be shared across various healthcare organizations, giving a more comprehensive picture of a patient's medical information than a paper chart or centralized system can. Each of the transactions is time-stamped and cannot be changed or deleted. Only the patient has access to his or her personal medical records. Cryptographic hashing of records guarantees data integrity. When healthcare providers, insurers or other healthcare stakeholders require information, they have to send a request to the patient along with details of the information they want to view. Only upon the patient authorizing access can the healthcare provider or stakeholder query the blockchain to provide them with the location of the information.

As blockchain data is encrypted with the private key of participants/nodes in the network and access to information is only possible when the sender shares his or her public key, we can be assured of confidentiality and privacy. With blockchain implementations for EHRs, members of a private, peer-to-peer network can share the relevant medical information with appropriate viewership permissions, while the original member maintains ownership of the data shared.

Current initiatives at play

- PhrOS (phros.io), jointly released by Taipei Medical University and Digital Treasury Corporation, aims at bringing transparency and interoperability by putting all the patients' medical records on the blockchain.
- MTBC (mtbc.com), a leading provider of cloud-based healthcare IT solutions and services has built an integrated EHR and Practice Management System that can not only store medical records but also enable the patients to have full control of their medical information. The patient has the ability to make available his/her records to one doctor and pass it on to another.

10.5.3 Claims and Billing Management

Medical billing mistakes are a common cause for lost profits. With rising medical costs, hospitals and clinics work in an environment where every penny counts. So it is no surprise that hospitals and medical facilities overcharge. This is not intentional but owing to the fact that insurance companies pay only a fraction of the actual costs or a pre-negotiated amount. If Claims section of the insurance company detects an error or discrepancy in billing, the claim is rejected. When quality patient care is the main priority, hospitals do not want to get into the administrative micro-management of billing reconciliation. Hence, many healthcare organizations have outsourced their billing process. Many have given the onus of insurance approvals on treatments and resolution denied claims back to the patient. Fixing an error and waiting for a period of time, sometimes months, to see whether the claim is accepted or rejected creates a financial dent in both the healthcare facility as well as the patient. This is not considering the health and welfare of the patient in question.

A bigger problem than unintentional billing errors is intentional cases of fraud. Some of the prevalent healthcare frauds are overcharging of the actual services, performing unnecessary services, claiming charges for the non-performed services, and misrepresenting non-covered medical services as services which are covered. Unfortunately, it is practised in many hospitals to cover financial loses. However, the patient is the real sufferer as he or she is negatively impacted by the loss of time, resources and health in between all the runaround. It is ironical that such complications arise, as the healthcare industry is all about providing quality healthcare to patients.

Blockchain can enable a smart payment system that can automate the required workflows and all the parties involved can share a single copy of the contracts and billing-related information. This elimination of back-and-forth data transfer increases transparency and efficiency, leading to lower administration costs, faster claims processing and less money lost.

Blockchain-run EHRs allow insurance companies to see a verifiable record of administered treatments, The result is lesser number of denied claims, less instances of billing fraud and lower healthcare costs allowing for better administration of care for all parties.

Current initiatives at play

- Veris Foundation Blockchain, a non-profit entity in the US, aims to reduce healthcare costs for patients by directly connecting the payers with the providers.
- Change Healthcare provides blockchain-based solutions for providers, payers and partners to manage claims and remittances, improve collection of patient payments, minimize denials and underpayments, as well as manage daily revenue cycles and business operations more effectively.

10.5.4 Drug Supply Chain Management

Distribution of counterfeit and fake drugs is a severe problem in the pharmaceutical space. The World Health Organization pegged global fake drug sales at $75bn in 2010, a 90% increase over five years. In many developing countries in Asia, Africa and South America, counterfeit drugs comprise between 10% and 30% of the total medicines on sale. Counterfeit drugs that may not have the same active pharmaceutical ingredients or dosage levels as the actual drug are known to cause deaths on a large scale. Pharmaceutical companies incur losses in millions annually due to counterfeit drugs.

To fully comprehend the issue, we need to understand the point(s) of failure. The journey of a product (drugs or medical devices) from manufacturer to consumer has multiple stages of transportation, handling, storage, redistribution and retail. Issues can arise at any of these stages from simple human error to malicious intent (fraud). Since the records of information are copious and stored centrally either on paper or in a database, it is difficult or near-to-impossible to identify all the issues with the drug supply chain.

The security feature of blockchain technology can be utilized to address some of the issues in drug traceability. Every new transaction added to a block is immutable and time-stamped. This ensures that the information inside the block cannot be altered and the product can be tracked. Hence, companies who are participants in the specific drug blockchain would have a proof that the drugs manufactured by them are authentic.

The pharmaceutical companies can have the right to choose who will be acting as miners in the supply chain – manufacturers, distributors or retailers. Based on their place in the supply chain, every participant can have different rights or accessibility options (refer Fig. 10.4). For example, while labs can register drugs, the wholesalers can have the permission to verify transactions. The drug or medical device can be easily tracked as it moves along the supply chain between different entities thus ensuring complete traceability.

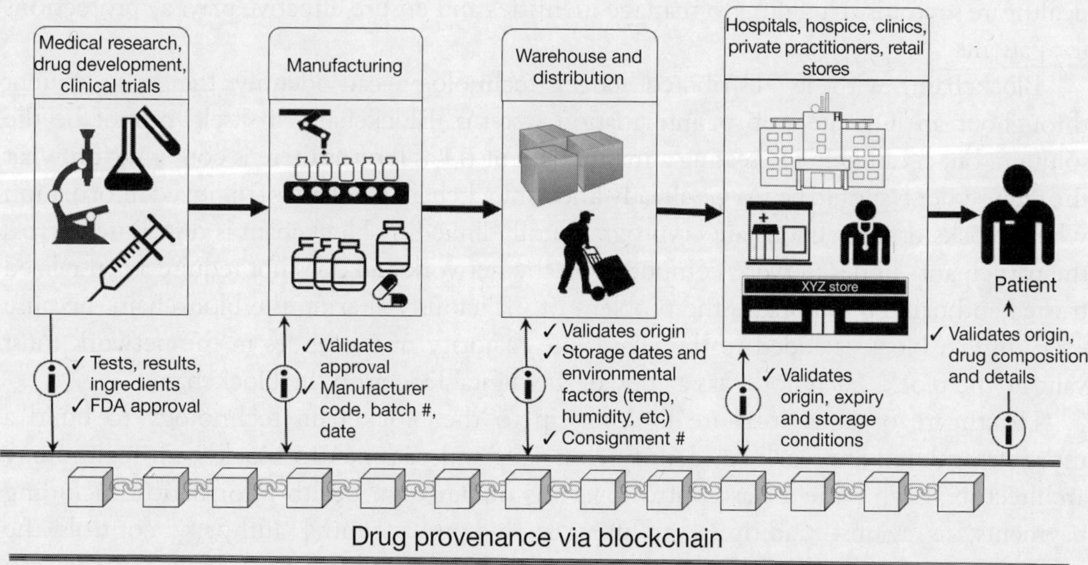

Figure 10.4: Drug traceability in blockchain

With blockchain ensuring data transparency, organizations can track the complete path of origin at a particular drug or product level, thus reducing the circulation of fake or counterfeit drugs.

Current blockchain projects in the making

- NITI (National Institution for Transforming India), a government think-tank, has partnered with U.S.-based information technology giant Oracle and India's Apollo Hospitals chain to implement a blockchain project with an aim to eliminate all channels of counterfeit medical products, including pharmaceuticals, by transferring the hospital chain's complete inventory to a blockchain-powered system.
- FarmaTrust aims to curtail counterfeit drugs through the integrated use of blockchain and AI based provenance systems. The supply chain tracking systems can prevent the counterfeit drugs from entering the system.

10.5.5 Patient and Provider Identity Management

According to Gartner, Identity and Access Management (IAM) is the security discipline that enables the right individuals to access the right resources at the right times for the right reasons. Identity management addresses the mission-critical need to ensure appropriate access to resources across increasingly heterogeneous technology environments. Stolen health records are highly profitable and have a longer shelf-life than financial data and is a major force driving the increase in healthcare breaches. Efficient and reliable patient identify management will be an essential element in making the benefits of integrated care a reality. However, with an assortment of electronic medical records (EMRs), the healthcare sector is struggling to manage identities and ensure effective privacy protections for patients.

Blockchain, with its distributed ledger technology, can identify the same patient throughout an intricate web of information systems. Blockchain, in itself, cannot be the solution, but it can help in creating a master patient ID as long as there is consensus on what the requirements should be. As we already know, blockchain is a chain of digitized information where blocks of transactions are cryptographically linked. A blockchain is distributed across the participant's nodes (servers/computers) in the network and does not require a centralized trusted authority. It eliminates the problem of maliciously altering the blockchain, because every time a block is added to the chain, the majority of the nodes in the network must validate the block. Each node has a copy of the digital ledger on the blockchain.

Healthcare organizations are looking up to the blockchain technology to build a more streamlined approach to identify clinicians and patients. In blockchain technology architecture, each patient has control over his or her own health information including payments, as against traditional architecture, where a central authority controls the information accesses and distribution. Blockchain-based identity management can eliminate the need to individually log in to every application to access different sets of information. It is secure because multiple checkpoints are required for any party to access the data. This method of validation in healthcare can cut down on login times as well as give patients the opportunity to be in control of their own health information.

Along with a master patient ID, the IDs for providers and other healthcare systems are required. Standardizing the data identifiers for all sources will require effort and time. The blockchain can be used to track the unique patient identifier if there is only one place to track. Currently, either a human or system has to do the initial work of matching this data to the right patient record. That is not something blockchain can do. However, paired with a master patient ID solution that identifies which record the data belongs in, blockchain could enrich the patient record. In short, with a combination of decentralized blockchain protocol and a comprehensive system for verifying identity, a unique digital ID can be created for each online user, be it a patient or provider, that can ascertain the truth of individual identity and prevent identity theft.

Current initiatives at play

- Trusted Key Solutions Inc., a blockchain-based secure digital identity company, in partnership with healthcare consortium NH-ISAC has built a solution that provides customers with a single, re-usable digital identity to access all of their healthcare services, such as signing up for health insurance, registering with healthcare providers and fulfilling prescriptions online.
- Hashed Health (hashedhealth.com) launched a provider credentialing solution that leverages blockchain to securely exchange/share credentialing information of healthcare professionals.

10.5.6 Clinical Trials and Medical Research Management

Considerable research, analysis and clinical trials are conducted to test the effectiveness of a drug or medicine or procedure before it is released to the public. To conduct these experiments and research, a massive amount of data is required by the researchers for analysis and further decision-making. This means that all the data collected and recorded needs to be secure, unbiased and transparent. In many instances, organizations and pharmaceutical companies modify or hide the data in order to record an outcome that is beneficial to them. Clinical research also requires real-life medical data to test their hypothesis in real-world scenarios. To ensure high accuracy in outputs of clinical research and trials, scientists require large amounts of de-identifiable raw data. Polls have shown that people are open to sharing their medical records for research purposes on the assurance that their privacy is maintained and the data is secure.

Blockchain technology can add credibility and authenticity to clinical trials through its distributed ledger and transaction engine. It can provide a permanent, valid, and immutable ledger of record for all research findings since it is time-stamped. With the smart contract layer, people can directly control their medical data and grant specific access rights to organizations. This helps research organizations to get direct and secure access to a vast repository of holistic and accurate clinical data.

Another arena of healthcare that can be exploited is the funding of research and medical trials. Any breakthrough research into new treatments, technologies or disease prevention needs funding that is currently either centralized or limited. Security of data being provided through encryption on a shared ledger can facilitate information sharing. A research-based blockchain healthcare ledger could potentially aid in the funding of clinical studies and research using crypto tokens. There are investors and like-minded people who are open to funding research that has merit. They will be able to view real-time information stored on the ledger and select the studies and research they would like to fund. The trustless nature of the ledger would minimize the chance of academic fraud and duplication of work. This will also enable faster regulatory compliance and approvals.

Current initiative at play

– Ambrosus, a multinational Internet of Things (IoT) supply chain manufacturer, aims to expand into pharmaceutical industry leveraging blockchain and IoT technologies to streamline processes for private clinical trials, help with the traceability of drug ingredients and conditions, and help government and related agencies with country-specific regulatory compliance, thus enabling smooth and faster flow of goods across borders.

10.6 Blockchain in Real-estate

Did you know?

"Blockchain offers an open source, universal protocol for property buying, conveyancing, recording, escrow, crowd funding, and more. It can reduce costs, stamp out fraud, speed up transactions, increase financial privacy, internationalize markets, and make real estate a liquid asset."

– *International Blockchain Real Estate Association (IBREA)*

The real-estate (RE) market currently faces several issues related to housing affordability, rising rate and economy, many of which cannot be addressed with blockchain. However, it cannot be denied that blockchain does have a quite a few applications that can bring measurable value to the real-estate sector (refer Fig. 10.5).

Blockchain-based smart contracts can potentially transform the real-estate sector by simplifying property transactions like purchase, sale, financing, leasing, and management. Though there can be some level of disintermediation, few trusted intermediaries would still be required, such as for the assessment of the building to understand renovation requirements.

Over time, blockchain adoption can have a broader impact when it is linked with IoT, AI and big data for public utility services such as smart parking, waste/water/energy billing, and also enable data-driven city management.

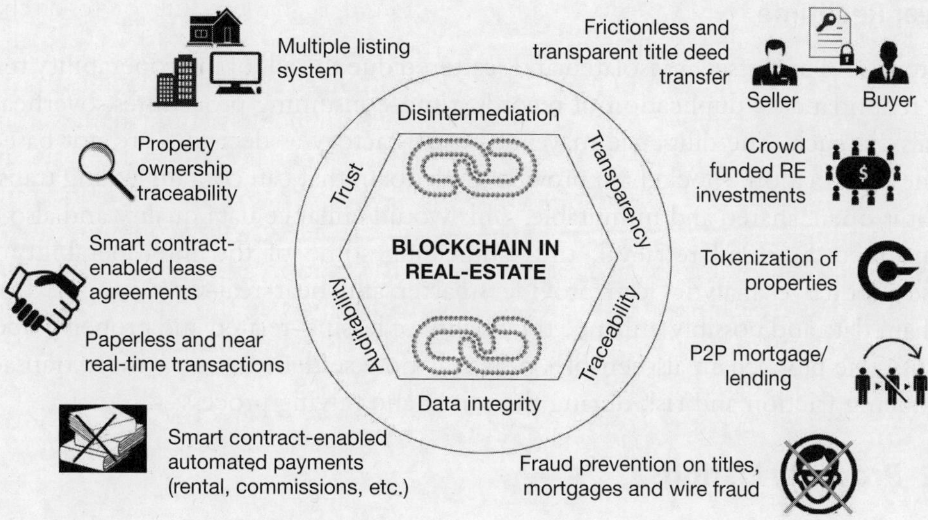

Figure 10.5: Potential blockchain applications in real-estate sector

10.6.1 Challenges in Real-estate

Blockchain technology is not the solution to all the existing inefficiencies in RE. However, it can cater to some of the challenges that the real-estate sector is facing:

1) Public Record Keeping

A lot of documentation and records change hands in a real-estate deal like title deeds, title insurance, warranty insurance, contractual agreements, and other related documents. Record keeping is still paper-based in most registrar or government offices that are wrought with bureaucracy and inadequate filing/storage facilities.

In blockchain data/transaction is stored as ledger, distributed across a peer-to-peer network in which every transaction is immediately, automatically, and permanently recorded and accessed only by other permissioned participants of the blockchain. As data stored on the blockchain is secure, immutable, and can be made public, it becomes the perfect platform to record property sales and leasing transactions.

2) Trust Is a Must

There are many players in a real-estate leasing cycle like the owners, tenants, real-estate agents, brokers, notaries, banks and mortgage brokers, to name a few. The different participants, more often than not, have pre-existing relationships, which results in mistrust. Since blockchain technology is based on consensus and cryptographic proof, it allows for two parties to transact directly without the need for a third party like notaries or brokers. Since every transaction made can be verified, risks associated with fraud, abuse, and manipulation of transactions are mitigated.

3) Near Real-time

RE systems and processes are isolated and scattered due to lack of interoperability resulting in data redundancies, duplication of records, time-consuming procedures, overheads and lost deals. As such, due diligence may not be satisfactory as decisions are not based on a real-time information. Blockchain provides a platform that can contain leasing transaction data that is open, shared and immutable. This would enhance data quality and also enable real-time recording and retrieval, thus addressing some of the interoperability issues. The use predictive analytics can provide smarter and near real-time insights from the blockchain data and possibly enhance the quality of leasing-related and property operating decisions. The blockchain also enables near real-time settlement of payment transactions, thus reducing friction and risk during the rental and buying process.

10.6.2 Property Listings

Selling and buying of property is synonymous with real-estate agents or realtors. Realtors access multiple listing service (MLS) to access property data such as property features, location, and rental rates. These platforms are typically subscription-based and command high access fees from users. Due to lack of standardized process, the details of property-level information is subjective and entirely dependent on the data entered by the brokers. This may result in inaccurate, dated, or incomplete listing information.

Using blockchain ledger for property listings would mean opening up the access to all available property for anyone to view. The listing could also include any terms or conditions that would need to be met for a successful sale. This will provide a more precise and accurate view of the property enabling faster decision-making and also save time.

Current initiatives at play

- REX (Real Estate Exchange Inc.) is one of the first property brokers to launch a blockchain platform aimed at creating a decentralized global database of properties that is easily and cheaply accessible to anyone who is looking for a property.
- Ubitquity (ubitquity.io) is a blockchain agnostic platform that is built on bitcoin but compatible with ethereum, hyperledger and multi-chain. It provides Banking-as-a-Service (BaaS) platform that allows individuals, businesses and municipalities to store and track properties securely.

10.6.3 Tokenization of Properties

Real-estate is one of the most reliable investment assets in the world due to consistent returns and value appreciation. Yet, many are hesitant to invest in it due to cross-border investing difficulties, high investment costs, lack of access or visibility to quality investment properties and lack of investment liquidity.

To "tokenize" real-estate means to associate a blockchain token or cryptocurrency with real property. Property owners can issue blockchain-based tokens, which represent shares in an asset. Investors can buy such tokens and become fractional owners of the underlying asset and participate in the asset's appreciation and cash flows.

Blockchain technology and security tokens leverage the advantages of real-estate investing and mitigate the downside such as high costs, inaccessibility and illiquidity.

Current initiatives at play

- RealtyReturns, a real-estate tokenization marketplace allows investors to easily co-own quality investment properties without worrying about cross-border investment problems and provides potential investment liquidity through token exchanges.
- Atlan (atlan.io), a real-estate platform build over ethereum, aims to offer services connected to tokenized ownership of real properties and blockchain-based P2P rentals.

10.6.4 Frictionless Transactions

One of the primary friction points of global real-estate markets is the complex and lengthy contract procedures associated with the property changing hands. The contract is reviewed and executed by a series of intermediaries such as real-estate agents and lawyers who interpret the language of the legal documents, verify whether the predetermined conditions have been met, and also check if there are any open legal cases pertaining to the property that are pending in court. This adds additional layers of costs and delays to the transaction.

Using blockchain applications, the execution of a legal transaction can be algorithmically encoded so that it happens almost instantaneously. The required basis for trust among parties is reduced to a mathematical formula that is error-free and eliminates the need for humans to establish trust through time-consuming traditional means. In short, the contractual terms of a transaction could be written inside the blockchain in the form of smart contracts. This will enable buyers and sellers to access pre-defined, legally valid real-estate contracts, tailor them to suit their specific property sale, and get them executed when the parties attest that the agreed-upon conditions are all satisfied.

Current projects in the making

- SmartRealty (smartrealty.io), a blockchain start-up, provides a smart contract platform that is customizable enabling buyers and sellers to access pre-defined, legally valid real-estate contracts that can be modified to suit any specific property sale. The contract automatically executes when the named parties agree and the pre-agreed conditions are all satisfied. This removes the intermediaries and the complicated legal process which creates friction and expenses in real-estate transfer and payments.
- Propy Inc. is a blockchain platform that allows for deed registration, title registry and real-estate transactions.

10.6.5 Peer-to-Peer Mortgages

Applying for and getting a mortgage is a lengthy and complicated process that could take more than two months. There are numerous intermediaries involved, increasing the turnaround time. The fact that the procedure is almost entirely paper-based involving bank statement, tax returns, personal ID, and other related information does not help the situation either. Another factor that has introduced complexity is the market conditions, where the mortgage lenders have to be aware of and comply with the changing regulations like legal approvals, tax information and other related information. Gone are the days when the mortgage used to be managed solely by the bank using funds generated by the deposits of its clients. With the advent of the internet, people can now obtain financing through peer-to-peer (P2P) lending that allows borrowers to obtain a loan from a group of individual lenders for both mortgage and mortgage refinancing needs. The downside to P2P lending is the long processing time and steep collection fees for default payments. Besides, there is a distance between the borrowers and lenders as both the borrower and lender are contracted to the P2P platform providers. The life-span of a mortgage could be up to more than 30 years causing a strain to both lender and borrower especially if there is bankruptcy or delinquency.

A peer-to-peer blockchain-based mortgage system can enable a lender and borrower to transact directly with each other. Borrower, lender and seller can exchange private and secure information without the need for a central party or authority. Blockchain-based P2P mortgage lending will help those who are unbanked and reduce costs as hidden fees mostly associated with banks are not applicable on this platform. Smart contracts can execute deals and facilitate payments. Scores can be attached to lenders and borrowers who pay on time. Late payments or defaulters could have negative score that is tracked in the blockchain. The transparency and trust of the blockchain promotes confidence between the borrowers, lenders and sellers while also providing a check on defaulters and fraudulent activities, thereby promoting honest transactions.

Current projects in the making

- Homelend (homelend.io), a mortgage crowd-funding platform, aims at building a blockchain-based, peer-to-peer mortgage lending platform.

10.6.6 Smart Contract Property Management

Any real-estate transaction requires enormous amounts of paperwork and hours of coordination between the bank, broker, seller, buyer, and local government, often making the contracting process slow, expensive, and still open to interpretation. This burden could be alleviated if smart contracts handled rental agreements, commercial real-estate tenancy, and other ongoing types of real-estate transactions. Blockchain technology in real-estate could create trust between parties on a level that does not currently exist.

The property owner can build the leasing conditions, i.e., rent, property management fee, payment frequency and reinstatement particulars into a new smart contract. Upon online review, if the tenant agrees to conditions, the concerned parties digitally sign the smart contract using a key that denotes their identity, which then turns into a legally binding digital smart contract stored on a blockchain.

At every agreed payment period, the smart contract digitally transfers funds from the tenant's account to the property owner's account. Utility payments can be calculated and data fed into the blockchain from Internet of Things devices that transparently record energy, utilities, and more.

Current blockchain initiative at play

– The Midasium Blockchain uses smart contracts that can replace the traditional contracts like mortgage agreements, contracts of sale and tenancy agreements. The smart contracts automatically execute the full payment life-cycle based on pre-agreed timelines, and can be used for the payment of rent to owner/landlord, proportional payment to maintenance/contract workers and refund of security deposit back to tenant on termination of lease.

10.7 Blockchain in Supply Chain

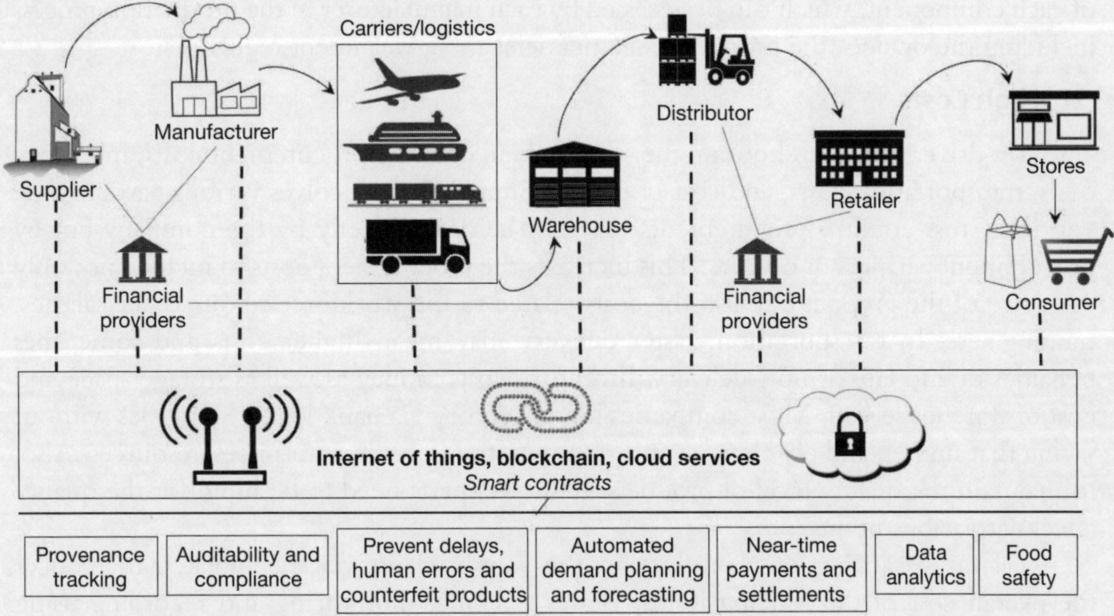

Figure 10.6: Benefits of blockchain in supply chain management

According to the Council of Supply Chain Management Professionals (CSCMP), supply chain management encompasses the planning and management of all activities involved in sourcing, procurement, conversion, and logistics management. Supply Chain Management (SCM) integrates supply and demand management within and across companies into a high-performance business model that drives competitive advantage. Due to the varied nature of the entities and data, manual paperwork, reconciliation issues and middlemen/intermediaries, it is difficult to identify points of failure. If any of the components fails, the brand bears the consequences. Blockchain technology offers the stakeholders a permanent and auditable digital record of the state of the product at each crucial step along the chain. The transparency can also help streamline the supply chain operations. Retailers and suppliers could benefit from a greater insight into the manufacturing process of the goods and ensure that all the required standards have been met, such as health and safety requirements (refer Fig. 10.6).

10.7.1 Challenges in Supply Chain

Some of the key challenges in SCM that can be addressed by blockchain technology are:

1) Provenance Tracking

When something goes wrong within a complex machinery system like an aircraft or freighter or lifesaving drug, it is critical to know the provenance, through supply chain management, of each component, down to the manufacturer, production date, batch, and even the manufacturing equipment program. Blockchain can hold the complete provenance details of each component, which can be accessed by each manufacturer in the production process including the owners, the producers, maintainers, and government regulators.

2) High Costs

The cost drivers that can shoot up the supply chain costs are procurement costs, inventory costs, transportation costs, and cost of quality. Supply chain involves various participating entities. Cross-country procurements may not be dealt directly by the company but by procurement partners or officers. This increases the procurement costs to include not only the price of the product but also the costs related to the workforce resource like salaries, training, etc. In transportation, faster delivery may mean higher costs and sometimes penalties due to late or non-delivery. Inventory costs money as well as storage space and cost of warehouse staff. Most companies buy inventory via bank loans. So it goes without saying that unused and obsolete inventory puts a strain on the enterprises' resources. Also, to maintain the quality level of products, trained experts need to be hired for the quality checks at various points.

Real-time tracking of a product in the supply chain via the blockchain reduces the overall cost of the moving items. Better tracking, monitoring and recording using blockchain technologies means more accurate planning in transportation needs, avoidance of delays, faster payments and settlements, better inventory control, assessments and quality checks recorded on the blockchain, and removal of intermediaries that are an overhead on the supply chain.

3) Trust Issues

Establishing trust within a complex supply chain system is critical for smooth operations. All participants are trusted to do their part honestly. The manufacturer must have confidence in the supplier to follow factory standards. Some of the common issues faced are specific suppliers getting preferential treatment due to personal relationship, market monopoly, underhand deals between procurement and supplier staff, and sub-standard products in circulation. Lack of transparency in current SCM system makes it extremely difficult to investigate illegal or unethical practices.

With the supply chain process/network on the blockchain, there is increased trust because no single authority controls the provenance information and frauds and corruption can be easily detected.

4) Process and Time Delays

Current supply chain processes handle massive data set, thus making the complex process slow; primarily as some supply chain nodes rely entirely on paper, e.g., shipping. Lack of transparency and provenance makes it challenging to locate and rectify faults in the best time possible as well as avoid recurring issues. Defective products at destination lead to total recall of the full consignment leading to reputation risk and costs.

With supply chain on the blockchain, the product transactions are time-stamped and can be tracked from the point of origin, including the parts or ingredients used in its manufacture or production. Identified defects can lead to specific recalls rather than the full set. The increased efficiencies in process can lead to reductions in time taken to diagnose and remedy a fault, thereby improving system utilization. The study of the full SCM life-cycle on the blockchain can lead to continuous process improvement.

10.7.2 Supply Chain Financing

Supply Chain Finance (SCF) is a financing process that links the buyers and suppliers with financial service providers like banks to unlock the working capital for both the buyer and the seller, thus providing liquidity and improving business efficiency.

Companies need working capital to run their businesses or strategic initiatives. In the real world, there are, more often than not, delays in payments over and above the payment terms. If the invoice is substantial with the promise of more business to come, the supplier may continue to deal with the large buyer and accept the strain on its business caused due to the impact of late payments. This is where supply chain financing can help.

In a standard SCF model, the supplier can sell the approved invoice at a discounted rate of invoice amount to the buyer's bank or financial lender for immediate payment to run its business. The bank will later collect the invoice amount from the buyer on the due date. Suppliers are then not subjected to lengthy payment terms and can control their cash flow for business growth and forecasting. As the suppliers have a strong financial base, they can continue to provide consistent service to the buyer. Buyers have the potential to extend payment terms without impacting the buyer–supplier relationship.

Though this seems a win-win for both buyer and supplier, SCF is not without its challenges. One of the reasons is that it is buyer that sets up the SCF facility based on the buyer's business strength/credit rating. Hence the system can work well only if the buyer is an established corporate. Also there has to be an organizational adoption that is not easy in a complex supply chain environment. Besides, the various financing models involve substantial paperwork and assessments making approvals time-consuming. Further, if the buyer defaults on goods received either due to poor financial management or fraud, it poses a risk to the SCF model.

Building the intricacies of SCF onto the blockchain can simplify the process by providing a single source of truth at critical points in the supply chain, such as purchase order receipt and approval, invoice receipt and approval, inventory delivery and quality control. With transparency, immutability and authentication measures built in every point, there is an inherent trust between the buyer, supplier and financial institution. Paper-based verification is substantially reduced, finance institutions are better equipped to take financing decisions and fraud is curbed.

By incorporating IoT tracking to goods delivery or inventory, the blockchain-based supply chain via smart contracts can automate transactions and payments including refunds for damaged or delayed goods.

Current initiative at play

- Mahindra and IBM announced, in 2017, the building of the first supply chain finance blockchain-enabled project in India. The aim is to create a common platform for Mahindra Finance's supplier-to-manufacturer transactions, allowing all parties to view the transactions in real time, driving trust and transparency through the supply chain.

10.7.3 Blockchain Logistics

Logistics is defined by the CSCMP (Council of Supply Chain Management Professionals) as "part of the supply chain process that plans, implements and controls the efficient, effective forward and reverse flow and storage of goods, services and related information between the point of origin and the point of consumption in order to meet the customer's requirements".

According to the International Chamber of Shipping, an estimated 90% of world trade is carried out by the international shipping industry every year with over 50,000 merchant ships trading internationally, transporting every kind of cargo. Other carrier services, involving domestic trade as well, would include the trucking companies, airlines, railways and others. Trust and collaboration is the key in logistics to optimize the flow of goods and the complex flow of information.

With huge number of stakeholders, each working with disparate systems that often leads to unstandardized processes, low transparency, conflicting interests and incorrect interpretation/assumptions in contractual terms, there is often a lack of trust and potential for disputes. Smart contracts powered by blockchain can help limit disputes by automating the contractual terms. Blockchain can bring about efficiencies through improved data sharing, permission-based transparency, disintermediation and collaboration between carrier/shipping partners. Blockchain can be an aid to streamline the complex processes, automating them wherever relevant, making them leaner and error-free. This can lead to fewer lost, stolen or damaged goods.

The required capacity can be monitored within the blockchain, and optimal carrier services can be made available in advance. Suppliers could even track factors such as temperature and humidity throughout the shipping journey. The visibility and predictability into the logistics operations can provide time-bound services leading to cheaper rates and customer satisfaction.

Current initiatives at play

- TradeLens (tradelens.com), a JV between Maersk and IBM, is a global blockchain-based supply chain platform for digitizing trade workflows and end-to-end shipment tracking. The blockchain enables the stakeholders to securely track shipments real-time, view the bills of lading, customs and other documents. With over 100 organizations on the blockchain including carriers, ports, 3PL, terminal operators and freight forwarders, it has the potential to significantly reduce delays and fraud, promote savings amounting to billions of dollars and revolutionize the logistics industry.
- CargoX (cargox.io) has piloted the first blockchain-based Smart Bill of Lading. While paper bills of lading can take up to ten days to process, CargoX was able to cut processing time down to four minutes with the use of blockchain and smart contracts.

10.7.4 Supply Chain Traceability

The current communication framework within the food ecosystem makes traceability a time-consuming task as many of the involved parties still track information on paper. Blockchain can replace paper-based tracking by allowing each player in the supply chain to generate and securely share data points that create an accountable and traceable system. Vast data points with labels that clarify ownership can be recorded promptly without any alteration. As a result, the record of a product's journey, from origin to destination, can be made available to monitor in real-time.

Blockchain can disrupt the automotive supply chain by connecting the whole sequence of manufacturing and allow for the seamless order or selling of units and parts along with the tracking, payment and maintenance of the said items. IoT devices can be used to maintain the whole history of the vehicle, including repairs and maintenance, on the blockchain.

Traceability and transparency in the retail industry can boost customer loyalty if the retailer can prove to the customer that the products they are purchasing are fresh and meet the required health and safety standard. Immutable digital recording at every stage of the retail supply chain process can help ensure authenticity and reduce the risk of counterfeiting that is often prevalent in the pharmaceuticals, luxury goods, arts, antiques and electronics sectors. This can bring about a new level of trust to the brand-customer relationship.

With the provenance and immutability of blockchain, consumers can be confident that their purchase is original, ethically sourced, and preserved in the right conditions.

Current initiatives at play

- Diamond giant De Beers uses blockchain technology to track stones from the point they are mined to the point they are sold to consumers. This ensures that the company avoids 'conflict' or 'blood diamonds' and assures the consumers that they are buying the genuine article.
- ADNOC (Abu Dhabi National Oil Company), in collaboration with IBM, has successfully launched a blockchain supply chain system pilot program to track oil from the well to customers, while simultaneously automating transactions along the way.

10.7.5 Food Safety

Fresh natural nourishing diet is essential for a healthy society. Currently there is an inherent lack of trust towards the food industry. The E. coli outbreak of 2018 linked to Romaine lettuce has caused the US farm to recall green and red lettuce and cabbages that were harvested between 27-30 November of the year. The domino effect resulted in their suppliers recalling sandwiches and products that used their produce. The 2008 milk scandal rocked the Chinese food industry when milk, infant formula and other food materials was found to be adulterated with melamine, a chemical that gives the appearance of higher protein content. A 2012 study conducted by the Food Safety and Standards Authority of India (FSSAI) across the country found that milk in India was adulterated with water, detergent, fat and even urea.

Fruits and vegetables are known to be adulterated with carcinogenic substances like oxytocin, saccharin, wax, calcium carbide and copper sulphate to bring about early ripening, brighter color, sweetness and shine. Overuse of pesticides and fertilizers means the food is contaminated from the field itself. Infections linked to beef products are identified and reported every year.

The global food supply chain consists of a network of farmers, distributers, wholesalers, retailers, warehouses, factories, shopkeepers and many others. It involves a massive amount of information flow between various entities and it would be challenging to track the origins of contaminated food or faulty records to its source. However, if all the information were to be captured on the blockchain, traceability would not be an issue. Human error or food tampering can be identified and rectified immediately at source before it reaches the supermarkets or consumers. Curbing the fraudulent activities can help reduce the food scandals that occur worldwide. A simple QR scan can let the consumer know all the details related to be product like the country and place of origin, ingredients used, actual date of manufacture and other details that can lead customers to make informed decisions on the product they wish to purchase.

Another advantage of food traceability is the ability to trace back contaminated or diseased food products that have entered the food supply chain. Tests carried out by Walmart and IBM demonstrated that the origin of certain foods was identified in 2.2 seconds using blockchain, as opposed to 6 days using manual verification. This would mean that with the adoption of blockchain, retailers could quickly identify and remove the contaminated food and avoid full recalls and overhead costs. Identifying the exact point of contamination can also allow for the fixing/repair of the process or equipment/technology to ensure non-recurrence.

Adopting blockchain technology in the food supply chain can also prevent fake and counterfeit products; for example, cheap wine sold in bottles with top-quality labels, cheap chocolates and confectionary packaged as top brands, and sub-standard meat of various animals sold as high-quality beef. With the distributed nature of blockchain, such kind of frauds would no longer be viable as the exact point of fraudulent activity can be identified and culprits brought to justice.

However, implementing blockchain in a complex food supply chain system is not easy. The integrity of the system is as good as the data entered. There are chances of counterfeit certificates on quality or farm inspections being introduced by malicious players. Many farmers in the network may not have the means or the technology to adopt blockchain. Food safety cannot to assured by blockchain alone. Blockchain is just a ledger of immutable information. It needs to work in cohesion with IoT, AI and other technologies to ensure that the data is as accurate as possible with least amount of manual intervention, e.g., the quantity and quality of pesticides, fertilizers and water introduced in the farms, quality and amount of animal feed provided.

Research is already on the way to study how to implement cheaper and cost-effective means of food safety for farmers, producers and consumers. An enormous responsibility falls on large corporates to pilot and test the potentials of blockchain leveraged supply chain to promote food safety.

Current initiatives at play

- Retail giant Carrefour has launched blockchain information for 20 items including chicken, eggs, raw milk, oranges, pork and cheese, and 100 more items with a focus on areas where consumers want reassurance, like baby and organic products. Customers can scan a QR barcode on a fruit with their smartphone and find out the date of harvest, farm location and owner, when it was packed and how long it was transported.
- Waltonchain (WTC) is a Smart Supply Chain ecosystem that is created through a combination of the RFID and blockchain technologies. It has recently launched a new initiative, WTC-Food, that aims at providing origin and destination tracking for food

products. The platform is primarily targeted at countries like China that lack robust food tracking systems.

– Bext360 (bext360.com) utilizes a combination of IoT, blockchain, machine learning, and artificial intelligence to build a fully transparent food supply chain Software-as-a-Service (SaaS) platform for farmers and different players along the value chain for sharing data and tracking progress.

10.8 The Blockchain and IoT

The Internet of Things (IoT) is a network of physical objects like home appliances, watches, vehicles, and other gadgets that contain embedded technology (electronics, software actuators and connectivity) to sense and communicate with their internal states or the external environment. Simply put, IoT are physical devices that are connected via the internet, either collecting or receiving information (or both) that can be acted upon. A typical example is in farming. Sensors can collect information on the moisture of the soil and automatically turn on the irrigation system. Other sensors can collect information on temperature and air quality and combine it with weather information from the internet on possibility of rain, and can decide whether to turn on the irrigation system or not, without the need for the farmer to intervene.

Did you know?

Sir Timothy John Berners-Lee, the inventor of the World Wide Web, founder of the World Wide Web Foundation and the original architect of the internet, believed in the philosophy that the Web needed to be open source and not controlled by a few central authorities or corporations who could potentially interfere or snoop on internet traffic, thus compromising fundamental human network rights.

He imagined a distributed system of computers communicating with each other directly, with each user owning a bit of responsibility for maintaining their contribution to the network. This is quite similar to the principle of Blockchain Technology. Web 3.0, the 3rd generation of internet, promises to hand control back to the users by architecting a web that protects individual property and privacy through a range of P2P technologies like the blockchain.

Source: What the Heck Is Web 3.0 Anyway? - forbes.com https://www.forbes.com/sites/juttasteiner/2018/10/26/what-the-heck-is-web-3-0-anyway/

With IoT devices, we are seeing the buildup of Smart Homes, Smart Cities, and Smart Factories all across the globe. Gartner, a research and advisory company, estimates that the number of connected IoT devices will reach 27.1 billion by 2021.

Though there are numerous benefits attributed to IoT like lifestyle comforts and efficiency, connectivity, revenue generation, cost reduction and better decision-making, these are not without its challenges. Lack of security and privacy, connectivity issues, and need for ample data storage and retrieval are real difficulties to reckon with.

By the very nature of the distributed ledger technology, blockchain is possibly the solution to address the challenges of the IoT market. Blockchain and IoT are currently trending technologies, both in their infancy. However, it is believed that blockchain can maximize the use of IoT devices enabling not just machine-to-human interaction but also machine-to-machine communication. Smart contracts will play a crucial role in ensuring that each party fulfills its part of the working relationship.

10.8.1 Advantages of Blockchain with IoT

The union of the two technologies can provide the following benefits:

1) Shared Platform

Present-day IoT architecture is centralized. Devices are identified, authenticated and connected via cloud servers and processing services. There is no single platform that connects all devices and there is no guarantee that cloud services offered by different manufacturers are interoperable and compatible. This needs a complex infrastructure network to transmit, collect and collate the myriad of data that is produced. Blockchain can provide a decentralized platform for inter-device communication to transfer all kinds of data securely and with permanence. We already know that once the data is accepted at the nodes, it cannot be changed. Blockchain can maximize the use of IoT devices. With smart devices communicating with other smart devices on a ledger-based platform operating on preset smart contracts, we are looking at a fully distributed reliable digital infrastructure.

2) Build Trust

IoT is all about the storing, sharing, exchanging, analyzing and monetizing of data. There are no central IoT standards and development of smart devices or projects are traditionally not done with IoT data security or data privacy as a key factor. As such, there is no way we can verify that the data has not been tampered with at the central database before being sold or used by other parties. Thus systems are prone to virus attacks, hackers and sometimes just simple human error. These could potentially cause not only losses but also huge casualties based on the sector it is applied. If we cannot fully trust the data at various source points, then we cannot trust the output.

However, data once stored in blockchain is immutable, transparent and traceable, thus making the IoT device data tamper-proof. This combination of secure IoT devices and blockchain introduces a real verifiable source of IoT data, which can be trusted to operate as required.

3) Reduce Costs

Any interaction between machines or devices is recorded in the blockchain, increasing the system's efficiency, accuracy and reducing costs. IoT can benefit from the reduction of the operational and maintenance overheads by cutting out the intermediaries.

Adopting a standardized peer-to-peer communication model to process the hundreds of billions of transactions between devices will significantly reduce the high infrastructure and maintenance costs associated with the traditional centralized model and improve efficiencies.

10.8.2 Applications of Blockchain in IoT

According to the World Bank, more than 54% of the global population resides in cities and the number is expected to grow to 67% by 2050. Urbanization has affected our physical environment with over-crowding, climate change, resource scarcity, pollution and negative impact on health. The ever-increasing population and people longevity jeopardizes the basic needs such as a clean habitat, water, energy and infrastructure. Hence, it is no surprise that governments are embracing the Smart City concept to address the socio-economic challenges.

One of the basic building blocks for modernization of cities is IoT. IDC, a global provider of market intelligence and advisory services, forecasts that the IoT network will grow to 41.6 billion by 2025. IoT utilizing the blockchain technology can enhance urban living.

Some of the Blockchain IoT usages are:

1) Identity

The challenges of IoT-based identity management systems related to storage, authentication and scalability issues can be addressed by blockchain. Decentralized Identity Management systems using blockchain can provide a secure mechanism for storing and validating user identities, thereby reducing identity thefts and related frauds. Foe example, Civic, a self-sovereign and secure identity management company has designed Secure Identity Platform (SIP) for multi-factor authentication without the need for passwords or usernames and relies instead on biometric information verified by the blockchain ledger.

2) Energy/Water/Waste Management

With IoT sensors and other technological devices, data from each of the utilities can be recorded and stored on interoperable systems. Data can be analyzed with AI prediction modeling to form smarter predictions on usage and consumption patterns and thereby improve efficiencies. For example, LO3 Energy blockchain-powered smart meter helps monitor utilities usage and distribution; WaterChain is creating a decentralized water funding platform with the help of blockchain technology and leading water innovators to radically improve the quality of water worldwide.

3) Transportation Management

Public transportation has increased 26% over the past 20 years. The emergence of smart cities and their highly localized economies is predicted to incentivize greater reliance on public transit. With blockchain, there can be a single point of payment for the various types of public transport, including bus, train, subway, taxis, and bikes with preloaded funds on a card. The blockchain can also enable a P2P mode of transportation. Malta has partnered with Omnitude, a UK-based Multi-Enterprise Blockchain Middleware Platform, to create a blockchain infrastructure for better public transit management.

4) Payments and Money Transfers

All public payments can be enabled on a blockchain-based solution, including those for municipality, welfare, payroll, and others. Blockchain can also aid fund transfer directly and securely without intermediaries who slow down transactions and charge fees for the service.

The Dubai Department of Finance in UAE recently launched a blockchain-powered payment system to enable real-time payments within and between government entities. The aim is to optimize an ethical and transparent governance process across the organizations.

5) Universal Data Storage

The massive amounts of data collected in smart cities via sensor data, smart grid data, smart vehicle data and others cannot be compromised by storing them on a centralized database that can be hacked. Blockchain technology can create a secure, encrypted, decentralized storage solution providing both interoperability and security. Interplanetary File System (IFPS) is a peer-to-peer hypermedia protocol that aims to function as a universal file system for all computing devices.

6) Keyless Signature Interface (KSI) for Hosting Government Services and Records

KSI is a form of blockchain established in Estonia for the government. The solution provides a high level of transparency and permanence for government records. The combination of data accuracy, permanence, privacy and invulnerability to passkey theft makes it the ideal storage solution for health records, land permits, court records, and all other government-related records that will be necessary to maximize the potential of smart communities.

Guardtime, a security software company, implements keyless signature infrastructure for federal and local agencies.

7) Smart Health

Blockchain-based Internet of Medical Things (IoMT) improves reliability, security and accountability, and prevents sensitive patient data and records from hackers. With this unique combination of IoT, AI and digital ledger technology, healthcare organizations, vendors, doctors as well as patients are interconnected, thus reducing healthcare costs and enhancing the overall patient experience. For example, a remote patient monitoring system with sensors can detect blood and glucose levels in patients, and send the data via the blockchain to medical professionals for analysis and diagnosis. Analysis of all of the data collected by sensors can be used to prescribe highly personalized treatments and medications for patients.

8) Smart City

Imagine a city that incorporates Smart Energy, Smart Health, Smart Agriculture, Smart Buildings, Smart Transportation, Smart Government and other Smart initiatives. This is slowly but surely becoming a reality. In a Smart City, IoT sensors can monitor faults instantaneously and alert the city agency's AI for an engineer or personnel. The engineer can analyze and fix the problem using AR or Smart Glasses. Any required parts could be developed or designed using a 3D printer. All the transactions at various points are recorded on the blockchain. Once repairs are completed or faults rectified, payments can be settled via the blockchain.

Estonia, the most advanced digital country in the world, has been using decentralized digital technology since 2012 with its foray into healthcare, ID management, government services and many others. Dubai has started a Smart City initiative, part of which is to have a 100% paperless government and storage system. India's Smart Cities Mission aims at developing 100 smart cities. Projects include affordable housing, integrated multi-modal transport, creation and preservation of open spaces, and waste and traffic management. Many of the smart-city initiatives will be implemented on a blockchain for enhanced security, immutability, and transparency.

The next step from a Smart City could potentially lead to a Smart Country by incorporating the cutting-edge technologies of blockchain, AI, big data, robotics, augmented reality and other IoT devices across cities.

Summary

A blockchain-based IoT solution has the potential to revolutionize businesses. It enhances customer experience and addresses all the global adaptability challenges of scalability, single point of failure, time-stamping, record, privacy, trust and reliability of digital devices and processes. However, unless we can build, deploy and run the IoT-based solution, the scheme just looks good on paper. To this end, it is an encouraging sign that with the amalgamation of IoT, AI, robotics, blockchain, nanotechnology and 3D-printing, with each technology supporting the other, we can look forward to the Fourth Industrial Revolution.

Banks and financial sectors have been slow to adopt new technologies. However, with technological innovation, we see the emergence and acceptance of financial technologies or FinTech companies within the user community. This means that the financial services sector needs to adapt rapidly with the times. Banks are looking at blockchain to improve settlements. Settlements that currently take around 7–10 business days are expected to be settled near-instantly with blockchain, thus removing the need for a lot of middle- and back-office staff.

However, despite the foreseen benefits, blockchain technology adoption is seen to be slow. Finance organizations may start implementing private blockchains where the financial institutions and their partners will have "permission" to participate. In this scenario, the benefits of a secure ledger, streamlined documentation process flow, faster response times and related cost saves may be realized. However, it will not be drastically different from the current modus operandi. Over time, when trust is built, one can look forward to a quicker and proactive adoption.

The adoption of blockchain technology is still in its infancy in Healthcare as well. However, we can expect to see healthcare solutions focused on providing more secure and integrated care to the patients in the near future. Though an emerging technology, blockchain cannot be viewed as a panacea for all the challenges hitting the healthcare industry. Enterprise systems are still centralized, and one cannot disregard the massive organizational change that goes into adopting a new distributed data culture entirely by implementing blockchain. While blockchain technology has its own set of challenges, it can still play a significant role in addressing existing interoperability and security challenges of the healthcare industry.

Lifelong learning is more important than ever. As there is a constant need to improve our skill-sets, education systems and recruitment processes need to be more adaptive and efficient to cater to the various needs. This is why blockchain has great potential in the education sector. Imagine a world where, instead of accumulating paper certificates, we can record them in the blockchain thereby enabling us to build up a secure, verifiable digital record of formal qualifications, experience and soft skills gained over our lifetime. Imagine we have sole control and custody of it and can make it available to employers anywhere in the world. Imagine the time, energy and money saved. This is Blockchain.

However, if not implemented systematically and correctly, blockchain can even increase costs. Having said that, it cannot be denied that blockchain does have a quite a few applications that can bring measurable value to the supply chain, real-estate and energy sector.

Blockchain technology is highly beneficial in the power and utility sector, which manages large networks, requires maintenance of data, faster transactions, and assurance of data security. Blockchain-enabled applications offer disintermediation, improved transparency and tamper-proof transactions, energy trading and independent energy trading platforms for decentralized energy. The smart grid is expected to balance the mismatch between supply and demand. The decentralized capability of this system could incentivize consumers and producers to trade capacity through a blockchain-enabled marketplace. Blockchain enables traceability, secures transactions between participants and keep everyone's trading activity and data private.

Practically, anything of value can be tracked and transacted on a blockchain network, reducing risk and cutting costs for all involved. Though quite promising, a lot more research into business needs, trials, project data and analysis is required into the scalability and performance of blockchain application to confirm if the technology is commercially viable to be adopted into the mainstream.

However, considering the potential benefits that have been realized in pilot studies, it goes without saying that blockchain is here to stay and will revolutionize how society operates in the next couple of years.

EXERCISES

Multiple Choice Questions

1. **KYC stands for:**

   A. Key in Your Crypto
   B. Know Your Customer
   C. Know Your Crypto
   D. Know Your Consensus

   Answer: B

   Explanation: Know Your Customer (KYC) is a regulatory and legal requirement of banks and financial institutions (FIs) to verify the identity of their customers. According to a report published by Burton Taylor International Consulting, global AML/KYC spending was estimated at a record $749 million in 2018 with Asia accounting for $301.1 million. Credit organizations need to perform KYC when processing applications.

2. **IoT stands for:**

   A. Internet of Tomorrow
   B. Internet of Things
   C. Internet of Technology
   D. Industries of Technology

   Answer: B

   Explanation: By the very nature of the distributed ledger technology, blockchain is possibly the solution to address the challenges of the Internet of Things (IoT) market. Blockchain and IoT are currently trending technologies, both in their infancy. However, it is believed that blockchain can maximize the use of IoT devices enabling not just machine-to-human interaction but also machine-to-machine communication. Smart contracts will play a crucial role in ensuring each party fulfills its part of the working relationship.

3. **On an average, cross-border transaction takes ____________ days**

   A. 20 to 40 days

B. 10 to 15 days
C. 3 to 5 days
D. 3 to 5 business days
Answer: D
Explanation: As per McKinsey, cross-border payments accounted for around 40% of global payments of transactional revenues with payment flows of more than $135 trillion during 2016. However, the current cross-border payment process suffers certain shortcomings. On an average a cross-border transaction takes 3–5 business days. International payments lack visibility and leave the beneficiary guesstimating on the delivery time and payment amounts as the exchange rate is not fixed until the arrival of the funds.

4. **__________ led by JPMorgan Chase and other banks uses blockchain technology for participating banks to transfer US dollars across borders and institutions.**
A. Global banking payments network
B. Interbank Information Network (IIN)
C. SWIFT
D. GPI Link payments
Answer: B
Explanation: IIN (Interbank Information Network) led by JPMorgan Chase and other banks uses blockchain technology for participating banks to transfer US dollars across borders and institutions.

5. **Global banking payments network SWIFT plans to launch a PoC with:**
A. Ripple
B. Hyperledger
C. Bitcoin
D. R3 Corda
Answer: D
Explanation: Global banking payments network SWIFT plans to launch a PoC with blockchain consortium R3 Tech to trial its GPI Link payments standard via R3's Corda platform.

6. **We.Trade (we-trade.com), a consortium powered by __________ blockchain, built a cross-border trade finance platform for tracking, managing and protecting trade transactions amongst SMEs.**
A. Ripple
B. Hyperledger Fabric
C. Bitcoin
D. R3 Corda
Answer: B

Explanation: We.Trade (we-trade.com), a consortium powered by the Hyperledger Fabric blockchain, built a cross-border trade finance platform for tracking, managing and protecting trade transactions amongst SMEs.

7. **____________ is all about owning one's own data and identity.**
   A. Self-sovereignty
   B. Immutability
   C. Disintermediation and Collaboration:
   D. Transparency and Trust
   Answer : A
   Explanation: Self-sovereignty is all about owning one's own data and identity. Our identity is our personal property and it is our sovereign right to keep it private and have control over the management of our personal data. Blockchain offers a robust infrastructure to identity attestations.

Short-answer Questions

1. **How data integrity issue in finance domain can be tackled in blockchain?**
   Blockchain offers the potential to create a single source of truth around customer identity. The immutability and transparency of blockchain solutions can improve data accuracy and security, reduce delays from paperwork and the risk of fraud, and show compliance through an audit trail.

2. **How efficiency related issues in finance domain can be tackled in blockchain?**
   Research by KPMG suggests a 40% increase in efficiency by leveraging blockchain for data processing and maintaining a single version of truth. Blockchain technology can create a ledger that is continuously synchronized throughout the network, thus eliminating the need for reconciliations.

3. **How cost-related issues in finance domain can be tackled in blockchain?**
   With blockchain consensus mechanisms and smart contracts, automatic transfer of funds can be triggered when the agreed set of conditions are met. This reduces the amount of time the capital is tied up in a transaction. Blockchain can also eliminate some transaction fees by reducing reliance on third parties.

3. **What is KYC?**
   Know Your Customer (KYC) is a regulatory and legal requirement of banks and financial institutions (FIs) to verify the identity of their customers. According to a report published by Burton Taylor International Consulting, global AML/KYC spending was estimated at a record $749 million in 2018 with Asia accounting for $301.1 million.

   Credit organizations need to perform KYC when processing applications.

4. **How blockchain can help in self-sovereignty in educational sector?**
   Self-sovereignty is all about owning one's own data and identity. Our identity is our personal property and it is our sovereign right to keep it private and have control

over the management of our personal data. Blockchain offers a robust infrastructure to identity attestations. This could give control to the students over their educational portfolio and avoid dependency on educational institutions for the storing of personally identifiable information (PII). A person can create a self-sovereign identity on a blockchain which is fully controlled and maintained personally by the individual. This could address the issue of identity theft prevalent on the traditional identity management system.

5. **How blockchain can help in immutability in educational sector?**

   Certificates can be faked, credentials can be weak, and education may not be accredited. With blockchain, once transaction records like transcripts, attendance sheets, assessment records and results, certificates, credits and other records of achievement are created, they are permanently stored and cannot be modified or deleted. A block in the blockchain contains the cryptographic hash of the previous block, as well as a time-stamp and the transaction data. This means that once transaction records like transcripts, attendance sheets, assessment records and results, certificates, credits and other records of achievement are created, they are permanently added to the blockchain in chronological order as a block. These blocks cannot be modified or deleted, once they are added to the blockchain.

6. **How blockchain can help in disintermediation and collaboration in education sector?**

   Professor John Dominique, Director at Knowledge Media Institute, an Innovation Lab at Open University in UK, has stated that the centralized model of present-day education is no longer sustainable. Blockchain technology allows for total disintermediation and disaggregation of education. Eliminating the need for a central controlling authority for transaction and record maintenance moderates the number of player interactions, saves time and cost and, best of all, promotes collaboration between students, teachers, and employers. Students can engage with top-quality educators anywhere in the world directly, provided they all exist in the blockchain ecosystem.

7. **How blockchain can help in learning new concepts?**

   Students can be rewarded with blockchain token to incentivize them to complete their courses. These tokens can be used to fund further studies or courses or converted to cryptocurrency and cashed. Blockchain-based learning systems allow people to access education around the world, and without the barrier of high costs. We could potentially be looking at a global learning model where students gain access to courses across the globe, attend courses in parallel that are tailored to the global/country requirements and all the while competing with other learners for token points and credits.

8. **How blockchain can help to overcome Energy inefficiencies.**

   The transmission over long distances causes power losses estimated to be between 8–15%. Transmission and distribution losses vary from country to country as well. In some countries, the loss is between 30–60%, attributed to electricity thieves and other factors. These losses not only hit our electricity bill but also indirectly contribute to carbonization as approximately 40% of global CO_2 emissions are emitted from electricity generations through the combustion of fossil fuels. In addition, if the energy supply and demand is not balanced in the grid, it can damage the infrastructure and cause safety issues like fire and human causalities. Blockchain is expected to help make enhanced, modern and optimized power grids.

9. How blockchain can help in environment and climate changes?

The global energy sector is responsible for two-thirds of all the greenhouse gases emitted into the atmosphere as reported by the International Energy Agency. From unstable climate change to poor water and air quality, pollution and disease, our natural habitat is under threat. It is time for our energy system to evolve in response to these environmental challenges. Blockchain technology can help in generating and distributing electricity more efficiently by reducing both the amount of fuel needed to generate electricity and, as a result, the amount of greenhouse gases and other air pollution emitted.

10. How blockchain can help in reducing Energy Inequality?

Energy retailers purchase energy (natural gas or electricity) in wholesale markets and sell it on to end-users at a fixed price. They charge consumers a premium for this service along with a provision for ancillary services, which is reflected in the energy bill. Thus energy retail companies make profit by marking up the price of energy the customer consumes. This system is inefficient and unfair to the consumers. Blockchain technology can help reduce energy inequality and inefficiency by empowering consumers to buy and sell energy from other consumers directly or buy directly from the supplier, thus removing the need for intermediaries/middlemen.

11. How blockchain can help in managing interoperability of health sector?

Interoperability means different things to different people. In healthcare, it is the process of freely exchanging healthcare information among electronic systems that ensures delivering the highest-quality, most effective, and most efficient care to patients. This means that any and all electronic health records (EHR) that are stored in disparate systems should be integrated for improving patient care by making the best information available at the point of care. Blockchain technology can address the challenge of varying standards by accessing data through application program interfaces (APIs). Data transfer through APIs achieves standardization of data formats. Thus, data can be transmitted irrespective of the capabilities of EHRs to communicate with different HL7 versions (HL 7 is the messaging standard for electronic data exchange in the clinical domain, widely implemented in the healthcare sector).

12. How blockchain can help accessibility in healthcare sector?

Universal accessibility is an important component of quality healthcare. More often than not, a pharmacy or a medical laboratory may need access to patient files for a particular or routine diagnosis. Providing relevant and specific patient information to the healthcare entity can be tricky. In majority of the cases, the whole patient file is shared or access to patient files is not possible and medical tests need to be redone. With blockchain-enabled smart contracts, the patient or authorized doctor can share the patient's electronic health records (EHR), even between organizations, say a medical clinic and an insurance company that do not have a business relationship to each other.

13. How blockchain can help in public record keeping in real-estate?

In blockchain, data/transaction is stored as a ledger, distributed across a peer-to-peer network in which every transaction is immediately, automatically, and permanently recorded and accessed only by other permissioned participants of the blockchain. As data stored on the blockchain is secure, immutable, and can be made public, it becomes the perfect platform to record property sales and leasing transactions.

14. What is tokenization in real-estate? How can blockchain be used for this?

To "tokenize" real-estate means to associate a blockchain token or cryptocurrency with real property. Property owners can issue blockchain-based tokens, which represent shares in an asset. Investors can buy such tokens and become fractional owners of the underlying asset and participate in the asset's appreciation and cash flows.

Blockchain technology and security tokens leverage the advantages of real-estate investing and mitigate the downside such as high costs, inaccessibility and illiquidity.

15. Write notes on peer-to-peer blockchain-based mortgage system?

A peer-to-peer blockchain-based mortgage system will enable a lender and borrower to transact directly with each other. Borrower, lender and seller can exchange private and secure information without the need for a central party or authority. Blockchain-based P2P mortgage lending will help those who are unbanked and reduce costs as hidden fees mostly associated with banks are not applicable on this platform. Smart Contracts can execute deals and facilitate payments. Scores can be attached to lenders and borrowers who pay on time. Late payments or defaulters could have negative score that is tracked in the blockchain. The transparency and trust of the blockchain promotes confidence between the borrowers, lenders and sellers while also providing a check on defaulters and fraudulent activities thereby promoting honest transactions.

16. What is provenance tracking in supply chain? How blockchain can help?

When something goes wrong within a complex machinery system like an aircraft or freighter or lifesaving drug, it is critical to know the provenance, through supply chain management, of each component, down to the manufacturer, production date, batch, and even the manufacturing equipment program. Blockchain can hold the complete provenance details of each component, which can be accessed by each manufacturer in the production process including the owners, the producers, maintainers, and government regulators.

17. How blockchain can help in reducing cost in the supply chain sector?

Real-time tracking of a product in the supply chain via the blockchain reduces the overall cost of the moving items. Better tracking, monitoring and recording using blockchain technologies means more accurate planning in transportation needs, avoidance of delays, faster payments and settlements, better inventory control, assessments and quality checks recorded on the blockchain, and removal of intermediaries that are an overhead on the supply chain.

18. How blockchain can help in tackling trust issues in supply chain sector?

Establishing trust within a complex supply chain system is critical for smooth operations. All participants are trusted to do their part honestly. The manufacturer must have confidence in the supplier to follow factory standards. With the supply chain process/network on the blockchain, there is increased trust because no single authority controls the provenance information and frauds and corruption can be easily detected.

19. How blockchain helps to reduce process and time delays in supply chain sector?

With supply chain on the blockchain, the product transactions are time-stamped and can be tracked from the point of origin, including the parts or ingredients used in its manufacture or production. Identified defects can lead to specific recalls rather than the full set. The increased efficiencies in process can lead to reductions in time taken to diagnose and remedy a fault, improving system utilization. The study of the full SCM life-cycle on the blockchain can lead to continuous process improvement.

20. How blockchain can help in supply chain financing?

Building the intricacies of SCF onto the blockchain can simplify the process by providing a single source of truth at critical points in the supply chain, such as purchase order receipt and approval, invoice receipt and approval, inventory delivery and quality control. With transparency, immutability and authentication built in at every point, there is an inherent trust between the buyer, supplier and financial institution. Paper-based verification is substantially reduced, finance institutions are better equipped to take financing decisions and fraud is curbed.

21. Write notes on blockchain and IoT.

By the very nature of the distributed ledger technology, blockchain is possibly the solution to address the challenges of the IoT market. Blockchain and IoT are currently trending technologies, both in their infancy. However, it is believed that blockchain can maximize the use of IoT devices enabling not just machine-to-human interaction but also machine-to-machine communication. Smart contracts will play a crucial role in ensuring that each party fulfills its part of the working relationship.

Essay-type Questions

1. Write an essay on applications of blockchain in banking sector.
2. What are challenges of the financial sector? List down how finance sector can leverage blockchain for the benefits.
3. How blockchain can help on KYC (Know your Customer) requirements?
4. How blockchain can help in cross-border payments?
5. How blockchain can help in trade finance?
6. How blockchain can help in stock trading?
7. How blockchain can help in insurance sector?
8. What are the challenges in educational sectors in which blockchain can play remediation role?
9. How blockchain can help in identity and maintaining student records?
10. How blockchain can help in students financing?
11. How blockchain can help in verifications of academic credentials?
12. How blockchain can help in new pedagogy?
13. What are the challenges in energy sector in which blockchain can play remediation role?
14. What is peer-to-peer trading? How blockchain can help?
15. Write an essay on smart grid. How Blockchain can help by making smart power grids?
16. How blockchain can help in energy trading?
17. What are the challenges in healthcare sector in which blockchain can play remediation role?
18. How blockchain can help in health records management?
19. How blockchain can help in claims and billing management?
20. How blockchain can help in drugs supply chain management?
21. How blockchain can help in patient and provider identity management?
22. How blockchain can help in clinical trials and medical research management?
23. What are the challenges in real-estate sector in which blockchain can play remediation role?
24. How blockchain can help in property listings area of real-estate sector?
25. How blockchain can help in tokenization of properties in real-estate sector?
26. How blockchain can help in peer-to-peer mortgages in real-estate sector?
27. What are the challenges in supply chain sector in which blockchain can play remediation role?
28. How blockchain can help in supply chain traceabilities?
29. How blockchain can help in food safety?
30. What are the advantages of blockchain with IoT?

CHAPTER 11

Limitations and Challenges of Blockchain

LEARNING OBJECTIVES

Blockchain has numerous benefits, such as decentralization, immutability, transparency, persistence, and stability. However, blockchain does not solve every problem. Only issues that need decentralization networks can be addressed using blockchain. At present, the use of blockchain has gained traction in the avenues of financial transactions and identity management. Solutions are being built to experiment with payments, supply chain management and provenance, identity management, property rights, and post-trade settlements.

This chapter provides an in-depth analysis of the various limitations and challenges of blockchain. Some of these have solutions in place for specific industries that are evolving continuously. With the growth in blockchain size, there are limitations related to software as well as hardware requirements and these need further research and investment.

11.1 Introduction

While blockchain has a very high potential of creating a bright future for internet systems and various industries like banking, insurance, healthcare, retail, media, etc., it also faces a lot of technical and business challenges. This niche technology has some constraints that have caused significant negative impact on its wider adoption, interoperability, technical and skill support, and regulatory requirements. Some of these limitations are contradictory in the sense that a few of blockchain's inherent features are to be compromised before it is put into meaningful use. The challenges that are foreseen in the application of blockchain are explained in detail in the subsequent sections.

11.2 Blockchain Implementation – Limitations

The chief obstacle for implementing blockchain technology is the lack of a universal protocol to make it universally accessible across all sectors. The major weaknesses of blockchain can be discussed broadly under the following headings:

11.2.1 Limited Scalability

Scalability is one of the significant limitations in the blockchain network since each of the nodes needs to verify the transactions processed in the system. Because of this, the speed of processing a transaction gets limited. Experts continue to work on distributed ledger technology, which is based on blockchain like hyperledger fabric to overcome scalability issues. There are three aspects of blockchain that have to be addressed – scalability, security, and decentralization. At a given time, you can only ascertain two out of the three properties. For example:

1. Security and decentralization can be taken up, but scalability will have to be compromised. Bitcoin blockchain is facing problems here.
2. Scalability and security can be taken up, but there is a need to compromise decentralization. BigTable and Cassandra are examples of such features, where they are parting with decentralization.
3. Decentralization and scalability can be taken up as the priority, but blockchain security will have to bear the brunt, which is not desirable for many industries. Private chains are examples of such features.

Since blockchains have a consensus algorithm where it is a must to verify all transactions, there is a limit on the number of transactions made in a given time. Nowadays, there are solutions like distributed ledger technology that provides the possibility of more number of transactions per second. However, the flip side is that it impacts the transaction rate or transaction speed of the blockchain negatively.

Since blockchain contains a distributed chain of blocks, with the size of blockchain growing in size, there can be severe concerns on storage requirements.

One of the most significant limitations of blockchain is scalability. Most of the public blockchain consensus protocols which operate in a decentralized way must compromise between the low network output and a high degree of centralization.

The scalability problem is defined as follows:

- Every single node on the blockchain network processes every transaction and stores a copy of the entire state.
- Because of inter-node latency that increases logarithmically with every additional node, the blockchain network gets less secure when more and more nodes are added to its system.
- As and when the size of the blockchain increases, there is a need for more storage capacity and computing power so that the network operates at its optimum.
- At some point, with the growth in the size of the blockchain, only a few nodes will be able to process a block resulting in the risk of going back to centralization again.

The difficulty in scaling a blockchain can be narrowed down to two primary factors:

1. As the blockchain itself inherently grows in size as more people use it, each node must maintain a more extensive storage history. As of March 2018, the bitcoin blockchain alone has a capacity of 163 Gigabytes. Every node must store this amount of information, which means anyone who starts a bitcoin node would have to have at least this much disk space. This number will only continue to grow.
2. There is no incentive to be a node that manages the ledger itself. The only people who have the incentive to maintain the network are the miners, who receive a bitcoin block reward for every block they 'create' and append to the chain.

These two problems mean that it becomes more unwieldy and costly for any one person to maintain a node. Miners might make the network run, but nodes are the ones who keep the actual ledger. This leaves any blockchain with a dilemma; they need to centralize the network to be able to scale it. There are many solutions offered by developers, blockchain companies and research organizations. One of the solutions which are commonly being used now is called Sharding.

Sharding is a kind of partitioning of database where vast databases are divided into smaller, faster, more easily manageable parts called data shards. When this concept is used in blockchain, the system can scale up as more and more nodes are added to it.

In the sharding process, every node need not process all the data. The entire end-to-end blockchain state can be split up into different shards and there will be no need for every node to store all the data. With shards, only a portion of the state would have to be saved by particular nodes. So now, when the transactions are initiated on the blockchain, they will be directed only to those depending on the shard they impact. Each shard will only process a part of the entire state instead of a complete state. However, a command mechanism needs to be there to establish seamless communication between the shards automatically.

Some projects are also working on the idea of nodes being run independently in standard off-the-shelf hardware, thus achieving true decentralization while also solving the scalability problem.

Experts are building an all peer-to-peer network using Shardus (a project). Thus the blockchain will benefit from sharding and auto-scaling to give high throughput, low latency, and immediate finality along with the assurance of security. This is supported by an algorithm, which is based on proof-of-quorum and Shardus Distributed Ledger. This is one that solves the problems of linear scaling and state sharding as the network increases in size.

Some other potential solutions for scalability problem are:

1. **Segwit (Bitcoin architecture-specific)** – Segregated witness (a.k.a. Segwit) is the solution to isolate transaction signatures (i.e., "witnesses") from the remaining transaction data, thus removing some extra weight off the blocks.
2. **Increasing the block size (Bitcoin architecture-specific)** – By enhancing the block size, it will be possible to have more transactions contained in each block, which in turn allows the network to handle more transactions per second.

3. **Off-chain state channels** – These state channels have mechanisms where the interactions happen within the blockchain instead of being conducted off (outside) the blockchain.
4. **Plasma** – This solution uses a cluster of smart contracts that executes on top of a root blockchain (i.e., the main Ethereum blockchain).
5. **Off-chain computations** – TrueBit is an example of a solution that uses off-chain calculations to enable scalable transactions among Ethereum smart contracts. Primarily, just like state channels, TrueBit uses an external layer of the blockchain to do the heavy lifting. Off-chain computation is a concept where there is an additional layer other than the blockchain in which all complex mathematical calculations are taken care of. This will not only take the burden from the Ethereum Blockchain but also reduce the transaction verification and processing costs. Not all the nodes in the blockchain participate in this computational model. Instead, only particular participants will perform computations along with a deposit, which is complicated. If the solution is right, the participant is given the award and also the deposit. If not, the deposit is forfeited. The off-chain computation does the verification using verifiers. By adding a separate layer to the blockchain, this mechanism ensures that processes that have slow transaction speed are executed separately. By moving the transactions off-chain, transactions can be processed in less than a second while providing a higher level of privacy of the information from public ledgers. One of the popular off-chain solutions is Raiden Network, which is being developed and gaining popularity.
6. **Proof-of-stake (PoS)** – Here, stakeholders vote with their money (in Ethereum's case, ether) instead of computing power. Bitcoins and Ethereum work on the PoW mechanism where complex mathematical equations have to be solved for verification and processing of transactions. Because of this, transaction speed becomes slow on Ethereum's blockchain. The PoW has its disadvantages, and to have higher transactional rates and higher levels of security, a new solution of proof-of-stake has been developed.

   The other main difference between the proof-of-work model and proof-of-stake is that in PoS, the validators will not earn ethers but instead, get a transaction fee for their effort. The transaction speed will be much higher in PoS.
7. **Transient blockchain** – Saito has an innovative solution to both problems. To solve the growing pains of keeping massive ledgers permanently, Saito has created a transient blockchain which deletes itself at specific intervals known as 'Genesis Periods.' A user who wants to retain any information on-chain can do so by paying a small fee, essentially paying for the actual cost of storage that it uses. The second solution of Saito is to provide an incentive to the nodes who supply storage capacity to upgrade the amount of data they can hold. For providing more storage capacity, nodes essentially become miners themselves by having a chance to claim new Saito tokens as they are issued. The more storage capacity they provide, the better chance they have of being compensated with a large sum of money. Therefore, this is similar to the probabilistic method of how Bitcoin miners spend money to make money.

The difference is that Saito, performing work for the network, helps it scale.

Although these solutions have provided limited improvement until now, the main limitations are due to the limited storage and computational abilities of current systems.

Hardware scalability: With the increase in popularity of blockchain, the cost of hardware is on the rise. This is also raising the price of entry to mining, thereby making it centralized.

Apart from this, there are also energy and cost issues.

11.2.2 Limited Privacy

Even though the identities are anonymous in a distributed ledger, looking at the transaction address, it is possible to link the user identity with that address and can get information about the user. Blockchain transactions are not aligned with one's status as anyone can create a new wallet anonymously and transact through that wallet. Identity verification data like security numbers cannot be openly stored in public smart contracts. Credential management is another factor that is not managed in an open, ultimately unsecured smart contract.

11.2.3 Lack of Technical Knowledge

Even as the blockchain is becoming increasingly popular, most of the investors are finding it difficult to understand the different technical terms, as there is no proper documentation available to help the users. Because of this, investors are not able to get their queries resolved. Also, many investors do not know much about Initial Coin Offering (ICO), which is famous for raising capital in the non-equity market.

Any niche technology will need time for the developers to learn, experiment, and pilot it in projects. Also, educational institutes need time to introduce the courses on the new technology and faculties have to be trained on the subject. Because of all this, there is a severe shortage of skilled developers of blockchain. This lack of experienced developers and workforce impacts innovation on the blockchain. Experts are made because of their experience in any technology. Statistically, there is a high demand for blockchain experts, which in turn also has raised the salary expectations of the few available in the market.

11.2.4 Security Concerns and Flaws

Blockchain is a network of people. If more than 50% of the people in this network are corrupt and manipulate the data, then the truth will be treated as a lie, and this could become one of the primary reasons for the complete network failure. So, members of the network need to carefully observe the transactions to avoid misusing data and to ensure security.

With the blockchain growing in size in terms of nodes or blocks, corruption among the participants compromises the blockchain security. If more than half of the nodal computers in a blockchain network validates something, then that is considered to be true. Hence, if more than half the nodes in a network approve a fraudulent transaction (lie), then the myth will be treated as truth by the entire blockchain network. This is called the " 51 percent attack ", which is a critical security flaw in blockchain and its applications.

The biggest security challenges of Blockchain are as follows:

1. **Spoofing of payment information:** In some instances, malware can replace the crypto wallet address with another one, when a user is transferring money. Not everyone is observant enough to double-check an address, which is usually a lengthy jumble of alphanumeric characters and symbols.
2. **Phishing:** Exchange users can be falsely directed to a phishing website through which their crypto wallet credentials can be misappropriated. For example, users can receive a fake message in their inbox providing information about a significant update for their crypto wallet. The message could specify that the user must sync his or her wallet with a recently hard-forked network; if ignored, the user will not be able to send or receive coins.
3. **Crypto wallet theft:** A majority of investors or users store their cryptocurrency wallet files on their personal computers, which leaves them susceptible to malware attacks.

11.3 Blockchain Implementation – Challenges

The principal challenge associated with blockchain is a lack of awareness of the technology, especially in sectors other than banking, and a widespread confusion on how it works. In addition, there are also the following factors that need to be addressed.

11.3.1 Transaction Processing Speed

One of the other challenges faced by blockchain is the scalability in the speed of transaction processing. Unlike what people believe, the most popular cryptocurrency Bitcoin can process only an average of 7 transactions per second. At the same time, Visa can process 24000 transactions per second, and PayPal can handle about 200 transactions per second. Bitcoin Cash can transact about 50 to 60 transactions per second, but it is still low compared to Visa. Among all the cryptocurrencies, Ethereum is relatively slow and can negotiate only about 20 transactions per second. With the growing size and complexity of Ethereum Blockchain, the transaction speed has been impacted, and sometimes, transactions take hours to get verified or are not verified at all.

Directed Acyclic Graph (DAG) has a solution to address the issue of low transaction speed. DAG has solved this by confirming transactions on a per-transaction basis rather than on a block-to-block base. Instead of waiting for miners to process a large block of transactions, DAG makes it possible for users to handle each other's transactions on a microscale, thereby reducing transaction costs and making it faster.

11.3.2 Complexity

The underlying technology of blockchain is very complicated for beginners to understand since the code consists of a lot of mathematical calculations. The technology of blockchain is at its infancy in terms of maturity; this has many new processes, standards, and niche technical aspects. Also, another aspect of blockchain is that it relies on decentralization and cryptography as the main backbone, which further adds to its complexity. Because of this, various challenges arise in its functionalities, such as the speed of transaction, the process of verifying the transaction, and the limits on data that can be sent or received. Another dimension to the complexity of blockchain technology is the size of its network. Just like any decentralized system, blockchain is susceptible to fraudulent actors who may be able to corrupt the network. To have a stable blockchain network, it needs to have many users and nodes connected with a robust network algorithm.

11.3.3 Implementation and Operation Cost

Even though blockchain benefits users with low transaction costs, the underlying technology of blockchain and its implementation cost are not small.

Four inputs predominantly impact the costs of both public and private blockchain solutions and depend entirely on the use case and the objectives an organization aims to accomplish. There are additional inputs to consider, but the most influential are:

- Transaction volume
- Transaction size
- Node hosting methods
- Consensus protocol.

Each of these uses high computing power with an increase in the size of blockchain with additional nodes and external integrations added. These all add to the operating costs. The development of blockchain solutions usually follows the agile process, which also adds to the costs with participation from different stakeholders of Business and IT.

11.3.4 Storage Constraints

Under the blockchain, the data is stored by every full node in the network indefinitely since the database in blockchain is append-only and immutable. Therefore, storage remains a considerable challenge for practical applications that are built on the blockchain.

To achieve fast transaction speed in blockchain, the data and the load on the nodes have to be cut down, which is the main challenge. As of now, each node has to store the entire data. So if the blockchain can have a mechanism that can allow the nodes to store only the data that is mostly used or locally relevant data for processing transactions, we can process transactions faster. Such solutions can make blockchain more efficient, and there is a need to research on such solutions to overcome the storage constraints.

11.3.5 Lack of Governance and Standards

With the evolution of more than 2000 cryptocurrencies and thousands of projects which use distributed ledger technology, numerous blockchain networks have come to the market. Since there are no global standards for interaction between these blockchain networks, these networks work in silos. That means there are many protocols, coding languages, and consensus mechanisms to support these networks, thereby creating inter-operability issues between blockchain networks.

Some of the solutions offered to solve the inter-operability issues are Ark (which works on Smartbridges architecture) and Cosmos, which makes use of the Interblockchain Communication (IBC) protocol.

Decentralized blockchain platforms do not have any centralized controlling authority, and it is a completely trustless, open and permissionless system where no one is responsible for setting and maintaining standards. No common regulatory standards are available globally now for blockchain applications. If there are standards, overall cost reduction, more efficient consensus mechanisms, and better interoperability can be achieved in the blockchain. This challenge further impacts the issue of onboarding new developers and blockchain security requirements.

Since blockchain technology is based on a distributed ledger, centralized databases will always be faster than blockchain networks. Just like a regular database, the network has to run all the processes whenever a block is added to it. But it also has to take care of new processes to ensure that performance is not compromised. Some of the processes that have to be carried out are:

1. **Signature verification** – In the blockchain, each transaction will have to be signed digitally, and a public or private key is used to validate the same. The whole process of this verification is time-consuming and complicated.
2. **Redundancy** – In a blockchain network, to process each node, it has to travel and process every intermediate node separately to get to the target node. In a centralized database system on the other hand, nodes can be processed in parallel without relying on any other nodes. This redundancy has a direct impact on the performance of the blockchain.
3. **Attaining consensus** – It is essential that in a blockchain, there should be a common consensus reached for every transaction made in the blockchain network. Based on the network size and the number of blocks or nodes in a blockchain, the to-and-fro communications involved to attain a consensus can consume a considerable amount of time and resources.
4. **Alternative consensus algorithms** – The two alternative consensus mechanisms possible are delegated proof-of-stake (DPoS) and practical byzantine fault tolerance (pBFT). These two alternative consensus mechanisms are more suitable for non-financial blockchains. For other use cases, these algorithms need further research and development, as of now.

5. **Delegated proof-of-stake (DPoS)** – In this mechanism, there will be a small number of delegates who maintain the ledger, and users vote for them using a real-time algorithm. Because of the reduced active nodes, the throughput of the network is much better. Since each node gets paid through inflation, it is justifiable for maintaining a large data centre for network support. But the negative side of the DPoS is that it functions in a centralized manner. That means more control on the network and less ability to be open and transparent. So if a user can buy more votes, it is easy to control and change the network rules to their benefit.

In the pBFT algorithm, every node takes turns for validating the state of the block. pBFT can perform many functions very efficiently, giving higher throughput and lower cost of resource consumption. The flip side of pBFT is that it is easily prone to Sybil attacks since it is insignificant to create new network nodes. Fraud actors may be able to create more nodes to reach 33% of the network and attack the complete chain. Compared to the Bitcoin network, which needs 50% of the network to gain control, pBFT needs only 33% of the network to gain control, which makes the pBFT very susceptible to be taken over.

11.3.6 Lack of Formal Contract Verification

Formal verification of smart contracts is a major unsolved problem since it needs an understanding of what a "formal proof" is. Formal verification is used to determine whether the program behaves as per the specification or not.

11.3.7 Energy and Resource Consumption

With additional nodes being added to the blockchain, it grows in size and the number of transactions also increases, which in turn results in the need for more resources and infrastructure to run. Miners need to validate every block in a blockchain. With an increase in the size of the blockchain, the time and effort required from miners to verify the network also drastically increases. To validate transactions, miners need to come up with many mathematical solutions, and they require substantial units of computing power to do that. Since each node has extreme limits of fault tolerance to ensure zero downtime, with high levels of security of network and data, the average cost of running the blockchain is also high based on the robust hardware infrastructure needed. All this means heavy energy consumption.

11.3.8 Simplified Mining

Mining for a typical blockchain cryptocurrency, for example, bitcoin, has led to an oligopoly issue. The majority of the mining power is with a limited number of mining pools that share profits. This directly contradicts the main principle of decentralization in bitcoin and cryptocurrency.

Ethereum operates on the PoW basis. Only when the validation rules are satisfied, the blocks will be created and appended into Ethereum's blockchain. This approach ensures the validity of each transaction. This PoW model secures the transactions as well as transfers for sure, but this still has its drawbacks. This method does not benefit the miners in any way, and there is no way to detect any fraudulent activity or transactions in the blockchain with this method. DAG is trying to solve this issue by defining key roles for the transactions in the blockchain, but this still has a long way to go. With this method, miners do not have power at their hands to manage the blockchain.

11.3.9 Human Errors

Errors in manual data entry will lead to outdated information or may also result in inconsistent data in the database. So essentially, data has to be validated before entry to the system and verified after the entry into the system. In the context of the blockchain, people use the phrase 'garbage in, garbage out' while referring to data.

Human error is also a common reason for issues in the system. Though we can consider the entire blockchain as a single database and each block in the blockchain as a storage stack, the date which is appended into the database is entered by humans in the first mile. If this data has to be clean, correct data should be fed at the entry point itself. However, there is no method or mechanism to verify the correctness and validity of the data, which is entered to be processed inside the blockchain network. Hence, there is a trust issue as false data entered into a blockchain can contaminate the data of the entire system.

Summary

Blockchain is a disruptive technology since it has empowered the global community with a path-breaking innovation that significantly alters the way that consumers and businesses operate. However, it comes with its own set of limitations/challenges, which all contribute to the slow path for mass adoption. Over time, active and measured steps are being taken to overcome these challenges and keep the course of blockchain adoption pointing upwards. Blockchain has some severe limitations which need to be fixed to utilize the revolutionary solutions it offers. Some of these problems will be fixed eventually, while some issues in its core architecture may still remain.

EXERCISES

Multiple Choice Questions

1. **This is the one of the major limitation in the blockchain network since all the transactions performed on the network needs to be verified by each of the nodes.**
   A. Immutability
   B. Privacy
   C. Technical Knowledge
   D. Scalability
   Answer: D
   Explanation: Scalability is the one of the major limitation in the blockchain network since all the transactions performed on the network needs to be verified by each of the nodes. Because of this, the speed of processing transaction gets limited. Experts continue to work on distributed ledger technology, which is based on blockchain like hyperledger fabric to overcome the scalability issues.
2. **In certain cases, malware can replace the crypto wallet address with another one, when a user is transferring money. This is called as:**
   A. Spoofing
   B. Phishing
   C. Resource Consumption
   D. Transfering Wallet
   Answer: A
   Explanation: Spoofing of payment information: In certain cases, malware can replace the crypto wallet address with another one, when a user is transferring money. Not everyone is observant enough to double-check an address, which is usually a lengthy jumble of alphanumeric characters and symbols.
3. **This is a kind of partitioning of database which divides very large databases into smaller, faster, more easily manageable parts. When this concept is used in blockchain, the system is able to scale as more and more nodes are added to it.**
   A. Breaking
   B. Dividing
   C. Partitioning
   D. Sharding
   Answer: D
   Explanation: Scaling blockchain is becoming one of the technology's great mysteries. Numerous solutions are being offered by various developers, researchers, and academics. One such solution being explored is known as sharding. Sharding is a kind of partitioning of database where very large databases are divided into smaller,

faster, more easily manageable parts called data shards. When this concept is used in blockchain, the system is able to scale up as more and more nodes are added to it.

4. **This has a solution to address the issue of low transaction speed. It solves this by confirming transactions on a per-transaction basis rather than a block-to-block basis.**
   A. DAG
   B. CAG
   C. Spoofing
   D. Sharding
   Answer A
   Explanation: Directed Acyclic Graph (DAG) has a solution to address the issue of low transaction speed. DAG solves this by confirming transactions on a per-transaction basis rather than a block-to-block basis. Rather than waiting for miners to process a large block of transactions, DAG allows users to process each other's transactions on a microscale, thereby reducing transaction cost and making it faster.

Short-answer Questions

1. **How is limited privacy an issue in blockchain?**
   In a distributed ledger, although identities are anonymous, it is possible to link the user identity with the address using the transaction patterns and can get information about the user. Blockchain transactions are not tied directly to user identity as anyone can create a new wallet anonymously and transact through that wallet. Identity verification data like security numbers cannot be openly stored in public smart contracts. Credential management is another factor which is not managed in an open, ultimately unsecured smart contract.

2. **What are the major security challenges of blockchain?**
   The biggest security challenges of blockchain are as follows:
   1. **Spoofing of payment information**: In certain cases, malware can replace the crypto wallet address with another one, when a user is transferring money. Not everyone is observant enough to double-check an address, which is usually a lengthy jumble of alphanumeric characters and symbols.
   2. **Phishing**: Exchange users can be falsely directed to a phishing website, through which their crypto wallet credentials can be misappropriated. For example, a user can receive a fake message in their inbox, providing information about a very important update for their crypto wallet. The message could specify that the user must sync his or her wallet with a recently hard-forked network; if ignored, the user will not be able to send or receive coins.
   3. **Crypto wallet theft**: A majority of investors or users store their cryptocurrency wallet files on their personal computers, which leaves them susceptible to malware attacks.

3. **How complexity is a limitation in blockchain?**

   Blockchain technology is not easy to understand by beginners as it involves a lot of mathematical calculations. Being a nascent technology, blockchain involves a lot of new processes and highly specialized terms and technologies. Blockchain has made cryptography and decentralization more mainstream, adding to its implementation complexities. Blockchain faces various challenges to its functionality such as transactional speeds, verification process, and data limitations.

   Another complexity of blockchain technology is its network size. Like all other distributed systems, blockchain is not 100 percent resistant against bad actors, who can corrupt the network. For the blockchain to remain stable and to avoid corruption in a network, it needs a huge set of users and nodes connected with a robust network.

4. **What are the factors that impact the costs of both public and private blockchain solutions?**

   Four inputs predominantly impact the costs of both public and private blockchain solutions, and these depend entirely on the use case and the objectives an organization aims to accomplish. There are additional inputs to consider, but the most influential are:

   - Transaction volume
   - Transaction size
   - Node hosting methods
   - Consensus protocol.

5. **How energy and resource consumption is a limitation for blockchain?**

   A blockchain network grows at an unprecedented pace, consuming heavy resources. Every block in a blockchain network needs to be thoroughly validated and mined by the miners. As the blockchain network grows, the need of miners to validate the blocks in a network also increases. Since the miners try many mathematical solutions to validate transactions, they utilize substantial units of computing power. Every node has extreme levels of fault tolerance, ensures zero downtime and makes data stored on the blockchain forever unchangeable and censorship-resistant. The average cost of a transaction is between 75 and 160 dollars and most of it goes for energy consumption. Every miner needs a supercomputer or a similarly powerful hardware resource to mine the blockchain. All these entail heavy energy consumption.

6. **How human errors affect the blockchain?**

   Human error is also a very common means of system failures. Manual errors can lead to outdated log information or even create a mismatching data while entering the data into the database. To make sure the data entered is correct, that data needs to be validated. Hence, the phrase 'garbage in, garbage out' is used in the context of the blockchain. The blockchain, as a whole, can be considered to be a database, and every block can be considered a storage container. However, the data that goes into the blockchain network is fed by humans. This data needs to be of good quality as there is no practical mechanism to monitor the data that is transmitted in a blockchain network. The data entered to a blockchain cannot be trusted and input of false data can contaminate the data of the entire network.

7. **What is spoofing of payment information?**

   In certain cases, malware can replace the crypto wallet address with another one, when a user is transferring money. Not everyone is observant enough to double-check an address, which is usually a lengthy jumble of alphanumeric characters and symbols.

8. **What is phishing?**

   In phishing, exchange users can be falsely directed to a phishing website, through which their crypto wallet credentials are misappropriated. For example, a user can receive a fake message in their inbox, which provides information about a very important for about their crypto wallet. The message could specify that the users must sync their wallet with a recently hard-forked network and that if ignored they will not be able to send or receive coins.

Essay-type Questions

1. Write an essay on limited scalability issues of blockchain.
2. How does lack of governance and standards affect the blockchain?
3. How does transaction processing speed affect the blockchain?
4. What are the potential solutions for the scalability problem of blockchain?

CHAPTER 12

Blockchain Case Studies

LEARNING OBJECTIVES

In this chapter, case studies of blockchain implementations for retail, banking, healthcare and energy industries have been discussed.

12.1 Case Study 1 – Retail

Industry Retail

Highlight Establishing unconditional transparency in the food supply chain using blockchain hyperledger fabric

Overview

Company ABC Retail is an American Multinational Retail Corporation established in the 1960s. ABC Retail operates a chain of hypermarkets, discount department stores and grocery stores, globally. It has wholly owned operations in Argentina, Chile, Canada, and South Africa.

In terms of revenue, ABC Retail is the world's largest company, with more than US$ 500 billion. It has more than 2 million employees across the globe. ABC Retail was the largest U.S. grocery retailer in 2019, and 65 percent of BC Retail's US$ 500 billion sales came from US operations.

Problem Statement

The company under study has access to fresh produce all year round and buys exotic food from worldwide markets; it is known to have more variety than any other food retail chain in the world. Their food is generally harmless to eat; nevertheless, they still occasionally have had cases where their consumers fell sick. Recently, there were at least 18 reported outbreaks of foodborne illnesses in the USA, including the E. coli found in romaine lettuce.

With consumers in recent times having access to agricultural produce across the world despite seasons, locations or the environment, they have expectations of trust on these products available to them. The retail chain has four significant challenges in assuring consumer expectations.

The first one is about food fraud, where there are a lot of possibilities of substituting, tampering, and faking the products, either at the production level or during the transit.

The other issue is about illegal production, where statistics show that 10% to 22% of the agricultural products are not reported or not regulated.

The major challenge faced is about foodborne illness, where it was proved in 2018 that 1 in 6 people fall sick from contaminated food or beverages per annum.

In such cases, the costs of food recall are very high, as much as $10 Million for ABC Retail, in addition to the loss in goodwill, loss of consumer confidence, and legal implication.

All the above challenges have resulted in high costs of the food chain in the following areas:

- Visibility into the status of goods as they move through the supply chain
- The cost of human health and life
- The cost of recalling contaminated food.

When there is an adverse event of a foodborne disease outbreak, it takes a long time, often in the order of days, to find the source of the disease. This is because the contamination could have occurred during the production, transit, or distribution channels. If there is better traceability possible, the companies can act fast, try to solve the issue and protect the farmers by removing the produce from the affected farms.

This, in retrospect, led to their interest in enhancing transparency and traceability throughout the food system. The company has tried many methods and approaches to solving this problem over the years, but none of them gave them the results they were looking for.

Approach

In 2016, ABC Retail established the Food Safety Collaboration Center in China, to develop its food provenance pilots using blockchain with a plan to invest about $25 million over the next five years to research global food safety.

The company looked into several blockchain technologies and eventually decided to use hyperledger fabric because it met most of their demands for blockchain technology. Hyperledger is an enterprise-grade blockchain technology. Hyperledger is also a permissioned blockchain. The team also found it essential to use an open-source-based blockchain so that it is vendor-neutral. The technology ecosystem has to be open since ABC Retail worked with many suppliers, distributors, direct competitors, and geographical collaborations.

ABC Retail decided to run two pilots concurrently to test the blockchain solution for traceability and transparency of the supply chain from end to end. Two use cases were tested – the traceability of mangoes, which are sold in the US stores of the company and the traceability of pork sold in their stores in China.

Life of ABC Retail mango typically consists of the following duration milestones:

- It takes 5 to 8 years for a mango tree to grow and bear fruits
- Usually, mangoes are produced by small farmers in Central or South America
- Once the mangoes are harvested, they are sent to a packing house for washing, drying, and packaging.
- The mango cartons are shipped to the US by air, sea or land (custom border)
- Then these mangoes are washed, peeled, sliced and put into convenient containers in a facility centre and then later shipped to ABC Retail Distribution centres to get refrigerated
- They are subsequently transported to one of the company's 11,200 stores, refrigerated and shelved.

In the course of an adverse event, it took ABC Retail an average of 6.75 days to track these mangoes within which time the damages would have already happened. However, after implementing the Blockchain solution, the 6.75 days tracking time of mangoes was reduced to 2.2 seconds. This created greater transparency across ABC Retail's food supply chain in 2017.

Solution

ABC Retail combined AI, IoT, and blockchain to improve the quality of the food supply chain from end to end. The Application logic of the system was hosted in the smart contract called "Chaincode" in the hyperledger. To put this system to test, the company created a food traceability system based on hyperledger fabric. The company roped in another technology partner, IBM Food Trust, and ran two proof-of-concept projects to test the system. The blockchain system built using hyperledger was designed to accurately record batch number, farm origin data, expiration dates, storage temperatures, and shipping details. ABC Retail used AI to predict and analyze the various patterns and trends of the retail policies that impact the supply chain. They also studied road traffic prediction and used the analysis to expedite the delivery. When the food products passed through the supply chain, sensors, and RFID tags supported by IoT technology were used to write the real-time data into the blockchain. These data helped ABC Retail to implement hazard management systems and critical control points for supporting the Food Safety Modernization Act and Inspection Service.

Every node in the blockchain network owns its data and will have control of those who can access the data elements based on the permission granted to share the pertinent records. The bolt-on adapters of the blockchain interfaced with the existing data stores like SAP Master Data of ABC Retail to make use of the business details like inventory, sale orders, and vendor master data. An API connector was built for the blockchain network administrators to manage the other complex environments and automatic web uploads of legacy data. The user was able to interact with the blockchain network either through the desktop or mobile interfaces or use the certifications module for uploading the regulation and inspection documents for supply partners. This solution also provided business-critical digital certificates needed for provenance verifications to ensure authenticity. The end-to-end traceability helped in fast product recall with real-time location and status of the food product or produce. To authorize users, predefined roles were created, such as given below:

- Account Owner
- Account Administrator
- Certifications Manager
- Food Safety Team Manager
- Onboarding Team Member.

Results

Thus, ABC Retail can now trace the origin of over 25 products from 5 different suppliers with the support of a system powered by hyperledger fabric. The company plans to span out the system to more products and categories, moving ahead in the future. It has recently announced that it will require all of its fresh leafy greens suppliers (like suppliers of salad and spinach) to trace their products using the system.

Now, ABC Retail will have the advantage to regain customer confidence because of the following:

- Food safety
- Food freshness
- Waste reduction
- Confidence and sustainability
- Traceability.

They collaborated with their technology partner to set up a blockchain, involving prominent players in the food industry, like Nestle and Unilever. ABC Retail is planning to extend this system in collaboration with Food Trust to more food products.

12.2 Case Study 2 – Banking and Financial Services

Industry Banking and Financial Services – Core Banking

Highlight Increasing transaction volume and network resilience while maintaining confidentiality requirements for real-time gross settlement among different banks of a central banking consortium

Overview

Banking institutions, central banks, and financial markets in the domestic and international space are going through a disruptive phase with the emergence of niche technologies and modern infrastructures. While some of the countries are still nascent in their strategy and planning to keep up with these advancements, the South African financial services industry has taken active steps to learn, grow and adapt to these opportunities. Having understood the broader implications and long-term benefits of automation, they were able to derive a contextual view of the finance sector in South Africa. South African Reserve Bank (SARB) required an innovative approach to enhance the resilience of interbank payment systems while ensuring the reduction of the overall cost of these systems.

The National Payment System Department (NPSD) executed the vision of the SARB by ensuring the overall effectiveness and integrity of the payment system.

The key aims of a properly functioning payments system are to:

- improve the stability of the financial system
- reduce the cost of the banking transaction
- encourage optimum use of financial resources
- enhance financial market liquidity
- facilitate the conduct of monetary policy.

Problem Statement

To understand the problem statement, it is essential to know how bank payments and settlements happen in South Africa.

As per the National Payment System Act 78 of 1998 in South Africa, settlement can be carried out using cash or through bank entries in the book of the SARB. For this to happen, the participants of the settlement system should have an account at the SARB through which interbank settlement takes place. This occurs on a pre-funded basis through the South African Real-time Gross Settlement System called SAMOS or South African Multiple Option System. The aim of this system is to reduce the risk of settlement failure by a participant in the payment system. Only if the issuing bank has sufficient funds, the transfer is carried out in the SAMOS. This reduces the risk of exposure for the participating banks as well as SARB. The funds transferred are final and irreversible.

With each bank in the country operating at different levels of maturity and business goals with their customers and vendors, inter-bank transactions are involved. They need to be highly secured for the customers to have a high degree of trust in them. Despite the design of the SAMOS system being aligned to SARB policy requirements, the system gradually became customized, slowly making the reconciliation payments inefficient. Also, the entire burden of suspicious transactions rested with the participating banks although it was not their fault.

Solution

Project Khokha is a project driven by the SARB, in collaboration with the consortium of South African settlement banks as well as participating technical and support partners. The prime goal of the Khokha was to build a proof of concept (PoC). With the modalities of the process being as vital as the outcome, it provided an opportunity to expand the distributed ledger technology (DLT) skills base in the South African banking industry. It also presented an opportunity to explore the approach of collaborative innovation, which could be a critical success factor for future developments. A distributed ledger was created between participating banks to have a payment system so that the participating banks can pledge, redeem, and track their balances and transactions of the tokenized rand. As part of Project Khokha, specific vital measurements such as scalability, privacy, security, and flexibility of the DLT were also assessed to know whether the DLT can be extended for other functions too. The RTGS system was built using tokens supported by funds held in the Central Bank. The tests were conducted for the performance of each node when each participant bank executed its code with different deployment models from separate locations.

To mobilize the project fast, Quorum technology was selected since many of the participating banks used Quorum, and the implementation partner had the necessary skills in Quorum.

Quorum is an enterprise version of Ethereum (developed by JPMorgan and EthLab) and works on an open-source platform that enables collaboration between entities. Quorum uses the Go Ethereum code. The main distinctions of Quorum are in-network and peer permissions management framework, enhanced transaction, contract privacy, and voting-based consensus mechanism.

Project Khokha followed an agile approach. The design and development of the RTGS DLT platform were executed through four iterations.

In Iteration 1, two banks (restricted to SARB) were enabled to transfer tokens and minting functions were designed, developed, and tested.

In Iteration 2, the payment approvals were made by SARB, with all participating banks being able to view the transactions in the whole network.

Iteration 3 involved shielding the amounts of transactions and balances using Pedersen commitments (these are commitment schemes that deal with "how the counterparty engages with value" – the surrendered value remains private and can be discovered only later when the employee finds the required parameter of the commitment process); access was given to SARB for opening the commitment, enabling it to verify and approve the payment.

Iteration 4 used range proofs and Pederson commitments to enhance the flexibility of the system. SARB nodes continue to have complete visibility of the Pederson commitments, and therefore, of the transactions. However, it does not have the need to verify them since the other nodes would have already performed this role. Quorum solution utilizes Whisper for private messaging, Pedersen commitments, and range proofs.

Results

With the blockchain implementation, the central bank consortium was able to surpass the transaction performance target of 70,000 transactions in much less than two hours. They were able to achieve 95% of block propagation time is less than 1 second and 99% propagation in less than 2 seconds. This proved that acceptable performance and service level is achievable, despite the geographical distribution of the banks' hardware.

The system was able to achieve privacy and information security while meeting the required transaction volumes.

The final results were positive, indicating that the Quorum platform can deliver the performance required, matching, and even exceeding the current performance criteria.

This was the first time that the banking consensus mechanism, legal commitments, and a set of legal proofs for confidentiality were used in tandem in an integrated enterprise. Together, all these essential elements delivered a combination of scalability, resilience, privacy, and settlement finality.

12.3 Case Study 3 - Healthcare

Industry Healthcare

Highlight Electronic health records and medical research data digitization

Overview

The first information technology-related changes in medical science were the digitization of medical files, now known as electronic health records (EHRs). The data contained in the EHR, other data sources, technology, and healthcare have the potential to transform medical practice by improving its overall performance. EHRs have produced a large volume of data.

Currently, for instance, most EHRs collect quantitative, qualitative, and transactional data, all of which could be collated, analyzed, and presented using sophisticated procedures and techniques.

Problem Statement

Electronic Healthcare Records (EHRs) were initially planned to manage distinct and straightforward medical records. As per their design, EHRs cannot maintain multi-institutional medical records of patients for their lifetime. Based on the various phases of the lives of patients who may have moved from one provider to another, the medical records will be scattered and not integrated. Some of the past data may be lost, since access to old data is controlled by the healthcare service provider who owns the medical history, rather than the patients. As per the HIPAA privacy rule, providers can take up to 60 days maximum to update or delete any record as per the patient's request.

In most cases, this 60-day maximum rule is also not complied with. Hence, beyond this time delay, maintaining the records with history becomes very challenging. Added to this is the complexity of the medical data being handled by different providers whom the patient may have approached during his or her lifetime. So we need a cohesive medical health record management solution that helps patients, providers, hospitals, and physicians/doctors to access the medical records in a hassle-free manner to prevent any adverse health-related events in a patient's life.

Solution

MedRec is a Blockchain product built for handling patient medical history. Since the permission management feature is enabled in blockchain, MedRec can allow patient-validated data exchange between medical jurisdictions and the record management system so that the physicians/doctors can have the needed access to these records. Now in this blockchain product, various stakeholders like physicians, patients and providers are thoroughly informed on the updated medical history and claims payment. They are also empowered to update, validate or delete the records based on their roles and responsibilities. Hence, the blockchain ledger has the convenience of auditable history with accurate traceability for root-cause analysis, if needed by the physicians, insurance regulators, or patients.

One of the key features built in this blockchain product is the robust fail-over model that depends on the many stakeholder entities, thereby preventing a single point of failure. Also, medical records are safe and secure from any cyber-attacks since the data is stored separately in provider and patient databases; authorization data is stored in each node of the network. With this, the raw medical data and log of the global authorization are distributed; this blockchain product is not a target to any content attack or data leak.

MedRec solution provides exhaustive patient agency across provider and treatment sites so that the citizens are empowered to make the right decisions of healthcare. This blockchain solution allows the patients to have a long-term confidential log of their medical information, current and historical, thereby enabling the patients to have predictive and preventive medicinal treatments. MedRec can also receive data from hardware devices such as Fitbit. Apple Watch, etc. to update the patient's daily health data.

Patients can create a holistic medical record and allow essential stakeholders to view or edit, receive a second opinion from physicians, and share it with guardians, family members, hospitals, or caretakers. MedRec also has modules of Predictive Analytics using which patients and physicians can learn from the family medical history and conditions, past medical care, the pattern of results, etc. so that this can be used for further treatments. An accurate Learning Health system can be created by deploying open APIs, machine learning, and AI on top of the data layers in MedRec. The modularity design also allows additional layers of functionality like disease surveillance, epidemiological monitoring, alerts to physicians if the patients consistently abuse prescription access such as in drug abuse, personal health dashboards, etc.

Another salient feature of MedRec is Community Model, which primarily helps medical research with evidence-based data collected from the medical records system residing on the blockchain. This insight from the model can facilitate research into comparative clinical effectiveness and enable a better understanding of treatment outcomes among similar medical issues. Medical research costs can be substantially reduced with a blockchain system like MedRec, where the records can be accumulated, organized, and expert insights received based on the reports that are available for analysis. Such medical research is otherwise very expensive due to the recruitment process cost and the lack of proper access to medical records.

EHRs are not directly stored on the Ethereum Blockchain but in a set of smart contracts that are used to locate the records. Based on the various agreements of patients, providers, and other users, MedRec has three types of contracts, namely, Registrar Contract, Patient–Provider Relationship Contract, and Summary Contract. These are briefly explained below:

Registrar Contract

The registrar contract ensures that participant IDs (providers, patients, and insurers) are mapped to their Ethereum address identity, which is treated as the public key. The rules and regulations are coded into the contract, so only authorized institutions can add data into the network. Hence, if there is new patient information to be added to the blockchain, the patient has to concede such information addition. Each ID is stored in a blockchain address referenced by the summary contract.

Patient–Provider Relationship Contract

This contract gives the relationship contract linking the two nodes in the system so that one node stores the data and manages the medical records for the other node. This kind of relationship exists typically between patient and provider but can be extended to different entities like insurers, physicians, etc. who have to access the records.

Summary Contract

Summary contract is like an indexed table of contents where each participant can locate the overall summary of their relationship with the other participants. This gives the list of references so that current and older engagements with other nodes can be referenced in the Patient–Provider contracts. This also provides a relationship "status" when the connection is established. The full control of the link, such as append, delete, and accept actions, is with the patient only. Thus summary contract gives a single point of dedicated location pointing to fragmented records of the patient while establishing the current and previous relationships.

The MedRec smart contract structure embodies as a single model of a "Healthcare Directory and Resource Location," secured with public-key cryptography and enabled with fundamental properties of provenance and data integrity. This blockchain directory model allows adding new participants and changing organizational relationships through additions to the smart contracts.

Whenever a patient wants to have access to a specific medical record, MedRec sends a request to the provider's database gatekeeper, part of the off-chain infrastructure of MedRec. The database gatekeeper executes the access interface call to the requester patient node's local database, based on the permissions stored on the blockchain. A server listening call is run to query requests, cryptographically signed by the issuer from clients on the network. The cryptographic signature is tallied to confirm identities. If it is a valid address, the question is run on the node's database, and the medical records are returned to the client; in this case, to the patient.

MedRec also gives the complete audit log information for the HealthIT ecosystem so that safety and security standards are taken care of.

Results

MedRec product strives to enable Precision Medicine and a holistic understanding of patient medical status without creating a centralized repository of data. Centrally stored data is challenged by the threat of cyber-attacks and data leaks. MedRec works on open APIs. Hence, it can be integrated into other applications of the Healthcare IT stack.

12.4 Case Study 4 – Energy and Utilities

Industry Energy and utilities

Highlight Renewable energy trading

Overview

In the current energy ecosystem, community and locally sourced energy projects and microgrids are becoming increasingly popular. Energy projects which are locally owned deliver benefits that are environmentally relevant for the communities involved.

In microgrids, distributed generators, storage devices, uncontrollable and controllable loads form an interconnected system that can operate in synchronization with the primary grid or in complete autonomy, if working in island-mode. Microgrids act as a central system which has clear electrical boundaries for the primary grid. Streamlining the control of supply and demand outside the physical and electrical boundaries can be done with a new concept of the virtual microgrid. Usage of microgrids helps in promoting localized energy production and utilization, which leads to a reduction in transmission and distribution losses.

Problem Statement

With energy systems becoming more complex, multi-agent, and decentralized, a higher degree of management control is needed with many stakeholders in the ecosystem. Advanced communication protocols, data exchanges within the global power network and operation of the central management need to be much smarter. Different models and techniques are required to facilitate these latest trends in the energy sector.

Solution

The solution developed is a blockchain-based, decentralized energy trading platform that will operate on a distributed P2P blockchain network, which allows users to sell and buy electricity harnessed from solar panels automatically. This is how it works: If a village has extra power that was captured during the day and is stored on a lithium-ion battery, it could automatically sell a specific predetermined portion to another village on the network. Without relying on big centralized power plants, microgrids will supply a small area with electricity consolidated from distributed sources such as diesel generators, solar power generators with battery storage, crowd-funded low-power source, etc. These localized microgrids will operate either separately or attached to the national grid so that small businesses, hospitals and other organizations can keep working without a break, in case large grids break down.

Brooklyn Microgrid is a blockchain-based P2P energy trading platform run by TransActive Grid, a partnership between LO3 Energy, Consensys, Siemens, and Centrica. The microgrid is situated in the Gowanus and Park Slope communities in Brooklyn, New York, and a three-month trial run of P2P energy trading between community members has been done using this blockchain platform. With this platform, now the prosumers can sell their surplus energy produced to other community members who are in need of energy. This is enabled on the Ethereum-based smart contracts implemented by a company called Tendermint. The first energy transaction was recorded in the first trial, where five prosumers and five neighbouring consumers were piloted. Specially designed smart meters measure the surplus energy and convert them into equivalent energy tokens, which are traded in the marketplace. Each token denotes solar panels that produce a specific unit of energy. These tokens can be transferred from the smart meter wallet of prosumer to end-consumers using blockchain technology. Whenever the consumer uses this unit of energy in his house, these tokens are deleted by the consumer's smart metering device and marked as being purchased by the consumer.

The users of Microgrid can specify their price preferences for selling or buying the electricity by interacting with the blockchain platform. Based on the pricing preferences entered in the system, the platform is designed to display real-time energy prices. The blockchain ledger will have a record of the transacting stakeholders, contract terms, pricing, energy traded and consumed, which are measured by the smart metering devices.

Additionally, all payments are automatically executed by the smart contracts. As defined by the access and authorization, each member of the community will have access to all previous transactions in the ledger and will be able to validate/verify the transactions themselves. Several houses and small businesses have signed up for the next phase of implementation, which has the goal of fully automated transactions. All of these transactions will be done in native ICO token.

In the Energy sector, business is looking for new paradigms like demand response, mechanisms for auctioning energy, and automatically scheduled power consumption based on predefined agreements between stakeholders. This is possible only if the underlying framework allows synchronous flow of information. Smart contracts built on top of the blockchain will provide these functionalities a robust, transparent, secure, and reliable transactional system.

Results

The Brooklyn Microgrid project has become the model for exploring innovative business models for new projects to promote better consumer participation in communities. With the help of localized energy trading, better potential for energy cost savings is possible.

With this implementation, now all the participants will be able to have partial ownership of the grid infrastructures represented by tokens in the system. Since the prosumers use a decentralized network instead of a third party, the cost is also substantially reduced. Because of the lower price, now small-scale companies will be able to participate; thus the end-users would be allowed to choose the best option of electrical power trading.

Summary

Blockchain is gaining popularity, and many of the industries have started implementing blockchain solutions in a phased manner rather than as a big-bang implementation. Although these implementations can be done within a span of 14 to 22 weeks, it is essential to have the right collaboration between various stakeholders in the system to ensure that the implementation is successful and gains mass adoption. That is the reason why industries are implementing blockchain in the pilot mode. The pilot mode helps the stakeholders to check the functionality of the new technology and helps them to keep up and gain more business, customer confidence, and scalability.

CHAPTER 13

Blockchain Platform using Go Language

LEARNING OBJECTIVES

Rapid technological breakthroughs like blockchain have opened up an avenue of innovative ideas. However, bringing an idea to reality has a whole new set of hurdles from research, design, experimentation, build, promotion and implementation. Perhaps choosing a suitable technology is the important challenge.

`GoLang` is an easy and efficient go-to language for coding any decentralised system including blockchain. `Golang` is an open-source general programming language that shows low steepness in the learning curve for developers. Cryptographic calculations, which are essentials of blockchain, are made easy in `Golang`. `Golang` also has several libraries and packages that are specific to blockchain. Hence, `Golang` is considered as one of the top five programming languages that can be used for writing blockchains. It has an easy and readable syntax and fast compilation time. `Golang` is a compiled language with execution time much faster than other similar languages. Because of these features, `Golang` has become a prominent platform for many blockchain applications.

13.1 Introduction

Before we start blockchain programming in Go Language, we need to take care of specific pre-requisites. Quite understandably, the first and foremost activity is to install Go Language and get started with the primary programming interface of Go Language, i.e., Go Language console. In this section, we shall follow a step-by-step guide for fulfilling these pre-requisites.

13.1.1 Install `Golang` in Your System

- Golang 1.13.3 or higher (https://golang.org/dl/)

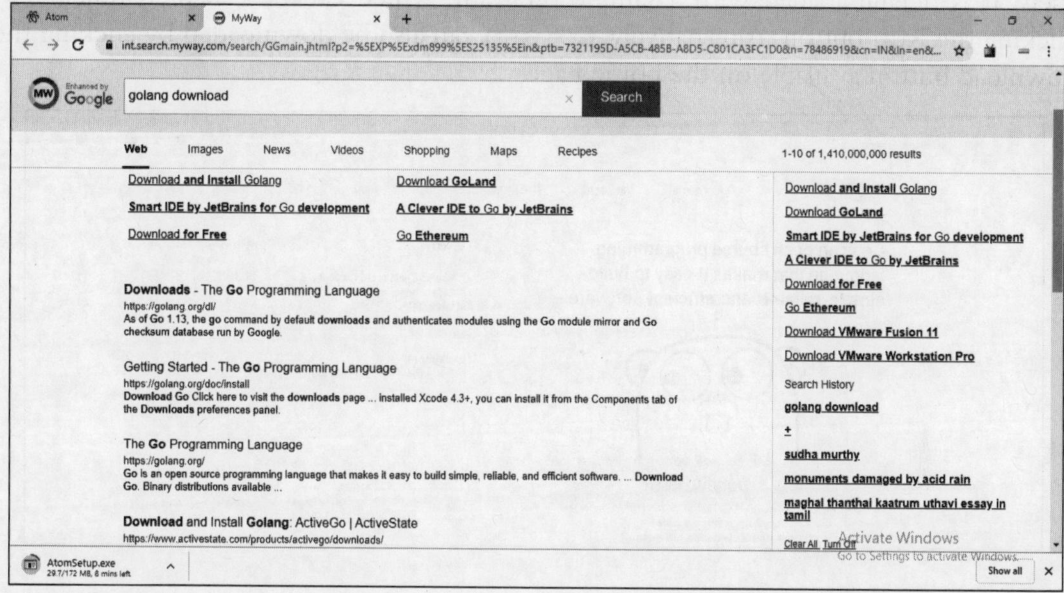

- Atom 1.41.0 or higher *l*, only if you want to leverage the advantage of using an IDE. Otherwise, `Golang` console is sufficient. (https://atom.io/)

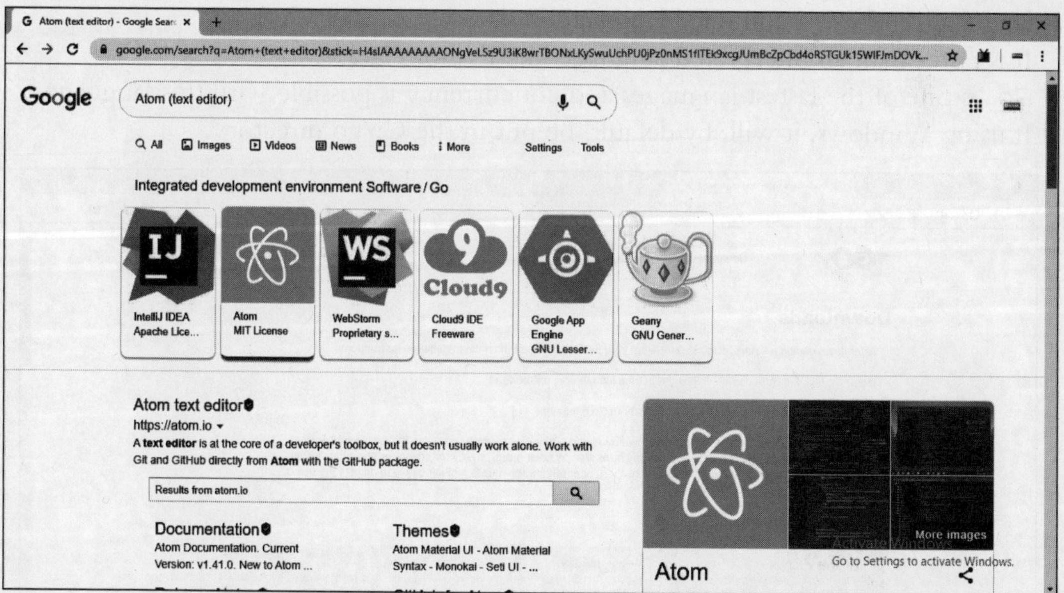

Note

`Golang` has its online web service to run its program called `Playground`. It can use most of the standard library functions of `Golang`. It runs in `Golang`'s sandbox environment, which is a cloud environment used for testing.

13.1.2 Learn How to Use `Golang Playground`

`Golang Playground` is available at the website https://golang.org on the home page itself. It is default loaded with a sample program, which can be executed quickly by clicking the `Run` button. Alternatively, `Golang` can also be downloaded by clicking the download button available on the home page.

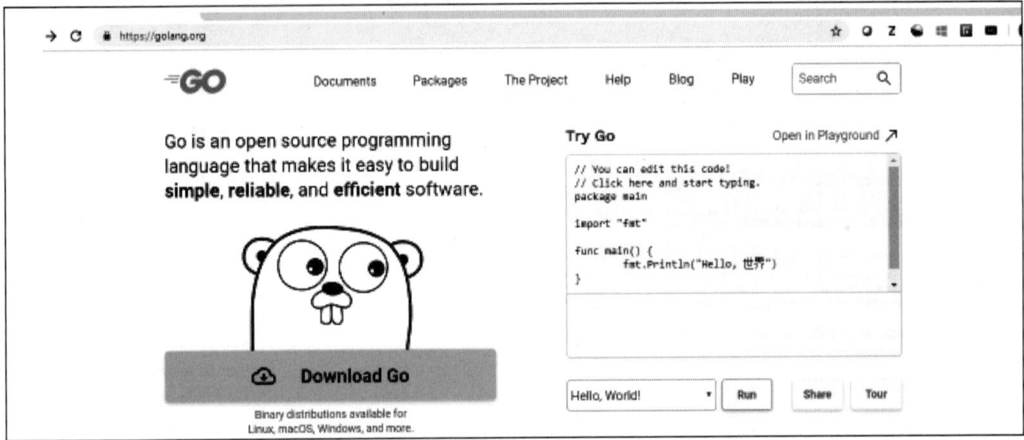

13.1.3 Learn How to Install `Golang` in Your System

- `Go` programming language, also called `golang/go` is a machine-level programming language, and it is a compiled language.
- It is faster than an interpreted language like Python/Java
- `Go` is one of the fastest languages and concurrency is possible with this language
- If using Windows, it will, by default, be put in the C:/go directory

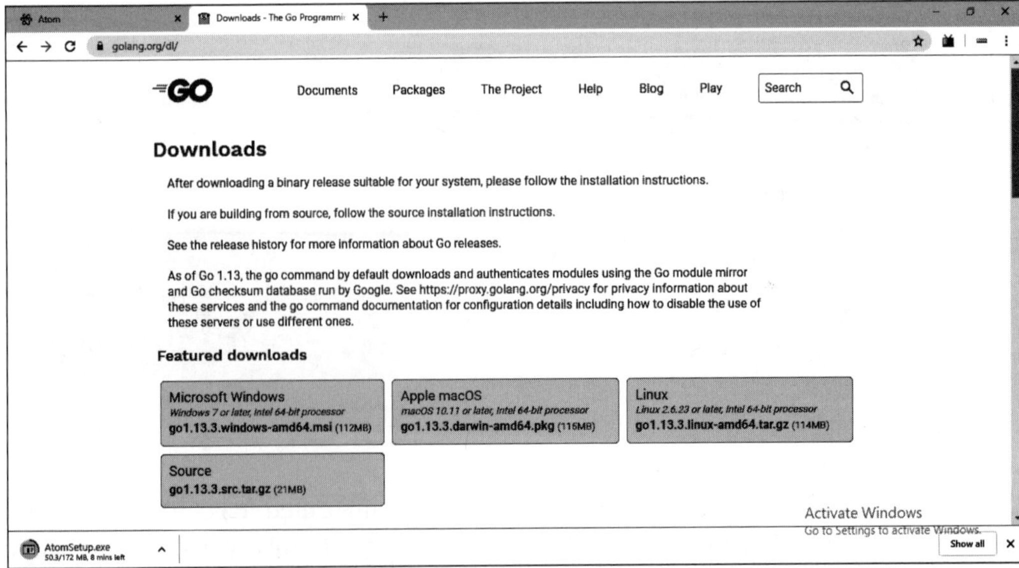

Windows MSI Installer and Zip Archive

- There are two options to install `Golang` in the Windows operating system, one is MSI Installer, and the other is the ZPI archive.

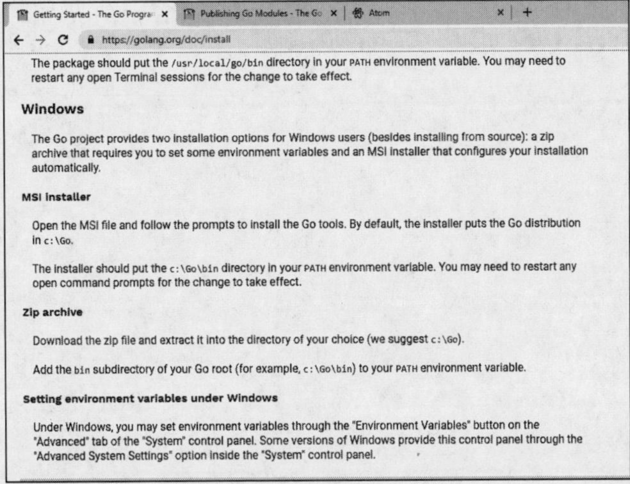

The package should put the /usr/local/go/bin directory in your PATH environment variable. You may need to restart any open Terminal sessions for the change to take effect.

Windows

The Go project provides two installation options for Windows users (besides installing from source): a zip archive that requires you to set some environment variables and an MSI installer that configures your installation automatically.

MSI installer

Open the MSI file and follow the prompts to install the Go tools. By default, the installer puts the Go distribution in c:\Go.

The installer should put the c:\Go\bin directory in your PATH environment variable. You may need to restart any open command prompts for the change to take effect.

Zip archive

Download the zip file and extract it into the directory of your choice (we suggest c:\Go).

Add the bin subdirectory of your Go root (for example, c:\Go\bin) to your PATH environment variable.

Setting environment variables under Windows

Under Windows, you may set environment variables through the "Environment Variables" button on the "Advanced" tab of the "System" control panel. Some versions of Windows provide this control panel through the "Advanced System Settings" option inside the "System" control panel.

- Setting Environment Variables under Windows: Right-click "This PC" folder. Click "Properties" and then click on advanced system settings. Select "Environment variable." Check the system variable path, and you can see that it added C:\go\bin. If not, it will have to be added manually.

13.1.4 Learn How to Install `Atom` in Your System

- `Atom` is a free and open-source integrated development environment (IDE) for multiple languages, which includes `Golang`.
- Atom is a sound editor for `Golang`, and it is an open-source community created editor.
- First, download the windows installer.

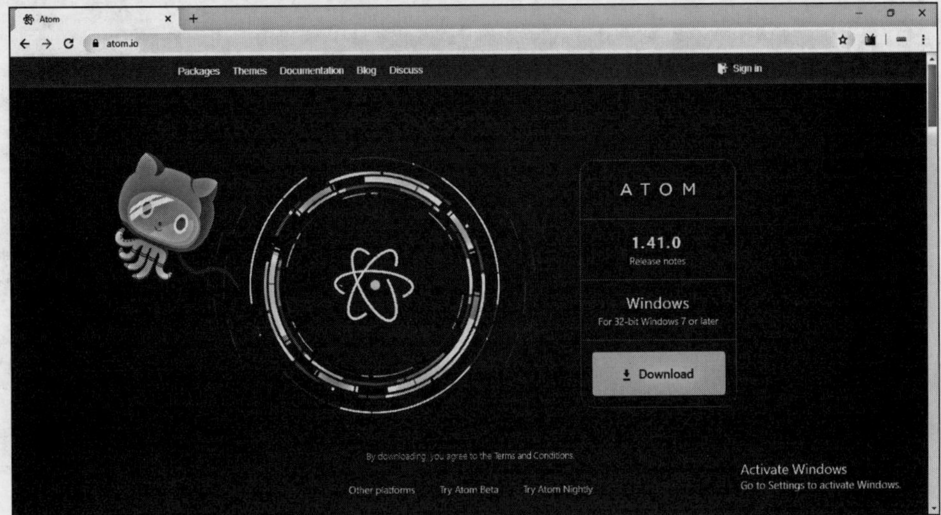

- After Atom installs, click on "install a package", open the Installer and then search for the package "Script," install it

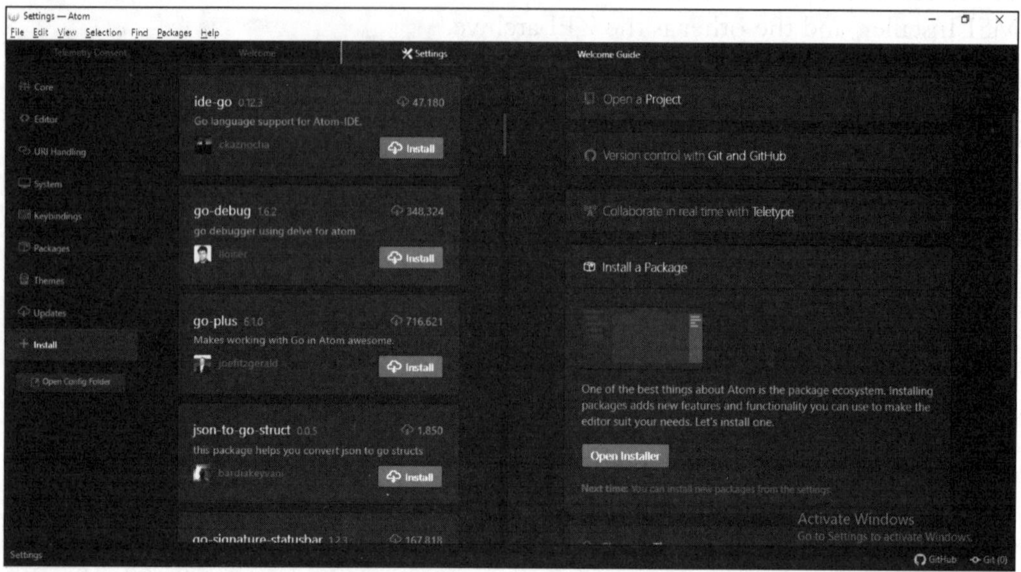

- Also, install the "go plus" package in Atom if required.

13.2 Learn How to Execute Your First Golang Program in Atom

- Write the first program called "HelloWorld.go"
- Run the program (in C:/go directory), in the command prompt, type: `go run HelloWorld.go`

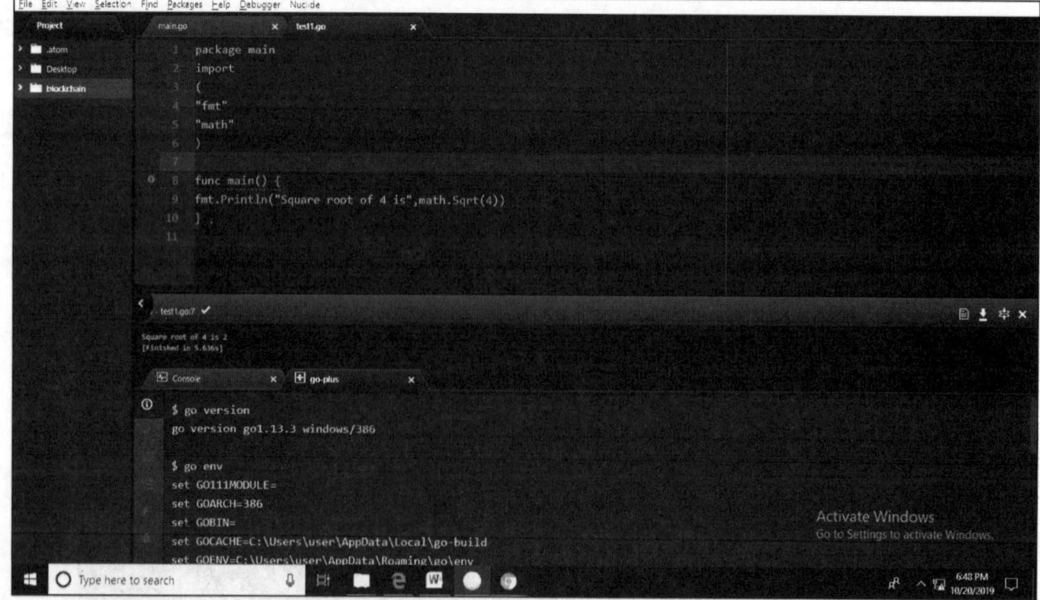

- There are no semicolons in Google Go at the end of each statement
- `Main` function indicates the start of the code and all the codes need to be written inside the main function.

13.3 Know How to Do Basic Programming Using Golang

In this section, we will discuss basic programming in `golang`, so that we set our foundation strong in defining blockchain using `golang`.

13.3.1 Basic Packages and Commands in Golang

- One of the basic package structures in `go` language is the "fmt" package. It consists of various functions, which include printing functions.

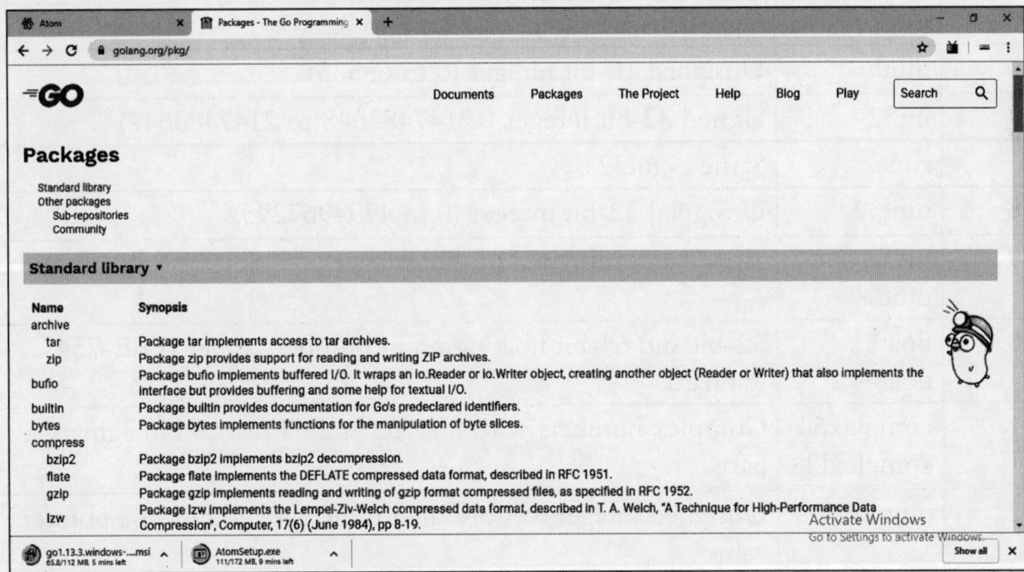

- Packages in `Golang` are imported using command **import "Package name."**

```
package main
import   "fmt"
func main() {
fmt.println("Hai, This is my first program")
}
```

- There is no semicolon at the end of each line.
- The main function is first executed while we run the package.
- Within a single package, there can be only one main function.
- We will look into other useful packages in Section 13.4

13.3.2 Data Types in `Golang`

Basic data types in `golang`: It is used for declaring data variables or functions. It also determines how much memory it will occupy to store data.
Note: Choosing an optimal data type is critical to utilize memory efficiently.

S. no	Data type	Remarks
1	int	Integer (32 bits or 64 bits)
2	uint	Unsigned integer
3	int8	Signed 8-bit integer (-128 to 127)
4	uint8	Unsigned 8-bit integer (0 to 255)
5	byte	Same as uint8
5	intl6	Signed 16-bit integer (-32768 to 32767)
6	uintl6	Unsigned 16-bit integer (0 to 65535)
7	int32	Signed 32-bit integer (-2147483648 to 2147483647)
	rune	Same as int32
8	uint32	Unsigned 32-bit integer (0 to 4294967295)
9	int 64 uint64	Signed and unsigned 64-bit integer
10	float32 float64	32-bit and 64-bit floating point numbers as per IEEE-754 standard
11	complex64 complexl28	Complex numbers with float 32 and 64 real and imaginary parts.
12	uintptr	Unsigned integer to store the un interpreted bits of a pointer value

Defining a Variable

1. `Golang` uses "var" keyword, and it is a specialized keyword
2. "var" keyword is followed by the "Name" of the variable
3. After the Name of the variable, "data type" of the variable (int, int16, int32, int64, byte, [] byte, bool, string, float, etc.) is given.

Example 1

var Difficulty int = 20
var name = "Hello"
We can also define variables of other types, such as folat, bool, string.

Example 2

var Price1 float32 = 8.25
var Price2 float32 = 10.25

Example 3

In the above declarations, Price1 and Price2 are of same type (float32). It can also be given as below:

var price1, price2 float32 = 8.25,10.25

The above statements declare that both price 1 and price 2 are of type float 32.

Example 4

price1, price2:= 8.25, 10.25

The above statement automatically declares the type based on the value given.

Example 5

constant difficulty int = 12

The value of the constant cannot be changed at a later point in time. If we try to assign some other value to the constant (difficulty), then it will throw an error. However, if it is just defined as a normal variable, then the value can be changed. For example, the below codes will work.

var difficult int = 12

Difficulty = 20

Arrays

It is used to store a homogenous group of elements.

Example: var b [20] int

It declares a variable named "b" as an array of twenty integers.

```
package main
import "fmt"
func main() {
        var s [2]string
        s[0] = "Hello"
        s[1] = "Students"
        fmt.Println(s[0], s[1])
        fmt.Println(s)

        finonacci := [12]int{0, 1, 1, 2, 3, 5, 8, 13, 21, 34, 55, 89}
        fmt.Println(fibonacci)
    }
```

```
Output:
Hello Students
[Hello Students]
[0 1 1  2 3 5 8 13 21 34 55 89]

Program exited.
```

Structure

While Array is a collection of a homogenous set of same elements, Structure is a group of various data types.

```
package main
Import "fmt"

type Block struct {
        Index    int
        Timestamp string
        BPM      int
        Hash     string
        PrevHash  string
        }


func main() {

        var Block1 = Block{100, "2019 October 31 11:00 AM", 0, "hash1", "hash0"}
        fmt.Println(Block1)

        }
```

```
Output:

{100 2019 October 31 11:00 AM 0 hash1 hash0}

Program exited.
```

13.3.3 Loops

For loop

```
Syntax:

 for variable in sequence {
(loop body)
}
```

```
Example:
package main
import "fmt"
func main() {
        sum := 0
        //Sum of all integers up to 5 (1+2+3+4+5)
        for i := 0; i <= 5; i++ {
                sum += i
                        }
        fmt.Printf("Sum of all integers upto 5 is %d",sum)
}
```

```
Output:
Sum of all integers upto 5 is 15
Program exited.
```

13.3.4 `If-Else` Statement

```
Syntax:
if (condition 1) {
Statement 1
} else if (condition 2){
Statement 2
} else {
Statement 3
}
```

```
Example:
package main

import (
  "fmt"
 )

func main() {
        var b = "BlockChain"
        var x = len(b)
                if (x >= 25){
                fmt.Println("Length of blockchain is greater than
25...and is", x)
        } else if (x > 20 &&  x<25) {
                fmt.Println("Length of blockchain is between 20 to
25...and is", x)

                } else if (x >5 &&  x<20) {
                fmt.Println("Length of blockchain is between 5 to
20....and is ", x)

                } else  {
                fmt.Println("Length is less than are equal to 5 ....and
is ", x)
        }
}
```

```
Output:
Length of blockchain is between 5 to 20...and is 10
Program exited.
```

13.3.5 Writing Functions

```
Syntax:
func function name (argument list) return value {
(function body)
}
```

```
Example: Function to calculate factorial of an input number n.

package main
import "fmt"
func factorial(a int)  int {
var f = 1
for i :=1; i<=a; i++ {
        f = f*i
        }
        return f
}
func main() {
        fmt.Println("Factorial of 3 =", factorial(3))
}
```

```
Output:
Factorial of 3 = 6
Program exited.
```

The same function can be used to return multiple values. In the below example, `addsub()` function returns the addition and subtraction values in the same function.

```
func addsub(a, b int) (int, int) {
        return a+b, a-b
}

func main() {

add, sub := addsub(4,5)
fmt.Printf("addition %d \n", add)
fmt.Printf("Subtraction %d \n", sub)
}
```

13.3.6 `Switch` Statements

```
i := 3

    switch i {

        case 1:  fmt.Println("One")
        case 2:  fmt.Println("Two")
        case 3:  fmt.Println("Three")
        default:  fmt.Println("Default value")
```

The `switch` statement is used to switch across various statements based on the value given. In the above case, the value of i is 3, and hence it will print "Three."

```
J := 10

    switch {

        case J<10:  fmt.Println("J value is less than 10")
        case J>=10 && J< 15 :  fmt.Println("J Value is between 10 to 15")
        default:  fmt.Println("Default value")

    }
```

In the above case, the value of J is 10, Case J > = 10 && J < 15 will get evaluated, hence it will print, "J Value is between 10 to 15".

13.4 Basic Packages in Golang

Packages are the basic building blocks of go language. There are functions/methods, which are part of these packages, which can be used in the code. These packages are imported into the system using the "import package name" syntax. Some examples of Go packages are given below.

- fmt
- log
- crypto/sha256
- time
- strconv
- io
- sync
- encoding/hex
- encoding/json
- net
- net/http
- github

We will discuss some of the above packages in this section.

13.4.1 `Fmt` Package

- Package `fmt` implements formatted Input and Output related functions. Below is the set of functions, which are most widely used in this package.

Function	Remarks	Example	Output
Errorf(String)	Used to create Error format	err := fmt.Errorf("user%q (id %d) not found", "John", 260) fmt. Println(err.Error())	user "John" (id 260) not found
Println(content)	Used to print the output in the console itself	const name, age = "John", 44 err := fmt.Errorf("user%q (id %d) not found", name, age) fmt. Println(err.Error())	age of "John" is 44
printf(content)	It prints according to format specifier. (%d, %s etc..)	const name, age = "John", 44 fmt. Printf("ageof %q is%d", name, age)	age of "John" is 44

- Formatters are used while printing variables in a particular format. For example: % d is used to print base 10 of integer variables.
- Below are the set of formatters used within the print functions, which are most widely used in this package (init).

Types	The verbs	Meaning
General	%v	Default Format
	%T	A Go-syntax representation of the type of the value
Integer	%b	Base 2 representation
	%c	Character representation of the value
	%d	Base 10 representation
	%o	Base 8 representation
	%x, %X	Base 16 representation
	%U	Unicode format: U+1234; same as "U+%04X"
Float	%b	Decimal less scientific notation with exponent a power of two of age: %b
	%f, %F	Decimal point notation of age: %f
String	%s	The uninterrupted bytes of the string or slice
	%q	A double-quoted string safely escaped with Go-syntax

- Below is an example of the `golang` program, which utilizes the package `fmt` along with associated functions.

```
package main

import "fmt"

func main() {

        const  age = 44
        //General representation

        fmt.Printf("The value in a default format of age: %v \n", age)
        fmt.Printf("A Go-syntax representation of the type of the value of age: %T \n", age)

        //Integer representation

        fmt.Printf("Base 2 representation of age: %b \n", age)
        fmt.Printf("The character representation of age: %c \n", age)
        fmt.Printf("Base 10 representation of age: %d \n", age)
        fmt.Printf("Base 8 representation of age: %o \n", age)
        fmt.Printf("Base 16 representation of age: %x \n", age)
        fmt.Printf("Unicode format of age: %U \n", age)

        //Float representation

        const height = 182.67

        fmt.Printf("Decimal less scientific notation with exponent a power of two of height: %b \n", height)
        fmt.Printf("Decimal point notation of height: %.1f \n", height)

        //String

        const name = "John Davis"

        fmt.Printf("The uninterrupted bytes of the string or slice of name: %s \n", name)
        fmt.Printf("A double-quoted string safely escaped with Go syntax of name: %q \n", name)

}
```

```
Output:

The value in a default format of age: 44
A Go-syntax representation of the type of the value of age: int
Base 2 representation of age: 101100
The character representation of age: ,
Base 10 representation of age: 44
Base 8 representation of age: 54
Base 16 representation of age: 2c
Unicode format of age: U+002C
Decimal less scientific notation with exponent a power of two of height: 6427129249466941p-45
Decimal point notation of height: 182.7
The uninterrupted bytes of the string or slice of name: John Davis
A double-quoted string safely escaped with Go syntax of name: "John Davis"


Program exited.
```

13.4.2 Log Package

```
func Handle(err error){
        if err != nil {
                log. Panic(err)
        }
}
```

The package log of golang implements a simple logging package, which defines a type called Logger and has methods for formatting output. It also has a predefined 'standard' Logger accessible through helper functions Print[f|ln], Fatal[f|ln], and Panic[f|ln], which are easier to use than creating a Logger manually.

13.4.3 Crypto/Sha256 Package

This package is used to create hash value; sum256 functions return the SHA256 checksum of the data.

Example: **hash := sha256.Sum256([]byte info1)**

In the above example, the contents of info1 are converted into hash bytes. This function is handy in blockchain to create the hash value.

13.5 Creating Simple Blockchain Using Golang

We assume the readers of this book know what a blockchain is and how to code programs using go language. Blockchain primarily uses a public database that is distributed across multiple peers, making it a revolutionary network. In a generic database, every node would be trustworthy. Every piece of data coming from each node has to be correct. With blockchain, however, one of the nodes could be producing incorrect data. A blockchain is composed of multiple blocks; each block contains the data that we want to pass around inside of our database as well as a hash, which is associated with the block itself.

13.5.1 Block Structure

To represent a block, create a struct, which has

1. A hash data field,
2. Previous hash field and
3. The data field.

```
type Block struct {
        Hash [] byte
        Data [] byte
        PrevHash [] byte
}
```

The hash field represents the hash of this block. The data represents the data inside the block. The data can be anything, ranging from ledgers to documents to images. The previous hash represents the last block's hash. This previous hash allows linking the blocks together to form a linked list.

13.5.2 Blockchain Structure

Blockchain is created by implementing a struct which contains one field that has an array of pointers to blocks.

```
type BlockChain struct {
blocks []*Block
}
```

We already know that a blockchain is a group of blocks; hence an array of pointers to blocks is formed within the blockchain.

13.5.3 Creating a Block

We already know that a block consists of three fields, namely: 1. A hash data field, 2. Previous hash field and 3. The data field.

```
func CreateBlock(data string, prevHash []byte) *Block {

block := &Block {[] byte{}, [] byte(data), prevHash}
block.CreateHash()
return block
}
```

`CreateBlock` function is called to create a block using the data to be included in the block and the previous hash. The current hash is formed by combining the data and the previous hash. `CreateBlock` function returns the newly formed block. After creating the block, it is added to the overall blockchain.

13.5.4 Creating Hash

Let us create a method (function) that allows us to create a hash based on the previous hash and the data. We shall use the bytes library. We create a variable called `info`. Data and PrevHash are elements within the Block (referred as b), which are passed as parameter to `CreateHash()` function.

b.Data refers to the data within the Block b.

b.PrevHash refers the previous hash within the Block b.

A temporary variable `info` is used to create an array of bytes (from b.Data and b.PrevHash). We create the actual hash by using the sum256 hashing function which comes from the sha-256 library. The resultant hash is pushed into the block using the command: b.Hash = hash[:]

```
func (b *Block) CreateHash() {

info := bytes.Join([][] byte{b.Data,
b.PrevHash}, [] byte{})

hash :=sha256.Sum256(info)
b.Hash = hash[:]
}
```

13.5.5 Adding Blocks to the Blockchain

Let us create a method to add a block to the blockchain. This method (Addblock) gets the pointer for the blockchain, and then it takes in a data string (data to be added). First, we need access to the previous block in our blockchain. We can do this by calling chain blocks and then passing in the length of our blockchain blocks minus one.

PrevBlock := chain.blocks[len(chain.blocks)–1]

We can create the current block by calling the create block function and passing in the data string from here to get our previous block hash and passing it on as the previous hash.

new := CreateBlock(data, prevBlock.Hash)

After creating the current block, we can append this new block to our original blockchain by using the `append` function passing in chain dot blocks.

chain.blocks = append (chain.blocks,new)

```
func (chain *BlockChain) AddBlock(data string){

PrevBlock := chain.Blocks[len(chain.Blocks)-1]
new := CreateBlock(data, PrevBlock.Hash)
chain.Blocks = append(chain.Blocks, new)
}
```

Since the pointers are passed here, the values are not returned, and the addition of the new block is done using the pointer itself.

13.5.6 Creating the First Block: Genesis Block

Let us create a function `Genesis()` to create the Genesis block.

```
func Genesis() *Block{
return CreateBlock("Genesis",
[]byte{})
}
```

Inside this function, call the `create` block with data that we want for the first block. In the current case, let us put in the string Genesis and then a previous empty hash, which is just a slice of bytes. With this `Genesis` block function, we can create a blockchain function that builds our first blockchain block. `Initblockchain()` function will call this `Genesis ()` function for forming the first block (Genesis) and initiating the blockchain.

13.5.7 Initiating the Blockchain

Inside of the `initblockchain` function, we can return a reference to our blockchain, and inside of it we can create an array of blocks with a call to our Genesis function as below:

```
func InitBlockchain() *BlockChain {

return &BlockChain{ [] *Block{Genesis()} }

}
```

13.5.8 `Main` Function of the Blockchain

In our `main` function, we first `init` the blockchain and add other blockchains. First, let us call the init blockchain and assign it to the chain variable. We can add some blocks (three blocks here) to the chain by merely calling `add` block and then passing in the data that we want to push into our blocks (a string in this case). The hashes for these blocks will be derived from the information inside of the block and the previous hash.

```
func main() {

chain := InitBlockchain()
chain.AddBlock("First Block")
chain.AddBlock("Second Block")
chain.AddBlock("Third Block")

}
```

To see the blockchain, let us run a `for` loop and then take out each block, one at a time. Then we print each of the fields by calling `fmt.printf` on each of the fields.

```
for _, block :=range chain.Blocks {

fmt.Printf("Previous Hash: %x \n",
block.PrevHash)

fmt.Printf("Data in Block: %s \n",
block.Data)

fmt.Printf("Hash: %x \n", block.Hash)
fmt.Printf("\n\n")
}
```

13.5.9 Running the Simple Blockchain Golang Program

Run: `simpleblockchain.go`

```
When we run the application, you can see the output:

****************************************************************************************
Previous Hash:
Data in Block: Genesis
Hash: 81ddc8d248b2dccdd3fdd5e84f0cad62b08f2d10b57f9a831c13451e5c5c80a5

****************************************************************************************
Previous Hash: 81ddc8d248b2dccdd3fdd5e84f0cad62b08f2d10b57f9a831c13451e5c5c80a5
Data in Block: First Block
Hash: 98b0453614d970a5e9c29de4d51e0557037f684ca8ea9ecc3c89ec89cb66a679

****************************************************************************************
Previous Hash: 98b0453614d970a5e9c29de4d51e0557037f684ca8ea9ecc3c89ec89cb66a679
Data in Block: Second Block
Hash: fb6e107ce4bf8cfde65da1b451fe2a8a0b28bf076e9d681fbc90d1e452631f73

****************************************************************************************
Previous Hash: fb6e107ce4bf8cfde65da1b451fe2a8a0b28bf076e9d681fbc90d1e452631f73
Data in Block: Third Block
Hash: 77303eab2ead49febe03c57b2ec3bf68672cfcee493022112d17ccb9719b5829

Program exited.
```

13.6 Creating Simple Blockchain with Proof-of-Work (PoW) Using Golang

This version of blockchain adds proof-of-work (PoW) algorithm on top of the simple blockchain created above (simpleblockchain.go).

13.6.1 Introducing Nonce into the Block

Hash of the current block is not only created using the hash of the previous block and the data (transaction). It can also include a nonce (random number) and the timestamp. In this example, we include only nonce into the block.

```
type Block struct {
        Hash     []byte
        Data     []byte
        PrevHash []byte
        Nonce    int
}
```

13.6.2 Proof-of-Work Struct

Proof-of-Work includes the newly created block for verification (which consists of the nonce number) and also the target number. The blockchain sets a target number, and the hash of the newly formed block needs to be lesser than this target for validation purpose.

```
type ProofOfWork struct {
        Block  *Block
        Target *big.Int
}
```

Creating Proof-of-Work Structure

Proof-of-work is a structure that is created by calling the function as shown below.

```
func CreatePOW(block *Block, target *big.Int) *ProofOfWork {

        pow := &ProofOfWork{block, target}
        return pow

}
```

13.6.3 Creating/Running Proof-of-Work

A new hash function is formulated using previous hash, data, nonce number (random number) and difficulty number in such a way that the resultant hash is lesser than the target set by the blockchain. We can secure our blockchain by forcing the network to add a block to the chain. When you hear about miners mining bitcoin, they are mostly doing this proof-of-work algorithm.

Miners sign the blocks on the blockchain and get the fees. They are essentially powering the actual network by running the proof-of-work algorithm. By doing this, miners make the actual blocks and the data inside of the blocks more secure. Generating proof of work is hard to do but checking the proof is relatively easy. As a first step, we use this algorithm takes the data from the block; then, along with that data, we create a counter or a nonce, which starts at 1 and increments upwards theoretically up to infinity. We then create a hash of the data plus the counter. And then, we check the resulting hash, to see if it meets a set of requirements. This is where it becomes difficult; if the hash meets the set of requirements, then we use that hash, and it signs the block. Otherwise, we go back and create another hash, and we repeat this process until we get a hash that does meet the set of requirements. According to the requirements, the first few bytes of the hash must contain zeros. In the original bitcoin proof-of-work specification, which is called hash cash, the first difficulty was to get 20 consecutive bits of the hash as zeros. This requirement got adjusted over time with more zeros added before the hash, and that is essentially the difficulty. That means that we must have more zeros preceding the hash for it to be valid.

const difficulty = 12

An arbitrary value of 12 is assigned to a constant called difficulty. In this algorithm, the difficulty is going to stay static. In a real blockchain, the algorithm would slowly increment this difficulty over a long period. The difficulty is determined based on the time taken by a miner to mine and also the computational power.

```
func (b *Block) RunPOW() {

        Found := false
        noncenumber := 1

        for Found == false {

                var Target *big.Int
                info := bytes.Join([][]byte{b.PrevHash, b.Data}, ToHex(int64(noncenumber)))
                Target = big.NewInt(1)
                Target.Lsh(Target, uint(256-Difficulty))
                info1 := bytes.Join([][]byte{info}, ToHex(int64(Difficulty)))
                hash := sha256.Sum256(info1)
                b.Hash = hash[:]
                sha256Value := hash
                var intHash big.Int
                intHash.SetBytes(hash[:])

                if intHash.Cmp(Target) == -1 {

                        b.Hash = sha256Value[:]
                        b.Nonce = noncenumber
                        b.Hash = sha256Value[:]
                        Found = true
                        pow := CreatePOW(b, Target)

                        if validate(pow) {
                                fmt.Println(" ")
                        }

                }

                noncenumber++
        }
}
```

In the above function, `intHash` is the integer value of the new hash formulated using previous hash, data, nonce number (random number), and difficulty number. The value of the nonce number is continuously changed to fit the criteria, which is that `intHash` should be lesser than the target (integer). Whenever a new block is created (`CreateBlock` function), the above `RunPOW` function is called to ensure that proof-of-work is thoroughly done before adding the block.

```
func CreateBlock(data string, prevHash []byte) *Block {

    block := &Block{[]byte{}, []byte(data), prevHash, 0}
    block.RunPOW()
    return block
}
```

13.6.4 Validating Proof-of-Work

```
func validate(pow *ProofOfWork) bool {

    info := bytes.Join([][]byte{pow.Block.PrevHash, pow.Block.Data}, ToHex(int64(pow.Block.Nonce)))
    info1 := bytes.Join([][]byte{info}, ToHex(int64(Difficulty)))
    hash := sha256.Sum256(info1)
    var intHash big.Int
    intHash.SetBytes(hash[:])
    return intHash.Cmp(pow.Target) == -1
}
```

Validating proof-of-work is easy; it just takes the proof-of-work structure and target. It forms the hash and ensures that the hash is lesser than the target value set by the blockchain.

13.6.5 Running a Simple Blockchain `Golang` Program with Proof-of-Work (PoW)

Run: `simpleblockchainPOW.go`

```
***********************************************************************************
*******
Previous Hash:
Data in Block: Genesis
Hash: 0000378f18108d5325f3a83535583efa7d96225662e850d0e7739571838571ba
Nonce number: 56254
PoW: true


***********************************************************************************
*******
Previous Hash: 0000378f18108d5325f3a83535583efa7d96225662e850d0e7739571838571ba
Data in Block: First Block
Hash: 000036cde4b85a1ab2aa387bd35ee4e5459dcc266e3714157a7dd04d3603b01c
Nonce number: 94791
PoW: true


***********************************************************************************
*******
Previous Hash: 000036cde4b85a1ab2aa387bd35ee4e5459dcc266e3714157a7dd04d3603b01c
Data in Block: Second Block
Hash: 00003acc75411c1d5b2b6f967bb64004680c2915ebec475cff275b02d6d73eaf
Nonce number: 9068
PoW: true


***********************************************************************************
*******
Previous Hash: 00003acc75411c1d5b2b6f967bb64004680c2915ebec475cff275b02d6d73eaf
Data in Block: Third Block
Hash: 000019cd06bd1b5b325572ec9c991a94dda05c961da9cdcee0d1dc4a156e6c7c
Nonce number: 285820
PoW: true


Program exited.
```

13.7 Connecting to Ethereum Using Golang

In this section, we will learn about connecting to Ethereum using `Golang` by dialing in the free ethereum client : ("https://mainnet.infura.io"). Visit: https://infura.io/ to know more information about infura.

13.7.1 Getting and Importing Github Packages to Connect to Ethereum Network

In this `golang` program, we will be using two GitHub packages (these are open-source online storage of packages and codes, which can be reused) to connect to Ethereum network, namely,

1. "github.com/ethereum/go-ethereum/common"
2. "github.com/ethereum/go-ethereum/ethclient"

To get into the directory where Go Langauge is installed (in my machine, it is installed in C:/Go directory), in the command prompt, run the below commands separately:

- `go get github.com/ethereum/go-ethereum/common`
- `go get github.com/ethereum/go-ethereum/ethclient`

By running the above commands, these two packages are automatically packaged in, and then it can be used within `Golang` program using "import" command.

13.7.2 Dialling Infura to Get the Connection of the Client

After importing the packages, the next step is connecting the Ethereum client using eth client package and dialing infura.

```
conn, err := ethclient.Dial("https://mainnet.infura.io")

if err != nil {
	log.Fatal("Whoops something went wrong!", err)
}

ctx := context.Background()
tx, pending, _ := conn.TransactionByHash(ctx, common.HexToHash("0xae9c3776de9ed6bf0e025704bbeced567b428c78e00330b59c25fe45e7ef87a9"))
if !pending {
	fmt.Println(tx)
}
```

Conn, err := ethclient.Dial(«https://mainnet.infura.io»)

The above statement is used to get the connection client by connecting infura using the dial package of ethClient, which we imported. After getting the client (conn), getting the context is important as it is passed as a parameter to receive the transaction details to establish the connection. To get the context, the "context" package is imported.

`ctx := context.Background()` is used to get the context.

13.7.3 Getting Transaction Details by Hash from the Connection

In the last section, we discussed getting the connection and the context. Once the connection is established, the next step is accessing the transaction and block details. Let us now look at how the transaction details are got from the connection if the hash is known.

```
tx, pending, _ := conn.TransactionByHash(ctx, common.HexToHash("0xae9c3776de9ed6bf0e025704bbeced567b428c78e00330b59c25fe45e7ef87a9"))
if !pending {
        fmt.Println(tx)
}
```

`Conn.TransactionByHash()` method is used to get the transaction (Tx) details based on the hash. For this function, the content and the hash are passed as parameters. We can print this transaction using `fmt.Println(tx)` method.

13.7.4 Getting the Latest Block Details

```
// Get the latest block
header, err := conn.HeaderByNumber(context.Background(), nil)
if err != nil {
        log.Fatal(err)
}

blockNumber := header.Number
fmt.Println("Latest block: " + blockNumber.String())

// Query a specific block by number
block, err := conn.BlockByNumber(context.Background(), blockNumber)
if err != nil {
        log.Fatal(err)
}
```

- We can get the header of the latest block using the below code: `header, err := conn.HeaderByNumber(context.Background(), nil)`
- Once we get the header, we can get the latest block number using the below command: `blockNumber := header.Number`
- The latest block details can be obtained by passing the latest block number. `block, err := conn.BlockByNumber(context.Background(), blockNumber)`

- We can get the details of the block using the block obtained above.

```
fmt.Println("Block difficulty: " + block.Difficulty().String())
fmt.Println("Block hash: " + block.Hash().String())
fmt.Println("Number of TXs: " + strconv.Itoa(len(block.Transactions())))
```

13.7.5 Query a Specific Block by Hash

```
// Query a specific block by hash
blockByHash, err := conn.BlockByHash(context.Background(), block.Hash())
if err != nil {
        log.Fatal(err)
}

//fmt.Println("Block time: " + blockByHash.Time().String())
fmt.Println("Block difficulty: " + blockByHash.Difficulty().String())
fmt.Println("Block hash: " + blockByHash.Hash().String())
fmt.Println("Number of TXs: " + strconv.Itoa(len(blockByHash.Transactions())))
```

Specific hash can be used to get the block numbers using the bleow command:
`blockByHash, err := conn.BlockByHash(context.Background(), block.Hash())`

```
Block difficulty: 2478239002127040
Block hash:
0xba1d02f8fa5dd2e86dfc7d38850ed98534601ac3bc4b1a2730bcd8001b8e272b
Number of TXs: 84
```

13.7.6 Getting the Transaction Details from the Block

Once the block details are known, then `block.Transactions()` can be used to get all the transaction details.

```
// Getting Transaction details from the block

block, err = conn.BlockByNumber(context.Background(), blockNumber)
if err != nil {
        log.Fatal(err)
}

for _, tx := range block.Transactions() {
        fmt.Printf("TX Hash: %s\n", tx.Hash().Hex())
        fmt.Printf("TX Value: %s\n", tx.Value().String())
        fmt.Printf("TX Gas: %d\n", tx.Gas())
        fmt.Printf("TX Gas Price: %d\n", tx.GasPrice().Uint64())
        fmt.Printf("TX Nonce: %d\n", tx.Nonce())
        //fmt.Printf("TX Data: %v\n", tx.Data())
        fmt.Printf("TX To: %s\n", tx.To().Hex())
        receipt, err := conn.TransactionReceipt(context.Background(), tx.Hash())
        if err != nil {
                log.Fatal(err)
        }

        fmt.Printf("Receipt Status: %d\n", receipt.Status)
        fmt.Println("---")
}
```

A sample output for the above code `snippet:aa` is shown here:

```
TX Hash: 0x634656e6a0d66a79a0c1ff6bc63804aad6e9e66d0644cf081377eef43a7e9b0a
TX Value: 201466361894076992
TX Gas: 50000
TX Gas Price: 15000000000
TX Nonce: 14072578
TX To: 0x414Db6757feE9a9797De7eE20eb76bafA1bDd1ad
Receipt Status: 1
---
TX Hash: 0x85c819a77c667f8c7c108772510961221l524ba80e34a30ed3e56857e1a9951e
TX Value: 500529593786228992
TX Gas: 50000
TX Gas Price: 15000000000
TX Nonce: 14072579
TX To: 0x44e8F70C86626e7F3EeF737D2524d8Cd9307f7f2
Receipt Status: 1
```

13.7.7 Getting the Specific Transaction Details from the Block Based on the Transaction Index and Block Hash

```
// Grab block by hash then iterate over transactions by index
blockHash :=
common.HexToHash("0xae9c3776de9ed6bf0e025704bbeced567b428c78e00330b59c25fe45e7ef87a9")
count, err := conn.TransactionCount(context.Background(), blockHash)
if err != nil {
        log.Fatal(err)
}

for idx := uint(0); idx < count; idx++ {
        tx, err := conn.TransactionInBlock(context.Background(), blockHash, idx)
        if err != nil {
                log.Fatal(err)
        }

        fmt.Printf("TX Hash: %s\n", tx.Hash().Hex())
}
```

The number of transactions in a block can be obatained using the below command.

```
count, err := conn.TransactionCount(context.Background(),
blockHash)
```

Sepcific transaction details based on the transaction id (idx) can be obtained using the below command.

```
tx, err := conn.TransactionInBlock(context.Background(),
blockHash, idx)
```

13.7.8 Getting the Specific Transaction Details Based on the Transacion Hash

```
// Grab a transaction by it's individual hash
txHash := common.HexToHash("0xae9c3776de9ed6bf0e025704bbeced567b428c78e00330b59c25fe45e7ef87a9")
tx, isPending, err := conn.TransactionByHash(context.Background(), txHash)
if err != nil {
        log.Fatal(err)
}

fmt.Printf("TX Hash: %s\n", tx.Hash().Hex())
fmt.Printf("Pending?: %v\n", isPending)
```

Specific transaction details based on the transaction hash can be obtained using the below command.

```
tx, isPending, err := conn.TransactionByHash(context.
Background(), txHash)
```

Summary

Packages in `Golang` are imported using command import "package name". There is no semicolon at the end of each line. The main function is first executed while we run the package. Within a single package, there can be only one main function.

Basic data types in `Golang`: It is used for declaring data variables or functions. It also determines how much memory it will occupy to store.

Note: Choosing an optimal data type is critical to utilize memory efficiently.

`Golang` uses "var" keyword, which is a specialized keyword

"var" keyword is followed by the "Name" of the variable

Arrays: It is used to store a homogenous group of elements.

Structure: While Array is a collection of a homogenous set of same elements, Structure is a group of various data types.

The same function can be used to return multiple values.

The `switch` statement is used to switch across various statements based on the value given.

Packages are the basic building blocks of `go` language. There are functions/methods, which are part of these packages that can be used in the code. These packages are imported into the code using the "import package name" syntax.

Package fmt implements formatted Input and Output related functions.

The package log of `golang` implements a simple logging package, which defines a type called Logger and has methods for formatting output.

Crypto/sha256 Package: This package is used to create hash value, Sum256 functions return the SHA256 checksum of the data.

A simple blockchain is created by implementing a struct which contains one field with an array of pointers to blocks.

Proof-of-work includes the newly created block for verification (which includes the nonce number) and also the target number. The blockchain sets a target number, and the new hash of the newly formed block needs to be lesser than this target for the purpose of validation.

Const difficulty: An arbitrary value of 12 is assigned to a constant called "difficulty". In this algorithm, the difficulty is going to stay static. In a real blockchain, the algorithm would slowly increment this difficulty over a significant period. The difficulty is determined based on the time taken by a miner to mine and also the computational power.

Free ethereum client : ("https://mainnet.infura.io").

- `go get github.com/ethereum/go-ethereum/common`
- `go get github.com/ethereum/go-ethereum/ethclient`

By running the above commands, the two packages are automatically packaged in, and then it can be used within `Golang` program using the "import" command.

```
Conn, err := ethclient.Dial("https://mainnet.infura.io")
```

This statement is used to get the connection client by connecting infura using the Dial method of ethClient, which was imported earlier.

`Conn.TransactionByHash()` method is used to get the Transaction (Tx) details based on the hash.

We can get the header of the latest block using the below code.

```
header, err := conn.HeaderByNumber(context.Background(),
nil)
```

Specific hash can be used to get the block numbers using the bleow command:

```
blockByHash, err := conn.BlockByHash(context.Background(),
block.Hash())
```

Number of transactions in a block can be obatained using the below command.

```
count, err := conn.TransactionCount(context.Background(),
blockHash)
```

Sepcific transaction details based on the transaction id (idx) can be obtained using the below command.

```
tx, err := conn.TransactionInBlock(context.Background(),
blockHash, idx)
```

EXERCISES

Practical Questions (Using Golang)

1. Program to print "Hello world".
2. Program to calculate the average of numbers in a given list.
3. Program to swap numbers without using a temporary variable.
4. Program to reverse a given number.
5. Program to print odd numbers within a given range.
6. Program to find the sum of the digits in a number.
7. Program to print all numbers in a range divisible by a given number.
8. Program to find the smallest divisor of an integer.
9. Program to count the number of digits in a given number.
10. Program to check whether the given number is palindrome.
11. Program to read a number *n* and print the series (1 + 2 + 3 + ... *n*).
12. Program to check if a number is an Armstrong number.
13. Program to find LCM of given two numbers.
14. Program to count the number of vowels in a string.
15. Program to count the number of lower case characters in a string.

16. Program to reverse a string.
17. Create a simple blockchain program using Struct and add three blocks to it (without PoW). (Refer: simpleblockchain.go)
18. Create a simple blockchain with proof-of-work (PoW) using `Golang`.
(Refer: simpleblockchainPOW.go)
19. Write a program to connect to Ethereum using `Golang` and GitHub (User infura). Get the latest block details.
20. Write a program to connect to Ethereum using `Golang` using GitHub (User infura). Get the specific transaction details from the block based on transaction index and block hash.

CHAPTER 14

Blockchain Ethereum Platform using Solidity

LEARNING OBJECTIVES

The Ethereum platform is the most commonly used decentralized blockchain ledger. This platform uses Solidity, which is today's language of choice for smart contract development. Solidity is designed to be easy to learn for those who are already familiar with one or more programming languages. In this chapter, we will look into different aspects of the language like the variable types, functions, interactions with other smart contracts, and how to send ether between the contracts.

14.1 Introduction

Ethereum is a cryptocurrency, which runs in a decentralized virtual machine (EVM). It allows transferring value in the form of ether back and forth to different wallets or addresses on the blockchain. It allows for creating smart contracts. Smart contracts are software codes deployed into the blockchain, and they also store states. You can create a smart contract that can take a bunch of ether, store it until a later date, and then give it back to specific addresses. The smart contract can also be used for voting from a valid address. Ethereum smart contracts are created in a programming language called solidity, which was designed specifically for the Ethereum virtual machine. It can also be used on many other blockchains such as Hyperledger. The syntax of solidity looks almost like Java Script. If you already know JavaScript, it is easy to pick up Solidity language. With the Remix browser, learning solidity is very easy. The solidity language has changed a lot between versions 0.4 and 0.5. In this chapter, we will look into different aspects of the language like the variable types, the functions, interactions with other smart contracts, and how to send ether between the contracts.

14.2 Remix IDE

You may want to check out the solidity documentation by the Ethereum team, at https://solidity.readthedocs.io/en/develop which breaks down the entire language and explains how to create smart contracts.

There are several different Integrated Development Environments (IDEs) and frameworks to create smart contracts. The necessary steps are as follows: first write the solidity code, compile it, and then deploy it. We may use the browser IDE called remix (refer Fig. 14.1) created by the Ethereum team. There is an integrated blockchain in the Remix browser where all accessories are pre-loaded and already connected. In addition, this integrated blockchain gives dummy ethers, which can be used to send a transaction to the network. Remix also uses a friendly graphical interface. We can write code, deploy, rewrite it, and deploy it to our content without spending any real money.

For further information about this browser, the reader may visit https://remix.ethereum.org or https://ethereum.github.io/browser-solidity

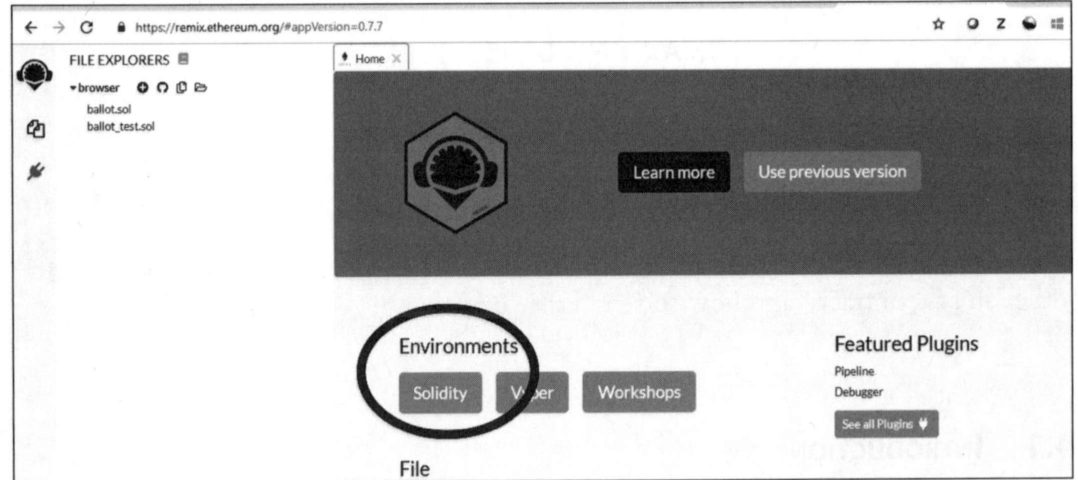

Figure 14.1: Home page of remix.ethereum.org

14.3 Structure of a Smart Contract Program

Like any other program, the Solidity program has its structure, which should be followed while writing the code for the Ethereum contract.

Pragma Statement

The first statement in a solidity file is the pragma statement, which instructs the compiler to compile with a specific version of solidity, denoted as "pragma solidity ^0.5.1;"

In this statement, `pragma` is the keyword. 0.5.1 indicates the version number of the solidity program (refer Fig. 14.2). The symbol ^ indicates that any version above this version is compatible. The middle number 5 (in 0.5.1) is significant. If a smart contract is written in a certain version of solidity, and compiled with another version which is incompatible, then solidity will throw errors (due to compatibility issues). For example, if the code is written in 0.5.0 version and compilation is attempted with the 0.6.0 (or) 0.4.0 version, it will not compile. However, it can be compiled with 0.5.1 version (or) higher versions like 0.5.2, etc. You should always write a smart contract with the latest version, which is the stable version of solidity. Immediately after the pragma statement, import statement can be used.

```
pragma solidity ^0.5.1;
        contract Hello {
                String name;
        constructor () public {
                name = "Benjamin" ;
        }
}
```

Figure 14.2: First solidity Ethereum program (contract Hello)

Contract

In solidity, everything is a contract, similar to a class (in Java). The contract is defined after the pragma statement (refer Fig. 14.2). Here, `contract` is the keyword. 'Hello' is the name of the contract. Open braces ({) indicate the start of the contract. Close braces (}) indicate the end of the contract. Codes of the contract are written between these braces. It is a good practice to name the smart contract with the same name as the file. Therefore, in the above example, the contract named Hello will be saved in a file called Hello. Sol, which indicates, the Hello program, is a solidity program. In a practical scenario, a single sol file of solidity can be used to write several smart contracts. For example, it is also possible for us to define another smart contract, namely Hello2, within the same file.

Variables

Variables names are defined at the beginning of the contract immediately after the open braces of the contract. Variables are easy to create in contracts. They just take a type. In this case, we have defined a string (refer Fig. 14.2) and named the variable "name." Variables within the blockchain contract are called state variables. In this example, the name is the state variable.

Functions

After defining the variable, you can define a function. A function is usually written to manipulate the state variable(s) or to get the value of the state variable(s). The syntax for the definition of a function is very similar to JavaScript (refer Fig. 14.2). A function can also be called from outside the smart contract.

Constructor

Solidity also has a constructor. As you may already know, a constructor is just a function that is called at the very beginning of the instantiation of the class (in Java). In solidity, a constructor is a function named "constructor" (refer Fig. 14.2). The only time this constructor is called is when you first deploy this contract to the blockchain and is never called again — deploying the contract to the blockchain is the same as instantiating the contract. In our example, in the constructor, we set the name to some default value. We set the name to "Benjamin." In solidity, all line statements end with a semicolon (;).

14.4 Using Remix to Write and Run a Solidity Program

14.4.1 Example 1

Step 1: First, ping the URL: https://remix.ethereum.org (refer Fig. 14.3).

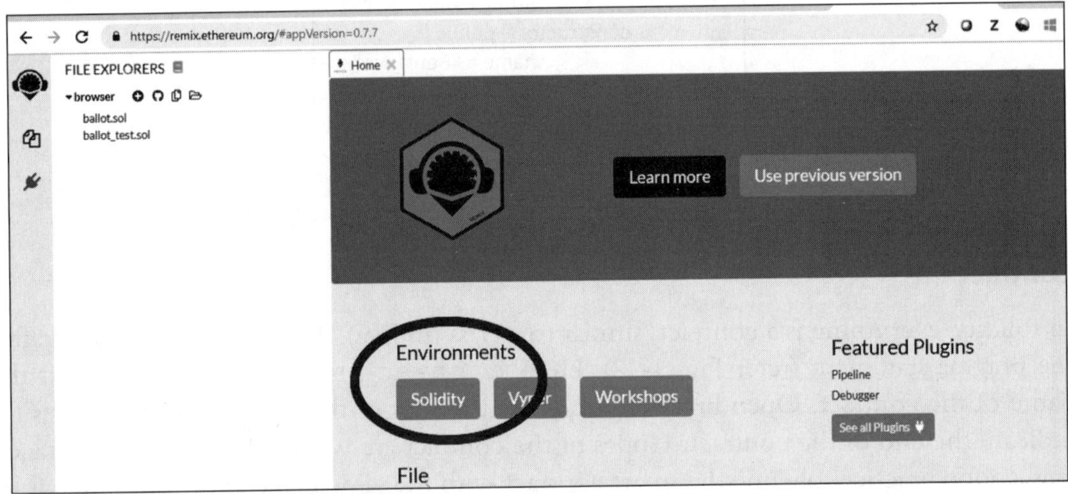

Figure 14.3: Home page of remix Ethereum

Step 2: Change the environment to "JavaScript VM."

First, change the environment to "JavaScript VM," indicating we will use JavaScript Virtual Machine for running the code (refer Fig. 14.4).

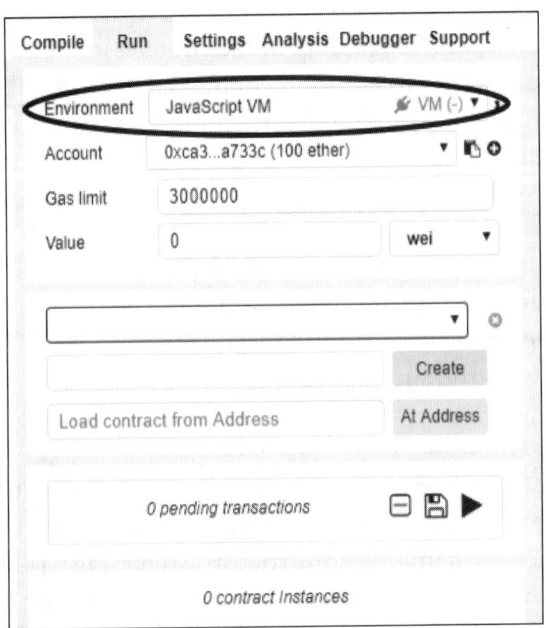

Figure 14.4: Changing environment to "JavaScript VM"

Step 3: Create a new solidity file

Let us create a new solidity file. Click on the + symbol on the left-hand top corner of the screen. Name the file as Helloword.sol. The extension "sol" indicates it as a solidity file (refer Figs 14.5 and 14.6).

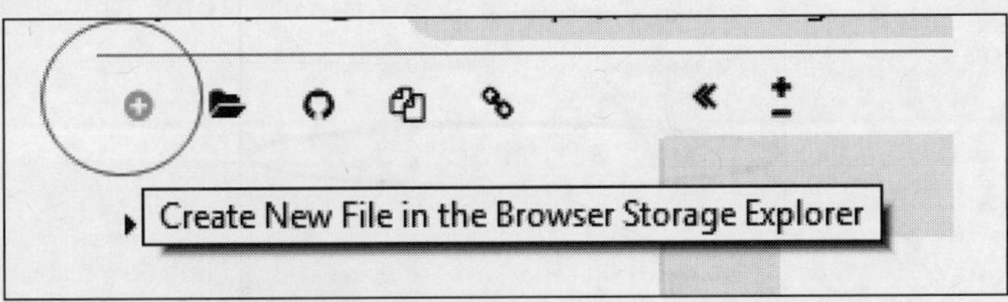

Figure 14.5: Creating a new file

Figure 14.6: Naming a new solidity file

Step 4: Write the solidity code.

Write the solidity program in the space given in the remix browser (refer Fig. 14.7).

```
pragma solidity ^0.5.1;
    contract Helloword {
    string public name;
    constructor() public {
        name = "Benjamin";
    }
   function sayhello() public view returns (string memory, string memory)  {
        return("hello", name);
    }
    function setName(string memory n) public {
        name = n;
    }
    function getName() public view  returns (string memory)  {
        return name;
    }
}
```

Figure 14.7: First solidity file

Step 5: In the "Compile" menu of the browser, click "Start to Compile" in order to compile the code written (refer Fig. 14.8). Remove warnings, if any.

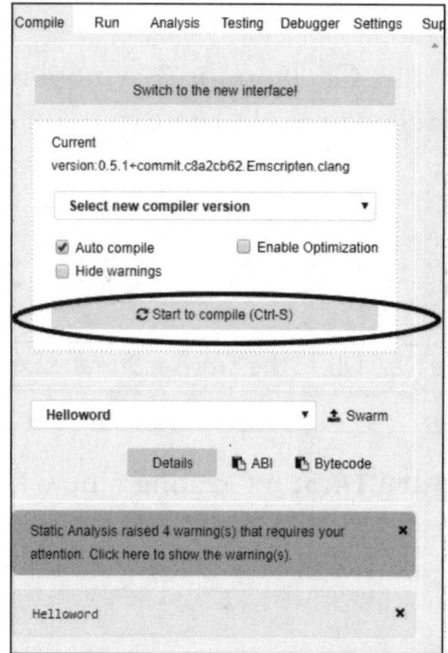

Figure 14.8: Compiling solidity code

Step 6: In the "Run" menu drop-down, select the contract name to deploy ("Helloword" in this case). Click Deploy. The contract will appear in the "Deployed Contracts" section (refer Fig. 14.9).

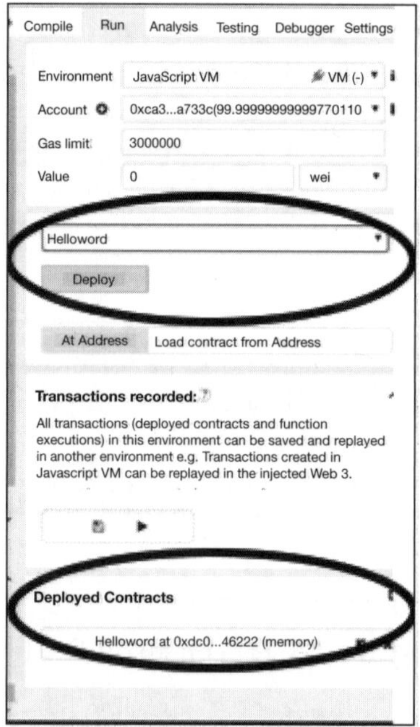

Figure 14.9: Deploying and running the solidity code

Step 7: Click on the contract name in the "Deployed Contracts." In our case, click on "Helloword" in the "Deployed Contracts."

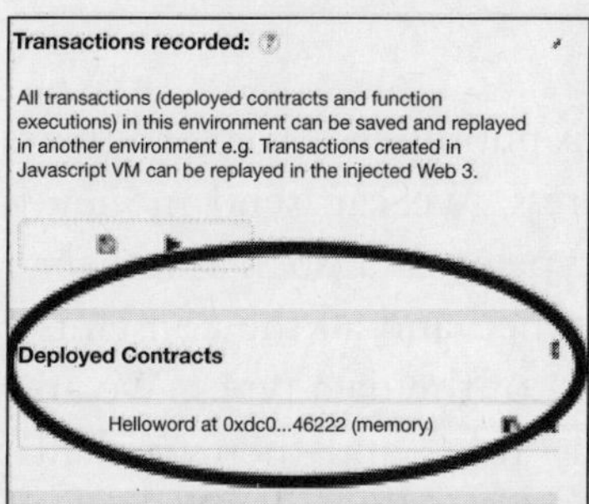

Figure 14.10: Click and run the deployed contracts

Step 8: You can now see the public function names and variable names (refer Fig. 14.11). In our case, 1. setName 2. getName 3. name 4. sayHello

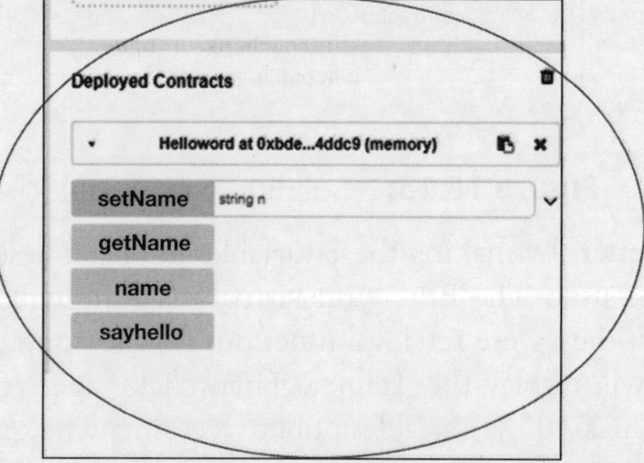

Figure 14.11: Functions and variables in the deployed contracts

Step 9 When the contract is created, the name is set to Benjamin. In "SetName," let us try to set the name to some other value. For example: "Tony Robbins." After that, click on "sayhello" again; it displays hello Tony Robbins (refer Fig. 14.12).

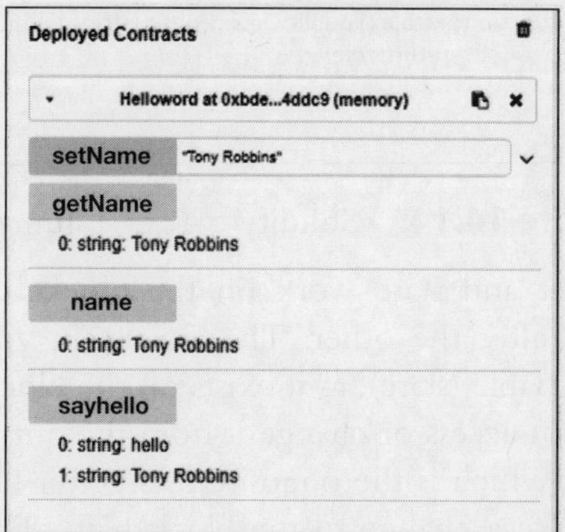

Figure 14.12: Helloword program output

14.4.2 Example 2

Every Ethereum contract that is published is part of a transaction on the blockchain. They also receive an Ethereum address. We can send money (ether) to it. State variables are strongly typed, meaning each type of data (such as integer, character, hexadecimal, packed decimal, and so forth) is predefined and all the constants or variables defined for a given program are described with one of the data types. We are going to declare two variables; the first one is a string called "role," and then we will set the role to "Trainer"(refer Fig. 14.13). Therefore, the Trainer is the role. Then, we are going to create an integer (unsigned integer). An integer can be either positive or negative. Unsigned integers are always positive. We will declare an unsigned integer called "age" with a value of 44. To make these variables accessible to the public, we use the keyword `public`.

```
pragma solidity ^0.5.1;
        contract Hello {
                string public role = "Trainer";
                uint public age = 44;
}
```

Figure 14.13: Solidity state variables

We have created getters by making these variables public. These getters are used to get the value from the end user. The Ethereum virtual machine will create this on its own. You can run them as if they were retrieval functions on the contract. Therefore, you can click the role, and it will display the Trainer. Then click "age," it will display 44. If the user does not want to make these variables public, access may be given to only the desired variables by creating a getter function. Thus, we create these variables as private again, and then we create a new function called get the role (refer Fig. 14.14).

```
pragma solidity ^0.5.1;
        contract Hello {
                string role = "Trainer";
                uint age = 44;
        function getrole () public view returns (string memory) {
                return role;
        }
}
```

Figure 14.14: Solidity `getter` function

The Ethereum contract and state work on the blockchain, which is a ledger of transactions, arranged one after the other. The Ethereum virtual machine returns the last known value of that variable (state), as it exists on the blockchain ledger. Since these values are private, we cannot access or change it from the outside. To change the values, we use the setter function, which is the opposite of a getter function. So, let us call this setrole function; it will make a change to the blockchain. Therefore, when we set the role, it will write a transaction to the current block, and that will be saved on the ledger permanently. Therefore, setters also take a variable with a type, but they do not return anything. We take in a string; we can call it "w," and set the role to the value of w. Because it is a string, we have to enclose it in quotes (while giving value in the Remix browser). We will set the role of "Coach." If we now hit the `getrole` again, it returns Coach (refer Fig. 14.15).

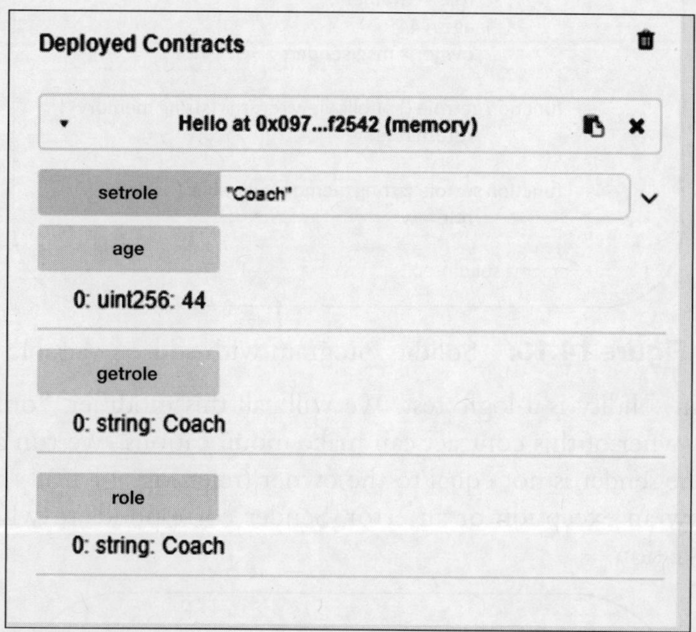

Figure 14.15: Solidity output for `setrole` and `getrole` function

A constructor is used for any sort of setup work such as setting variables to default values. Instead of initializing these variables, we create a constructor. We set the role to Trainer and age to 44. Let us say that we only want the owner or the creator of this contract to be able to make any sort of modifications to its internal states. In this case, we are going to use modifiers.

14.5 Modifiers

Let us create a variable of type address. The address is a rare type in the Ethereum virtual machine and is a memory location that stores Ethereum addresses (refer Fig. 14.16). We will just call this address owner, and it stands for the actual creator of a contract. In the constructor, we set the owner to message the sender.

Because creating a contract or deploying the contract is a transaction on the blockchain, it is sent from an Ethereum address. Every transaction on the blockchain has a corresponding message (msg). The message is like a container object, with a whole bunch of metadata. The sender address is the address of whoever initiated that transaction (msg. sender). By placing this in the constructor message, that sender corresponds to the address creating this contract. Now we have privately saved the address of the creator.

```
pragma solidity ^0.5.1;
        contract hello {
                string role;
                uint public age;
                address owner;
        constructor() public {
                role = "Trainer";
                age = 44;
                owner = msg.sender;
        }
        function getrole () public view returns (string memory) {
                return role;
        }
        function setrole (string memory w) public {
                role = w;
        }
}
```

Figure 14.16: Solidity program with address variable

A modifier in solidity is a logic test. We will call this modifier, "onlyowner," which means only the owner of this contract can make modifications. We run a test to check if the message of the sender is not equal to the owner (refer Fig. 14.17).

We also throw an exception or an error. Sender corresponds to whoever is sending the current transaction.

```
pragma solidity ^0.5.1;
        contract Hello {
                string role;
                uint public age;
                address owner;
        constructor() public {
                role = "Trainer";
                age = 44;
                owner = msg.sender;
        }
        modifier onlyOwner {
                if (msg.sender != owner) {
                        revert();
                }
         _;}
        function getrole () public view returns (string memory)
        {
                return role;
        }
        function setrole (string memory w) onlyOwner public {
                name = w;
        }
}
```

Figure 14.17: Solidity program with modifier function and revert

In this case, it is not necessary that the owner of the contract should send the message. It could be anybody executing the function. If the message is not from the owner of the contract, we shall want to revert and abort the transaction. However, if the message is from the owner, we will go ahead and execute. Accordingly, we modify the `setrole` function. Only the owner, the body of this function, will run this. Remix testing IDE gives the user a list of different addresses (Accounts) to play with. The first address is the address that created the contract (refer Fig. 14.8).

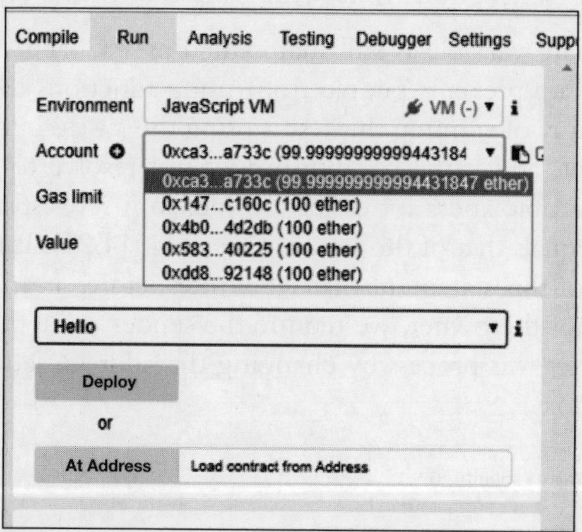

Figure 14.18: Working with Account 1 in solidity

We can now go ahead and set a role without any problem, because we shall be using the address that we created this contract with. Thus, if we set roles, everything looks okay, and when we use `getrole` we get Coach. Let us now use another address in the list (refer Fig. 14.19), and set the account back to some value. Let us see what happens now.

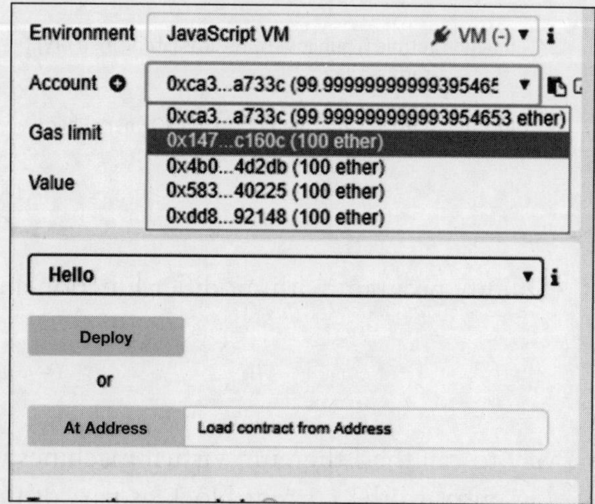

Figure 14.19: Working with Account 2 in solidity

You will see that an exception is thrown, and we will not be able to set the role (refer Fig. 14.20).

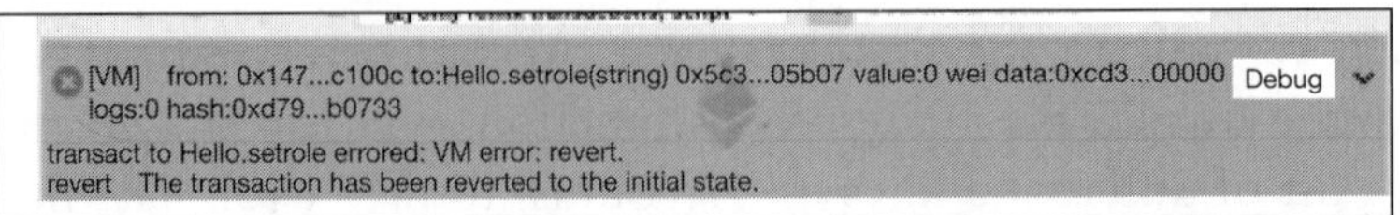

Figure 14.20: Exception while setting the role with a different account

We see that the state variable is still maintained as a Coach, which is the previous value. Thus, the contract prevents people from using functions they are not allowed to use. A more readable way of running this test is using the `require` keyword. `Require` is just an `if` statement, and if the condition does not resolve to true, it will throw an error. This is more readable and a lot easier to type; so it saves space and can be used to check if the message equals that of the sender (refer Fig. 14.21). It may be noted that it is actually the opposite of this statement that we wrote before, and instead of throwing an error if the sender is not the owner, we require the sender to be the owner, and then we continue. We can repeat this process by changing the address, setting Trainer to Coach, and we still get the error.

```
pragma solidity ^0.5.1;
        contract Hello {
                string role;
                uint public age;
                address owner;
        Constructor () public {
                role = "Trainer";
                age = 44
                owner = msg.sender;
        }
        modifier onlyOwner {
                require(msg.sender == owner);
        _;}
        function getrole () public view returns (string memory) {
                return role;
        }
        function setrole (string memory w) onlyOwner public {
                name = w;
        }
}
```

Figure 14.21: Solidity program with modifier function and `require`

14.6 Events

An event in solidity is a way to call the Ethereum virtual machines logging functionality. It also serves to write a log entry to the current block as part of the transaction. These logs are useful for decentralized applications. The Ethereum JavaScript SDK can listen to the Ethereum blockchain for events; so these events can be tied to JavaScript callback functions to run any kind of logic on the user interface or another piece of application. So, to write one of these log entries, an event has to be created.

With events, you can push data from the smart contract to outside consumers, such as the web front end. An event is a kind of function that takes parameters (refer Fig. 14.22). Let us create an event called "changed". This "changed" event will take an address, and in this case, we shall log the time when a change is made to the role variable. The timelog will help us to know if someone from a particular address has made a change to the contract.

```
pragma solidity ^0.5.1;
        contract Hello {
                string role;
                uint public age;
                address owner;
        constructor() public {
                role = "Trainer";
                age = 44;
                owner = msg.sender;
        }
        event Changed(address a);
        modifier onlyOwner {
                require(msg.sender == owner);
        _;}
        function getrole () public view returns (string memory) {
                return role;
        }
        function setrole (string memory w) onlyOwner public {
                name = w;
                emit Changed(msg.sender);
        }
}
```

Figure 14.22: Solidity program with `event` function

14.7 Arrays in Solidity

Arrays are a collection of similar kinds of data types, for example, Array of integers, Array of Boolean values. We cannot have integer values and Boolean values in the same array. The concept of arrays is used very often in smart contracts. In solidity, there are two kinds of arrays. The first one is the storage arrays. These arrays are stored inside the blockchain. Therefore, after executing the function, the array will stay in the memory of the blockchain. The second type of array is memory arrays. The memory array is just a temporary array – it exists while executing a function, but it disappears after that. First, let us look into the details on how to declare an Array. If we are declaring an array of, say, integers, we first mention the integer type, then use the square bracket, and then name the Array.

For example: `uint[] public myArray`

- `uint` is the type of variable.
- `[]` indicates it is an array type.
- `public` indicates that this array is accessible openly.
- `myArray` is the name of the array.

First, let us look at how an element may be added to an array. Let us assume that we are inside a function, because, we cannot manipulate arrays outside of functions (refer Fig. 14.23).

```
pragma solidity ^0.5.1;
            contract Hello {
                        uint[] public myArray;
            function push(uint i) public {
                        myArray.push(i);
            }
}
```

Figure 14.23: Solidity program with `Array`

The `push` command is used to push an array element to the top of the array. `Push` is the function name that makes the array public. Inside the function, we refer to the array using its name (`myArray`). To add a new element, we use the `.push ()` method. `Push` will always add the element at the top of the array.

To read an array element, we reference the array by its variable name, and we use the square bracket notation. To access the first element in an array named `myArray`, we use `myArray [0]`. The first element is referenced as `index 0`. The second element is referenced as `index 1` etc.

The next operation is to update an element in an array. Once again, we reference the variable name, and then the index at which we want to update. For example, if we want to update the first element, then the index is zero. The new value is assigned using equal to symbol. For example: `myArray [0] = 20`; this operation will update the first element to 20. If the array elements are read again, myArray of 0 will have a value of 20. If we try to access an element that does not exist, like myArray [50], then we shall get an error while running the smart contract.

The last operation that can be used in an array is the `delete` operation. We use the `delete` keyword (refer Fig. 14.24). For example: `delete myArray [index];` Note: Index of zero represents the first value. If an array's length is 10, then the index is numbered from 0 to 9 (0,1,2,3,4,5,6,7,8,9) indicating 10 elements of the array. The statement `delete` will not modify the length of the array. It just resets the value to its default type. For example, the default value for an integer is zero. For an array of Boolean, the default value will be false. For iterating an array, we can use it for a loop. It is also possible to specify the size of the array while declaring it. For example, `uint[10]` public myFixedSizeArray;. If we declare the size of the array, then the push method cannot be used. For example, myFixedArray.push(i) will throw an error. However, we can implement push method by specifying the index and the number to be added as myFixedSizeArray[index] = number;. The main difference between memory arrays and storage arrays is that memory arrays are not safe. Data cannot be saved with memory arrays. The memory array is just a temporary array. It is not saved inside the blockchain. The memory cannot have a dynamic array. We need to specify its length. To add a new value to it, we need to reference the index directly.

14.7.1 Array Example 1

```
pragma solidity ^0.5.1;
        contract ArrayContract {
                uint[] public myArray;
                uint[] public myArray2 = [1,2,3];
                uint[10] public myFixedSizeArray;

        function push(uint i) public {
                myArray.push(i);
        }

        function pop() public {
                myArray.pop();
        }

        function getLength()  public view returns(uint) {
                return myArray.length;
        }

        function getArray()  public view returns(uint[] memory)  {
                return myArray;
        }

        function remove(uint index) public {
                delete myArray[index];
        }
}
```

Figure 14.24: Solidity program with Array, Push and Pop

Push function is used to add value at the top of the stack. Pop function is used to take out value from the stack.

Outputs of the Program

Deployed Contracts
ArrayContract at 0x100...90e28 (memory)
pop
push 20
remove uint256 index
getArray
0: uint256[]: 20
getLength
0: uint256: 1
myArray uint256
myArray2 uint256
myFixedSizeArray uint256

Figure 14.25: Output of Solidity program with `Array`, `Push` and `Pop`

When we enter 20 in the push field of the output and click, it adds the number 20 into the array (refer Fig. 14.25). When we click `getLength`, it shows as 1 and `getArray` indicates that the value 20 is added into the array.

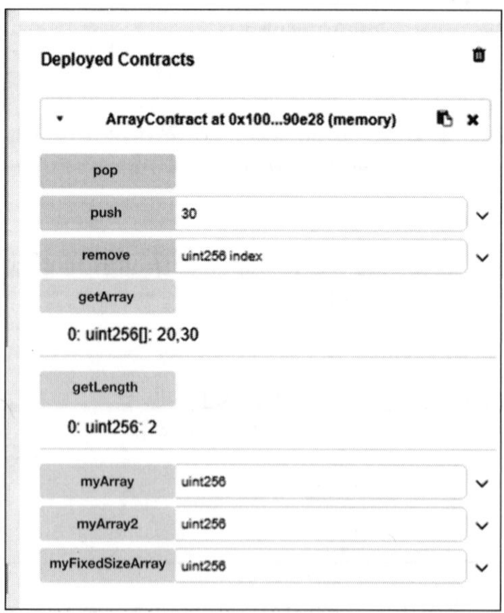

Figure 14.26: Output of Solidity program with `Array`, adding the second element

Next, when we enter 30 in the push field and click, it adds the number 30 into the array. When we click `getLength`, it shows as 2 and `getArray` indicates that the value 30 is also added into the already existing array (refer Fig. 14.26).

After that, when we enter 40 in the push field and click, it adds the number 40 into the array. When we click `getLength`, it shows as 3 and `getArray` indicates that the value 40 is also added into the already existing array (refer Fig. 14.27).

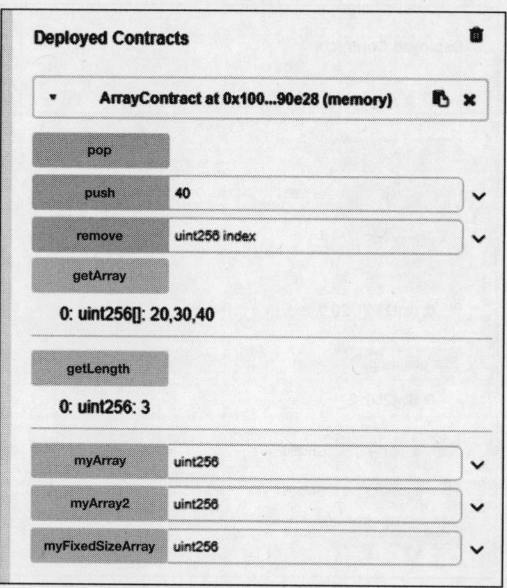

Figure 14.27: Output of Solidity program with `Array`, adding the third element

After adding the elements, when we click on the `Pop` button, it removes the number 40 from the array. When we click `getLength`, it shows as 2 and `getArray` indicates that the value 40 is removed from the already existing array (refer Fig. 14.28).

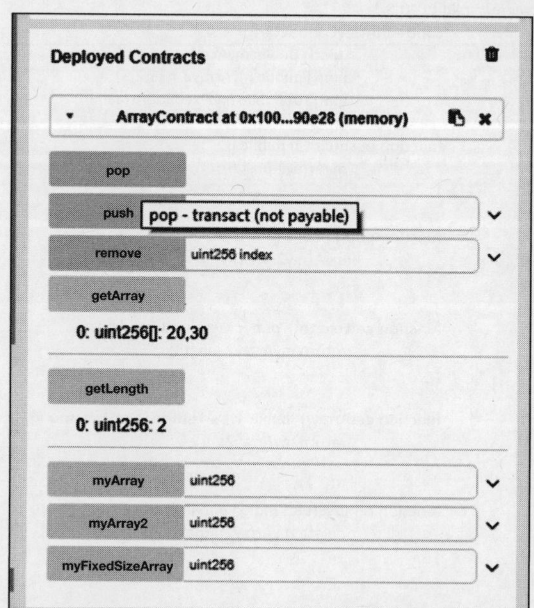

Figure 14.28: Output of Solidity program with `Array`, `Pop` button

Then, when we enter 1 and click on the `remove` button, it removes the index 1 element from the array. When we click `getLength`, it shows as 2, and `getArray` indicates that the value at index 1 is reset to 0 (refer Fig. 14.29).

Figure 14.29: Output of Solidity program with `Array`, `Pop` and `Remove` button

14.7.2 Array Example 2

```
pragma solidity ^0.5.1;
        contract ArrayContract2 {
                uint[] public myArray;
                uint[] public myArray2 = [1,2,3];
                uint[10] public myFixedSizeArray;

        function push(uint i) public {
                myArray2.push(i);
        }

        function pop() public {
                myArray2.pop();
        }

        function getLength() public view returns(uint) {
                return myArray2.length;
        }

        function getArray() public view returns(uint[] memory) {
                return myArray2;
        }

        function remove(uint index) public {
                delete myArray2[index];
        }
}
```

Figure 14.30: Array Example 2

When we run the above solidity program (refer Fig. 14.30) in the Remix browser, we get the same results we got for Example 1, except that the arrays are initialized with the values of 1, 2 and 3 representing the first three elements. **Note**: The array size is not fixed here. It is kept open.

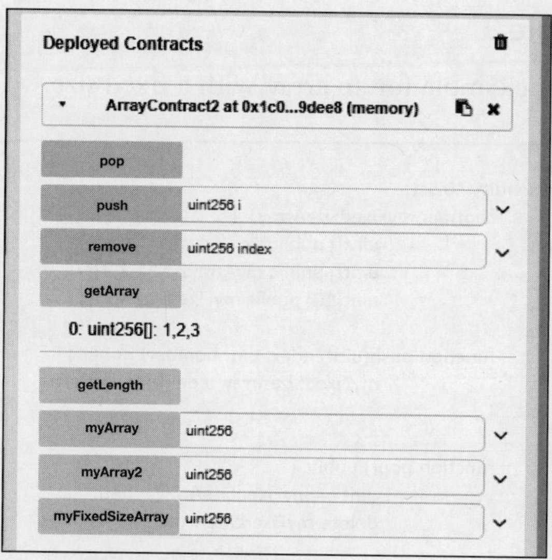

Figure 14.31: Array Example 2, `getArray` output

When we click `getArray`, we get the value as 1, 2, 3, which are the contents of Array 2 (refer Fig. 14.31).

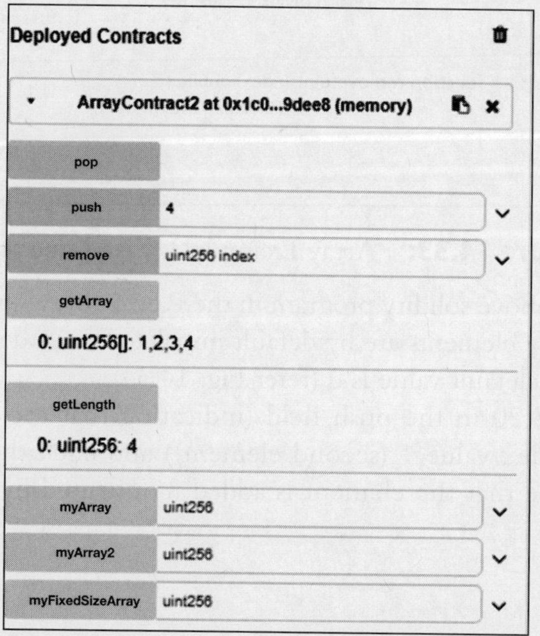

Figure 14.32: Array Example 2 output, adding the element (push)

When we enter the number 4 in the push field and click `push` (refer Fig. 14.32), the new element entered (4 here) is added to the already existing array, and this can be confirmed by clicking `getArray` button which displays 1, 2, 3 and 4. This can also be reconfirmed by clicking the `getLength` button, which displays 4.

14.7.3 Array Example 3

Let us look into another example for an array with a fixed size.

```
pragma solidity ^0.5.1;
        contract myFixedSizeArray1 {
                uint[] public myArray;
                uint[] public myArray2 = [1,2,3];
                uint[10] public myFixedSizeArray;

        function push(uint index, uint number) public {
                myFixedSizeArray[index] = number;
        }

        function pop() public {
                uint i = myFixedSizeArray.length - 1;
                delete myFixedSizeArray[i];
        }

        function getLength() public view returns(uint) {
                return myFixedSizeArray.length;
        }

        function getArray() public view returns(uint[10] memory) {
                return myFixedSizeArray;
        }

        function remove(uint index) public {
                delete myFixedSizeArray[index];
        }
}
```

Figure 14.33: Array Example 3: Fixed size array

When we run the above solidity program in the Remix browser (refer Fig. 14.33), we can see that all the array elements are by default initialized to its default value.

Note: For integer, the default value is 0 (refer Fig. 14.34).

When we enter 1, 20 in the push field (indicating that the array element 20 has to be added to the index value 1 (second element)) and click the `push` button (refer Fig. 14.35), we can see that the element is added (confirmed by clicking `getArray` button).

Deployed Contracts

myFixedSizeArray1 at 0x139...6f1ef (memory)

pop

push uint256 index, uint256 number

remove uint256 index

getArray

0: uint256[10]: 0,0,0,0,0,0,0,0,0,0

getLength

0: uint256: 10

myArray uint256

myArray2 uint256

myFixedSizeArray uint256

Figure 14.34: Array Example 3: Fixed size array output

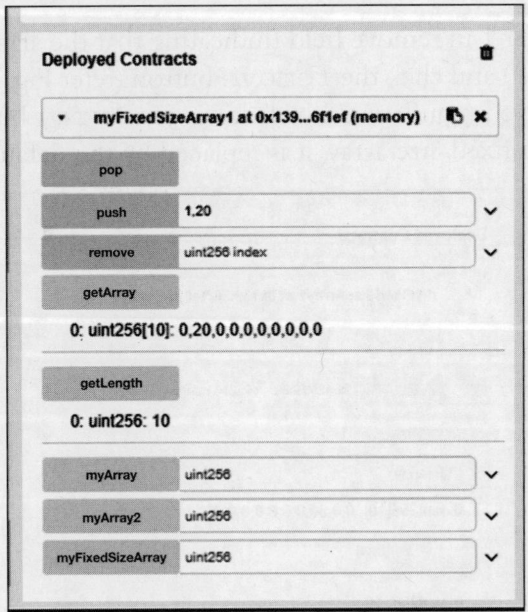

Figure 14.35: Array Example 3: Fixed size array output, push button

After that, when we enter 2, 30 in push field (indicating that the array element 30 has to be added to the index value 2 (third element)) and click the `push` button (refer Fig. 14.36), the second element is added (confirmed by clicking `getArray` button).

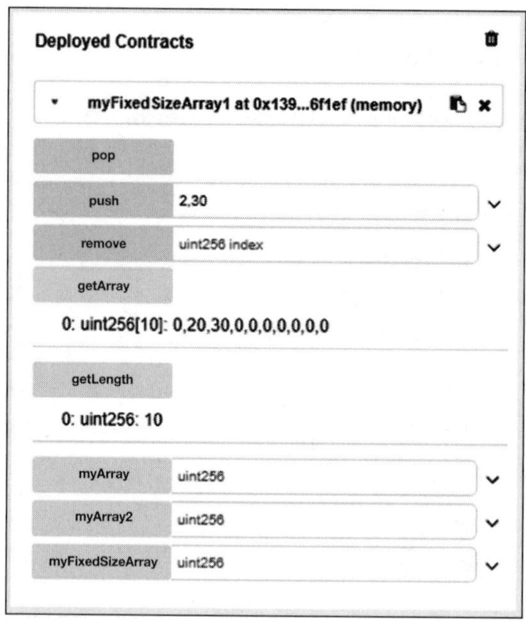

Figure 14.36: Array Example 3: Fixed size array output, push button adding the next element

Next, when we enter 1 in remove field (indicating that the array element at the index value 1 (second element)) and click the `remove` button (refer Fig. 14.37), we can see that the element 20 is removed (confirmed by clicking `getArray` button). When the array element is removed in a fixed-size array, it is replaced by the default value (0).

Deployed Contracts
myFixedSizeArray1 at 0x139...6f1ef (memory)
pop
push
uint256 index, uint256 number
remove
1
getArray
0: uint256[10]: 0,0,30,0,0,0,0,0,0,0
getLength
0: uint256: 10
myArray
uint256
myArray2
uint256
myFixedSizeArray
uint256

Figure 14.37: Array Example 3: Fixed size array output, remove button

When we enter 9,100 in the push field (indicating that the array element 100 to be added to the index value 9 (tenth element)) and click the `push` button (refer Fig. 14.38), we can see that the element is added (confirmed by clicking `getArray` button).

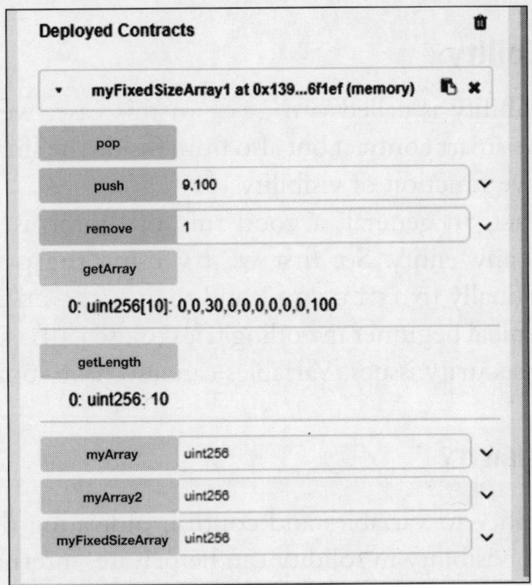

Figure 14.38: Array Example 3: Fixed size array output, push (9,100)

14.8 Function Visibility

Function visibility in solidity can be private, internal, public, and external. Visibility keywords allow defining who can have access to a particular function. The most restrictive visibility keyword is called `private`.

Private Function Visibility

When `private` keyword is used, we can only call the function from inside the smart contract. It is conventional to prefix the name of the function by an underscore to indicate a private function. It is just a convention; solidity does not mandate this. With private function, when we deploy the smart contract then we will not see `get` value because this is private, and it cannot be called from outside.

Internal Function Visibility

The next keyword is called internal. With the `internal` keyword, the function is still limited to be called from within a smart contract, but it can also be called from a smart contract that inherits from this smart contract. `Internal` is slightly less restrictive than `private`, but we still cannot call this function from outside the smart contract.

External Function Visibility

With the `external` keyword, we can only call the function from outside a smart contract. For an external function, we do not prefix them by an underscore.

Public Function Visibility

The last function in visibility is called `public`. In this case, we not only can call the function from outside the smart contract but also from inside the smart contract. Therefore, this is the most permissive function of visibility.

With a smart contract, in general, a good rule of thumb is to give the minimum amount of privilege to any entity. So, first we try using the `private` keyword and relax the restrictions gradually to `internal` and then `external`, before finally using `public`. However, a typical beginner in coding tries to use `public` for all the functions, which may create some security issues. Variables can also have some visibility keywords.

14.9 Variable Visibility

We can also assign visibility to variables and control, indicating the details of how they can be accessed. Variable visibility in solidity can be private, internal, and public. We first declare an integer variable and make it private; then we declare its name. This means the variable is of type integer, the visibility is `private`, and the name is `name`. A private variable can only be read from within the same smart contract.

When a variable is stored in the Ethereum blockchain, which is a public blockchain, we cannot have anything private. In the public blockchain, it is always possible to read the value of a private variable. In cases like this, we use `internal` keyword instead of `private`.

The internal variable can be read from inside of smart contract or another smart contract that is inherited from it. We cannot read it from the outside of the smart contract.

When `public` keyword is used, we can read the variable not only from within an inherited smart contract but also from outside of smart contracts. It gives the most access to solidity variable. By default, solidity considers a variable as a private variable.

14.10 Function Modifier Keyword

```
pragma solidity ^0.5.1;
        contract MyContract {
                uint value;
        function getValue() external view returns(uint) {
                return value;
        }
        function setValue(uint _value) external {
                value = _value;
        }
}
```

Figure 14.39: MyContract with `setValue` and `getValue` functions

When we deploy the above contract (refer Fig. 14.39), there are two functions represented by two different colored buttons, an orange and a blue. The button `getValue` is blue-colored and is a read-only function. The button `setValue` is orange and can modify data on the blockchain. When we use `view` keyword in a function, it means this is a read-only function. It does not modify the blockchain variables. When `view` keyword is used in a function signature, any attempt to modify a value inside the function body of the smart contract will throw an error as "Function declared as view, but this expression (potentially) modifies the state and thus requires non-payable (the default) or payable." There is another keyword that is quite similar to `view` and is called `pure`. With `pure` keyword, the function is read-only, but instead of just returning a value from the blockchain, it can do some computation.

14.11 How Funds Are Accepted

Let us now examine how funds are accepted into a contract from other Ethereum addresses. Let us create a contract that lets users buy concert tickets; the name of the contract is called "FunkConcert."

1. We shall allow people to send money to the contract to purchase a ticket
2. The address from which they send their ether shall be registered as of somebody who owns a ticket
3. We shall start with a set number of tickets and reduce that number every time a ticket is purchased until we have no more tickets.
4. Finally, we shall return all the funds for the purchased tickets to the creator of the contract.

The first thing we shall need is the owner's address, so that would be our address. The next thing we need is tickets to hold (number of tickets). Then we need a price for the tickets. This will be an unsigned integer constant. A constant is a variable that you cannot change once it is initialized. Price of one ticket defined as 1 ether here (refer Fig. 14.40).

Next, we create a registry to store the details of those who bought the tickets and how many tickets they own. We shall use mapping. The map is a variable that stores keys and a corresponding value to the keys. `Purchaser` is a mapping that uses an Ethereum address as the key and points to a value (an unsigned integer) representing the number of tickets that a particular purchaser has bought. We declare the purchasers variable as public so that we may check who has purchased the tickets and how many.

Let us also create a constructor in which we shall set the owner to message sender. Message sender in the constructor refers to the address creating or deploying the contract. We shall also set the number of tickets to five. Let us also create a function to purchase these tickets. We shall call this function "buyTickets." When we buy a ticket, we shall be able to specify the number of tickets and to make sure that this function can receive ether. We also need to make the function payable. Hence, we add `payable` as a keyword, at the end of our function.

```
pragma solidity ^0.5.1;
        contract FuncConcert {
                address payable owner;
                uint public tickets;
                uint constant price = 1 ether;
                mapping (address =>  uint) public purchasers;

        constructor() public {
                owner = msg.sender;
                tickets = 5;
        }
        function() external  payable{
                buyTickets(1);
        }
        function buyTickets(uint amount) public payable {

                if(msg.value != (amount *price) || amount > tickets)  {
                        revert();
                }
                purchasers[msg.sender]  = purchasers[msg.sender]  +  amount;
                tickets -=amount;

                if( tickets ==0) {
                        selfdestruct(owner);
                }
        }
}
```

Figure 14.40: Accepting funds

To send ether to a smart contract, we need to have a function with the `payable` keyword. The `payable` keyword specifies that the function can accept ether. Without this keyword, the function will reject the incoming transfer. The function should not be a `view` function. When someone purchases a ticket, we want him or her to specify the number of tickets. We also want them to send a valid amount of ether. Therefore, we first check if the purchaser has sent the right amount of ether for tickets that he or she wants to buy. This is done by verifying if the message value, which is a message table value, corresponds to the amount of ether that was sent to this function. If the message value is not equal to the number of tickets purchased times the price, then the program throws an error (shows an exception).

One other thing we have to check is whether or not we have enough tickets left. For this, let us set a key on the purchaser's mapping to denote the number of tickets bought.

This is done by referencing the index of the sender's address, which will be zero to start with. We can just add the number of tickets bought and take the message to the sender, which gives the address of whoever is trying to buy the ticket. We then add the number of tickets that the sender has successfully bought. In the next step, we remove that number of tickets from the total number of tickets that we have.

Let us now make this ticket available to the public as well to see how many tickets are left. Ethereum automatically creates getters for any sort of public variables. Therefore, these ticket getters are automatically created. We can already see that there are five tickets left. Let us select an address and try to buy a ticket. However, it might sometimes throw an error as shown in Fig. 14.41 resulting from transaction failure.

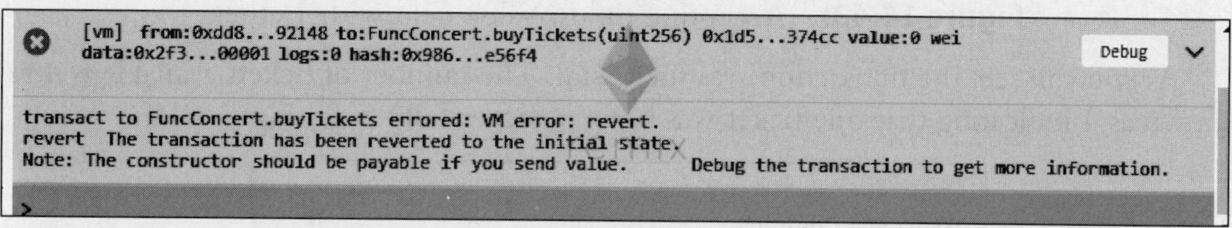

Figure 14.41: Accepting funds: Spinning failure

This happens when we do not spin any ether, i.e., we kept the value field as empty (refer Fig. 14.42)

Environment JavaScript VM VM (-)
Account 0xdd8...92148 (97.999999999914682!
Gas limit 3000000
Value 0 ether
FuncConcert
Deploy
or
At Address Load contract from Address

Figure 14.42: Accepting funds: Spinning failure – Value field with 0 value

Let us now send one ether (select value as 1) because that is the price of one ticket (refer Fig. 14.43).

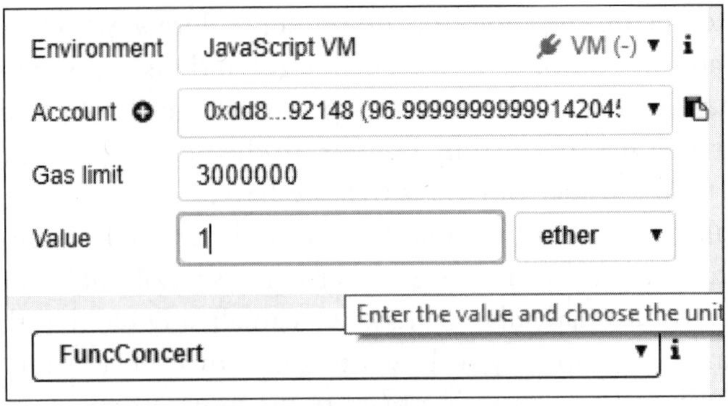

Figure 14.43: Accepting funds: Value field with 1 ether

As you can see, the transaction was successful. The number of tickets available is now shown as 4 indicating that one ticket was bought successfully (refer Fig. 14.44).

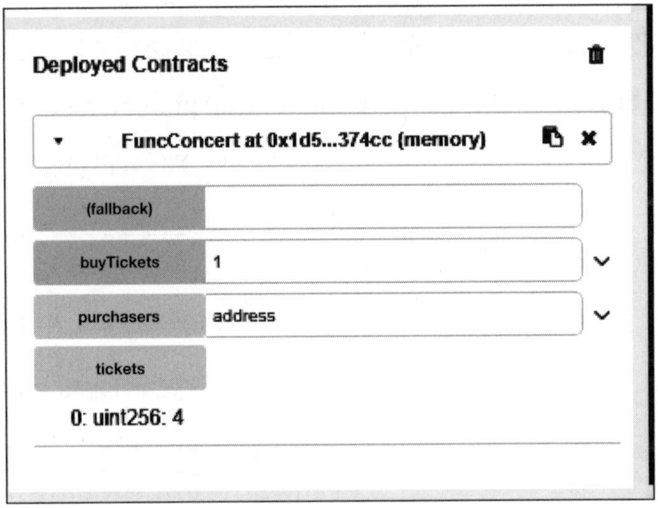

Figure 14.44: Accepting funds: Buying 1 ticket

When working with this remix IDE, if we try to run a function that is not payable, i.e., if we try to send ether to it, it is going to fail. Hence, we need to clear the value box and run tickets again.

We can also copy this address and look at the purchasers in the index. The index gives the address of whoever has purchased the tickets. It will show that we have purchased one ticket. We can try to buy two more tickets, with two ethers and go through the transaction once again (refer Fig. 14.45).

We need to first clear the value before checking how many tickets are left. It shows that two tickets are left. Also, under purchasers, we have three tickets in total.

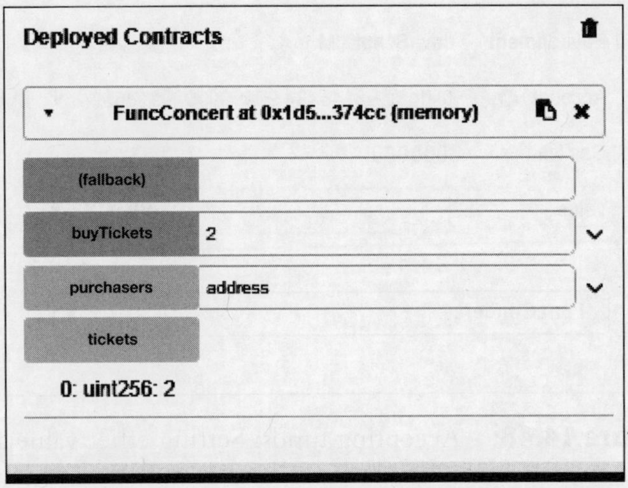

Figure 14.45: Accepting funds: Buying 2 tickets

Let us now see what happens when we try to buy a ticket at a wrong price. Let us give 1.5 ethers for one ticket. It fails (refer Figs 14.46 and 14.47).

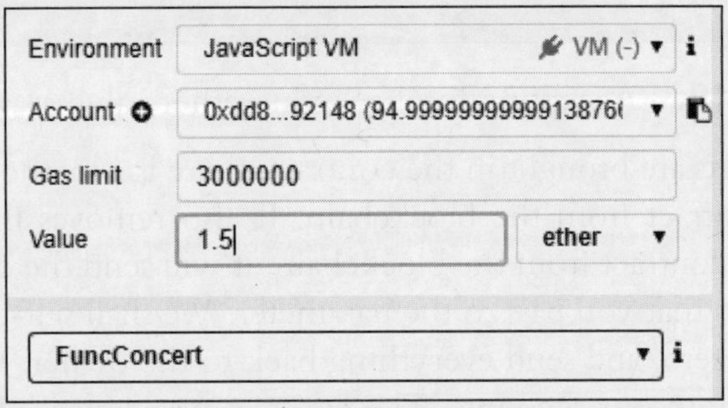

Figure 14.46: Accepting funds: Setting ether value as 1.5

Figure 14.47: Accepting funds: Setting ether value as 1.5 with error

This makes sure that we can transact with the correct amount of money. Ether spinning should be equal to the amount for the tickets bought. There are two tickets left. Let us try to buy three more tickets by giving the right amount of money (3 ethers) (refer Figs 14.48 and 14.49). We cannot buy 3 more tickets because there are only two tickets left.

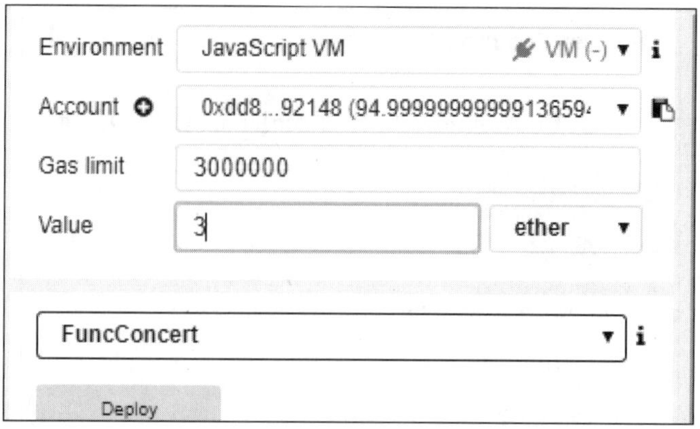

Environment JavaScript VM VM (-)
Account 0xdd8...92148 (94.999999999913659
Gas limit 3000000
Value 3 ether
FuncConcert
Deploy

Figure 14.48: Accepting funds: Setting ether value as 3

```
[vm] from:0xdd8...92148 to:FuncConcert.buyTickets(uint256) 0x1d5...374cc
value:3000000000000000000 wei data:0x2f3...00003 logs:0 hash:0xc63...424e1
transact to FuncConcert.buyTickets errored: VM error: revert.
revert   The transaction has been reverted to the initial state.
Note: The constructor should be payable if you send value.    Debug the transaction to get more information.
```

Figure 14.49: Accepting funds: Setting ether value as 3 with error

When all the tickets are brought in the contract, there is a method called self-destruct. This removes the contract from the blockchain. It also removes the address parameter. When it removes the contract from the blockchain, it will send the current balance of the contract to the address that you pass to the parameter. We shall self-destruct this contract, check if tickets equal zero, and send everything back to the owner. We just got five more ether deposited back into our accounts. The contract was destroyed and the balance was sent back to our address.

14.12 Fallback Function

Let us also learn about fallback function. Solidity allows us to create a function within a contract that has no name whatsoever. The fallback function is called if we try to call a function by a name that does not match anything else in the contract (refer Fig. 14.50). It is triggered when we try to call a function that does not exist.

```
function() external payable{
        buyTickets(1);
}
```

Figure 14.50: Fallback function with external payable

Use the `function` keyword and specify the `external` keyword. A fallback function does not have any name. We can only have one fallback function in a smart contract. There is no argument to a fallback function. The only visibility that is allowed in the fallback function is external. To accept ether, we need to add the `payable` keyword to the fallback function. One useful scenario for this is allowing people to send money from their wallet directly to the address of the contract. Sending ether directly to the wallet does not call a function by name. It is done by using this callback function. We allow ether to be sent straight to the contract by using the `payable` keyword just as we would with a normal function. This keyword is the fallback function and it allows people to send money straight to our contract to buy a ticket. Because the fallback function cannot take an argument, it can be used only to buy one ticket. Let us call buyTickets from within this function, send ether and click the fallback function. The transaction is seen to be successful (refer Fig. 14.51).

```
pragma solidity ^0.5.1;

contract FuncConcert {
address payable owner;
uint public tickets;
uint constant price = 1 ether;
mapping (address => uint) public purchasers;

constructor() public {
    owner = msg.sender;
    tickets = 5;
        }
function() external  payable{
    buyTickets(1);
        }
function buyTickets(uint amount) public payable {

    if(msg.value != (amount *price) || amount > tickets)  {
                revert();
                    }
    purchasers[msg.sender]  =   purchasers[msg.sender] +  amount;
    tickets -=amount;

    if( tickets == 0)   selfdestruct(owner);
    }


}
```

Figure 14.51: Sending ether using `payable` keyword

Note: While creating a contract, the Value field should be empty. In our case, the fallback function is a payable function, and so the Value field should be filled.

14.13 Contract Inheritance

Let us discuss the concept of contract inheritance (refer Fig. 14.52), how we can use existing contracts and build upon them and create entirely new contracts using the old functionality. To demonstrate this, we are going to use a fun concert example.

```
pragma solidity ^0.5.1;
        contract FuncConcert {
                address payable owner;
                uint public tickets;
                uint constant price = 1 ether;
                mapping (address => uint) public purchasers;
        constructor(uint t) public{
                owner = msg.sender;
                tickets = t;
        }
        function() external payable{
                buyTickets(1);
        }
        function buyTickets(uint amount) public payable {
                if(msg.value != (amount *price) || amount > tickets) {
                        revert();
                }
                purchasers[msg.sender] += amount;
                tickets -= amount;
                if( tickets ==0) {
                        selfdestruct(owner);
                 }
        }
        function website() public returns (string memory) ;
        }

        interface Refundable{
         function refund(uint numTickets)  external returns (bool);
        }

        contract AbstractFuncAttack is FuncConcert (10), Refundable {

                function refund(uint numTickets) public returns (bool) {
                        if (purchasers[msg.sender] < numTickets) {
                        return false;
                        }
                         msg.sender.transfer(numTickets * price);
                        purchasers[msg.sender] -= numTickets;
                        tickets +=numTickets;
                        return true;
                }

                function website() public returns (string memory) {
                        return "http://abstractfuncattack.com";
                }
        }
```

Figure 14.52: Contract inheritance program

We shall create a new contract that inherits from this contract. It shall be a FuncConcert contract but with a very different name. It is called as "AbstractFuncAttack," which is a FuncConcert. We can deploy this "AbstractFuncAttack" contract. Just like the FuncConcert contract, it has all the different features. Contract AbstractFuncAttack is FuncConcert {}

In the initial FuncConcert contract, we set the tickets to be five (5) every time. What if they are at a larger venue? They have more seats and so they have a larger number of tickets. In this case, we modify the main constructor to take an argument. When a new fun concert contract is created, it is supposed to take an argument on creation to set the number of tickets. In the child contract, we add the arguments to the contract that we are inheriting from (main Contract) as shown: contract AbstractFuncAttack is FuncConcert (10), Refundable {}. One other useful feature of inheriting from contracts is the ability to set an abstract function. An abstract function is a function that has no actual implementation, but it does define a signature. Let us say that we want every single contract that inherits from this funcconcert contract to have a function that displays the concert or the band's website. To do that, we need to create this abstract function. When we use the function to call a website, it returns a string. Function `website()` returns `(string)` (refer Fig. 14.53). We need to implement it in full in the child contract. The function has to have the same signature as the parent contract.

```
function website() public returns (string memory) {
        return "http://abstractfuncattack.com";
}
```

Figure 14.53: The `website` function

The abstract function may be used when we want to inherit or let someone know that a contract has a set of functions that follow a set of specific signature but would prefer not to go through the process of creating a completely new contract with a bunch of implemented functions along with some abstract functions. In other words, it is useful when we care only about signatures and not about implementation.

`Interface` Keyword

```
interface Refundable{
        function refund(uint numTickets) external returns (bool);
}
```

Figure 14.54: Refundable interface

Let us say that we want this particular concert to be able to refund tickets. For this we shall to create an interface, use the `interface` keyword, and call it refundable. The refundable interface has a function called `refund` (refer Fig. 14.54). It will take several tickets and refund that number of tickets to the caller. It will also return true or false based on whether or not it can refund. To return true or false, we use the bool or Boolean return type (refer Fig. 14.55).

AbstractFuncAttack, which inherits from FuncConcert, is also a refundable contract. We cannot deploy until we implement the refund function. So let us implement the refund function using the same signature, as in the interface. What we want to do is to check if whoever is calling the refund function has purchased tickets, and if they have the same number of tickets or more than what they are trying to refund before sending the money back.

```
function refund(uint numTickets) public returns (bool) {
            if (purchasers[msg.sender] < numTickets) {
                        return false;
            }
            msg.sender.transfer(numTickets * price);
            purchasers[msg.sender] -= numTickets;
            tickets +=numTickets;
            return true;
}
```

Figure 14.55: Refund function logic

14.14 Contract Communicating with Another Contract

Ethereum virtual machine (EVM) allows us to create contracts that can communicate with other contracts. This allows us to create complex decentralized applications. With EVM, we can have one contract to handle all the states for our application and another contract to handle the logic. These two contracts can communicate with each other through their various publicly accessible functions. Let us create contracts that talk to other contracts (refer Fig. 14.56).

```
pragma solidity ^0.5.1;
            contract Foo {
                        Calculator calc = new Calculator();
            function fourTimesSix() external view  returns(int) {
                        return calc.multiply(4,6);
            }
            function TwoPlusEight() external view  returns(int) {
                        return calc.add(2,8);
            }
}

            contract Calculator {
            function add (int a, int b) external pure returns (int) {
                        return a+b;
            }
            function multiply (int a, int b) external pure returns (int) {
                        return a*b;
            }
}
```

Figure 14.56: Contract Foo calling calculator contract

Let us create a calculator contract (refer Figs 14.57 and 14.58) that can add and multiply. If we publish this contract, we can add or multiply two numbers (refer Fig. 14.57).

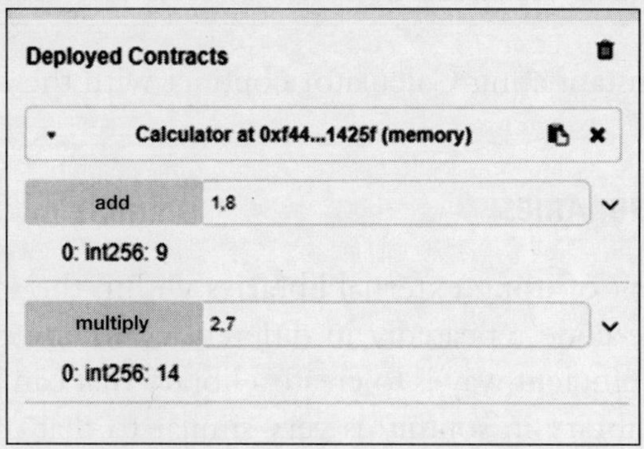

Figure 14.57: Calculator contract outputs

```
pragma solidity ^0.5.1;
        contract Calculator {
        function add (int a, int b) external pure returns (int) {
                return a+b;
        }
        function multiply (int a, int b) external pure returns (int) {
                return a*b;
        }
}
```

Figure 14.58: A simple program for calculator

When contract Foo is created (refer Figs 14.56 and 14.59), it instantiates another contract called "Calculator," which uses the commands Calculator calc = new Calculator(); After the instantiation, Foo can call the functions within Calculator contract. For example, multiply function within the Calculator contract can be called calc.multiply(4,6);

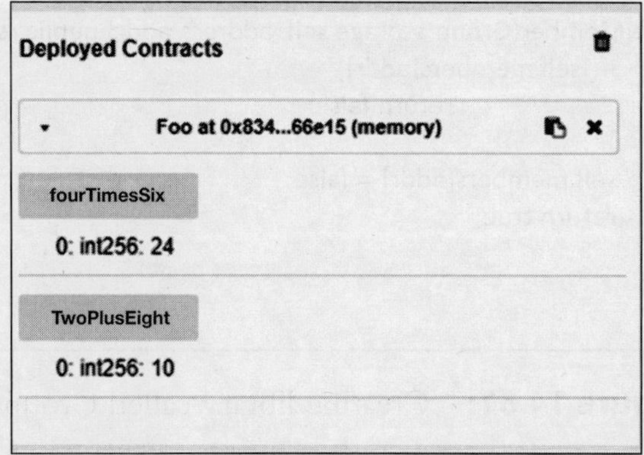

Figure 14.59: Foo contract output (calculator)

Another way to instantiate the calculator contract is through its Transaction Hash, if it is known. The syntax for this hash is given in Fig. 14.60.

```
Calculator calc = Calculator(0x19aec731c4c511d0408a2b19e033dd378257e6d763898b614bea09169b8e6428);
```

Figure 14.60: Instantiating Calculator contract with the transaction address

14.15 External Libraries

Let us discuss the concept of using external libraries within the solidity contracts. When we have to use the same code repeatedly in different contracts, one way is to copy and paste the code. Another efficient way is to create a library that can be imported into a new contract and reused. A library in solidity is very similar to that of a contract. The name of the file ends in .sol. Instead of defining a contract, we use the `library` keyword. Libraries are very useful for reusing codes. For example, if we have to spot contracts with the same code, then we can extract the common code into a library to avoid code duplication. The way to declare a library is with the `library` keyword. Then, we give a name to the library (refer Fig. 14.61) and write all functionalities that are to be reused within.

```
pragma solidity ^0.5.1;
library Groups {
        struct Group {
                mapping (address => bool) members;
        }
        function addMember( Group storage self, address addr) public returns (bool) {
                if(self.members[addr]){
                        return false;
                }
                self.members[addr]=true;
                return true;
        }
        function delMember(Group storage self, address addr) public returns (bool) {
                if(!self.members[addr]){
                        return false;
                }
                self.members[addr] = false;
                return true;
        }
}
```

Figure 14.61: Creating library called Groups.

First, we write a struct. A struct is a data type much like an integer or a string or an array, but it allows defining our type or objecting. In this case, the struct has one data type called mapping. However, it can have any other number of variables within the data type (refer Fig. 14.62).

```
struct Group {
            mapping (address => bool) members;
}
```

Figure 14.62: Creating structure and mapping

We have created a struct group. The mapping maps an address to a Boolean. We shall access these members to check whether the member exists or not.

```
function addMember(Group storage self, address addr) public returns ( bool ){
            if(self.members[addr]){
                        return false
            }
            self.members[addr]=true;
            return true;
}
```

Figure 14.63: The `addMember` function

This library manages groups, and the groups contain addresses. In the "addMember" function (refer Fig. 14.63), we shall add the address as a member if it is already not a member. If it (address) is already a member, the function will return false. This function takes the group as its first parameter. The `storage` keyword tells the Ethereum virtual machine (EVM) that when we pass in this group, we want to reference it by its actual memory address or storage location address. Since EVM is not a memory, this is stored on the blockchain; so we want to reference it by its actual location. Hence, we can modify its value rather than taking a copy of whatever is passed in and modifying it later. We do this because this library is external to any contract that we are using. This is the only way for the Ethereum virtual machine (EVM) to know that we are referencing the group passed in from the contract, which is using this library. The next parameter is an address, and it returns a Boolean, true or false. We check the Boolean to see if the address already exists. If it does, the value returns false and if it does not exist, then the value returns true, indicating that we have successfully added a new member to the group.

```
function delMember(Group storage self, address addr) public returns (bool) {
            if(!self.members[addr]{
                        return false;
            }
            self.members[addr] = false;
            return true;
}
```

Figure 14.64: The `delMember` function

The final function is `delMember` and it removes a member from the group (refer Fig. 14.64). This function also takes a group as the first parameter, referenced by its storage address. It passes an address as the second parameter and returns the Boolean as true or false. If the member does not exist, we cannot remove it, so the function returns false. If the member does exist, the function sets it to false, thereby removing it, and then return true, indicating that successful removal. Let us see how to use this library (Groups) in a new contract. We stored this library in a .sol file named Groups.sol (refer Fig. 14.65).

```
pragma solidity ^0.5.1;
import './Groups.sol';
        contract Libraryfun {
                Groups.Group admins;
                event success();
        modifier onlyAdmins() {
                require(admins.members[msg.sender]);
         _;}
          constructor() public{
                 Groups.addMember(admins,msg.sender);
          }
         function add(address addr) public  onlyAdmins{
                 if (Groups.addMember(admins,addr)){
                         emit success();
                 }
        }
        function remove(address addr) public onlyAdmins {
                if(Groups.delMember(admins,addr)){
                        emit success();
                 }
        }
}
```

Figure 14.65: Contract library function

Use the `import` keyword followed by the filename as given: import 'Groups.sol'. We need to import at the top of the file, outside of the contract and create a group using Groups.Group admins; we will reference this admin. We are going to add ourselves to the admins group using a constructor (refer Fig. 14.66).

```
constructor() public{
        Groups.addmember(admins,msg.sender);
}
```

Figure 14.66: Contract library function: Constructor

Next, we create a proxy function to the add member group.

```
function add(address addr) public onlyAdmins{
        if (Groups.addMember(admins,addr)){
                emit success();
        }
}
```

Figure 14.67: Contract library function: `add` function

This function takes an address. It calls `addMember` function within the Groups (refer Fig. 14.67), which in turn returns true or false. When it is successful, we shall call the event `success ()`. We also have created a modifier (refer Fig. 14.68), namely, "onlyAdmins." That will allow only admins to run this function. People who are added to the admins group can run the function.

```
modifer onlyAdmins() {
        require(admins.members[msg.sender]);
_;}
```

Figure 14.68: Contract Library Function: `onlyAdmins` function

Let us add one more function, called "remove" function (refer Fig. 14.69), which is a proxy to the delete member function. Again, only the admins can do this.

```
function remove(address addr) public onlyAdmins {
        if(Groups.delMember(admins,addr)){
                emit success();
        }
}
```

Figure 14.69: Contract library function: `Remove` function

14.16 ERC20 Token Transfer

Let us discuss how we can transfer ERC20 token in solidity. The 20 is a standard that describes how we could implement the token on the Ethereum blockchain. With 20 tokens, we can create our coin on the Ethereum blockchain. Most tokens represent some form of value, and a lot of them are traded in decentralized exchanges. First, we will write a contract named "Token". Only the import statement is written there. It is possible to import a solidity file directly from the web using an import statement (refer Fig. 14.70). In this program, we import ERC20.sol from GitHub. Contract Token in ERC20 indicates that the token contract is ERC20. We write another solidity file with two contracts, namely, Transfer Token and Owner.

```
pragma solidity ^0.5.1;
            import "https://github.com/OpenZeppelin/openzeppelin-contracts/contracts/token/ERC20/ERC20.sol" ;
contract Token is ERC20 {
}
```

Figure 14.70: Importing ERC20 solidity file from GitHub.

We have created a contract titled "Token" by inheriting ERC20. Contract Token in ERC20 represents the same. We are also importing ERC20.sol program from github.com, which has defined functions within it to transfer ERC20 tokens.

```
pragma solidity ^0.5.1;
        import "./Token.sol";
        contract TransferToken {
                function transfer() external {
                        //Token Address copied from Token.sol
                        Token token = Token(0xCA35b7d915458EF540aDe6068dFe2F44E8fa733c);
                        // use the below URL to convert address to checksum valid address
                        //https://etherscan.io/address/0xca35b7d915458ef540ade6068dfe2f44e8fa733c
                        token.transfer(msg.sender,100);
                }
                function transferFrom(address recipient,uint amount) external {
                        //Token Address copied from Token.sol
                        Token token = Token(0xCA35b7d915458EF540aDe6068dFe2F44E8fa733c);
                        //Transfer amount of token to msg.sender
                        token.transferFrom(msg.sender,recipient,amount);
                }
}

        contract Owner {
                function transfer(address recipient, uint amount) external {
                        //Token Address copied from Token.sol
                        Token token = Token(0xCA35b7d915458EF540aDe6068dFe2F44E8fa733c);
                        token.approve(0xdD870fA1b7C4700F2BD7f44238821C26f7392148,amount);
                        TransferToken transferToken = TransferToken(0xdD870fA1b7C4700F2BD7f44238821C26f7392148);
                        transferToken.transferFrom(recipient,amount);
                }
}
```

Figure 14.71: ERC20 solidity program

Let us examine how to use the transfer function of an ERC20 token. For this, we need to get into the code at GitHub. The name of the function is transferred with the following signatures: function transfer (address recipient, uint256 amount) public returns (bool) {}. There are two arguments for this function. We need to provide the recipient and the amount. Therefore, when we call this function, we want to send this many tokens (amount) to the given address. This will be taken out of our balance. The calling smart contract must have enough token. Otherwise, this operation will fail. We shall create a function to call this transfer function. Let us call our function as transfer, and inside that, we declare a variable of type token, ERC20 token (refer Figs 14.71 and 14.72).

```
function transfer() external {
        Token token = Token(0xCA35b7d915458EF540aDe6068dFe2F44E8fa733c);
        token.transfer(msg.sender,100);
}
```

Figure 14.72: ERC20 solidity program: `Transfer` function

We give the address in the parentheses of the token. For example, in order to send 100 tokens to msg.sender the statement is: token.transfer(msg.sender,100);

There is another function within ERC20 token called "transferFrom," which is used to send a certain amount from a sender to the recipient. TransferFrom function allows transfer of tokens on behalf of someone else (refer Fig. 14.73), and this is particularly useful in the case of decentralized exchanges. For instance, assume there is an owner and an operator. The operator is another address that has been approved by the owner to send a certain amount of token. In this case, we shall use the approve function.

```
token.approve(0xdD870fA1b7C4700F2BD7f44238821C26f7392148,amount);
TransferToken transferToken = TransferToken(0xdD870fA1b7C4700F2BD7f44238821C26f7392148);
transferToken.transferFrom(recipient,amount);
```

Figure 14.73: ERC20 solidity program: `TransferToken` contract

If we try to send more tokens than what was approved, the transaction will fail.

14.17 Error Handling in Solidity

An error can happen if it is triggered due to failure in business logic with one of the keywords or if solidity does not allow certain things like accessing an array, out of bound in an array etc. When this happens, the execution of the function is stopped and other instructions that are written after this statement will never be executed. Secondly, any state change is reverted. The state value will stay the same as it was before the call. When a transaction is sent to a smart contract, it has to be paid with gas. Depending on the nature of the error, either all the gas up to the point of execution will be consumed, or the whole gas that was provided for a transaction will be consumed.

The `throw` keyword is the first keyword to throw an error in solidity, and it was deprecated instantly in version 0.5. The latest editions (0.5) use the `revert` keyword. With the `revert` keyword, we can specify an explanation for the error; revert ('this is the reason for the error'). Also, if we want to do a conditional revert, we can do something as shown in Fig. 14.74.

```
if(Condition) {
        revert(Reason);
}
```

Figure 14.74: `revert` function

For example

```
if (a > 20) {
revert('Value of a is greater than 20');
}
```

If a particular value meets the given condition (a >20 here), then the `revert` statement is executed. A shorter way to do this is to use the `require` keyword. `Require` takes two arguments. First, it tests for a condition and next it gives the error message to display (refer Fig. 14.75).

```
require(condition, 'Something bad happened');

For example:

require(a > 20, 'Value of a is greater than 20');
```

Figure 14.75: `require` function.

If the value of a is greater than 20, then trigger an error. The `assert` statement uses a condition (refer Fig. 14.76) as a parameter. When an `assert ()` statement fails, it means something very wrong and unexpected has happened, and we need to fix our code.

```
assert(condition);

For example:

assert(a > 20);
```

Figure 14.76: `assert` function.

Let us write another example to test all three functions (refer Fig. 14.77).

```
pragma solidity ^0.5.1;
            contract ErrorTest {
            function TestRevert() pure external {
                        revert('Revert error statement');
            }
            function TestRequire() pure external {
                        require(true == false, 'Require error statement');
            }
            function TestAssert() pure external {
                        assert(true == false);
            }
}
```

Figure 14.77 `ErrorTest` contract program

When we click `TestRequire ()`, it displays the below error message (refer Fig. 14.78).

Figure 14.78: `ErrorTest` contract program: `TestRequire ()` function output

When we click `TestRevert()`, it displays the error message shown in Fig. 14.79.

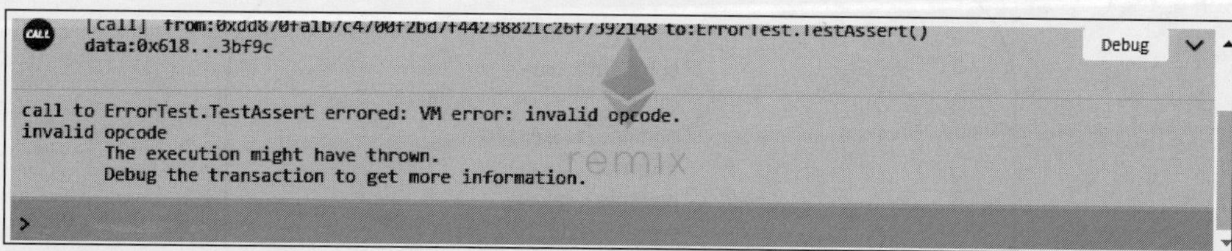

Figure 14.79: `ErrorTest` contract program: `TestRevert ()` function output

When we click `TestAssert()`, it displays the below error message shown in Fig. 14.80.

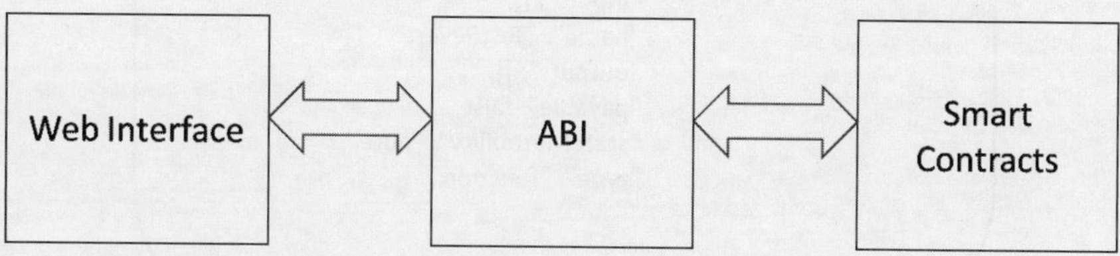

Figure 14.80: `ErrorTest` contract program: `TestAssert ()` function output

14.18 Application Binary Interface (ABI)

ABI is an interface between web applications and smart contracts. In other words, ABI can be used to connect a web application with the contract source code (byte code) that was deployed into the blockchain (refer Fig. 14.81). ABI is needed to connect to the byte code of the smart contract. It defines which functions can be invoked (called) and guarantees that the function will return data in the expected format.

Web Interface ⟺ ABI ⟺ Smart Contracts

Figure 14.81: ABI interface with smart contracts

For example, let us consider the smart contract written in Remix Browser Editor (refer Fig. 14.82).

```
pragma solidity ^0.5.1;
        contract ErrorTest {
        function TestRevert() pure external {
                revert('Revert error statement');
        }
        function TestRequire() pure external {
                require(true == false, 'Require error statement');
        }
        function TestAssert() pure external {
                assert(true == false);
        }
}
```

Figure 14.82: ABI interface example: Smart contracts

Let us try to create ABI for the above smart contract. ABI has only information about functions and events (refer Fig. 14.83). It has no information about state variables. If a web application needs to interact (connect) with a smart contract, it just needs ABI and the corresponding contract address.

```
[
        {
                "constant": true,
                "inputs": [],
                "name": "TestAssert",
                "outputs": [],
                "payable": false,
                "stateMutability": "pure",
                "type": "function"
        },
        {
                "constant": true,
                "inputs": [],
                "name": "TestRequire",
                "outputs": [],
                "payable": false,
                "stateMutability": "pure",
                "type": "function"
        },
        {
                "constant": true,
                "inputs": [],
                "name": "TestRevert",
                "outputs": [],
                "payable": false,
                "stateMutability": "pure",
                "type": "function"
        }
]
```

Figure 14.83: ABI interface example: Smart contracts' ABI code

Let us compare ABI function with the smart contract function to understand the structure of ABI. Let us take the function named "TestAssert" and its corresponding solidity code (refer Fig. 14.84).

Solidity code of "TestAssert":

```
function TestAssert() pure external {
        assert(true == false);
}
```

ABI of the function "TestAssert":

```
{
        "constant": true,
        "inputs": [],
        "name": "TestAssert",
        "outputs": [],
        "payable": false,
        "stateMutability": "pure",
        "type": "function"
}
```

Figure 14.84: ABI interface example: `TestAssert()` function

In the type (function) "ABI" indicates it specifies the details of a function (of solidity code). Function name (in solidity code) is referred to as name in ABI. The constant = `true` indicates that this function returns a value (error in this case). The parameter `"payable": false` indicates that this function is not payable. The parameter `"inputs": []` indicates that this function has no inputs. The parameter `"outputs": []` indicates that this function has no outputs. The parameter `"stateMutability": "pure"` indicates that the function is pure.

14.19 Swarm (Decentralized Storage Platform)

Swarm is a decentralized storage platform and content distribution service for Ethereum. It is designed to serve as a decentralized and redundant store of Ethereum's public record and also to store and distribute DApp code. Swarm is a peer-to-peer storage platform, which is maintained by the peers who contribute to the storage and bandwidth resources. Being a peer-to-peer system, the swarm has no single point-of-failure and is resistant to false and distributed denial-of-service or DDoS attacks. Swarm is also censorship-resistant, as any central authority does not control it. Swarm has a built-in incentive system for peers who pool in their storage and bandwidth resources.

14.20 Whisper (Decentralized Messaging Platform)

Whisper is a communication protocol that allows decentralized applications or DApps to communicate with each other. With a whisper, DApps can publish messages to each other. Whisper messages are transient in nature and have their time to live (TTL) set. Each message has one or more topics associated with it. The DApps running on a node inform the node about the topics to which they want to subscribe. Whisper uses topic-based routing where the nodes advertise their topics of interest to their peers.

Summary

Ethereum is a cryptocurrency, which runs in a decentralized virtual machine (EVM). It allows transferring value in the form of ether back and forth to different wallets or addresses on the blockchain. It allows the creation of smart contracts. Smart contracts are software codes deployed into the blockchain, and it also stores states.

The solidity documentation by the Ethereum team, which breaks down the entire language and how to create smart contracts, can be accessed at the web site https://solidity.readthedocs.io/en/develop

There are several different Integrated Development Environments (IDEs), frameworks, and environments to create smart contracts. The basic steps are to first write the solidity code, compile it, and then deploy it. We can use the browser IDE called a remix. We can write code, deploy it and rewrite the content without the risk of physical money.

More information about solidity can be gained by visiting
https://remix.ethereum.org and https://ethereum.github.io/browser-solidity

The first statement in a solidity file is the `pragma` statement, which instructs the compiler to compile with a certain version of solidity, denoted as "pragma solidity ^0.5.1;"

In solidity, everything is a contract, similar to a class (in Java). The contract is defined after the `pragma` statement.

Variables' names are defined at the beginning of the contract immediately after the open braces. They are easy to create in contracts and just take a type.

After defining the variable, we define a function. A function is usually written to manipulate the state variable(s) or to get the value of the state variable(s).

Solidity also has a constructor. As you may already know, a constructor is just a function that is called at the very beginning of the instantiation of the class (in Java). In solidity, a constructor is a function named "constructor." The only time this constructor is called is when we first deploy this contract to the blockchain and is never called again.

Step 1: First, ping the URL: https://remix.ethereum.org.

Step 2: Change the environment to "JavaScript VM."

Step 3: Create a new solidity file.

Step 4: Write the solidity code.

Step 5: Compile the code.

Step 6: Run the code.

Step 7: Click on the deployed contract.

Step 8: Give the relevant inputs.

Step 9: Check the outputs.

Every Ethereum contract that is published is part of a transaction on the blockchain. The contract also receives an Ethereum address. We can send money (ether) to the contract. State variables are strongly typed.

We create getters for the variables by making the variables public. The Ethereum virtual machine will create this on its own. We can run them as if they were retrieval functions on the contract.

The Ethereum contract and state work one after the other on the blockchain, which is a ledger of transactions. The Ethereum virtual machine returns the last known value of that variable (state), as it exists on the blockchain ledger.

The address is a very special type in the Ethereum virtual machine and is a memory location that stores Ethereum addresses.

Because creating a contract or deploying the contract is a transaction on the blockchain, it is sent from an Ethereum address. Every transaction on the blockchain has a corresponding message (msg). The message is like a container object, with a whole bunch of metadata.

A modifier in solidity is a logic test.

An event in solidity is a way to call the Ethereum virtual machine's logging functionality to write a log entry to the current block as part of a transaction. These logs are useful for decentralized applications. The Ethereum JavaScript SDK can listen to the Ethereum blockchain for events; so these events can be tied to JavaScript callback functions to run any kind of logic on the User Interface or another piece of your application.

Arrays are a collection of similar kinds of data types. In solidity, there are two kinds of arrays. The first one is the storage array. These arrays are stored inside the blockchain. After executing the function, the array will stay inside the memory of the blockchain. The second type of array is the memory array. The memory array is just a temporary array; it exists while executing a function, but it disappears after that.

The `push` command is used to push an array element at the top of the array. The push is the function name that makes the array public. The statement `delete` will not modify the length of the array. It just resets the value to its default type.

Function visibility in solidity can be private, internal, public, and external. Visibility keywords allow defining who can have access to a particular function. The most restrictive visibility keyword is called `private`. When a `private` keyword is used, we can only call the function from inside the smart contract. As a convention, we prefix the name of the function by an underscore to indicate a private function.

With the `internal` keyword, the function is still limited to be called from within a smart contract, but it can also be called from a smart contract that inherits from this smart contract. With an `external` keyword, we can only call the function from outside a smart contract. We do not prefix an external function by an underscore. In the case of the `public` keyword, we can call the function not only from outside the smart contract but also from inside the smart contract. Therefore, this is the most permissive function for visibility.

A private variable can only be read from within the same smart contract. A variable that is stored in the Ethereum blockchain cannot be private since it is a public blockchain. When public keyword is used, we can read the variable not only from within an inherited smart contract but also from outside of smart contracts. The `public` keyword gives the most permissive kind of access to the solidity variable.

A constant is a variable that cannot be changed once it is initialized.

To send ether to a smart contract, we need to have a function with the `payable` keyword. The `payable` keyword specifies that the function could accept ether. Without this keyword, the function will reject the incoming transfer.

Solidity allows us to create a function within a contract that has no name whatsoever. This function, known as fallback function, is called if we try to call a function by a name that does not match anything else in the contract. If we call a function that does not exist, then the fallback function is triggered.

Ethereum virtual machine (EVM) allows us to create contracts that can communicate with other contracts. It allows us to create complex decentralized applications. Using EVM, we can have one contract that handles all the states for our application and another contract that handles the logic. Both of those contracts can communicate with each other through their various publicly accessible functions.

One way to use the same code in different contracts is to copy and paste the code. Another efficient way is to create a library that can be imported into a new contract and reused. A library in solidity is very similar to that of a contract. The name of the file ends in .sol. Instead of defining a contract, we use the `library` keyword. Libraries are very useful if we want to reuse codes.

In ERC20 token transfer, the 20 is a standard that describes how we could implement the token on the Ethereum blockchain. With 20 tokens, we can create our coin on the Ethereum blockchain. Most tokens represent some form of value, and a lot of them are traded in decentralized exchanges.

An error can happen either because it is triggered (due to failure in business logic) with one of the keywords or because solidity does not allow certain things like accessing an array, out of bound in an array, etc. Therefore, when this happens, the execution of the function is stopped and other instructions that are written after this statement will never be executed.

ABI is an interface between web applications and the smart contract. In other words, using ABI, a web application can be connected with the contract source code (byte code) that was deployed into the blockchain.

Swarm is a decentralized storage platform and content distribution service for Ethereum. Swarm has been designed to serve as a decentralized and redundant store of ethereum's public record and also to store and distribute DApps code.

Whisper is a communication protocol that allows the decentralized applications or DApps to communicate with each other. With a whisper, DApps can publish messages to each other.

EXERCISES

Practical Questions (Using Ethereum Solidity)

1. Program to print "Hello world".
2. Program to calculate the average of a given set of 3 numbers (Get the numbers as inputs).
3. Program to swap numbers without using a temporary variable (Get the numbers as inputs).
4. Program to check if a number is an Armstrong number.
5. Program to find LCM of given two numbers.
6. Write an array program to implement stack functionalities (Last-in–First-out).
7. Write a calculator program for adding and subtracting two given numbers. Call this program from another contract. Write only single Solidity program.
8. Write a calculator program for adding and subtracting two given numbers. Call this program from another contract. Create and use Library.
9. Write a "CrowdSource" program to collect 2 ethers from Accounts until it reaches 20 ethers. Once it reaches 10, ethers transfer the fund to the owner.
10. Write a solidity program to transfer ERC20 token.

CHAPTER 15

Blockchain Platform using Python

LEARNING OBJECTIVES

Python is a language that has a gentle learning curve that makes it easier for blockchain developers to master it quite effortlessly. It is simple in that it does away with a lot of white spaces, codes and keywords making it popular among developers. It is a secure, performant and scalable language making it ideal for blockchain implementation.

15.1 Introduction

Python is an easy-to-learn and powerful programming language. It has high-level data structures and effective approach to object-oriented programming. Python is a perfect language for scripting and rapid application development (RAD) in many areas on most platforms including Blockchain. Python has free packages for writing blockchain programs.

15.2 Learn How to Use Python Online Editor

Python online editor is available at the website:
https://www.jdoodle.com/python3-programming-online/.

It is default loaded with sample programs, which can be executed easily by clicking the "Execute" button. Alternatively, Python language can also be downloaded into the local machine by clicking the download button available in the home page of python (https://www.python.org/). Python documentation is available in the website of python for easy reference (https://docs.python.org/3/).

15.3 Basic Programming Using Python

15.3.1 Introduction to Basic Python Commands

First, install the following basic libraries of Python:

- "os" for using operating system dependent functions
- "pandas" for extensive data manipulation functionalities

Anaconda Python, by default, contains all the basic packages required to start programming. However, in case, any extra package is needed, 'pip' is the recommended installer.

Command: `pip install`

Syntax: `pip install 'SomePackage'`

After installing, libraries need to be loaded by invoking the command:

Command: import <<library-name>>

Syntax: `import os`

Try out each of the following commands.

Sr #	Command	Purpose	Sample code with output
1	os.getcwd()	Getting the current working directory	>>> os.getcwd() C:\Users\book
2	os.chdir()	Setting the current working directory	>>> os.chdir('C:\\ Python programs')
3	os.listdir()	See directory content	>>> os.listdir('C:\\ Python programs') Out[16]: ['Blockchain_ Python.py']
4	python '<<filename>>'	Compile source file for execution	>>> python new2. py
5	print()	Command for basic user output	>>> print("Hello") Hello

(*Continued*)

Sr #	Command	Purpose	Sample code with output

6	input ()	Command for basic user input	>>> a = input("Give input: ")
			Give input: 15
			>>> print("Your input: ", a)
			Your input: 15

7	type()	Gives the type of an object	>>> x = 25
			>>> type(x)
			Out[28]: int

8	help(<<keyword>>)	Access help related to some function.	>>> help(print) >>> help(os.listdir)

15.3.2 Basic Data Types in Python

Python has five standard data types:

- Number
- String
- List
- Tuple
- Dictionary.

Though Python has all the basic data types, Python variables do not need explicit declaration to reserve memory space. The declaration happens automatically when you assign a value to a variable. The equal sign (=) is used to assign values to variables. The operand to the left of the = operator is the name of the variable and the operand to the right of the = operator is the value stored in the variable.

For example:

```
>>> var1 = 24
>>> var2 = 28.72
>>> var3 = "Blockchain"

>>> type(var1)
Out[38]: int
>>> print(var1)
24
```

```
>>> type(var2)
Out[39]: float
>>> print(var2)
28.72

>>> type(var3)
Out[40]: str
>>> print(var3)
Blockchain
```

Python variables can be easily type-casted from one data type to another by using its data type name as a function (actually a wrapper class) and passing the variable to be type-casted as a parameter. However, not all data types can be transformed into another. For example, converting a String to Integer/Number is impossible (That is obvious!).

```
>>> var2 = int(var2)
>>> print (var2)
28
>>> type (var2)
Out[42]: int
```

15.3.3 `for-while` Loops and `if-else` Statement

For loop

Syntax

```
for i in range(<lowerbound>,<upperbound>,<interval>):
    print(i*i)
```

Example: Printing squares of all integers from 1 to 4.

```
for i in range(1,5):
    print(i*i)
1
4
9
16

for i in range(1,20,5):
    print(i)
1
```

```
6
11
16
```

While loop

Syntax

```
while < condition>:
    <do something>
```

Example: Print cubes of all integers from 1 to 3.

```
i = 1
while i < 4:
    print(i*i*i)
    i = i + 1

1
8
27
```

`if-else` statement

Syntax

```
if < condition>:
    <do something>
else:
   <do something else>
```

Example: Print the squares of a number if it is a negative number. If it is a positive number, then print that number.

```
i = -5
if i < 0:
    print(i*i)
else:
    print(i)

25
```

15.3.4 Writing Functions

Writing a function (in a script):

Syntax

```
def functionname(parametername,...):
(function_body)
```

Example: Function to calculate factorial of an input number n.

```
def factorial(n):
    fact = 1
    for i in range(1,n+1):
        fact = fact*i
    return(fact);

k = factorial (4)
print (k)

24
```

15.3.5 Mathematical Operations on Data Types

Vectors

```
>>> n = 20
>>> m = 15
>>> n + m    #addition
Out[53]: 35

>>> n - m  #subtraction
Out[55]: 5

>>> n * m    #multiplication
Out[56]: 300

>>> n / m    #division
Out[57]: 1.333333
```

15.4 Python Packages for Blockchain

The high-level language Python may be utilized to create Blockchain software. Python language can interact with the public ledger API of Blockchain using 'blockchain.

info',which is a bundle offered in GitHub. The commands given below are used for installing Python Blockchain from GitHub.

```
$ git clone https://github.com/blockchain/api-v1-client-python
$ cd api-v1-client-python
$ python setup.py install
```

15.4.1 Blockchain Package Modules

This package consists of seven modules, as below:

Module Names

1. blockexplorer
2. create wallet
3. exchange rates
4. pushtx
5. v2.receive
6. Statistics
7. Wallet

A module can be imported into Python using the below commands:

```
from blockchain import blockexplorer
from blockchain import createwallet
from blockchain import exchangerates
from blockchain import pushtx
from blockchain import v2.receive
from blockchain import statistics
from blockchain import wallet
```

15.4.2 Blockchain Block Explorer Module

This module is an important package, which has ten functions for operating the blockchain.

Function Name

1. get_block(Str - block Hash)
2. get_tx(Str- Transaction Hash)
3. get_address(Str- Block Address)
4. get_block_height(Block number)
5. get_xpub()
6. get_multi_address()
7. get_unspent_outputs()
8. get_latest_block()
9. get_unconfirmed_tx()
10. get_blocks()

1. Blockexplorer Module

15.4.3 Create Wallet Module

This module helps to create wallet based on multiple parameters.

createWallet Module

password	:	str	password for the new wallet. At least 10 characters.
api_code	:	str	API code with the create wallets permission
service_url	:	str	URL to an instance of service-my-wallet-v3 (with trailing slash)
priv	:	str	private key to add to the wallet (optional)
label	:	str	label for the first address in the wallet (optional)
email	:	str	email to associate with the new wallet (optional)

example:-

```
wallet = createwallet.create_wallet('675password', '32ts91klsop', 'http://localhost:3002/', label = 'demo wallet')
```

15.4.4 Exchange Module

This module helps to get the ticket and bitcoin exchange values.

exchangerate Module

```
ticker = exchangerates.get_ticker()
btc_amount = exchangerates.to_btc('INR', 400000)
```

15.4.5 Pushtx Module

This module helps in broadcasting the transaction.

pushtx Module

```
tx :        str   (hex encoded)
api_code :  str   (optional)
example:-
pushtx.pushtx('10sdh47eb239rh39sk4jdudg534kfk565msdfu89fgfhtu67856y47efbhdfn4u3487hcb')
```

15.4.6 V2.Receive Module

This module has two functions, one to receive automatic bitcoin payments and the second as a call back function.

V2.receive Module: receive Function

```
xpub      :   str   the xpub address
callback  :   str   call back address
api_code :    str   your api key
example:-
resp = receive.receive('1sdyh35sdgsdf6fgrt535gtrfg423gbcd2A', 'http://my.demo.com?invoice=634125', 'your api key'
```

V2.receive Module: callback Function

```
callback  : str
api_code : str
example:-
logs = receive.callback_log('http://my.demo.com?invoice_id=634125', 'your api key')
```

15.4.7 Statistics Module

This module is used to get network statistics.

statistics Module

api_code : str (optional)

example:- stats= statistics.get()

15.4.8 Wallet Module

Wallet module is used to send money to wallets.

Wallet Module

Identifier	: str	
password	: str	
service_url	: str	URL to an instance of service-my-wallet-v3 (with trailing slash)
second_password	: str	(optional)
api_code	: str	(optional)

constructor example;

wallet = Wallet('7i2gh5dv-5t1n-54j1-nges-947243m4zx12', 'password987', 'http://localhost:3001/')

Wallet Module: send Function

to	: str	receiving address
Amount	: int	amount to send (in satoshi)
from_address	: str	specific address to send from (optional)
fee	: int	transaction fee in satoshi. Must be greater than default (optional)
Note	: str	public note to include with the transaction if amount >= 0.005 BTC

example:-

payment = wallet.send('9ght53sdvmi98r6dxcvnjjtd45hmmi75erdcbj876', 5000000, from_address='5jkhd378kjhtd57nh55cvmo73svl98gce4090mnv')

15.4.9 Blockchain Exercise

Write a python program to check all the above functions.

Summary

This chapter summarises the basic commands of Python language and presents the commands for installing blockchain package of Python from Github.

EXERCISES

Practical Questions (Using Python)

1. Program to print "Hello world".
2. Program to calculate the average value of a given set of 3 numbers (get the numbers as inputs).
3. Program to swap numbers without using a temporary variable (get the numbers as inputs).
4. Program to check if a number is an Armstrong number.
5. Program to find LCM of given two numbers.
6. Write an array program to implement Stack functionalities (Last-in–First-out).
7. Write a python program for blockchain using Github package.

CHAPTER 16

Blockchain Platform using Hyperledger Fabric

LEARNING OBJECTIVES

Hyperledger fabric is intended to serve as a foundation for developing applications and solutions in blockchain. Its modular design has been found to be acceptable among several industry use cases. Because of its interoperability and ability to handle cross-chain and cross-application data flow, hyperledger has, of late, attracted significant attention. In this chapter, we shall examine the components and codes of Hyperledger before discussing its applications in the blockchain platform.

16.1 Introduction

Linux Foundation declared the creation of the Hyperledger Project (open source) in December 2015. In early 2016, Hyperledger accepted code from Digital Assets, Blockstream, and IBM; this has evolved into an enterprise-grade solution: Hyperledger Fabric.

16.2 Components of Hyperledger Fabric Network

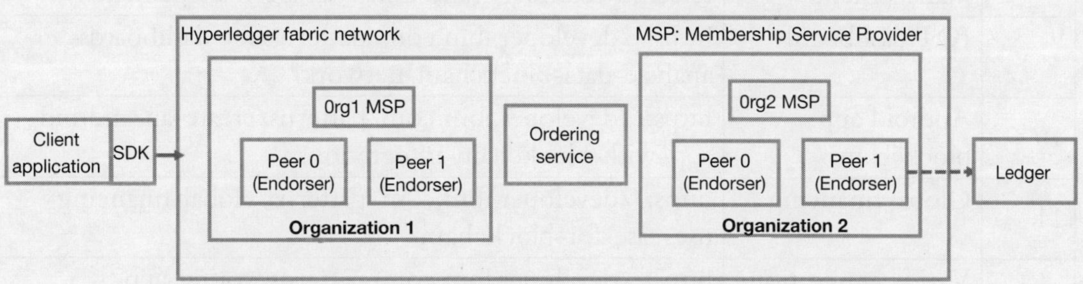

Figure 16.1: Hyperledger fabric network components

Figure 16.1 depicts the hyperledger fabric framework along with its components. It shows two organizations (Org1 and Org2). Within organization, there are peers (computer nodes). Peers can act as a Certifying Agent (CA) depending on the configuration. The chaincode written in Hyperledger Fabric Code (usually `Golang`) is deployed into the Hyperledger Fabric network. The chaincode deployed can be assessed using a Client Application Software Development Kit (SDK). All the nodes (peers) will have the copy of the ledger (blockchain).

16. 3 Chaincodes from Developer.ibm.com

IBM has shared codes for hyperledger fabric, blockchain application in its website, which can be reused/modified/deployed.

Table 16.1 Developer.ibm.com link for reference

S. no	Project name	Link for reference
1	Fabric Java SDK	https://developer.ibm.com/patterns/
2	Interacting with a blockchain	https://developer.ibm.com/patterns/interacting-with-a-blockchain-network/
3	Asset transfer application	https://developer.ibm.com/patterns/deploy-an-asset-transfer-app-using-blockchain/
4	Fitness club rewards points	https://developer.ibm.com/patterns/fitness-club-rewards-points-iot-and-retail-integration/
5	Car auction network	https://developer.ibm.com/patterns/car-auction-network-hyperledger-fabric-node-sdk-starter-plan/
6	IOT asset tracking application	https://developer.ibm.com/patterns/develop-an-iot-asset-tracking-app-using-blockchain/
7	Securing art	https://developer.ibm.com/patterns/securing-art-using-blockchain-digital-certificates/
8	Telecom fraud management	https://developer.ibm.com/patterns/blockchain-for-telecom-roaming-fraud-and-overage-management/
9	IOT dashboard	https://developer.ibm.com/patterns/iot-dashboards-analyze-data-blockchain-network/
10	Android app blockchain	https://developer.ibm.com/patterns/create-an-android-app-with-blockchain-integration/
11	Global financing	https://developer.ibm.com/patterns/global-financing-use-case-for-blockchain/
12	Voting application	https://developer.ibm.com/patterns/voting-app-hyperledger-ethereum/
13	Loyalty points fabric EVM	https://developer.ibm.com/patterns/loyalty-points-fabric-evm/

Table 16.1 lists the various applications (with free codes) available in developer.ibm.com website. Links to download the corresponding chaincodes (GitHub) are also available in the above URLs along with a demo link (YouTube) for the given program. The URLs also provide a quick summary, description (about the program), flow and instructions to use the chaincode program

16.4 Blockchain Application Using Fabric Java SDK

Now, let us discuss/deploy/run the first Project titled "Blockchain Application using Fabric Java SDK". More details about this project are available in the link: https://developer.ibm.com/patterns. The source code of the program is available in the link: https://github.com/IBM/blockchain-application-using-fabric-java-sdk. Readers are encouraged to visit the website and read the README.md file available. Please refer to the second paragraph of the README file: this paragraph explains that though the blockchain network is a purely back-end system in the real world, the information in the network is to be shared with users of network. The SDK has been provided by IBM to achieve this. The code provided by IBM uses a Java-based SDK, which will connect to the blockchain network (written in `golang`). The Java SDK offers various features for Java developers. It is recognized that many programmers in the industry use Java as their primary skill and many enterprise applications are built in Java; hence, it is a natural choice to use Java sdk for fabric. Java SDK for fabric offers API that helps manage the lifecycle of hyperledger channels and chaincodes. The pre-requisites and steps are given in the website as below (refer Fig. 16.2).

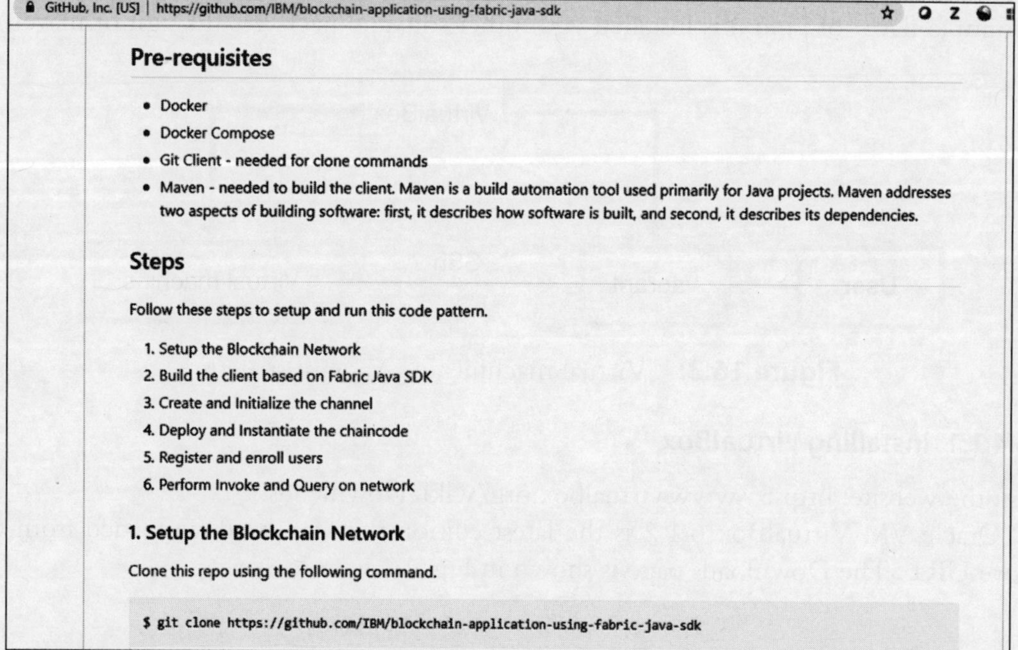

Figure 16.2: Prerequisites and the steps for fabric Java SDK chaincode

More Pre-requisites

For this exercise, the user needs to have a PC with following characteristics:

- Windows operating 64-bit OS. (Windows 8 or above is advised).
- RAM 8 GB
- Internet connection (Broadband internet is advised).

High-level Objectives

The high-level objective of this exercise is to get familiarized with the hyperledger fabric. It helps to

- experience the hyperledger channel.
- experience the hyperledger chaincode
- query hyperledger for users and assets.

16.4.1 Creating Hyperledger Environment on Windows

Hyperledger fabric code works with Linux environment. VirtualBox (software) allows installation of Multiple Operating Systems (OS) within the same box. In this section, we are going to install a VirtualBox on Windows and install a Linux flavor. VirtualBox is a mechanism to handle virtualization and enterprises solutions such that users can continue using Windows OS and Linux OS and the underlying applications. A tool named Vagrant can be used for managing virtual machines environment (in our case VirtualBox) using a single workflow mechanism (refer Fig. 16.3). Vagrant allows setting up development environment, and allows consistent environment using Vagrant file, across multiple environments. For our project, Vagrant provides an environment where Ubuntu (a flavor of Linux) is installed with other software packages (like git or maven).

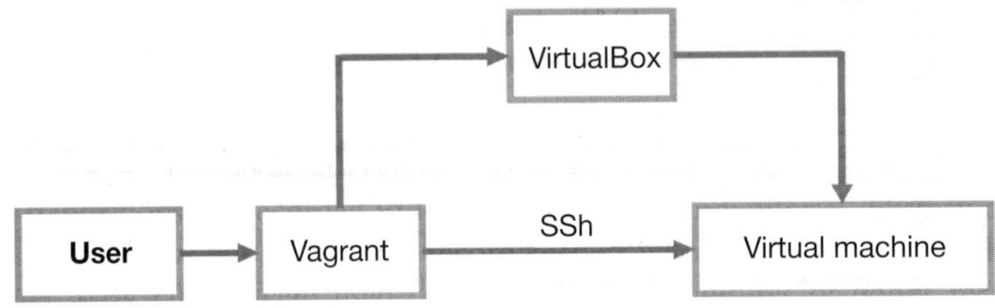

Figure 16.3: Virtual machine and Vagrant structure

16.4.1.1 Installing VirtualBox

Visit the website: https://www.virtualbox.org/wiki/Downloads.

Oracle VM VirtualBox 6.1.2 is the latest edition that can be downloaded from the above URL. The Downloads page is shown in Fig. 16.4.

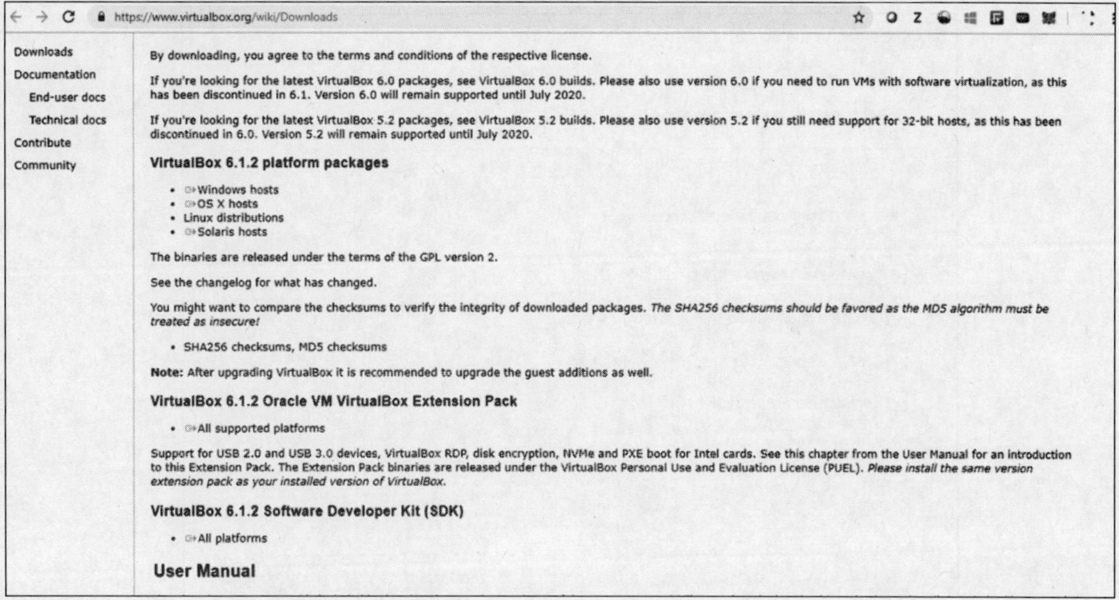

Figure 16.4: VirtualBox website home page

After downloading the file, click on EXE to install "Oracle VM VirtualBox 6.1.2". (refer Fig. 16.5).

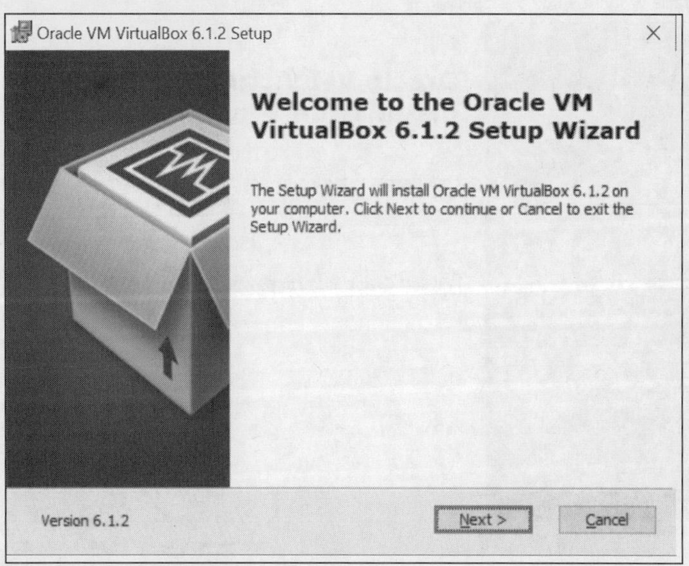

Figure 16.5: VirtualBox installation: Step 1

While installing the file, it will ask to select the features to be installed. Select all the figures. (refer Fig. 16.6). Click "Next" and "Yes".

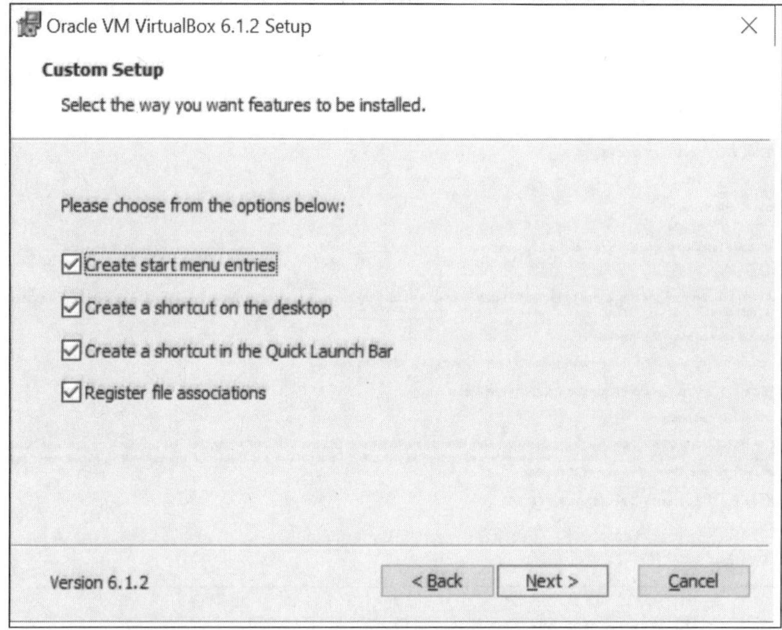

Figure 16.6: VirtualBox installation: Step 2

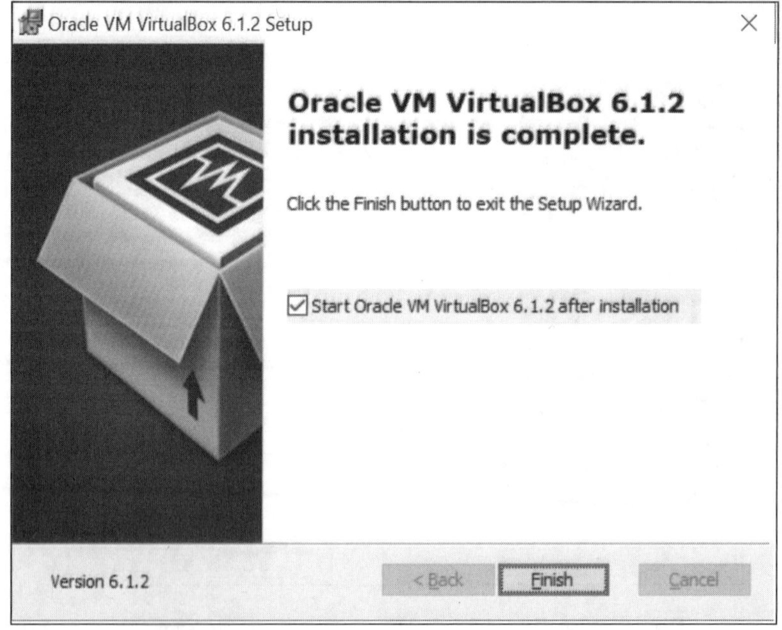

Figure 16.7: VirtualBox installation: Step 3

After installing the complete package, click "Finish" button (refer Fig. 16.7). After the installation, you can see the VirtualBox icon on your desktop.

When you click "VirtualBox" icon on the desktop, the screen will appear as in Fig. 16.8.

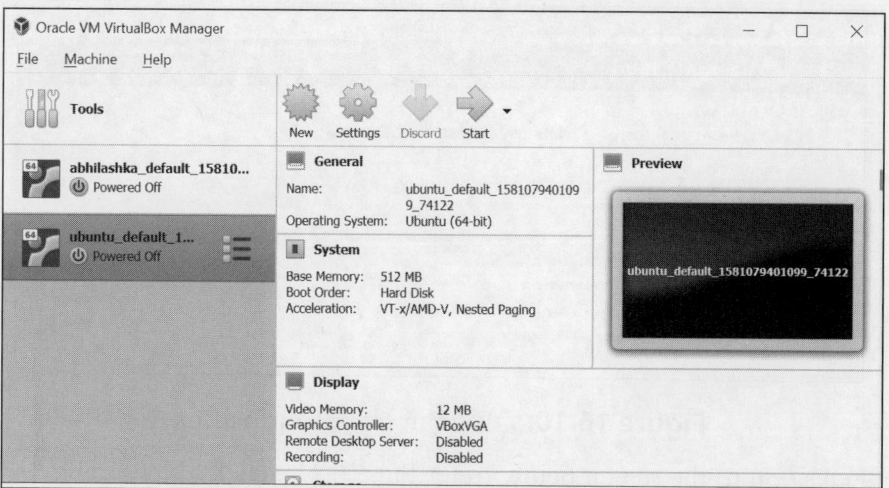

Figure 16.8: VirtualBox installation: Step 4

16.4.1.2 Installing Vagrant

Visit the website of Vagrant (https://www.vagrantup.com/). Click "Download 2.2.7" button (refer Fig. 16.9).

Figure 16.9: Vagrant home page screen

From the Vagrant page, select 64-bit Windows. From the downloaded installer, install Vagrant to the machine. Go to command prompt and create a folder called vagrantUbuntu1804 (you could create a folder as per your wish), the folder is named vagrantUbuntu1804 to clearly distinguish between other flavors of Linux in Vagrant.

Again, go to home page of Vagrant (refer Fig. 16.9). Click on the button "Find boxes". We are now installing Ubuntu Linux for 64-bit, please find box for ubuntu/ubuntu-18.04. Hit on the url against the bento/ubuntu-18.04 (refer Fig. 6.10).

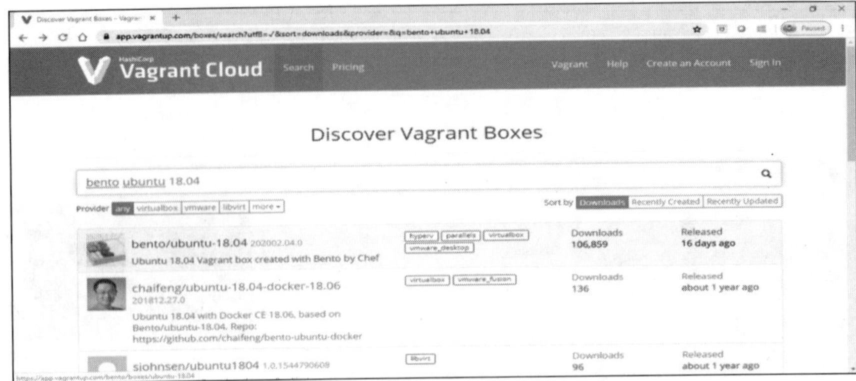

Figure 16.10: Vagrant cloud: VirtualBox

It will take you to the screen below (refer Fig. 16.11), click "New" tab.

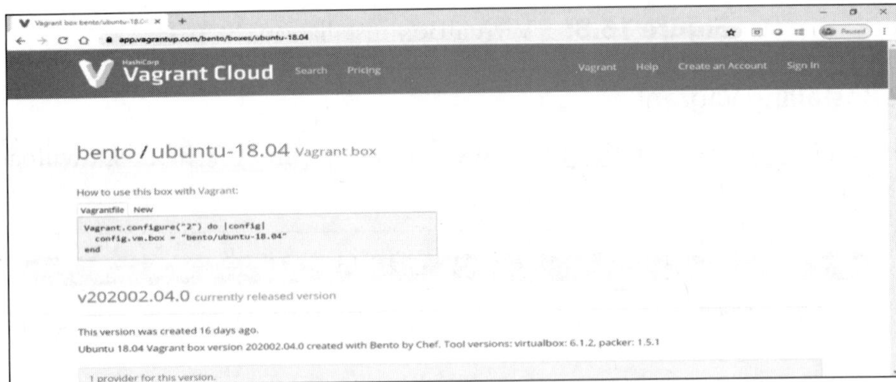

Figure 16.11: Vagrant cloud: New tab

Copy the code in the new tab (refer Fig. 16.12).

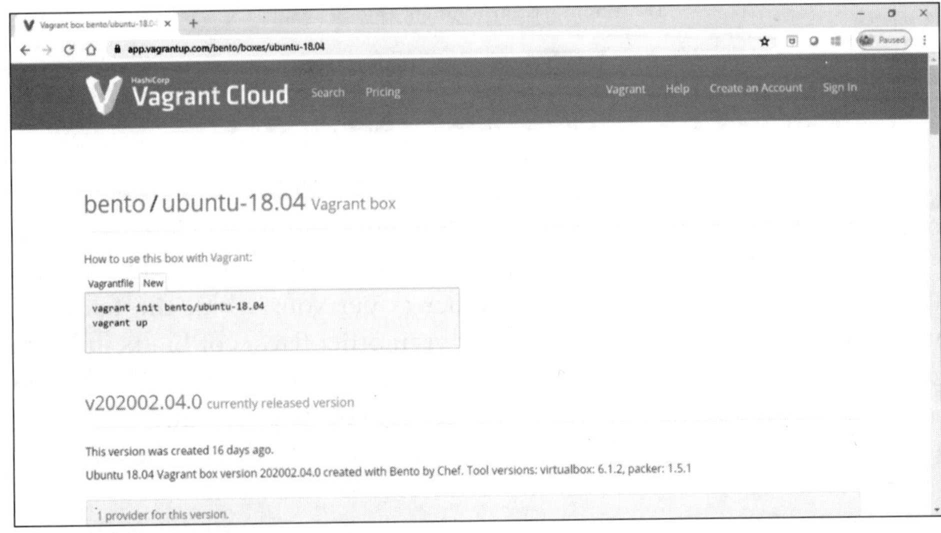

Figure 16.12: Vagrant cloud: New tab contents (commands)

Come back to command prompt on your machine. Go to the directory named "vagrantUbuntu1804". Fire the command: vagrant init Ubuntu/trusty64. Once init is done properly using the command, you could see a vagrant file in the folder. (refer Fig. 16.13). A Vagrant file is created in the directory (vagrantUbuntu1804).

```
C:\Users\abhil\vagrantUbuntu1804>dir/p
 Volume in drive C is Windows
 Volume Serial Number is 46D7-71AA

 Directory of C:\Users\abhil\vagrantUbuntu1804

11-02-2020  22:02    <DIR>          .
11-02-2020  22:02    <DIR>          ..
11-02-2020  22:02             3,095 Vagrantfile
               1 File(s)          3,095 bytes
               2 Dir(s)  863,301,681,152 bytes free
```

Figure 16.13: Vagrant file creation after the init command

Now, run Vagrant up command (refer Fig. 16.14).

```
vagrant@vagrant: ~

C:\Users\abhil\vagrantUbuntu1804>vagrant up
Bringing machine 'default' up with 'virtualbox' provider...
==> default: Box 'bento/ubuntu-18.04' could not be found. Attempting to find and install...
    default: Box Provider: virtualbox
    default: Box Version: >= 0
==> default: Loading metadata for box 'bento/ubuntu-18.04'
    default: URL: https://vagrantcloud.com/bento/ubuntu-18.04
==> default: Adding box 'bento/ubuntu-18.04' (v202002.04.0) for provider: virtualbox
    default: Downloading: https://vagrantcloud.com/bento/boxes/ubuntu-18.04/versions/202002.04.0/providers/virtualbox.bo
x
    default: Download redirected to host: vagrantcloud-files-production.s3.amazonaws.com
    default:
==> default: Successfully added box 'bento/ubuntu-18.04' (v202002.04.0) for 'virtualbox'!
==> default: Importing base box 'bento/ubuntu-18.04'...
==> default: Matching MAC address for NAT networking...
==> default: Checking if box 'bento/ubuntu-18.04' version '202002.04.0' is up to date...
==> default: Setting the name of the VM: vagrantUbuntu1804_default_1581440691351_12840
==> default: Fixed port collision for 22 => 2222. Now on port 2200.
==> default: Vagrant has detected a configuration issue which exposes a
==> default: vulnerability with the installed version of VirtualBox. The
==> default: current guest is configured to use an E1000 NIC type for a
==> default: network adapter which is vulnerable in this version of VirtualBox.
==> default: Ensure the guest is trusted to use this configuration or update
==> default: the NIC type using one of the methods below:
==> default:
==> default:   https://www.vagrantup.com/docs/virtualbox/configuration.html#default-nic-type
==> default:   https://www.vagrantup.com/docs/virtualbox/networking.html#virtualbox-nic-type
==> default: Clearing any previously set network interfaces...
==> default: Preparing network interfaces based on configuration...
```

Figure 16.14: Vagrant up command

Once Vagrant is up, we can connect to Ubuntu downloaded using the command **vagrant ssh** (refer Fig. 16.15).

```
vagrant@vagrant: ~
    default:
    default: Guest Additions Version: 6.1.2
    default: VirtualBox Version: 5.2
==> default: Mounting shared folders...
    default: /vagrant => C:/Users/abhil/vagrantUbuntu1804

C:\Users\abhil\vagrantUbuntu1804>vagrant ssh
Welcome to Ubuntu 18.04.4 LTS (GNU/Linux 4.15.0-76-generic x86_64)

 * Documentation:  https://help.ubuntu.com
 * Management:     https://landscape.canonical.com
 * Support:        https://ubuntu.com/advantage

 System information disabled due to load higher than 1.0.

 * Multipass 1.0 is out! Get Ubuntu VMs on demand on your Linux, Windows or
   Mac. Supports cloud-init for fast, local, cloud devops simulation.

     https://multipass.run/

0 packages can be updated.
0 updates are security updates.


This system is built by the Bento project by Chef Software
More information can be found at https://github.com/chef/bento
vagrant@vagrant:~$ uname -a
Linux vagrant 4.15.0-76-generic #86-Ubuntu SMP Fri Jan 17 17:24:28 UTC 2020 x86_64 x86_64 x86_64 GNU/Linux
vagrant@vagrant:~$
```

Figure 16.15: Vagrant ssh command to connect Ubuntu

You are advised to note the version of Ubuntu from the screen; also, you could find version using the following command: lsb_release –a (refer Fig. 16.16).

```
vagrant@vagrant:~$ lsb_release -a
No LSB modules are available.
Distributor ID: Ubuntu
Description:    Ubuntu 18.04.4 LTS
Release:        18.04
Codename:       bionic
vagrant@vagrant:~$ uname -a
Linux vagrant 4.15.0-76-generic #86-Ubuntu SMP Fri Jan 17 17:24:28 UTC 2020 x86_64 x86_64 x86_64 GNU/Linux
vagrant@vagrant:~$
```

Figure 16.16: lsb_release –a command to see the version of Ubuntu

16.4.2 Blockchain Prerequisites Installation

Now that we have done the virtualization part, let us go back to the exercise of blockchain application. It has the following pre-requisites:

- Git installation
- Maven installation
- Docker installation
- Docker-compose installation

16.4.2.1 Installing Git on Environment

Give the below commands in the command prompt for installing Git on the environment. Git helps to write distributed code (like blockchain).

- sudo apt-get install git

```
vagrant@vagrant: ~
vagrant@vagrant:~$ sudo apt-get install git
Reading package lists... Done
Building dependency tree
Reading state information... Done
git is already the newest version (1:2.17.1-1ubuntu0.5).
0 upgraded, 0 newly installed, 0 to remove and 10 not upgraded.
vagrant@vagrant:~$
```

Verifying the installation of git ($ git – Version): It displays the version of the git installed.

16.4.2.2 Installing Maven on Environment

This helps to create Java-based projects. This is a time consuming process and would take a few minutes. The below command is used to install Maven.

➢ sudo apt-get install maven

```
vagrant@vagrant: ~
vagrant@vagrant:~$ vagrant@vagrant:~$ sudo apt-get install maven
Reading package lists... Done
Building dependency tree
Reading state information... Done
The following additional packages will be installed:
  ca-certificates-java default-jre-headless fontconfig-config fonts-dejavu-core java-common libaopalliance-java
  libapache-pom-java libasound2 libasound2-data libatinject-jsr330-api-java libcdi-api-java libcommons-cli-java
  libcommons-io-java libcommons-lang3-java libcommons-parent-java libfontconfig1 libgeronimo-annotation-1.3-spec-java
  libgeronimo-interceptor-3.0-spec-java libguava-java libguice-java libhawtjni-runtime-java libjansi-java
  libjansi-native-java libjpeg-turbo8 libjpeg8 libjsr305-java liblcms2-2 libmaven-parent-java libmaven-resolver-java
  libmaven-shared-utils-java libmaven3-core-java libnspr4 libnss3 libpcsclite1 libplexus-cipher-java
  libplexus-classworlds-java libplexus-component-annotations-java libplexus-interpolation-java
  libplexus-sec-dispatcher-java libplexus-utils2-java libsisu-inject-java libsisu-plexus-java libslf4j-java
  libwagon-file-java libwagon-http-shaded-java libwagon-provider-api-java libx11-6 libx11-data libxcb1 libxext6 libxi6
  libxrender1 libxtst6 openjdk-11-jre-headless x11-common
Suggested packages:
  default-jre libaopalliance-java-doc libasound2-plugins alsa-utils libatinject-jsr330-api-java-doc libservlet3.1-java
  libcommons-io-java-doc libcommons-lang3-java-doc libasm-java libcglib-java libjsr305-java-doc liblcms2-utils
  libmaven-shared-utils-java-doc liblogback-java pcscd libplexus-cipher-java-doc libplexus-classworlds-java-doc
  libplexus-interpolation-java-doc libplexus-sec-dispatcher-java-doc libplexus-utils2-java-doc junit4 testng
  libcommons-logging-java liblog4j1.2-java libnss-mdns fonts-dejavu-extra fonts-ipafont-gothic fonts-ipafont-mincho
  fonts-wqy-microhei | fonts-wqy-zenhei fonts-indic
The following NEW packages will be installed:
  ca-certificates-java default-jre-headless fontconfig-config fonts-dejavu-core java-common libaopalliance-java
  libapache-pom-java libasound2 libasound2-data libatinject-jsr330-api-java libcdi-api-java libcommons-cli-java
  libcommons-io-java libcommons-lang3-java libcommons-parent-java libfontconfig1 libgeronimo-annotation-1.3-spec-java
  libgeronimo-interceptor-3.0-spec-java libguava-java libguice-java libhawtjni-runtime-java libjansi-java
  libjansi-native-java libjpeg-turbo8 libjpeg8 libjsr305-java liblcms2-2 libmaven-parent-java libmaven-resolver-java
  libmaven-shared-utils-java libmaven3-core-java libnspr4 libnss3 libpcsclite1 libplexus-cipher-java
  libplexus-classworlds-java libplexus-component-annotations-java libplexus-interpolation-java
```

Verifying the version of maven:

➢ mvn --version

The above command will display the version of the maven, which is an indication that the program has been installed.

16.4.2.3 Install Docker on the Environment

The below command is used to install the Docker in the environment.

➢ sudo apt-get install docker.io

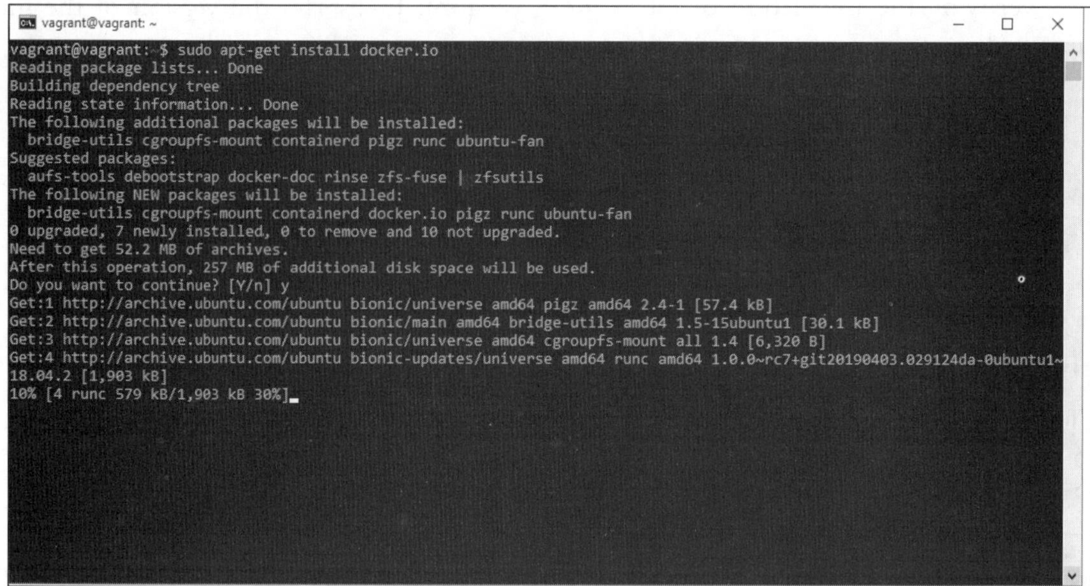

Verifying the version of Docker (This is used to check the Docker installation):

➢ docker -v

Also try the following command to verify the docker installation:

➢ docker version

```
vagrant@vagrant:~$ docker version
Client:
 Version:           18.09.7
 API version:       1.39
 Go version:        go1.10.1
 Git commit:        2d0083d
 Built:             Fri Aug 16 14:20:06 2019
 OS/Arch:           linux/amd64
 Experimental:      false

Server:
 Engine:
  Version:          18.09.7
  API version:      1.39 (minimum version 1.12)
  Go version:       go1.10.1
  Git commit:       2d0083d
  Built:            Wed Aug 14 19:41:23 2019
  OS/Arch:          linux/amd64
  Experimental:     false
vagrant@vagrant:~$
```

Please note that there are two sets of data, one for client and one for server.

16.4.2.4 Install Docker-Compose on the Environment

➢ sudo curl -L "https://github.com/docker/compose/releases/download/1.24.0/docker-compose-$(uname -s)-$(uname -m)" -o /usr/local/bin/docker-compose

```
vagrant@vagrant:~$ sudo curl -L https://github.com/docker/compose/releases/download/1.17.0/docker-compose-`uname -s`-`uname -m` -o /usr/local/bin/docker-compose
  % Total    % Received % Xferd  Average Speed   Time    Time     Time  Current
                                 Dload  Upload   Total   Spent    Left  Speed
100   617  100   617    0     0    745      0 --:--:-- --:--:-- --:--:--   745
  0     0    0     0    0     0      0      0 --:--:--  0:00:06 --:--:--     0
```

➢ sudo chmod +x /usr/local/bin/docker-compose

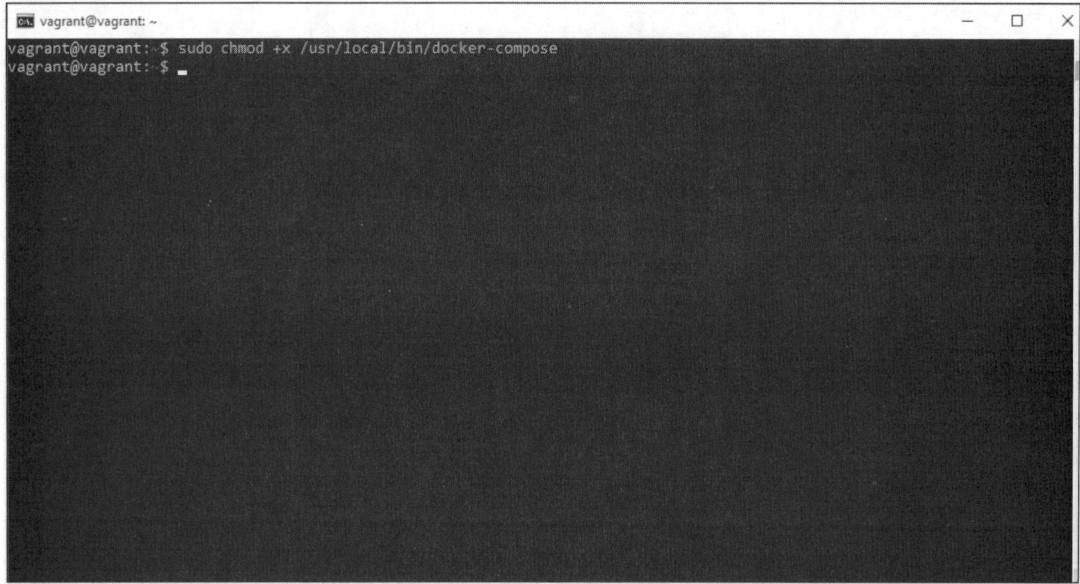

➢ docker-compose -version

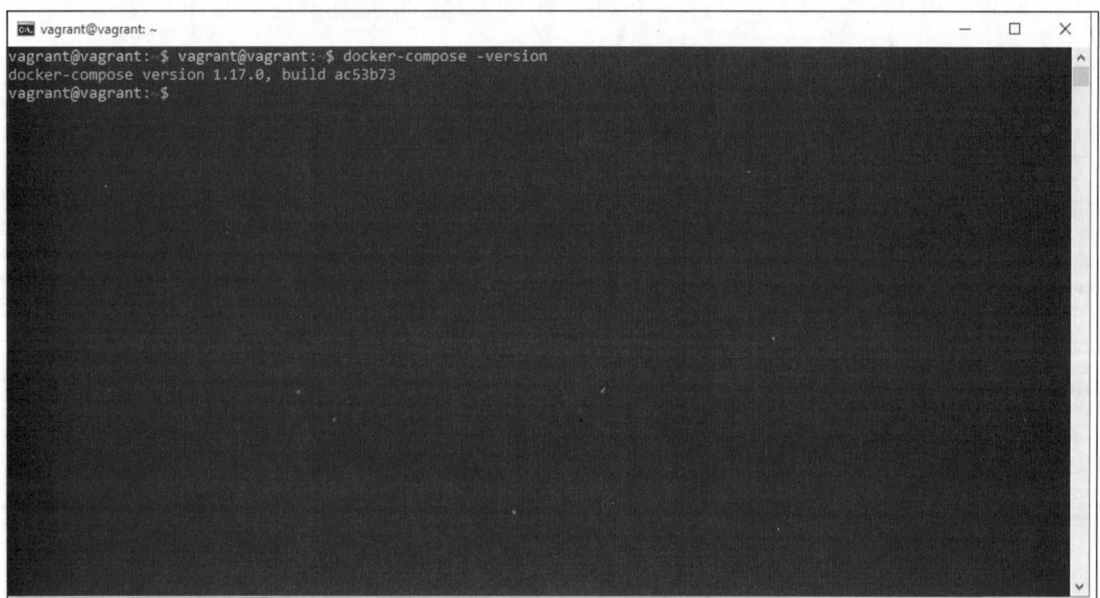

Now, we have installed all perquisites for the solution.

16.4.3 Setup the Java SDK Blockchain Network

Next, we shall set up the blockchain network using Java SDK.

16.4.3.1 Clone the Contents from Git

➢ git clone https://github.com/IBM/blockchain-application-using-fabric-java-sdk

(**Note:** This step will be repeated for different projects based on the source of the projects mentioned in Table 16.1)

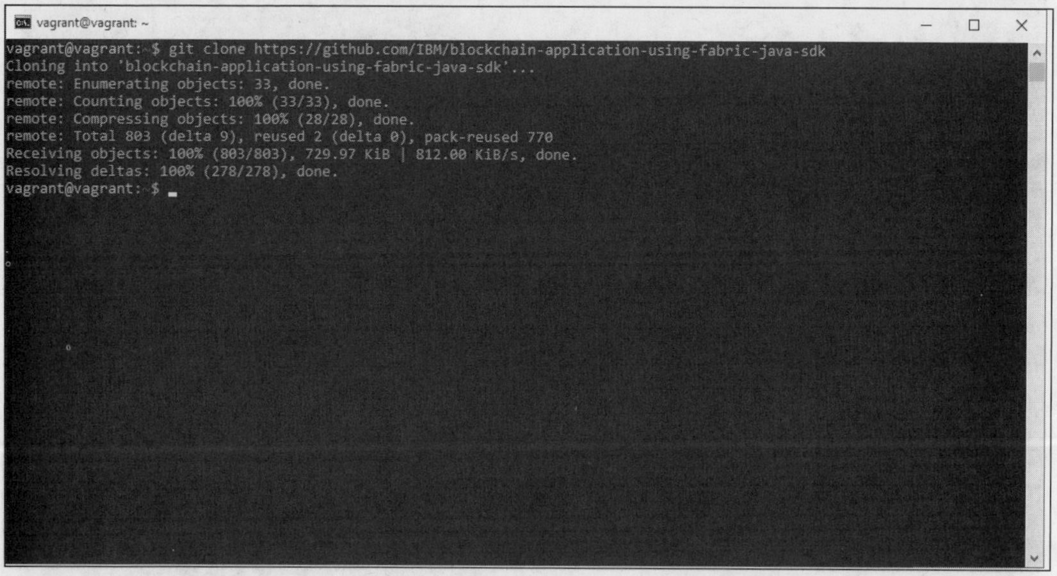

```
vagrant@vagrant: ~
vagrant@vagrant:~$ git clone https://github.com/IBM/blockchain-application-using-fabric-java-sdk
Cloning into 'blockchain-application-using-fabric-java-sdk'...
remote: Enumerating objects: 33, done.
remote: Counting objects: 100% (33/33), done.
remote: Compressing objects: 100% (28/28), done.
remote: Total 803 (delta 9), reused 2 (delta 0), pack-reused 770
Receiving objects: 100% (803/803), 729.97 KiB | 812.00 KiB/s, done.
Resolving deltas: 100% (278/278), done.
vagrant@vagrant:~$
```

Verify the contents of the clone:

➢ ls

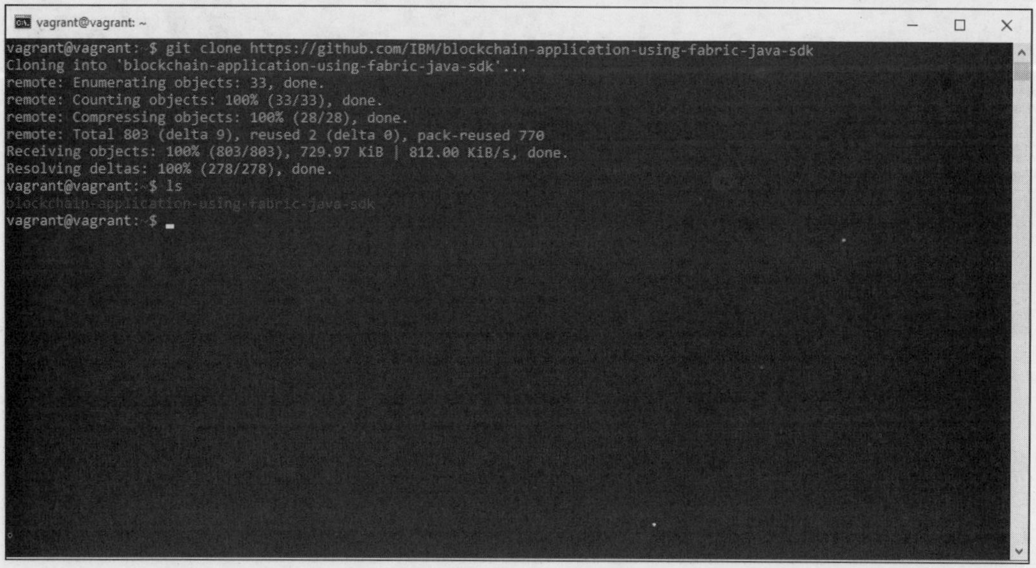

```
vagrant@vagrant: ~
vagrant@vagrant:~$ git clone https://github.com/IBM/blockchain-application-using-fabric-java-sdk
Cloning into 'blockchain-application-using-fabric-java-sdk'...
remote: Enumerating objects: 33, done.
remote: Counting objects: 100% (33/33), done.
remote: Compressing objects: 100% (28/28), done.
remote: Total 803 (delta 9), reused 2 (delta 0), pack-reused 770
Receiving objects: 100% (803/803), 729.97 KiB | 812.00 KiB/s, done.
Resolving deltas: 100% (278/278), done.
vagrant@vagrant:~$ ls
blockchain-application-using-fabric-java-sdk
vagrant@vagrant:~$
```

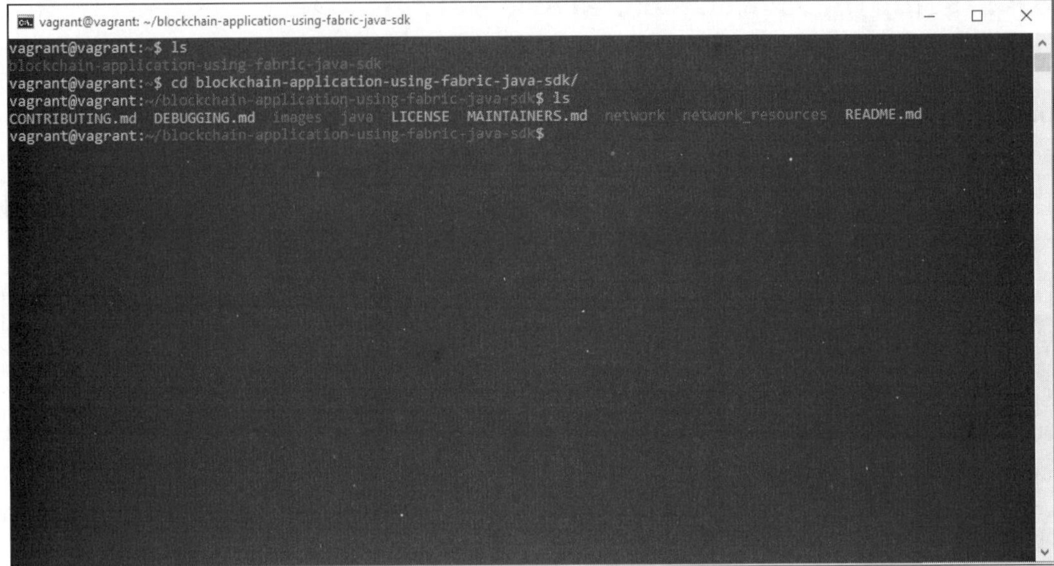

16.4.3.2 Run the Network

➢ cd network

```
vagrant@vagrant: ~/blockchain-application-using-fabric-java-sdk/network
vagrant@vagrant:~$ ls
blockchain-application-using-fabric-java-sdk
vagrant@vagrant:~$ cd blockchain-application-using-fabric-java-sdk/
vagrant@vagrant:~/blockchain-application-using-fabric-java-sdk$ ls
CONTRIBUTING.md  DEBUGGING.md  images  java  LICENSE  MAINTAINERS.md  network  network_resources  README.md
vagrant@vagrant:~/blockchain-application-using-fabric-java-sdk$ cd network
vagrant@vagrant:~/blockchain-application-using-fabric-java-sdk/network$ ls
build.sh  configtx.yaml  crypto-config.yaml  docker-compose.yml  stop.sh  teardown.sh
vagrant@vagrant:~/blockchain-application-using-fabric-java-sdk/network$
```

➢ chmod +x build.sh

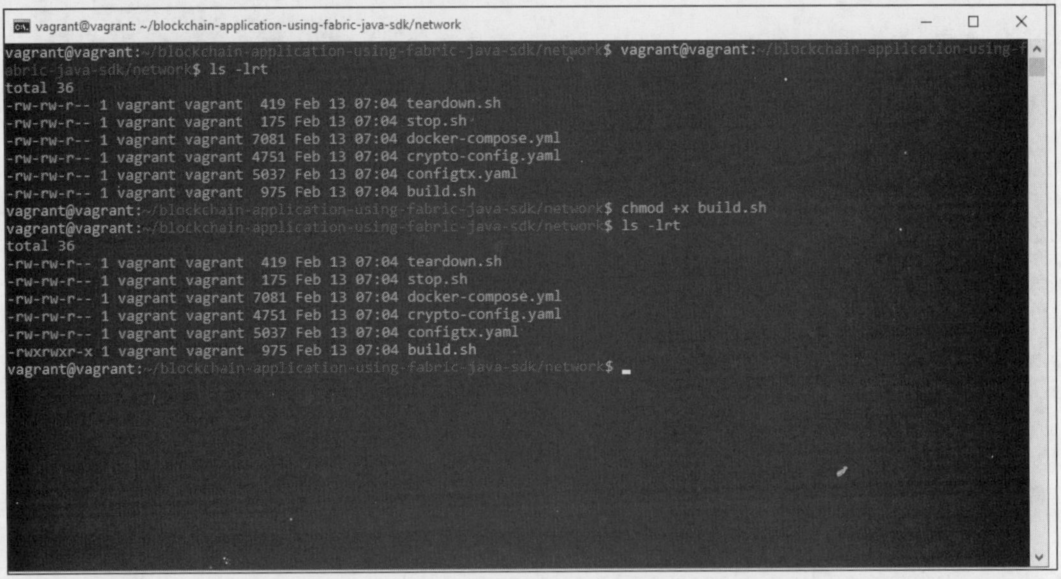

```
vagrant@vagrant: ~/blockchain-application-using-fabric-java-sdk/network
vagrant@vagrant:~/blockchain-application-using-fabric-java-sdk/network$ vagrant@vagrant:~/blockchain-application-using-f
abric-java-sdk/network$ ls -lrt
total 36
-rw-rw-r-- 1 vagrant vagrant  419 Feb 13 07:04 teardown.sh
-rw-rw-r-- 1 vagrant vagrant  175 Feb 13 07:04 stop.sh
-rw-rw-r-- 1 vagrant vagrant 7081 Feb 13 07:04 docker-compose.yml
-rw-rw-r-- 1 vagrant vagrant 4751 Feb 13 07:04 crypto-config.yaml
-rw-rw-r-- 1 vagrant vagrant 5037 Feb 13 07:04 configtx.yaml
-rw-rw-r-- 1 vagrant vagrant  975 Feb 13 07:04 build.sh
vagrant@vagrant:~/blockchain-application-using-fabric-java-sdk/network$ chmod +x build.sh
vagrant@vagrant:~/blockchain-application-using-fabric-java-sdk/network$ ls -lrt
total 36
-rw-rw-r-- 1 vagrant vagrant  419 Feb 13 07:04 teardown.sh
-rw-rw-r-- 1 vagrant vagrant  175 Feb 13 07:04 stop.sh
-rw-rw-r-- 1 vagrant vagrant 7081 Feb 13 07:04 docker-compose.yml
-rw-rw-r-- 1 vagrant vagrant 4751 Feb 13 07:04 crypto-config.yaml
-rw-rw-r-- 1 vagrant vagrant 5037 Feb 13 07:04 configtx.yaml
-rwxrwxr-x 1 vagrant vagrant  975 Feb 13 07:04 build.sh
vagrant@vagrant:~/blockchain-application-using-fabric-java-sdk/network$ _
```

➢ ./build.sh

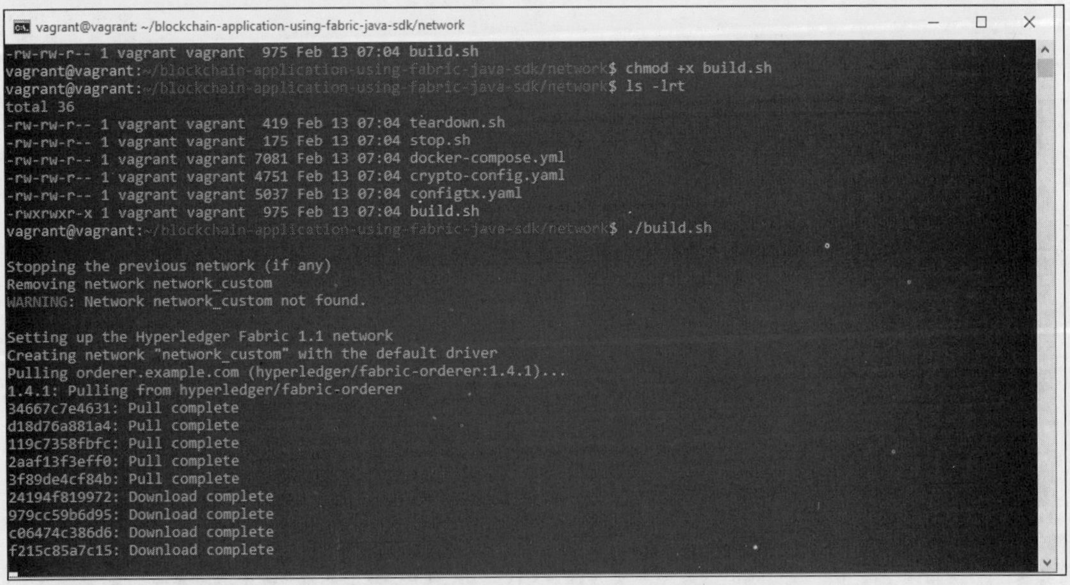

```
vagrant@vagrant: ~/blockchain-application-using-fabric-java-sdk/network
-rw-rw-r-- 1 vagrant vagrant  975 Feb 13 07:04 build.sh
vagrant@vagrant:~/blockchain-application-using-fabric-java-sdk/network$ chmod +x build.sh
vagrant@vagrant:~/blockchain-application-using-fabric-java-sdk/network$ ls -lrt
total 36
-rw-rw-r-- 1 vagrant vagrant  419 Feb 13 07:04 teardown.sh
-rw-rw-r-- 1 vagrant vagrant  175 Feb 13 07:04 stop.sh
-rw-rw-r-- 1 vagrant vagrant 7081 Feb 13 07:04 docker-compose.yml
-rw-rw-r-- 1 vagrant vagrant 4751 Feb 13 07:04 crypto-config.yaml
-rw-rw-r-- 1 vagrant vagrant 5037 Feb 13 07:04 configtx.yaml
-rwxrwxr-x 1 vagrant vagrant  975 Feb 13 07:04 build.sh
vagrant@vagrant:~/blockchain-application-using-fabric-java-sdk/network$ ./build.sh

Stopping the previous network (if any)
Removing network network_custom
WARNING: Network network_custom not found.

Setting up the Hyperledger Fabric 1.1 network
Creating network "network_custom" with the default driver
Pulling orderer.example.com (hyperledger/fabric-orderer:1.4.1)...
1.4.1: Pulling from hyperledger/fabric-orderer
34667c7e4631: Pull complete
d18d76a881a4: Pull complete
119c7358fbfc: Pull complete
2aaf13f3eff0: Pull complete
3f89de4cf84b: Pull complete
24194f819972: Download complete
979cc59b6d95: Download complete
c06474c386d6: Download complete
f215c85a7c15: Download complete
_
```

```
vagrant@vagrant: ~/blockchain-application-using-fabric-java-sdk/network
1.4.1: Pulling from hyperledger/fabric-peer
34667c7e4631: Already exists
d18d76a881a4: Already exists
119c7358fbfc: Already exists
2aaf13f3eff0: Already exists
3f89de4cf84b: Already exists
24194f819972: Already exists
979cc59b6d95: Already exists
8eb49214b84e: Pull complete
0e108783b4cb: Pull complete
Digest: sha256:05315d05b2892d34b4ed48f6502d28fe15a71090c36a39c97022a44475a984ad
Status: Downloaded newer image for hyperledger/fabric-peer:1.4.1
Creating ca_peerOrg2 ...
Creating orderer.example.com ...
Creating ca_peerOrg1 ...
Creating ca_peerOrg2
Creating orderer.example.com
Creating orderer.example.com ... done
Creating peer0.org2.example.com ...
Creating peer0.org1.example.com ...
Creating peer1.org1.example.com ...
Creating peer1.org2.example.com ...
Creating peer1.org1.example.com
Creating peer0.org1.example.com
Creating peer0.org2.example.com
Creating peer0.org1.example.com ... done

Network setup completed!!

vagrant@vagrant:~/blockchain-application-using-fabric-java-sdk/network$
```

16.4.3.3 Build the Client Based on Fabric Java SDK

Check maven:

➢ mvn -v

```
vagrant@vagrant: ~/blockchain-application-using-fabric-java-sdk/network
vagrant@vagrant:~/blockchain-application-using-fabric-java-sdk/network$ mvn -v
Apache Maven 3.6.0
Maven home: /usr/share/maven
Java version: 11.0.6, vendor: Ubuntu, runtime: /usr/lib/jvm/java-11-openjdk-amd64
Default locale: en_US, platform encoding: UTF-8
OS name: "linux", version: "4.15.0-76-generic", arch: "amd64", family: "unix"
vagrant@vagrant:~/blockchain-application-using-fabric-java-sdk/network$
```

➢ mvn install

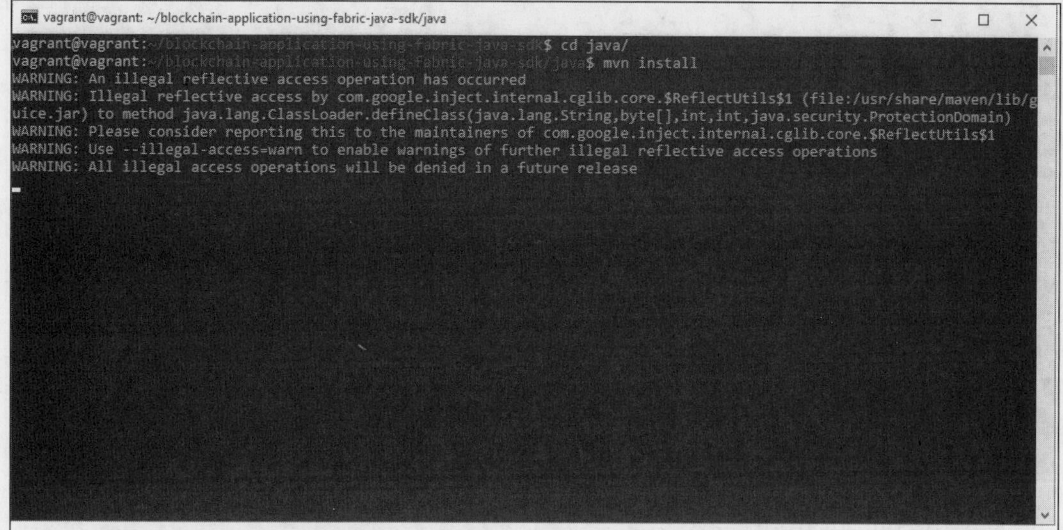

Note: This is a time consuming process.

```
vagrant@vagrant: ~/blockchain-application-using-fabric-java-sdk/java
Downloading from central: https://repo.maven.apache.org/maven2/org/codehaus/plexus/plexus-components/1.1.7/plexus-components-1.1.7.pom
Downloaded from central: https://repo.maven.apache.org/maven2/org/codehaus/plexus/plexus-components/1.1.7/plexus-components-1.1.7.pom (5.0 kB at 13 kB/s)
Downloading from central: https://repo.maven.apache.org/maven2/org/codehaus/plexus/plexus-utils/3.0.5/plexus-utils-3.0.5.jar
Downloading from central: https://repo.maven.apache.org/maven2/org/codehaus/plexus/plexus-digest/1.0/plexus-digest-1.0.jar
Downloaded from central: https://repo.maven.apache.org/maven2/org/codehaus/plexus/plexus-digest/1.0/plexus-digest-1.0.jar (12 kB at 31 kB/s)
Downloaded from central: https://repo.maven.apache.org/maven2/org/codehaus/plexus/plexus-utils/3.0.5/plexus-utils-3.0.5.jar (230 kB at 246 kB/s)
[INFO] Installing /home/vagrant/blockchain-application-using-fabric-java-sdk/java/target/blockchain-java-sdk-0.0.1-SNAPSHOT.jar to /home/vagrant/.m2/repository/blockchain-java-sdk/blockchain-java-sdk/0.0.1-SNAPSHOT/blockchain-java-sdk-0.0.1-SNAPSHOT.jar
[INFO] Installing /home/vagrant/blockchain-application-using-fabric-java-sdk/java/pom.xml to /home/vagrant/.m2/repository/blockchain-java-sdk/blockchain-java-sdk/0.0.1-SNAPSHOT/blockchain-java-sdk-0.0.1-SNAPSHOT.pom
[INFO] Installing /home/vagrant/blockchain-application-using-fabric-java-sdk/java/target/blockchain-java-sdk-0.0.1-SNAPSHOT-jar-with-dependencies.jar to /home/vagrant/.m2/repository/blockchain-java-sdk/blockchain-java-sdk/0.0.1-SNAPSHOT/blockchain-java-sdk-0.0.1-SNAPSHOT-jar-with-dependencies.jar
[INFO] ------------------------------------------------------------------------
[INFO] BUILD SUCCESS
[INFO] ------------------------------------------------------------------------
[INFO] Total time:  03:48 min
[INFO] Finished at: 2020-02-13T16:15:49Z
[INFO] ------------------------------------------------------------------------
vagrant@vagrant:~/blockchain-application-using-fabric-java-sdk/java$
```

Check the output from maven:

➢ Look for the new folder target, files within target folder.

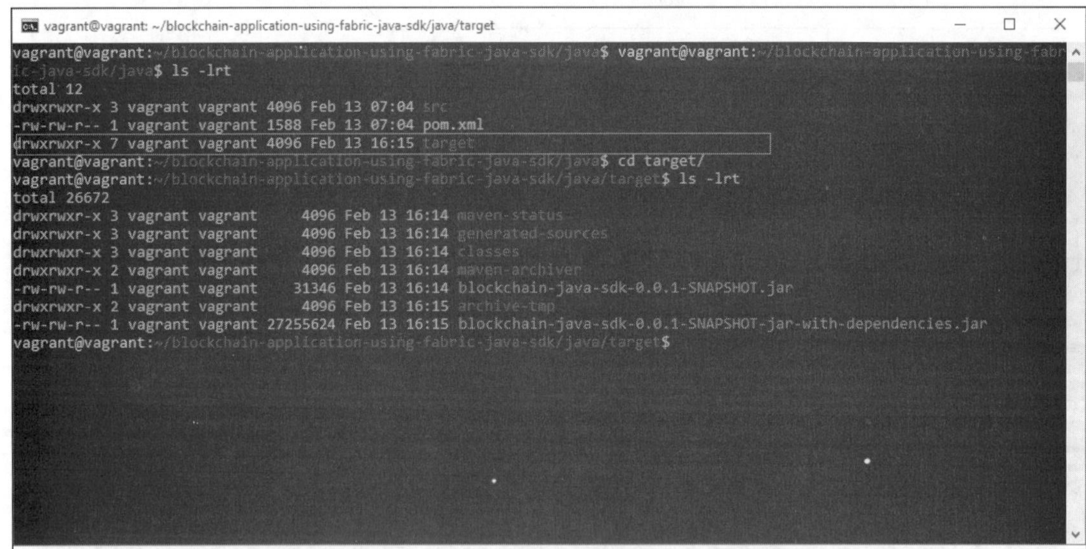

Copy the jar:

➢ cp blockchain-java-sdk-0.0.1-SNAPSHOT-jar-with-dependencies.jar blockchain-client.jar

```
vagrant@vagrant:~/blockchain-application-using-fabric-java-sdk/java/target$ ls -lrt
total 26672
drwxrwxr-x 3 vagrant vagrant     4096 Feb 13 16:14 maven-status
drwxrwxr-x 3 vagrant vagrant     4096 Feb 13 16:14 generated-sources
drwxrwxr-x 3 vagrant vagrant     4096 Feb 13 16:14 classes
drwxrwxr-x 2 vagrant vagrant     4096 Feb 13 16:14 maven-archiver
-rw-rw-r-- 1 vagrant vagrant    31346 Feb 13 16:14 blockchain-java-sdk-0.0.1-SNAPSHOT.jar
drwxrwxr-x 2 vagrant vagrant     4096 Feb 13 16:15 archive-tmp
-rw-rw-r-- 1 vagrant vagrant 27255624 Feb 13 16:15 blockchain-java-sdk-0.0.1-SNAPSHOT-jar-with-dependencies.jar
vagrant@vagrant:~/blockchain-application-using-fabric-java-sdk/java/target$ cp blockchain-java-sdk-0.0.1-SNAPSHOT-jar-wi
th-dependencies.jar blockchain-client.jar
```

Copy jar into new location:

➢ cp blockchain-client.jar ../../network_resources

16.4.3.4 Create and Initialize the Channel.

cd ../../network_resources

➢ java -cp blockchain-client.jar org.example.network.CreateChannel

```
vagrant@vagrant: ~/blockchain-application-using-fabric-java-sdk/network_resources
vagrant@vagrant:~/blockchain-application-using-fabric-java-sdk/network_resources$ java -cp blockchain-client.jar org.exa
mple.network.CreateChannel
log4j:WARN No appenders could be found for logger (org.hyperledger.fabric.sdk.helper.Config).
log4j:WARN Please initialize the log4j system properly.
log4j:WARN See http://logging.apache.org/log4j/1.2/faq.html#noconfig for more info.
Feb 13, 2020 4:24:11 PM org.example.util.Util deleteDirectory
INFO: Deleting - users
WARNING: An illegal reflective access operation has occurred
WARNING: Illegal reflective access by com.google.protobuf.UnsafeUtil (file:/home/vagrant/blockchain-application-using-fa
bric-java-sdk/network_resources/blockchain-client.jar) to field java.nio.Buffer.address
WARNING: Please consider reporting this to the maintainers of com.google.protobuf.UnsafeUtil
WARNING: Use --illegal-access=warn to enable warnings of further illegal reflective access operations
WARNING: All illegal access operations will be denied in a future release
Feb 13, 2020 4:24:15 PM org.example.network.CreateChannel main
INFO: Channel created mychannel
Feb 13, 2020 4:24:15 PM org.example.network.CreateChannel main
INFO: peer0.org2.example.com at grpc://localhost:8051
Feb 13, 2020 4:24:15 PM org.example.network.CreateChannel main
INFO: peer1.org1.example.com at grpc://localhost:7056
Feb 13, 2020 4:24:15 PM org.example.network.CreateChannel main
INFO: peer1.org2.example.com at grpc://localhost:8056
Feb 13, 2020 4:24:15 PM org.example.network.CreateChannel main
INFO: peer0.org1.example.com at grpc://localhost:7051
vagrant@vagrant:~/blockchain-application-using-fabric-java-sdk/network_resources$ _
```

The extract from the above output indicating successful initialization of the channels will be as shown:

```
INFO: Deleting - users
INFO: Channel created mychannel
INFO: peer0.org2.example.com at grpc://localhost:8051
INFO: peer1.org1.example.com at grpc://localhost:7056
INFO: peer1.org2.example.com at grpc://localhost:8056
INFO: peer0.org1.example.com at grpc://localhost:7051
```

➢ java -cp blockchain-client.jar org.example.network.DeployInstantiateChaincode

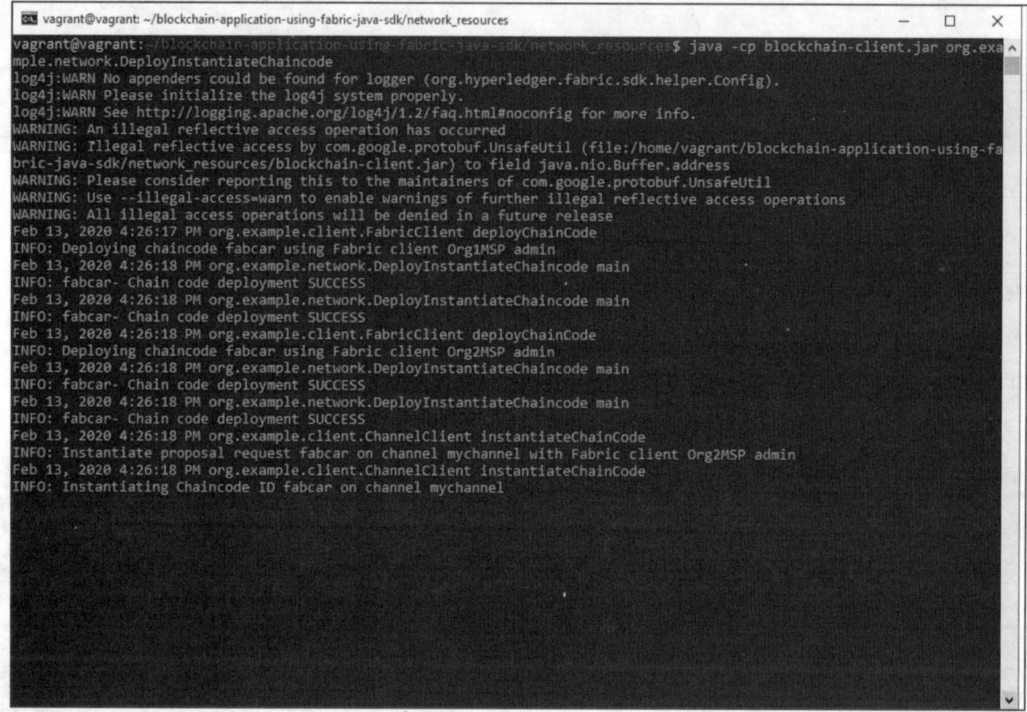

16.4.3.5 Deploy and Instantiate the Chaincode

➢ java -cp blockchain-client.jar org.example.network.DeployInstantiateChaincode

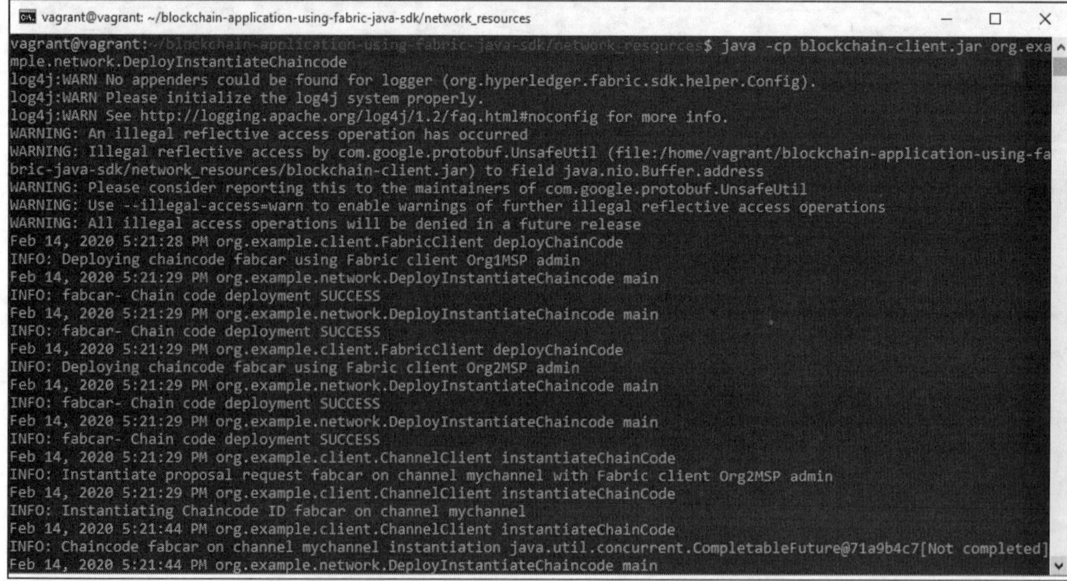

```
vagrant@vagrant: ~/blockchain-application-using-fabric-java-sdk/network_resources
Feb 14, 2020 5:21:29 PM org.example.network.DeployInstantiateChaincode main
INFO: fabcar- Chain code deployment SUCCESS
Feb 14, 2020 5:21:29 PM org.example.client.FabricClient deployChainCode
INFO: Deploying chaincode fabcar using Fabric client Org2MSP admin
Feb 14, 2020 5:21:29 PM org.example.network.DeployInstantiateChaincode main
INFO: fabcar- Chain code deployment SUCCESS
Feb 14, 2020 5:21:29 PM org.example.network.DeployInstantiateChaincode main
INFO: fabcar- Chain code deployment SUCCESS
Feb 14, 2020 5:21:29 PM org.example.client.ChannelClient instantiateChainCode
INFO: Instantiate proposal request fabcar on channel mychannel with Fabric client Org2MSP admin
Feb 14, 2020 5:21:29 PM org.example.client.ChannelClient instantiateChainCode
INFO: Instantiating Chaincode ID fabcar on channel mychannel
Feb 14, 2020 5:21:44 PM org.example.client.ChannelClient instantiateChainCode
INFO: Chaincode fabcar on channel mychannel instantiation java.util.concurrent.CompletableFuture@71a9b4c7[Not completed]
Feb 14, 2020 5:21:44 PM org.example.network.DeployInstantiateChaincode main
INFO: fabcar- Chain code instantiation SUCCESS
Feb 14, 2020 5:21:44 PM org.example.network.DeployInstantiateChaincode main
INFO: fabcar- Chain code instantiation SUCCESS
Feb 14, 2020 5:21:44 PM org.example.network.DeployInstantiateChaincode main
INFO: fabcar- Chain code instantiation SUCCESS
Feb 14, 2020 5:21:44 PM org.example.network.DeployInstantiateChaincode main
INFO: fabcar- Chain code instantiation SUCCESS
vagrant@vagrant:~/blockchain-application-using-fabric-java-sdk/network_resources$
```

The key extract from the above output will be as shown, indicating successful chaincode instantiation.

```
INFO: Deploying chaincode fabcar using Fabric client Org1MSP admin
INFO: fabcar- Chain code deployment SUCCESS
INFO: fabcar- Chain code deployment SUCCESS
INFO: Deploying chaincode fabcar using Fabric client Org2MSP admin
INFO: fabcar- Chain code deployment SUCCESS
INFO: fabcar- Chain code deployment SUCCESS
INFO: Instantiate proposal request fabcar on channel mychannel with Fabric client Org2MSP admin
INFO: Instantiating Chaincode ID fabcar on channel mychannel
INFO: Chaincode fabcar on channel mychannel instantiation
INFO: fabcar- Chain code instantiation SUCCESS
INFO: fabcar- Chain code instantiation SUCCESS
INFO: fabcar- Chain code instantiation SUCCESS
INFO: fabcar- Chain code instantiation SUCCESS
```

16.4.3.6 Register and Enrol Users

➢ java -cp blockchain-client.jar org.example.user.RegisterEnrollUser

```
vagrant@vagrant: ~/blockchain-application-using-fabric-java-sdk/network_resources
vagrant@vagrant:~/blockchain-application-using-fabric-java-sdk/network_resources$ vagrant@vagrant:~/blockchain-applicati
on-using-fabric-java-sdk/network_resources$
vagrant@vagrant:~/blockchain-application-using-fabric-java-sdk/network_resources$
vagrant@vagrant:~/blockchain-application-using-fabric-java-sdk/network_resources$ java -cp blockchain-client.jar org.exa
mple.user.RegisterEnrollUser
Feb 14, 2020 5:24:22 PM org.example.util.Util deleteDirectory
INFO: Deleting - users
log4j:WARN No appenders could be found for logger (org.hyperledger.fabric.sdk.helper.Config).
log4j:WARN Please initialize the log4j system properly.
log4j:WARN See http://logging.apache.org/log4j/1.2/faq.html#noconfig for more info.
Feb 14, 2020 5:24:27 PM org.example.client.CAClient enrollAdminUser
INFO: CA -http://localhost:7054 Enrolled Admin.
Feb 14, 2020 5:24:27 PM org.example.client.CAClient registerUser
INFO: CA -http://localhost:7054 Registered User - user1581701067112
Feb 14, 2020 5:24:27 PM org.example.client.CAClient enrollUser
INFO: CA -http://localhost:7054 Enrolled User - user1581701067112
vagrant@vagrant:~/blockchain-application-using-fabric-java-sdk/network_resources$
```

The key extract of the above output will be as shown, indicating successful registration and enrolment of user.

```
INFO: Deleting - users
INFO: CA -http://localhost:7054 Enrolled Admin.
INFO: CA -http://localhost:7054 Registered User - user1524459395783
INFO: CA -http://localhost:7054 Enrolled User - user1524459395783
```

16.4.3.7 Deploy and Instantiate the Chaincode

➢ java -cp blockchain-client.jar org.example.chaincode.invocation.InvokeChaincode

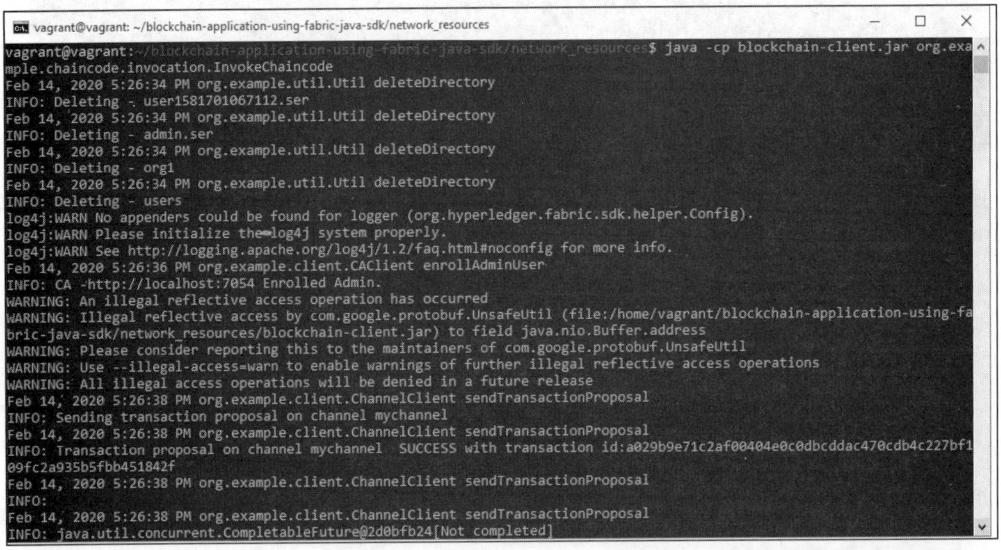

```
vagrant@vagrant: ~/blockchain-application-using-fabric-java-sdk/network_resources
vagrant@vagrant:~/blockchain-application-using-fabric-java-sdk/network_resources$ java -cp blockchain-client.jar org.exa
mple.chaincode.invocation.InvokeChaincode
Feb 14, 2020 5:26:34 PM org.example.util.Util deleteDirectory
INFO: Deleting - user1581701067112.ser
Feb 14, 2020 5:26:34 PM org.example.util.Util deleteDirectory
INFO: Deleting - admin.ser
Feb 14, 2020 5:26:34 PM org.example.util.Util deleteDirectory
INFO: Deleting - org1
Feb 14, 2020 5:26:34 PM org.example.util.Util deleteDirectory
INFO: Deleting - users
log4j:WARN No appenders could be found for logger (org.hyperledger.fabric.sdk.helper.Config).
log4j:WARN Please initialize the log4j system properly.
log4j:WARN See http://logging.apache.org/log4j/1.2/faq.html#noconfig for more info.
Feb 14, 2020 5:26:36 PM org.example.client.CAClient enrollAdminUser
INFO: CA -http://localhost:7054 Enrolled Admin.
WARNING: An illegal reflective access operation has occurred
WARNING: Illegal reflective access by com.google.protobuf.UnsafeUtil (file:/home/vagrant/blockchain-application-using-fa
bric-java-sdk/network_resources/blockchain-client.jar) to field java.nio.Buffer.address
WARNING: Please consider reporting this to the maintainers of com.google.protobuf.UnsafeUtil
WARNING: Use --illegal-access=warn to enable warnings of further illegal reflective access operations
WARNING: All illegal access operations will be denied in a future release
Feb 14, 2020 5:26:38 PM org.example.client.ChannelClient sendTransactionProposal
INFO: Sending transaction proposal on channel mychannel
Feb 14, 2020 5:26:38 PM org.example.client.ChannelClient sendTransactionProposal
INFO: Transaction proposal on channel mychannel  SUCCESS with transaction id:a029b9e71c2af00404e0c0dbcddac470cdb4c227bf1
09fc2a935b5fbb451842f
Feb 14, 2020 5:26:38 PM org.example.client.ChannelClient sendTransactionProposal
INFO:
Feb 14, 2020 5:26:38 PM org.example.client.ChannelClient sendTransactionProposal
INFO: java.util.concurrent.CompletableFuture@2d0bfb24[Not completed]
```

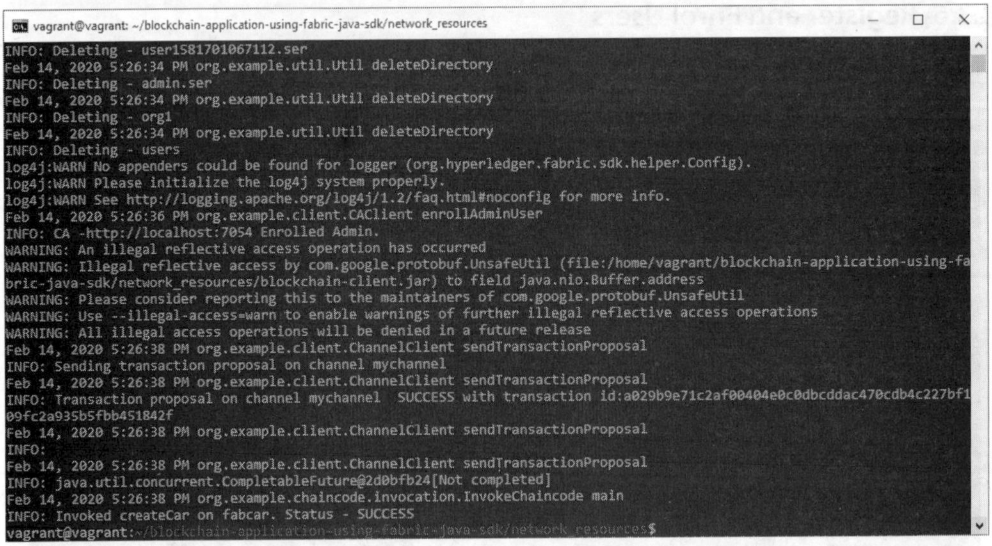

```
vagrant@vagrant: ~/blockchain-application-using-fabric-java-sdk/network_resources
INFO: Deleting - user1581701067112.ser
Feb 14, 2020 5:26:34 PM org.example.util.Util deleteDirectory
INFO: Deleting - admin.ser
Feb 14, 2020 5:26:34 PM org.example.util.Util deleteDirectory
INFO: Deleting - org1
Feb 14, 2020 5:26:34 PM org.example.util.Util deleteDirectory
INFO: Deleting - users
log4j:WARN No appenders could be found for logger (org.hyperledger.fabric.sdk.helper.Config).
log4j:WARN Please initialize the log4j system properly.
log4j:WARN See http://logging.apache.org/log4j/1.2/faq.html#noconfig for more info.
Feb 14, 2020 5:26:36 PM org.example.client.CAClient enrollAdminUser
INFO: CA -http://localhost:7054 Enrolled Admin.
WARNING: An illegal reflective access operation has occurred
WARNING: Illegal reflective access by com.google.protobuf.UnsafeUtil (file:/home/vagrant/blockchain-application-using-fa
bric-java-sdk/network_resources/blockchain-client.jar) to field java.nio.Buffer.address
WARNING: Please consider reporting this to the maintainers of com.google.protobuf.UnsafeUtil
WARNING: Use --illegal-access=warn to enable warnings of further illegal reflective access operations
WARNING: All illegal access operations will be denied in a future release
Feb 14, 2020 5:26:38 PM org.example.client.ChannelClient sendTransactionProposal
INFO: Sending transaction proposal on channel mychannel
Feb 14, 2020 5:26:38 PM org.example.client.ChannelClient sendTransactionProposal
INFO: Transaction proposal on channel mychannel  SUCCESS with transaction id:a029b9e71c2af00404e0c0dbcddac470cdb4c227bf1
09fc2a935b5fbb451842f
Feb 14, 2020 5:26:38 PM org.example.client.ChannelClient sendTransactionProposal
INFO:
Feb 14, 2020 5:26:38 PM org.example.client.ChannelClient sendTransactionProposal
INFO: java.util.concurrent.CompletableFuture@2d0bfb24[Not completed]
Feb 14, 2020 5:26:38 PM org.example.chaincode.invocation.InvokeChaincode main
INFO: Invoked createCar on fabcar. Status - SUCCESS
vagrant@vagrant:~/blockchain-application-using-fabric-java-sdk/network_resources$
```

The key output of the above command indicating success, will be as shown:

```
INFO: CA -http://localhost:7054 Enrolled Admin.
INFO: Sending transaction proposal on channel mychannel
INFO: Invoked createCar on fabcar. Status - SUCCESS
```

16.4.3.8 Perform Invoke and Query on Network

Blockchain network setup has now been completed and is ready to use. Now, we can test the network by performing `invoke` and `query` on the network. The `chaind code named (fabcar)` allows creating a new asset (which is a car). For the purpose of testing, `invoke` operation is performed to create a new asset (car) in the network and query operation is performed to list down the asset of the network. It is done by performing the following steps:

➢ java -cp blockchain-client.jar org.example.chaincode.invocation.InvokeChaincode

```
vagrant@vagrant: ~/blockchain-application-using-fabric-java-sdk/network_resources
vagrant@vagrant:~/blockchain-application-using-fabric-java-sdk/network_resources$ java -cp blockchain-client.jar org.exa
mple.chaincode.invocation.QueryChaincode
Feb 14, 2020 5:30:01 PM org.example.util.Util deleteDirectory
INFO: Deleting - admin.ser
Feb 14, 2020 5:30:01 PM org.example.util.Util deleteDirectory
INFO: Deleting - org1
Feb 14, 2020 5:30:01 PM org.example.util.Util deleteDirectory
INFO: Deleting - users
log4j:WARN No appenders could be found for logger (org.hyperledger.fabric.sdk.helper.Config).
log4j:WARN Please initialize the log4j system properly.
log4j:WARN See http://logging.apache.org/log4j/1.2/faq.html#noconfig for more info.
Feb 14, 2020 5:30:04 PM org.example.client.CAClient enrollAdminUser
INFO: CA -http://localhost:7054 Enrolled Admin.
WARNING: An illegal reflective access operation has occurred
WARNING: Illegal reflective access by com.google.protobuf.UnsafeUtil (file:/home/vagrant/blockchain-application-using-fa
bric-java-sdk/network_resources/blockchain-client.jar) to field java.nio.Buffer.address
WARNING: Please consider reporting this to the maintainers of com.google.protobuf.UnsafeUtil
WARNING: Use --illegal-access=warn to enable warnings of further illegal reflective access operations
WARNING: All illegal access operations will be denied in a future release
Feb 14, 2020 5:30:05 PM org.example.chaincode.invocation.QueryChaincode main
INFO: Querying for all cars ...
Feb 14, 2020 5:30:05 PM org.example.client.ChannelClient queryByChainCode
INFO: Querying queryAllCars on channel mychannel
Feb 14, 2020 5:30:05 PM org.example.chaincode.invocation.QueryChaincode main
INFO: [{"Key":"CAR1", "Record":{"make":"Chevy","model":"Volt","colour":"Red","owner":"Nick"}}]
Feb 14, 2020 5:30:15 PM org.example.chaincode.invocation.QueryChaincode main
INFO: Querying for a car - CAR1
Feb 14, 2020 5:30:15 PM org.example.client.ChannelClient queryByChainCode
INFO: Querying queryCar on channel mychannel
Feb 14, 2020 5:30:16 PM org.example.chaincode.invocation.QueryChaincode main
```

```
vagrant@vagrant: ~/blockchain-application-using-fabric-java-sdk/network_resources
INFO: Deleting - users
log4j:WARN No appenders could be found for logger (org.hyperledger.fabric.sdk.helper.Config).
log4j:WARN Please initialize the log4j system properly.
log4j:WARN See http://logging.apache.org/log4j/1.2/faq.html#noconfig for more info.
Feb 14, 2020 5:30:04 PM org.example.client.CAClient enrollAdminUser
INFO: CA -http://localhost:7054 Enrolled Admin.
WARNING: An illegal reflective access operation has occurred
WARNING: Illegal reflective access by com.google.protobuf.UnsafeUtil (file:/home/vagrant/blockchain-application-using-fa
bric-java-sdk/network_resources/blockchain-client.jar) to field java.nio.Buffer.address
WARNING: Please consider reporting this to the maintainers of com.google.protobuf.UnsafeUtil
WARNING: Use --illegal-access=warn to enable warnings of further illegal reflective access operations
WARNING: All illegal access operations will be denied in a future release
Feb 14, 2020 5:30:05 PM org.example.chaincode.invocation.QueryChaincode main
INFO: Querying for all cars ...
Feb 14, 2020 5:30:05 PM org.example.client.ChannelClient queryByChainCode
INFO: Querying queryAllCars on channel mychannel
Feb 14, 2020 5:30:05 PM org.example.chaincode.invocation.QueryChaincode main
INFO: [{"Key":"CAR1", "Record":{"make":"Chevy","model":"Volt","colour":"Red","owner":"Nick"}}]
Feb 14, 2020 5:30:15 PM org.example.chaincode.invocation.QueryChaincode main
INFO: Querying for a car - CAR1
Feb 14, 2020 5:30:15 PM org.example.client.ChannelClient queryByChainCode
INFO: Querying queryCar on channel mychannel
Feb 14, 2020 5:30:16 PM org.example.chaincode.invocation.QueryChaincode main
INFO: {"make":"Chevy","model":"Volt","colour":"Red","owner":"Nick"}
vagrant@vagrant:~/blockchain-application-using-fabric-java-sdk/network_resources$
```

Below is the key extract from the output indicating successful invoking:

```
INFO: Querying for all cars ...
INFO: Querying queryAllCars on channel mychannel
INFO: [{"Key":"CAR1", "Record":{"make":"Chevy","model":"Volt","colour":"Red","owner":"Nick"}}]
INFO: Querying for a car - CAR1
INFO: Querying queryCar on channel mychannel
INFO: {"make":"Chevy","model":"Volt","colour":"Red","owner":"Nick"}
```

Summary

This chapter summarises the basic components of Hyperledger Fabric network and walks the reader through its installation. It illustrates the writing of a sample Blockchain application using fabric Java sdk and also gives Chaincode references from Developer.ibm.com to write other programs using Hyperledger Fabric.

EXERCISES

Practical Questions (Using Hyperledger Fabric Platform)

1. Write a chaincode to interact with a blockchain network, using the below link: https://developer.ibm.com/patterns/interacting-with-a-blockchain-network/
2. Write a chaincode for an asset transfer app, using the below link: https://developer.ibm.com/patterns/deploy-an-asset-transfer-app-using-blockchain/
3. Write a chaincode for fitness club rewards points, using the below link: https://developer.ibm.com/patterns/fitness-club-rewards-points-iot-and-retail-integration/
4. Write a chaincode for a car auction network, using the below link: https://developer.ibm.com/patterns/car-auction-network-hyperledger-fabric-node-sdk-starter-plan/
5. Write a chaincode to develop an IoT asset tracking, using the below link: https://developer.ibm.com/patterns/develop-an-iot-asset-tracking-app-using-blockchain/
6. Write a chaincode for securing art using blockchain digital certificates, using the below link:
   https://developer.ibm.com/patterns/securing-art-using-blockchain-digital-certificates/
7. Write a chaincode for telecom roaming fraud and management, using the below link:
   https://developer.ibm.com/patterns/blockchain-for-telecom-roaming-fraud-and-overage-management/

8. Write a chaincode for IoT dashboards to analyze data blockchain network, using the below link:
https://developer.ibm.com/patterns/iot-dashboards-analyze-data-blockchain-network/
9. Write a chaincode to create an android app with blockchain integration, using the below link:
https://developer.ibm.com/patterns/create-an-android-app-with-blockchain-integration/
10. Write a chaincode for global financing use case for blockchain, using the below link:
https://developer.ibm.com/patterns/global-financing-use-case-for-blockchain/
11. Write a chaincode for voting app hyperledger ethereum, using the below link:
https://developer.ibm.com/patterns/voting-app-hyperledger-ethereum/
12. Write a chaincode for loyalty points fabric EVM, using the below link:
https://developer.ibm.com/patterns/loyalty-points-fabric-evm/

APPENDIX A

Connecting Remix with Ganache

A.1 Introduction

Ganache (previously known as TestRPC) is a personal blockchain for Ethereum, which can be used to deploy contracts, develop applications and run tests. Both desktop applications and command-line tools are available on various operating systems such as Windows, Mac and Linux.

You can visit https://www.trufflesuite.com/docs/ganache/overview to know more details about Ganache. Ganache GUI version can be downloaded by clicking Download button in the home screen as below (refer Fig. A.1).

Figure A.1: Home page of Ganache website

Table A.1 compares the features of Remix and Ganache.

Ganache GUI can be efficiently utilized for creating the user accounts and monitoring the execution (Blocks, Transaction and Accounts). We are going to connect Remix with Ganache. We will execute Steps 1 to 5 in Remix and Step 6 in Ganache.

Table A.1 Comparing Remix and Ganache

S. No	Activities	Remix	Ganache
1	Develop the contract	+	–
2	Compile the contract	+	–
3	Creating user account	+	+
4	Deploy the contract	+	–
5	Run the contract	+	–
6	Monitor the execution	–	+

A.2 Features of Ganache

Ganache comes up with a predetermined set of accounts (refer Fig A.2), which we can work with while writing and testing Ethereum solidity codes. All these accounts are pre-loaded with 100 ethers (ETH) to start with. Each account is also connected with its own address and private key to facilitate transactions. Rich look and feel is the main advantage of Ganache.

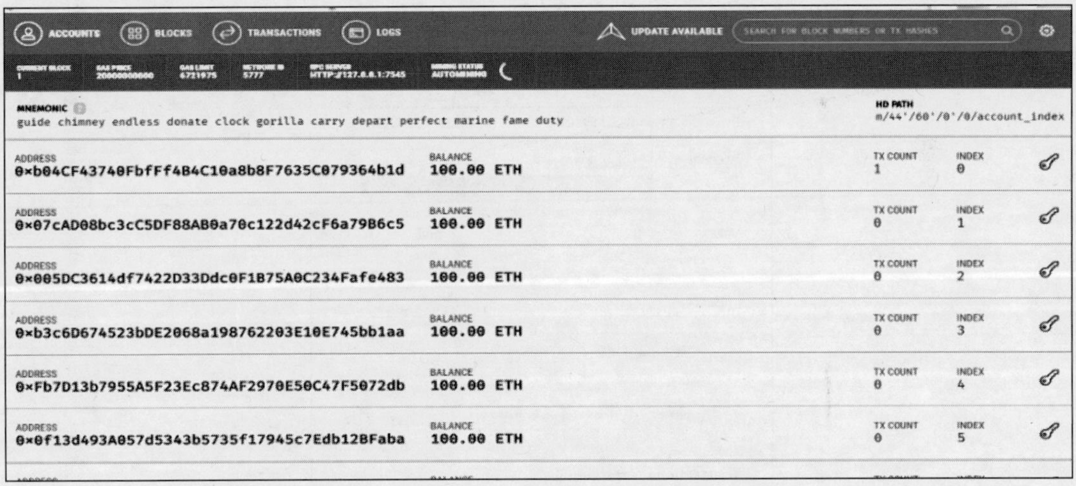

ACCOUNTS BLOCKS TRANSACTIONS LOGS UPDATE AVAILABLE

MNEMONIC
guide chimney endless donate clock gorilla carry depart perfect marine fame duty

HD PATH
m/44'/60'/0'/0/account_index

ADDRESS	BALANCE	TX COUNT	INDEX
0×b04CF43740FbfFf4B4C10a8b8F7635C079364b1d	100.00 ETH	1	0
0×07cAD08bc3cC5DF88AB0a70c122d42cF6a79B6c5	100.00 ETH	0	1
0×005DC3614df7422D33DdcOF1B75A0C234Fafe483	100.00 ETH	0	2
0×b3c6D674523bDE2068a198762203E10E745bb1aa	100.00 ETH	0	3
0×Fb7D13b7955A5F23Ec874AF2970E50C47F5072db	100.00 ETH	0	4
0×0f13d493A057d5343b5735f17945c7Edb12BFaba	100.00 ETH	0	5

Figure A.2: Ganache pre-loaded accounts with 100.00 ETH

Ganache has other features (refer Fig. A.3), which include advanced mining controls and configuring advanced mining options in a single click. It also allows a feature (link) to examine all blocks and transactions, which enables developers to check whether their code is working correctly as per the expectation. Vital debugging options are also available as part of this tool.

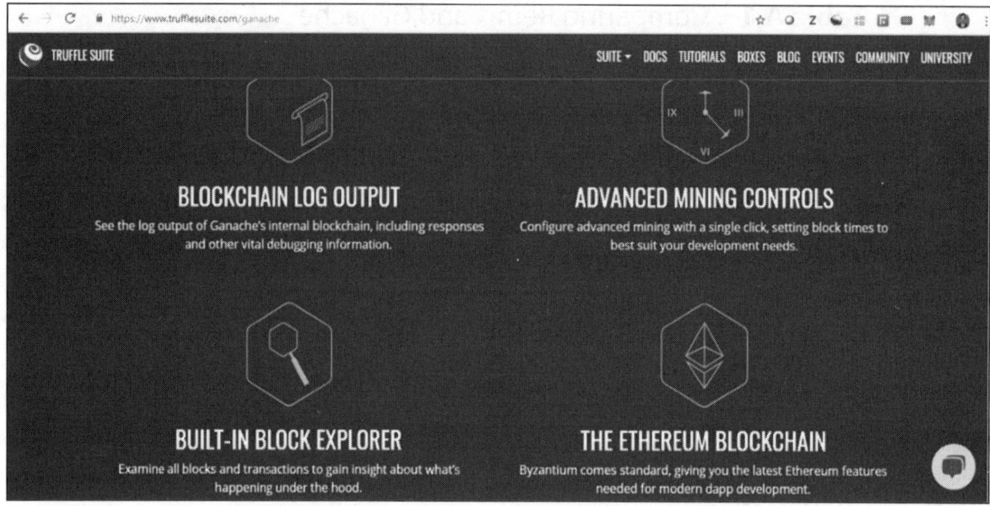

Figure A.3: Advance features of Ganache

A.3 Connecting Ganache from Remix

In Remix browser, in the environment column, the third option, namely, "Web3 Provider" needs to be selected (refer Fig. A.4).

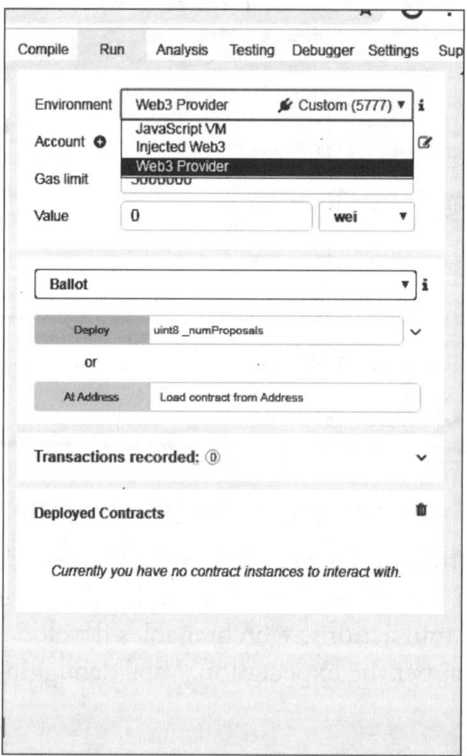

Figure A.4: Changing environment set up in Remix browser

After that, it will ask the URL to connect Ganache (refer Fig. A.5). In the box (in Remix browser), enter the RPC Server URL copied (refer Fig. A.6) from the top of the Ganache Accounts home page (HTTP://127.0.0.1:7545).

Figure A.5: URL to connect Ganache

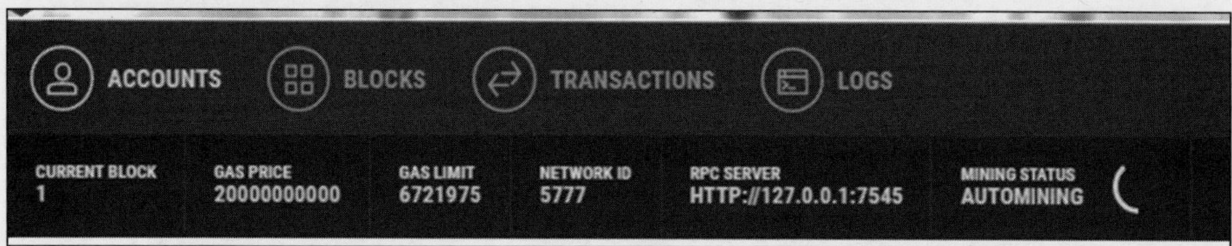

Figure A.6: RPC Server URL from Ganache accounts page

A.4 Confirming Connection Between Ganache Remix

After connecting Ganache environment from Remix through its Environment column, we need to check on whether the connection was properly established. This can be checked by ensuring that all account details from Ganache are exported into the Remix browser account details.

Environment Web3 Provider Custom (5777)
Account 0xb04...64b1d (99.9999999999997431(
0xb04...64b1d (99.99999999999743107 ether)
0x07c...9b6c5 (100 ether)
0x005...fe483 (100 ether)
0xb3c...bb1aa (100 ether)
0xfb7...072db (100 ether)
0x0f1...bfaba (100 ether)
0xb33...d6fac (100 ether)
0x91c...a97a4 (100 ether)
0x041...3bce4 (100 ether)
0xd4f...c2c73 (100 ether)
Gas limit
Value
AcceptingP
Deploy
or
At Address Load contract from Address

Figure A.7: Imported accounts from Ganache

In the Account section of Remix (refer Fig. A.7), we can see the accounts from Ganache (refer Fig. A.8) imported. The first account displayed in Remix is similar to the below account from Ganache. Hence, we can establish that connection was established between Remix and Ganache.

Figure A.8: Checking Ganache accounts and Remix accounts

A.5 Transaction from Remix to Ganache

As we know, every smart contract in Ethereum is connected to the blockchain and a Transaction ID is created while we run the Smart Contracts. We can run any smart contract program to check the establishment of the connection and creation of the transaction (refer Fig. A.9).

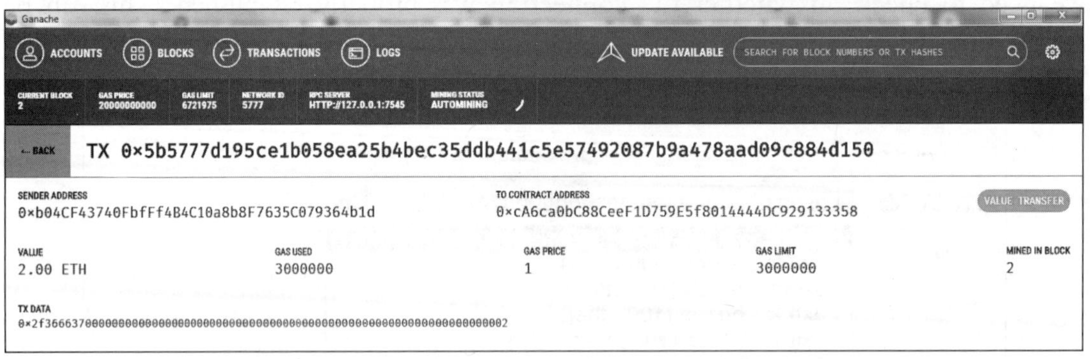

Figure A.9: Sample solidity program for checking the transaction

The above figure (refer Fig. A.9) indicates 2 blocks were created (in Ganache) and transactions established based on the contracts were executed in Remix.

APPENDIX B

Connecting MyEtherWallet (MEW) with Ganache

B.1 Introduction

Ganache (previously called as TestRPC) is a personal blockchain for Ethereum, which can be used to deploy contracts, develop applications and run tests. Both desktop application and command-line tools are available on various operating systems such as Windows, Mac and Linux.

You can visit https://www.trufflesuite.com/docs/ganache/overview to know more details about Ganache. Ganache GUI version can be downloaded by clicking Download button in the home screen as below (refer Fig. B.1).

Figure B.1: Home Page of Ganache website

Table B.1 compares the features of Remix, Ganache and MyEtherWallet (MEW).

We will use Remix for developing the contract (first option) and compiling the contract (second option). Creating the user accounts (third option) and monitoring the execution (sixth option) are taken care of by Ganache efficiently. We are going to deploy (fourth option) and run (fifth option) the same contract in MEW (resulting in transaction, monitoring the results with Ganache).

Table B.1 Comparing Remix, Ganache and MyEtherWallet(MEW)

S. No	Activities	Remix	Ganache	MEW
1	Develop the contract	+	–	–
2	Compile the contract	+	–	–
3	Creating user account	+	+	+
4	Deploy the contract	+	–	+
5	Run the contract	+	–	+
6	Monitor the execution	–	+	–

B.2 Features of Ganache

Ganache comes up with predetermined set of accounts (refer Fig. B.2), which we can work with while writing and testing Ethereum solidity codes. All these accounts are pre-loaded with 100 ethers (ETH) to start with. Each account is also connected with its own address and private key for doing transactions. Rich look and feel is the main advantage of Ganache.

Figure B.2: Ganache pre-loaded accounts with 100.00 ETH

Ganache has other features (refer Fig. B.3) which include advanced mining controls and configuring advanced mining options in a single click. It also allows a feature (link) to examine all blocks and transactions, which enable developers to check whether their code is working correctly as per the expectation. Vital debugging options are also available as part of this tool.

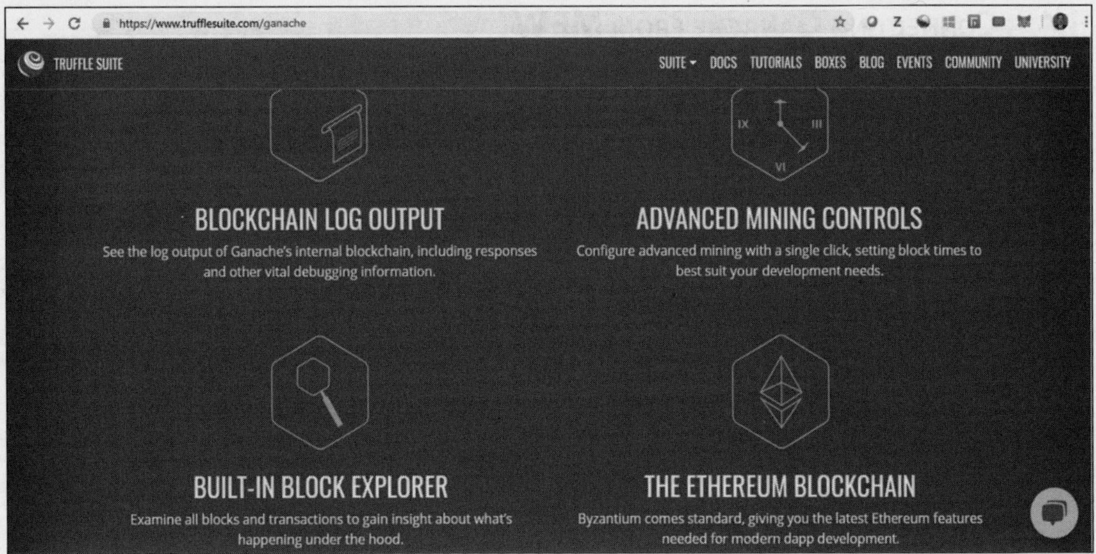

Figure B.3: Advanced features of Ganache

B.3 Writing and Compiling the Code in Remix

We write the actual solidity code in Remix and compile it (refer Table B.1: Steps 1 and 2). Remix generates ABI codes and Bytecodes for the same (refer Fig. B.4). Simple calculator code is written (for adding and subtracting two numbers), with the contract name as "Calculator".

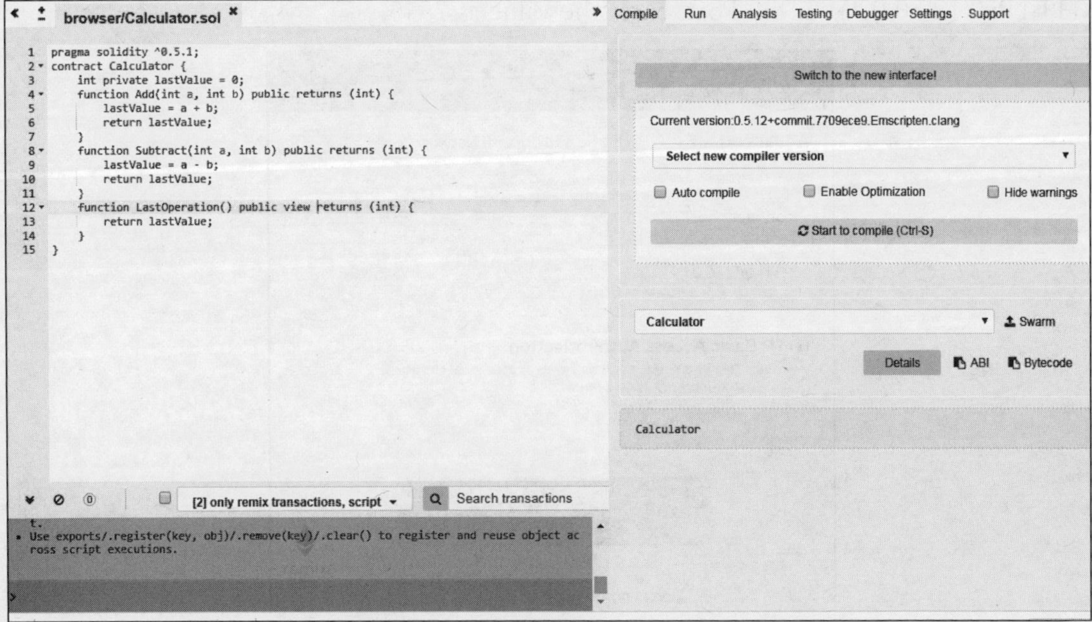

Figure B.4: Remix code and compilation

B.4 Connecting Ganache from MEW

In MEW (myetherwallet.com/interface/dashboard), click the network column, the last option namely: "Custom Networks" needs to be selected (refer Fig. B.5).

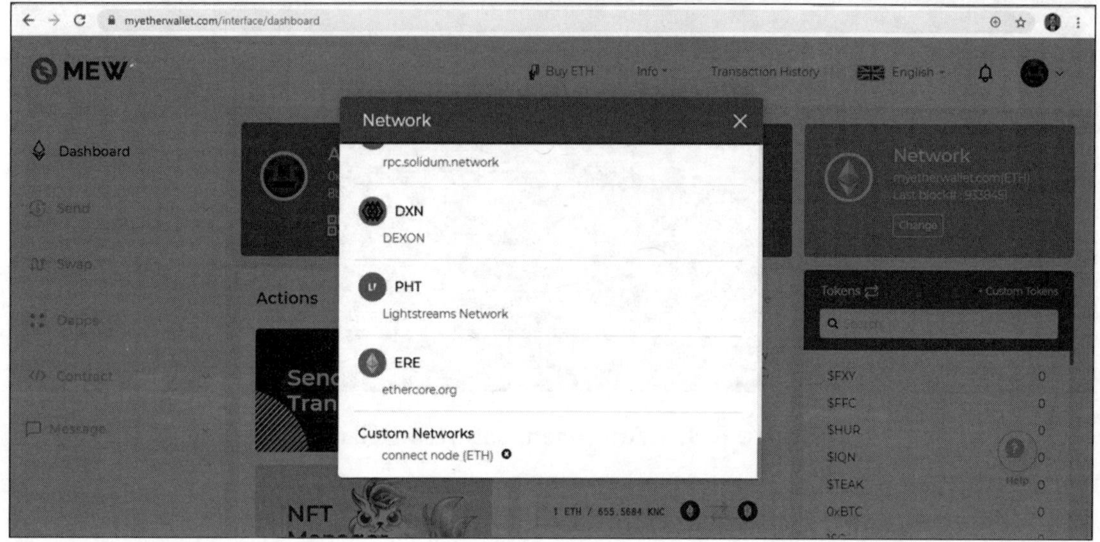

Figure B.5: Connecting custom network (Ganache) in MEW

After that, it will ask the URL to connect Ganache (refer Fig. B.6). In the box (Network), give a name to the node (it is given as "Connect node"), enter the RPC Server URL copied (refer Fig. B.7) from the top of the Ganache Accounts home page (HTTP://127.0.0.1). In the Port, enter the value as 7545.

Network

Advanced — Add Custom Network/Node

connect node | ETH - Ether

HTTP://127.0.0.1 | 7545

HTTP Basic Access Authentication

Warning: This information will be saved to your local storage, make sure your computer is secure.

Save

Don't know what it is? Help Center

Figure B.6: Network details to connect Ganache

Figure B.7: RPC server URL from Ganache accounts page

B.5 Connecting Ganache Accounts from MEW

In MEW (myetherwallet.com/access-my-wallet), click the link titled "Software" (refer Fig. B.8).

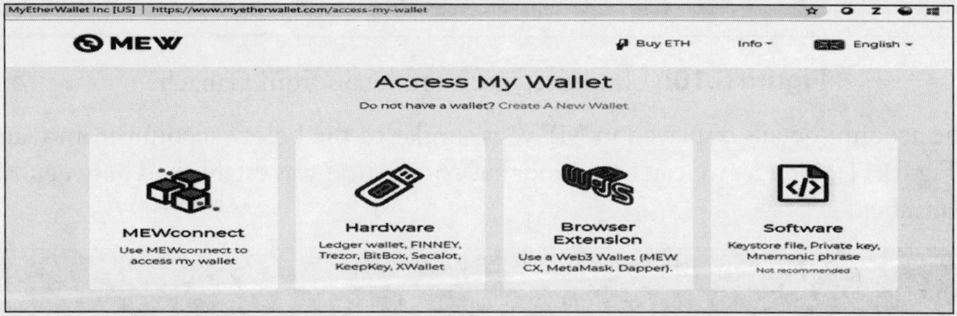

Figure B.8: Access My Wallet in MEW (Software option)

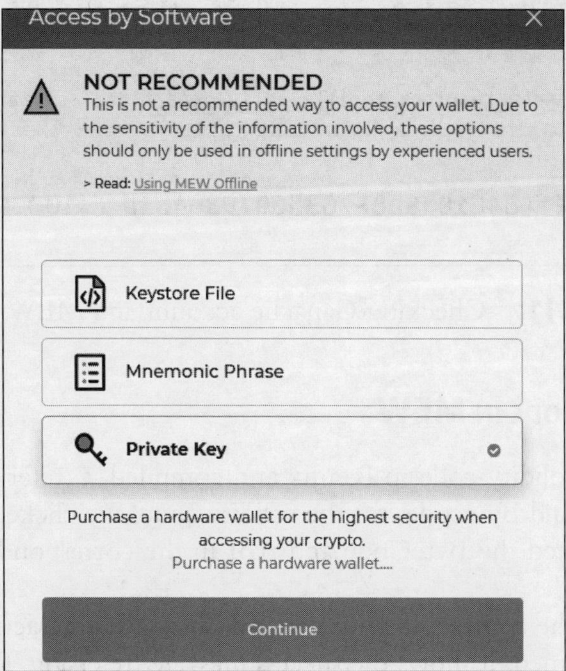

Figure B.9 Private Key connect to access My Wallet

In Access by Software screen, select Private Key (refer Fig. B.9), Click Continue. Copy the private key (from Ganache) for an account and enter that value here to connect to an account.

After connecting to Ganache environment from MEW, we need to check on whether the connection was properly established. This can be checked by ensuring that the account details from Ganache are exported into the MEW. Account balance (initial) shows as 100 ETH (refer Fig. B.10). This is automatically imported from Ganache.

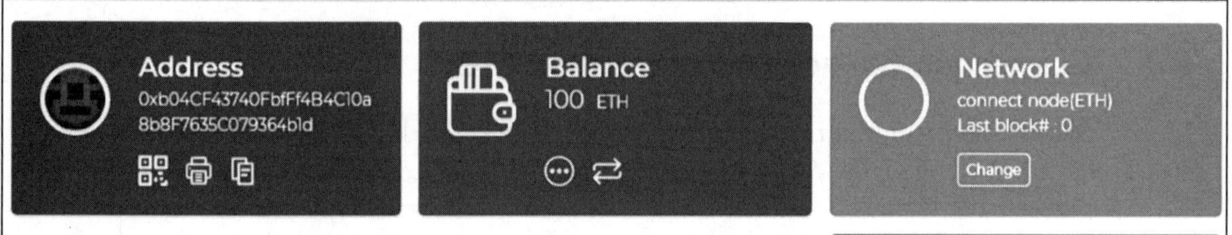

Figure B.10: Imported account details from Ganache

The account details displayed in MEW is similar to the below account from Ganache (refer Fig. B.11). Hence, we can conclude that connection was established between MEW and Ganache.

Figure B.11: Checking Ganache account and MEW account

B.6 Deploying Code in MEW

We wrote the actual solidity code in Remix and compiled it (refer Section B3). Remix generated ABI codes and Bytecodes for the solidity code. We clicked "Deploy Contract" link in MEW and pasted the ByteCode and ABI in the corresponding boxes (refer Fig. B.12).

While deploying the contract in MEW, it will show the transaction details (refer Fig. B.13) and once confirmed, it shows the success message (refer Fig. B.14).

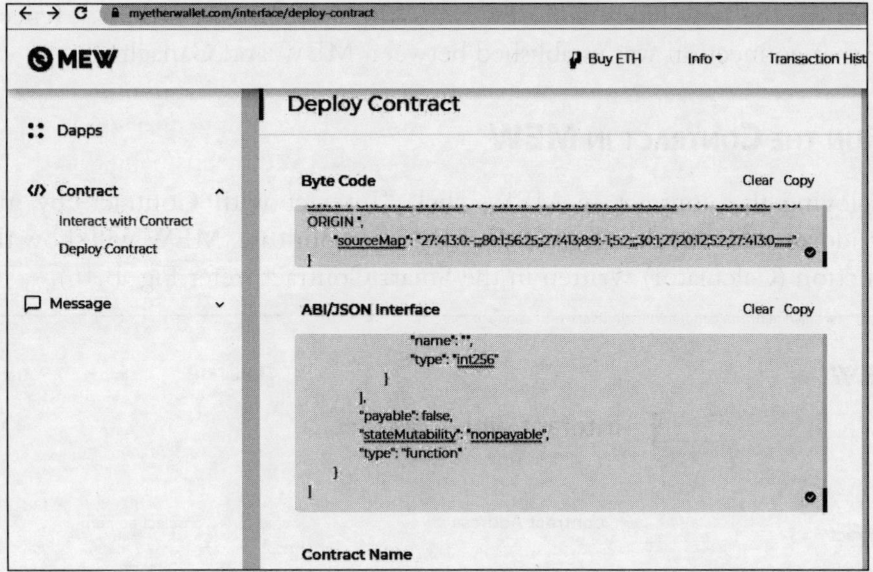

Figure B.12: Deploying the contract in MEW

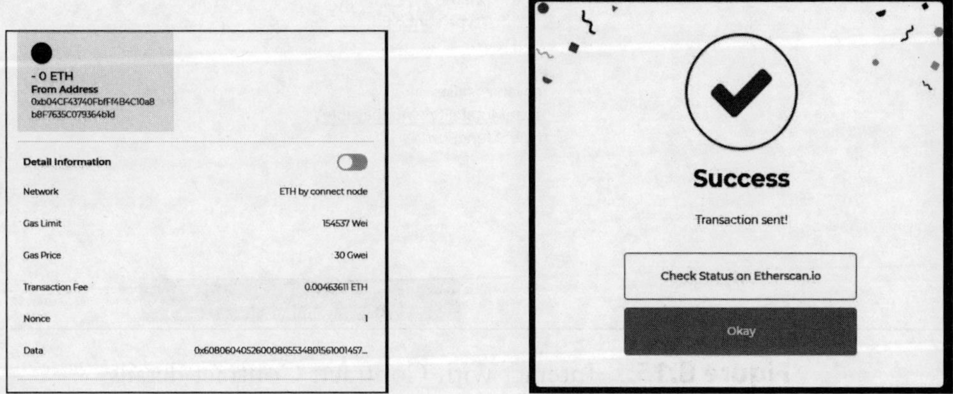

Figure B.13: Transaction details in MEW **Figure B.14:** Transaction Success details

We can see the new block and new transaction details in Ganache. Hence, we can conclude that connection was established between MEW and Ganache.

B.7 Run the Contract in MEW

After deploying the contract in MEW, click "Interact with Contract" by giving the Contract address and ABI details (refer Fig. B.15). In our case, MEW will show the details of the function (Calculator) written in the Smart Contract (refer Fig. B.16).

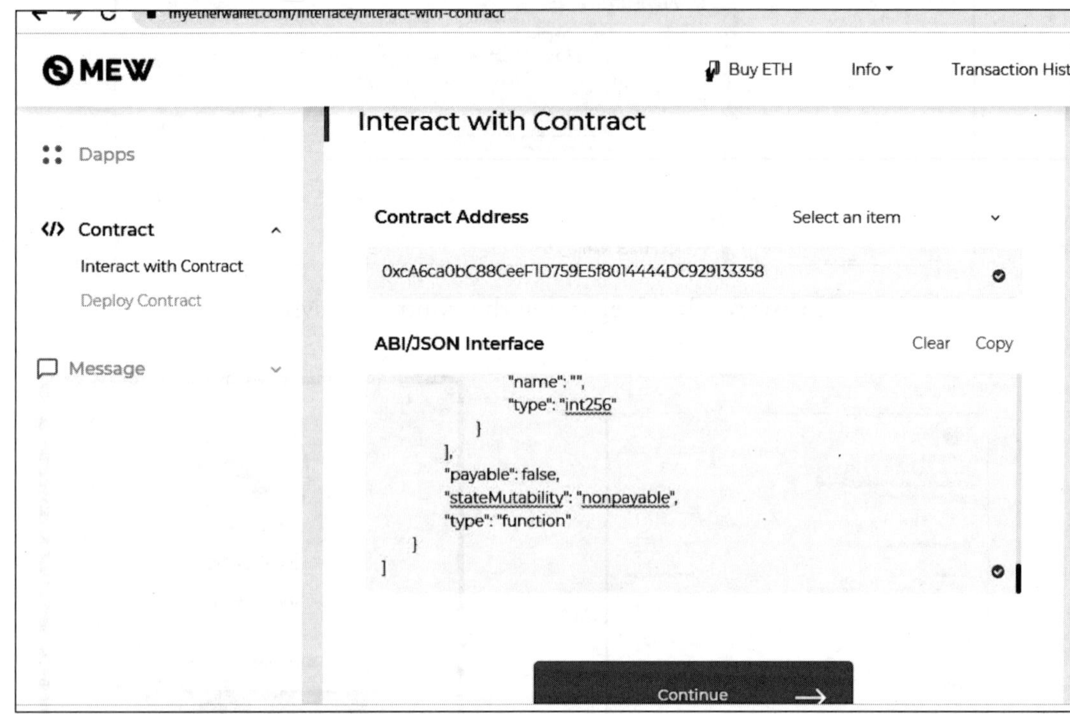

Figure B.15: Interact with Contract: Contract details

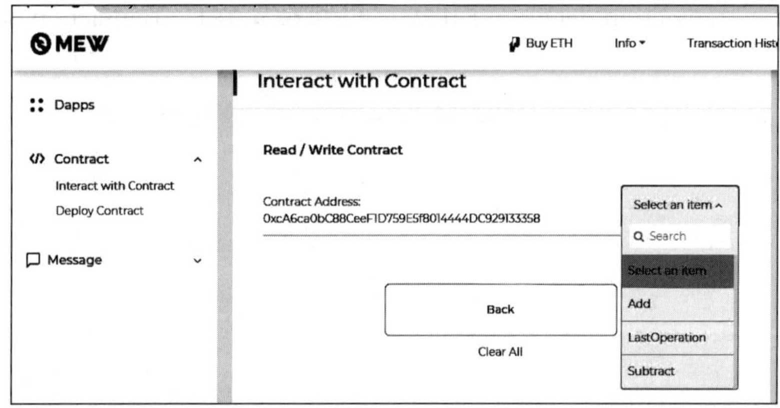

Figure B.16: Interacting with Contract: Contract details with functions

Connecting Remix with Metamask

C.1 Introduction

Metamask is a browser plugin that holds Ethereum wallet and connects the computer with the Ethereum network. Metamask can connect and network with other providers who offer free ethers to Accounts created in wallets of Metamask. For example: An account created in Metamask can be connected with an external network named Ropsen Test Network, which can then inject free ETH to the corresponding account.

Visit https://metamask.io/ to know more details about Metamask (refer Fig. C.1).

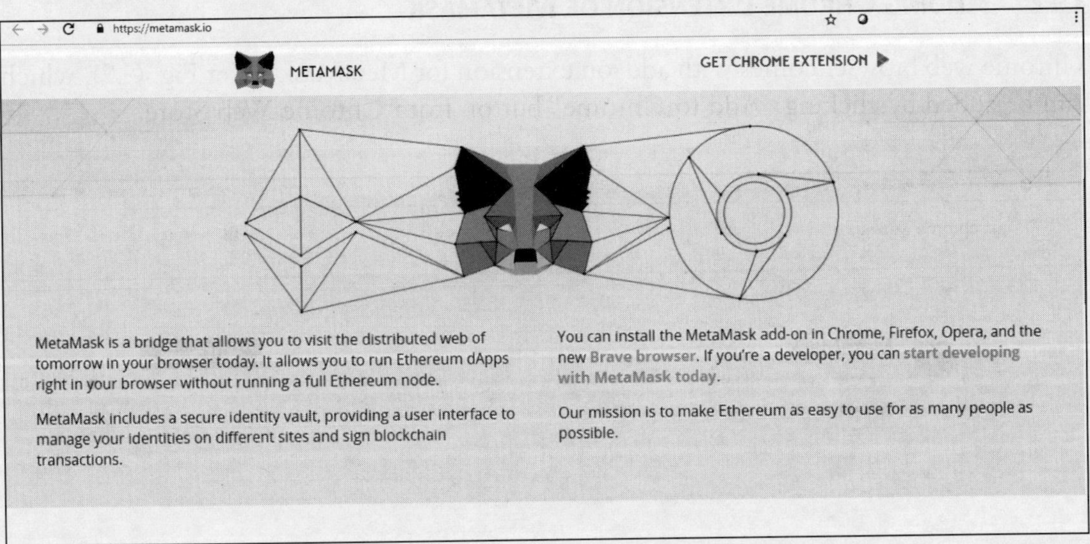

Figure C.1: Home page of Metamask website

Table C.1 compares the features of Remix and Metamask.

We will use Remix to develop the contract, compile, deploy and run the Contract (the first, second, fourth and fifth option respectively). Creating the user accounts (third option) and monitoring the execution (sixth option) are taken care of by Metamask.

Table C.1 Comparing Remix and Metamask

S. No	Activities	Remix	Metamask
1	Develop the contract	+	–
2	Compile the contract	+	–
3	Creating user account	+	+
4	Deploy the contract	+	–
5	Run the contract	+	–
6	Monitor the execution	–	+

C.2 Adding Chrome Extension of Metamask

Chrome web browser comes with add-on extension for MetaMask (refer Fig. C.2), which can be added by clicking "Add to Chrome" button from Chrome Web Store.

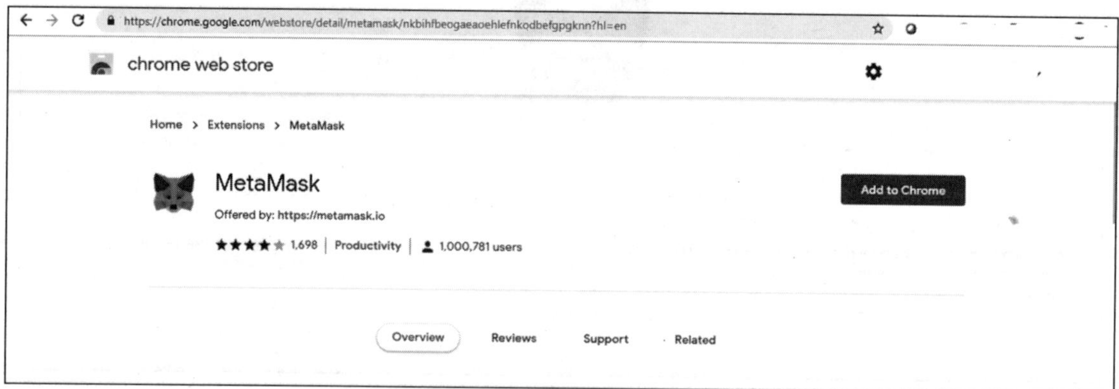

Figure C.2: Chrome browser add-on for MetaMask

While adding MetaMask, after the welcome page (refer Fig. C.3), there will be a page with options to either use already existing wallet or create a new wallet (refer Fig. C.4).

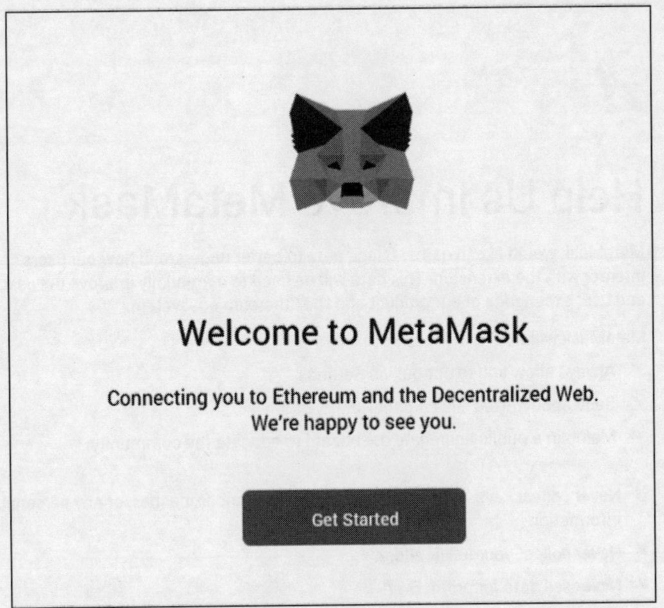

Figure C.3: MetaMask welcome message

Wallets already created using MetaMask can be imported using the first option. The second option is selected (Create a wallet) to create a wallet from the scratch.

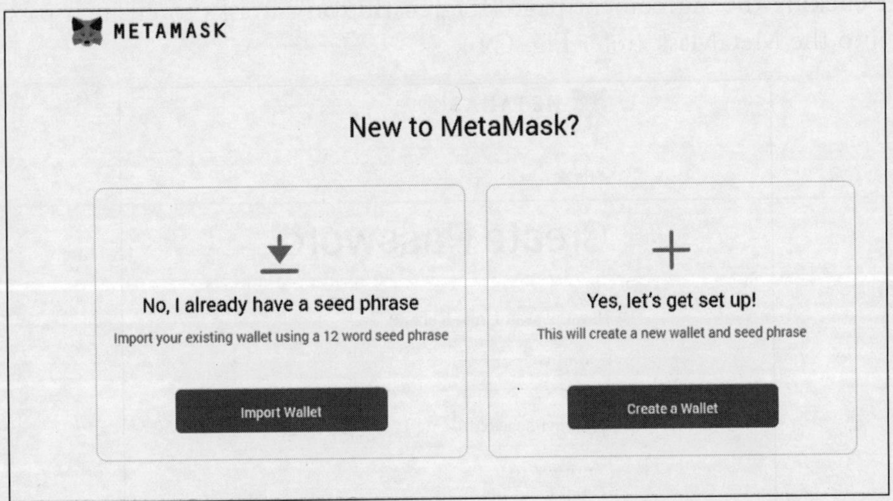

Figure C.4: MetaMask: Import Wallet (or) Create a Wallet option

An Agreement page appears (refer Fig. C.5) before creating the wallet in the MetaMask. It also gives some advice on how to use MetaMask, in the format of Do's and Don'ts.

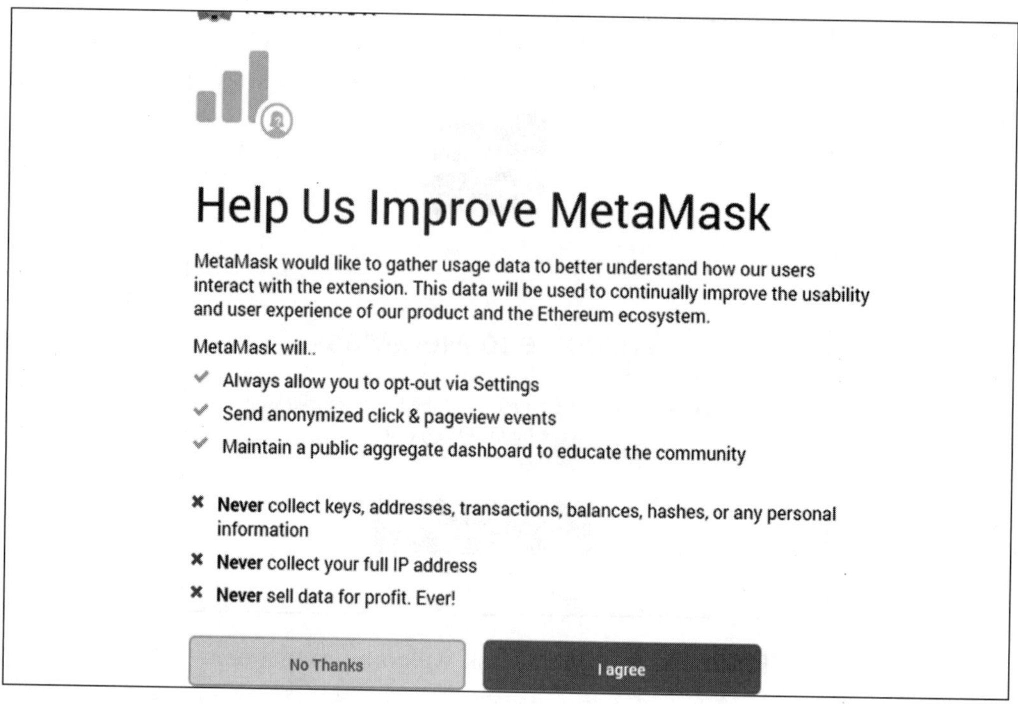

Figure C.5: Agreement page before creating the Wallet

After clicking the Agreement page (I agree Button), it asks to create a password for logging into the MetaMask (refer Fig. C.6).

METAMASK

< Back

Create Password

New Password (min 8 chars)

Confirm Password

I have read and agree to the Terms of Use

Create

Figure C.6: Password creation page of MetaMask

After creating the password, it displays a secret backup phrase (12 different words in order). It asks us to download the secret backup phrase and store it safely in a hard drive (refer Fig. C.7).

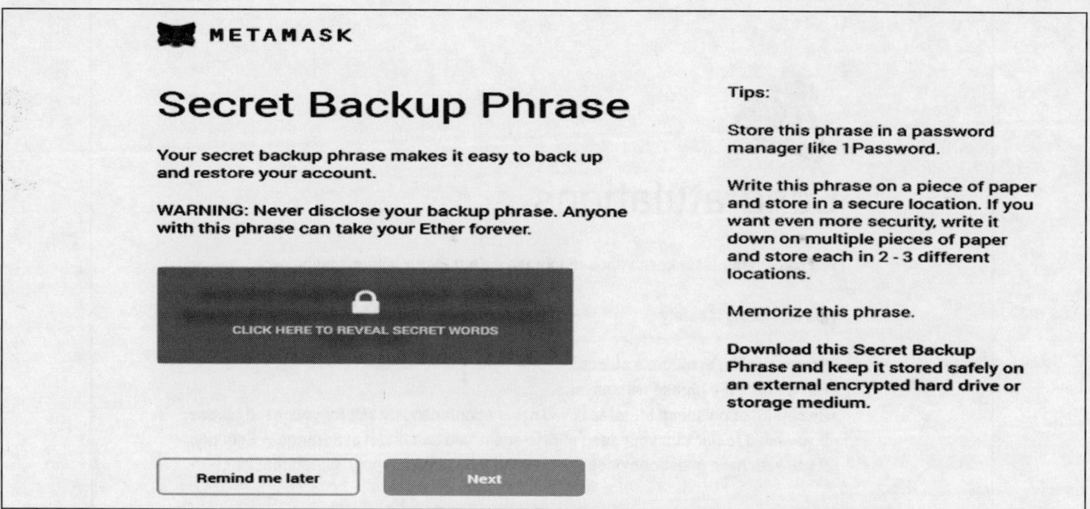

Figure C.7: Secret backup phrase page with warnings

After showing the secret backup phrase (12 different words in order), it asks us to enter the same phrase in the given order for confirmation (refer Fig. C.8).

< Back

Confirm your Secret Backup Phrase

Please select each phrase in order to make sure it is correct.

around	extend	sock	top
appear	crawl	rival	umbrella
exact	beyond	toss	train

Confirm

Figure C.8: Secret backup phrase page entry for confirmation

After the confirmation, it shows a Congratulations message indicating "all done" (refer Fig. C.9).

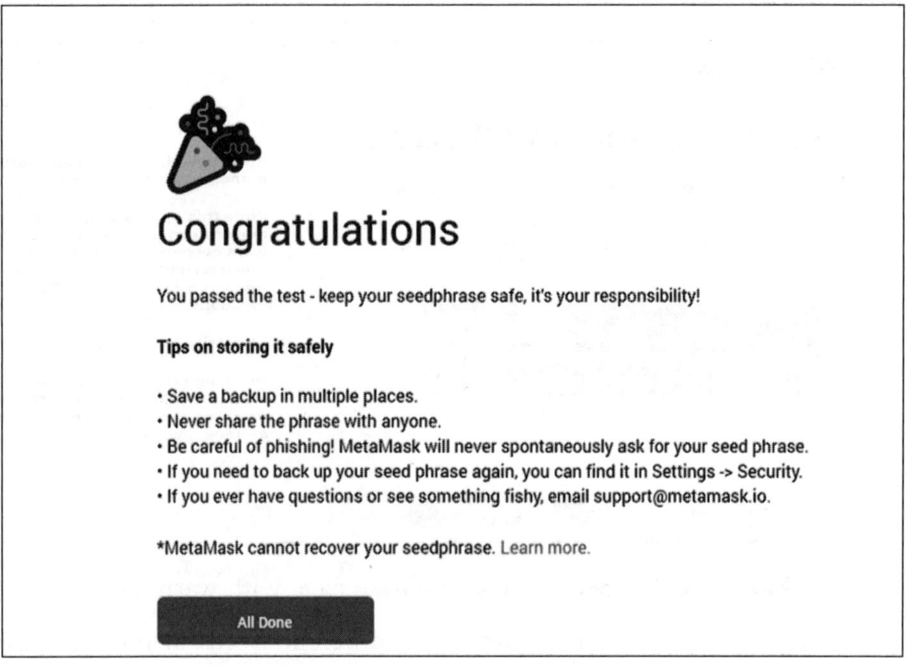

Figure C.9: Congratulation page before creating the wallet

While creating the wallet, it creates a default account (Account 1) with 0 ETH (refer Fig. C.10). This will be the default account created.

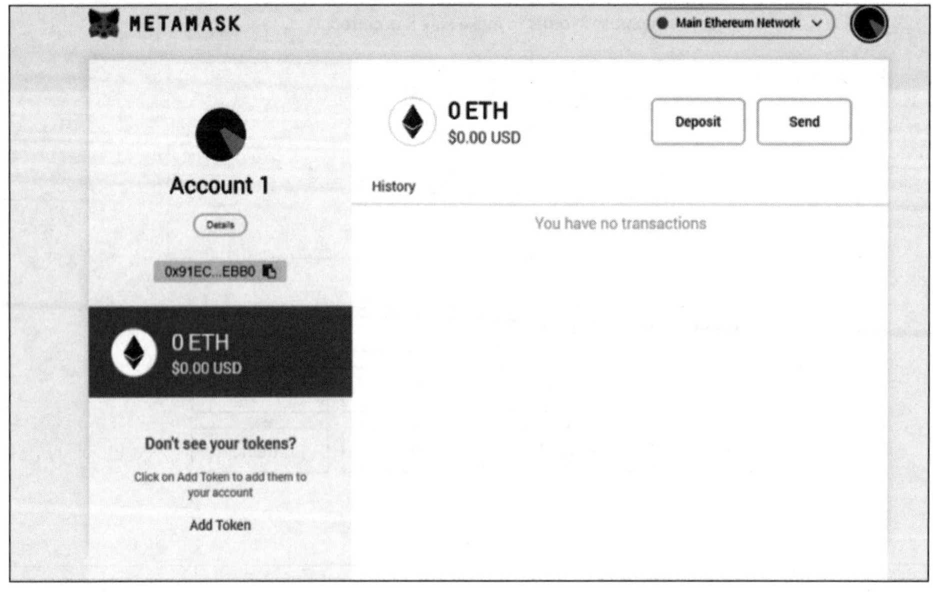

Figure C.10: Default Account 1 with zero ETH in MetaMask

C.3 Connecting Metamask to Ropsten Test Network

After getting the default account (Account 1) in MetaMask, connect to other test networks to get ETH for the accounts. In order to be connected to "Ropsten Test Network" to get free ETH for the accounts, Click Networks link and select "Ropsten Test Network" (refer Fig. C. 11).

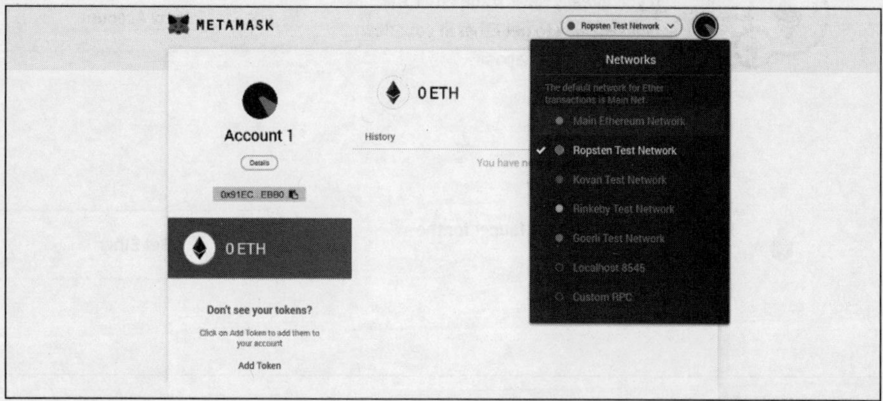

Figure C.11: Connecting MetaMask to Ropsten network

After connecting the "Ropsten Test Network", work on depositing some ETH from the Ropsten network by clicking Deposit button (refer Fig. C. 12).

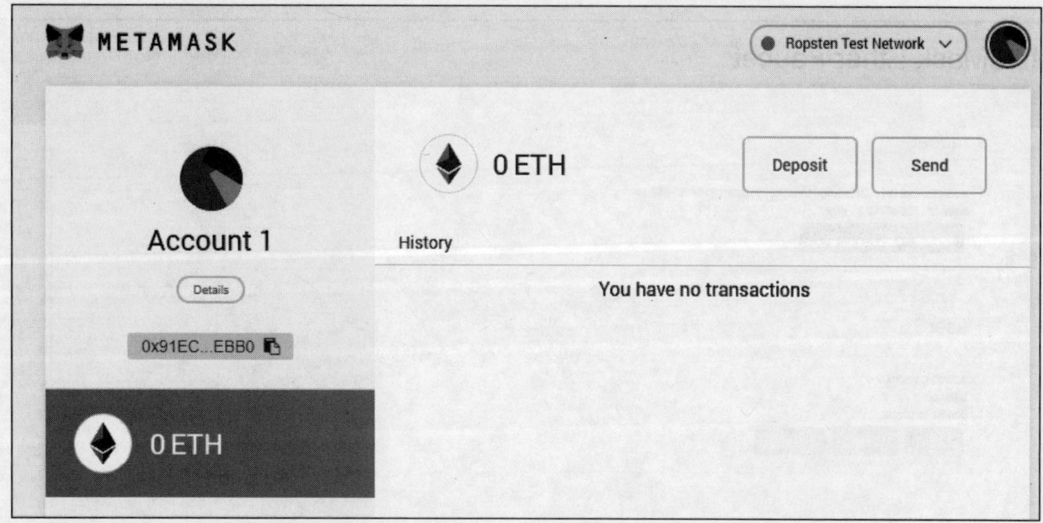

Figure C.12: Depositing ETH from Ropsten network

Two options are available to deposit ether from Ropsten network to Metamask account. The first option is "Directly Deposit Ether" by way of direct deposit (from one account to another). Another option is connecting to the "Test Faucet" to get the Ether for testing by clicking "Get Ether" button (refer Fig. C.13). Click the second option.

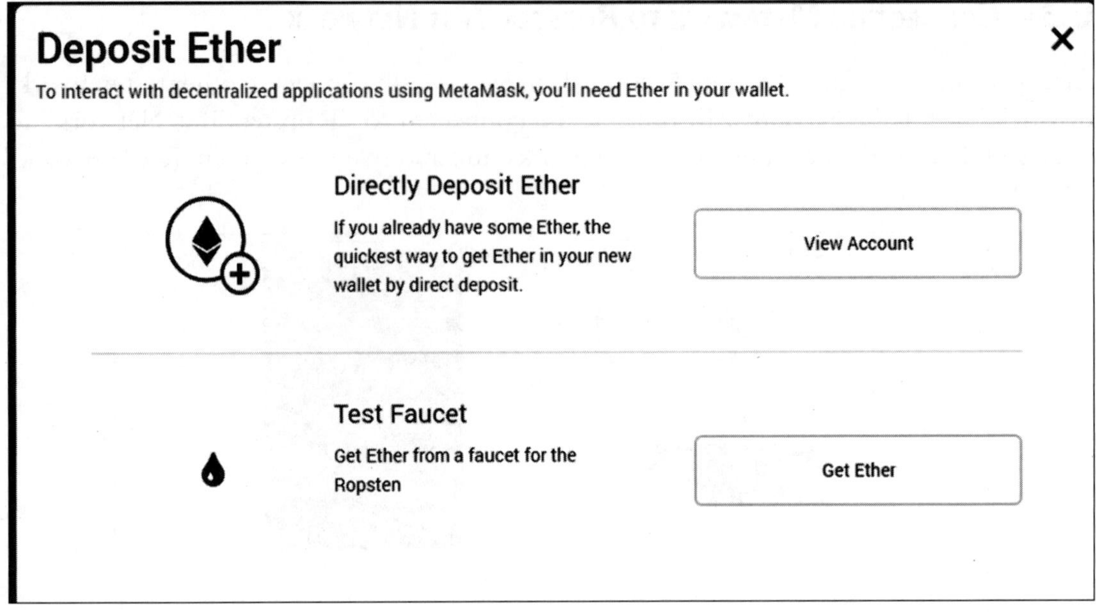

Figure C.13: Two options for depositing ETH: Directly Deposit Ether (or) Test Faucet

Clicking "Test Faucet" button to get the ether, will lead to a page to deposit one ETH to the account (Account 1) from the Test Faucet (refer Fig. C.14).

MetaMask Ether Faucet

faucet

address: 0x81b7e08f65bdf5648606c89998a9cc8164397647
balance: 92880718.35 ether
request 1 ether from faucet

user

address: undefined
balance: ...
donate to faucet:
1 ether 10 ether 100 ether

transactions

Figure C.14: Requesting one ether from Test Faucet

After clicking "request 1 ether from faucet" button, it will take us to MetaMask Notification page to confirm the connect request, giving permission in MetaMask to Connect to Test Ether Faucet (refer Fig. C.15). Click "Connect" for confirmation.

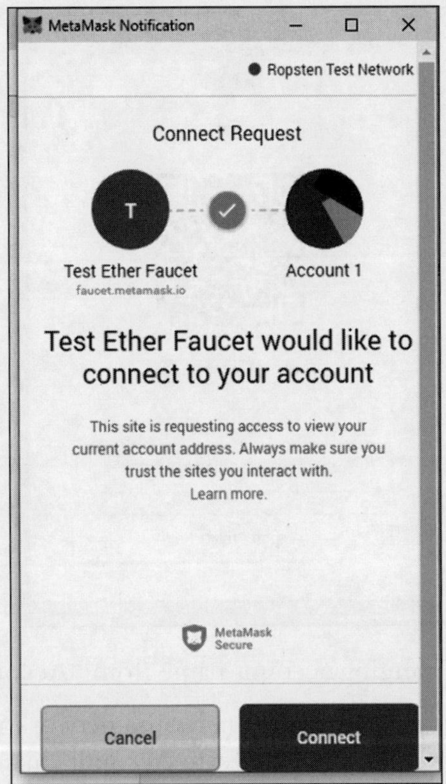

Figure C.15: Giving permission in MetaMask to connect to Test Ether Faucet

After clicking "Connect", one ETH is deposited into the Account 1 of the MetaMask from the Ropsten Test Network, which can be confirmed by visiting MetaMask (refer Fig. C.16).

Figure C.16: Account 1 deposited with 1 ETH from Ropsten Test Network

Account Name can be changed from "Account 1" to any other name (For example: John) by clicking on the details link under the Account (Account1) (refer Figs C.16 and C.17).

Figure C.17: Changing account name from "Account 1" to "John"

By clicking on the "My Accounts" link (after the network name), it is possible to add any number of test accounts by clicking "Create Account" link. Two accounts are added, namely, John and Tiger (refer Fig. C.18).

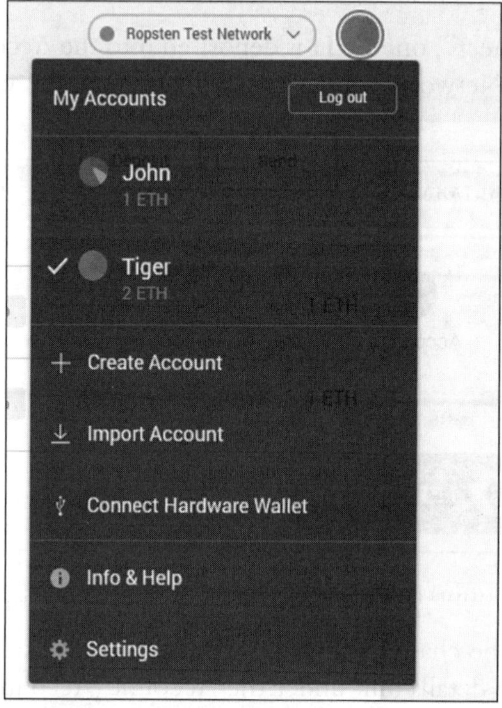

Figure C.18: Creating new accounts in MetaMask

C.4 Writing and Compiling the Code in Remix

We write the actual solidity code in Remix and compile it (refer Fig. C.19). A simple Hello Contract is written with functions for setting and getting the name.

browser/hello.sol

```
pragma solidity ^0.5.1;

contract Hello {
string public name;

    constructor() public {
        name = "Benjamin";
    }

    function sayhello() public view returns (string memory, string memory)  {
        return("hello", name);
    }

    function setName(string memory n) public {
        name = n;
   }

    function getName() public view  returns (string memory)  {
        return name;
    }

}

```

Figure C.19: Remix code and compilation

After writing and compiling the code (Contract Hello) in Remix, the environment is chosen as "Injected Web3" in Run menu (Remix) to connect to MetaMask (refer Fig. C.20).

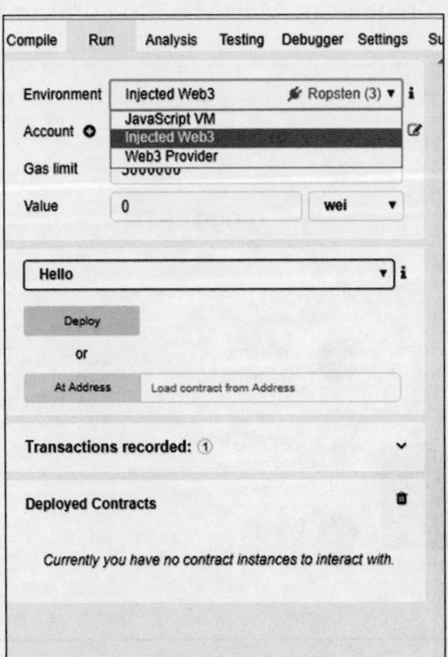

Figure C.20: Changing the environment to "Injected Web3" to connect to MetaMask

While the code is deployed in Remix (by clicking the Deploy button), it asks for confirmation (refer Fig. C.21) to connect Remix Ethereum to MetaMask Account (John) through MetaMask notification page. Click "Connect" button to confirm the deployment. The contract is created.

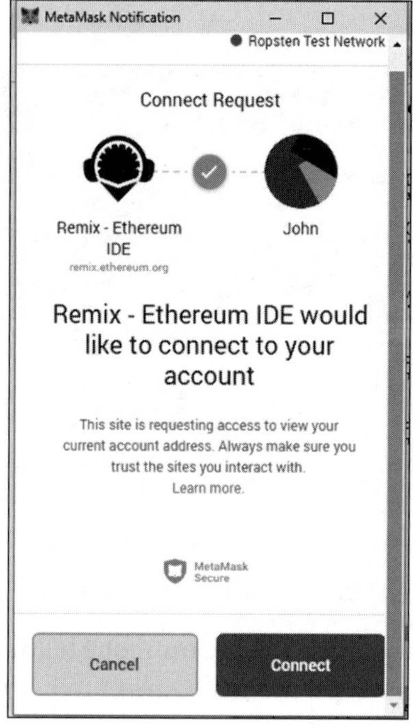

Figure C.21: Connect request in MetaMask to connect Remix Ethereum IDE

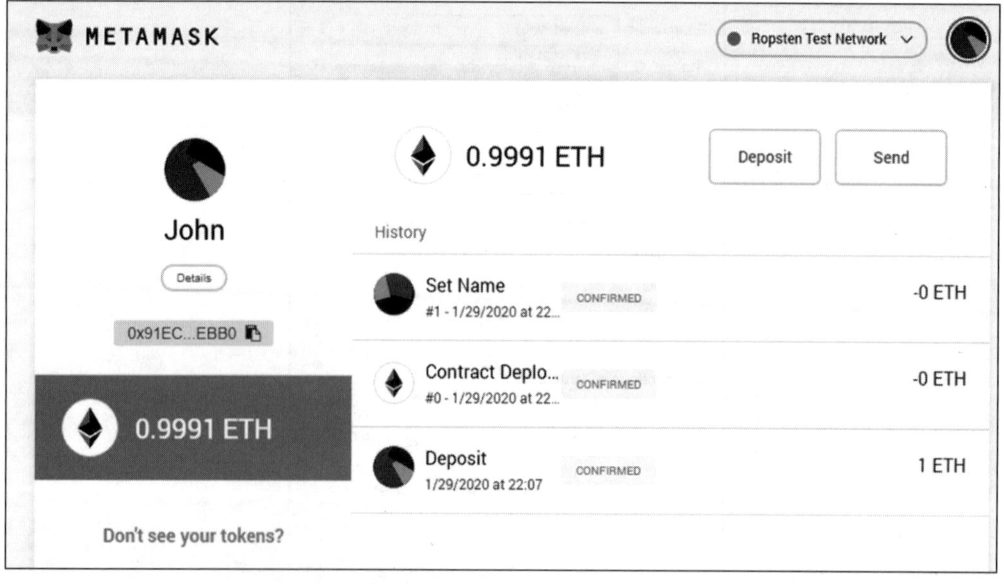

Figure C.22: Reflection of the Contract deployment and execution in MetaMask

When the code is deployed in Remix IDE, a corresponding entry (Contract Deployment) is done in MetaMask. Similarly, while the code is run (Set Name), a corresponding entry (Set Name) is done in MetaMask against the corresponding account (refer Fig. C.22).

C.5 Confirming the Transaciton in Etherscan

The transactions executed in Remix (by running Contract Hello) and the corresponding account level changes in MetaMask can be confirmed by visiting etherscan pages (https://ropsten.etherscan.io) and entering the corresponding address of the account (refer Fig. C.23).

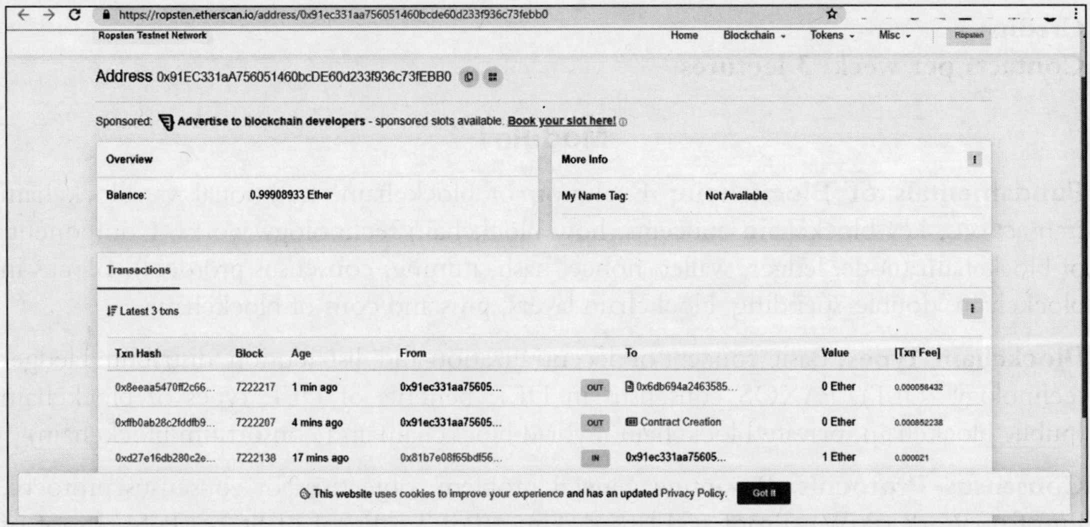

Figure C.23: Confirming the transactions of Ropsten account in Etherscan page

It can be seen that Etherscan pages reflect the Remix Contract (deployment and Run) connected to a MetaMask Account, which in turn is connected to Ropsten Network.

Model Syllabus for Blockchain Technology

Credits: 5
Contacts per week: 3 lectures

Module I

Fundamentals of Blockchain: Evolution of blockchain, traditional vs. blockchain transactions, key blockchain concepts, how blockchain technology works, Components of blockchain (node, ledger, wallet, nonce, hash, mining, consensus protocol), Blocks in blockchain, double-spending, blockchain layers, pros and cons of blockchain.

Blockchain Types: Basic concept of decentralization and distribution, distributed ledger technology (DLT), PAXOS consensus in DLT, benefits of DLT, types of blockchain (public blockchain, private blockchain, hybrid blockchain and consortium blockchain)

Consensus Protocols: Byzantine general problem, objectives of consensus protocol, Proof of Work (PoW), Proof of Elapsed Time (PoET), Proof of Stake (PoS), Proof of Authority (PoA), Practical Byzantine Fault Tolerance (pBFT), RAFT.

Module II

Cryptocurrency: Introduction, evolution of currency, birth of bitcoin, characteristics of cryptocurrencies, cryptocurrency wallets, altcoins, tokens, popular coins and tokens, ecosystem players, crypto mining, airdrop, coin burning, investing and trading, cryptocurrency safety, regulations around cryptocurrencies.

Public Blockchain Systems – Bitcoin: Characteristics of public blockchain, blockchain layers, bitcoin blockchain, bitcoin common terminologies, bitcoin mining, Proof of Work (PoW) and hashcash in bitcoin, block propagation and relay, transaction in the bitcoin network, bitcoin script.

Public Blockchain Systems: Ethereum: Ethereum components, mining in ethereum, Merkel Patricia tree, architecture of ethereum, workflow of ethereum, comparison of bitcoin and ethereum.

Consortium Blockchain: Key characteristics, why we need consortium blockchain, hyperledger platform, hyperledger frameworks, hyperledger tools, hyperledger fabric, multi-version concurrency control, channels in hyperledger fabric, overview of Ripple and Corda.

Module III

Smart Contracts: What is smart contract, smart contract in traditional sales agreement, how smart contracts work, characteristics of smart contract, types of smart contracts, oracles, types of oracles, different platforms for smart contracts, smart contracts in ethereum, smart contracts in industries.

Private Blockchain Systems: Key characteristics of private blockchain, why we need a private blockchain, e-commerce site example, smart contract in private environment, design limitations in a permissioned environment, CAP theorem, BASE theory, state machine replication mechanisms, smart contract in a state machine, applications of state machine replication.

Different Algorithms of Permissioned Blockchain: Common types of faults in distributed system, PAXOS algorithm, handling different failures in PAXOS, RAFT consensus algorithm, handling different failures in RAFT, Byzantine fault, three Byzantine, four Byzantine, Lamport–Shostak–Pease algorithm, introduced practical Byzantine Fault Tolerance (pBFT), multichain, streams in multichain.

Module IV

Initial Coin Offerings: What is ICO, ICO vs. IPO, important ICO terms, launching an ICO, investing in ICO (why invest in ICO, how to invest in ICO, understanding the white paper), pros and cons of ICO, successful ICO (pillars of ICO, examples of successful ICOs), evolution of ICO (ICO variants, regulation aspects).

Security in Blockchain: Security aspects in bitcoin, security and privacy challenges of blockchain in general (majority attack, deterministic transactions), inherent security attributes of blockchain (CIAR): mixing/bartering transactions, anonymous signatures, advanced encryption techniques, homomorphic encryptions, briber attack, hashing and Merkle tree, collusion attack, transaction malleability attack, reputation-based attack, the Finney attack, The vector76 attack, manipulation-based attacks, eclipse attacks, performance and scalability, replay attack, impersonation attack, Sybil attack, time-jacking, regulatory compliance and assurance, safeguarding blockchain smart contract (DApp), security aspects in hyperledger fabric.

Module V

Applications of Blockchain: Blockchain in banking and finance, blockchain in education, blockchain in energy, blockchain in healthcare, blockchain in real-estate, blockchain in supply chain, blockchain and IoT.

Limitations and Challenges: Blockchain limitations in general, limited scalability, limited security, lack of technical knowledge, security concerns and flaws, transaction processing speed, complexity, implementation and operational cost, storage constraints, lack of governance and standards, energy and resource consumption, simplified mining, human errors.

Practical: Generic blockchain using Golang, ethereum blockchain using solidity, multiple connectors of ethereum blockchain, blockchain using python, hyperledger blockchain.

Index

H

I

K

L

M

N

O

P